Advances in Intelligent Systems and Computing

Volume 1220

The series "Advances in Intelligent Systems and Computing" contains publications on theory, applications, and design methods of Intelligent Systems and Intelligent Computing. Virtually all disciplines such as engineering, natural sciences, computer and information science, ICT, economics, business, e-commerce, environment, healthcare, life science are covered. The list of topics spans all the areas of modern intelligent systems and computing such as: computational intelligence, soft computing including neural networks, fuzzy systems, evolutionary computing and the fusion of these paradigms, social intelligence, ambient intelligence, computational neuroscience, artificial life, virtual worlds and society, cognitive science and systems, Perception and Vision, DNA and immune based systems, self-organizing and adaptive systems, e-Learning and teaching, human-centered and human-centric computing, recommender systems, intelligent control, robotics and mechatronics including human-machine teaming, knowledge-based paradigms, learning paradigms, machine ethics, intelligent data analysis, knowledge management, intelligent agents, intelligent decision making and support, intelligent network security, trust management, interactive entertainment, Web intelligence and multimedia.

The publications within "Advances in Intelligent Systems and Computing" are primarily proceedings of important conferences, symposia and congresses. They cover significant recent developments in the field, both of a foundational and applicable character. An important characteristic feature of the series is the short publication time and world-wide distribution. This permits a rapid and broad dissemination of research results.

Indexed by DBLP, INSPEC, WTI Frankfurt eG, zbMATH, Japanese Science and Technology Agency (JST).

All books published in the series are submitted for consideration in Web of Science.

More information about this series at http://www.springer.com/series/11156

Zhengtao Yu · Srikanta Patnaik · John Wang ·
Nilanjan Dey
Editors

Advancements in Mechatronics and Intelligent Robotics

Proceedings of ICMIR 2020

Springer

Editors
Zhengtao Yu
School of Information Engineering and Automation
Kunming University of Science and Technology
Kunming, China

John Wang
Department of Information Management and Business Analytics
Montclair State University
Montclair, NJ, USA

Srikanta Patnaik
Department of Computer Science and Engineering
SOA University
Bhubaneswar, India

Nilanjan Dey
Department of Computer Science and Engineering
JIS University
Kolkata, India

ISSN 2194-5357 ISSN 2194-5365 (electronic)
Advances in Intelligent Systems and Computing
ISBN 978-981-16-1842-0 ISBN 978-981-16-1843-7 (eBook)
https://doi.org/10.1007/978-981-16-1843-7

This Springer imprint is published by the registered company Springer Nature Singapore Pte Ltd.
The registered company address is: 152 Beach Road, #21-01/04 Gateway East, Singapore 189721, Singapore

Preface

On behalf of the International Conference on Mechatronics and Intelligent Robotics-2020, we welcome all the participants to the International Conference on Mechatronics and Intelligent Robotics (ICMIR2020) held in Kunming, China, during December 11–13. ICMIR2020, like previous years, is being organized by Interscience Research Network, an international professional body, and International Journal of Computational Vision and Robotics, published by Inderscience Publishing House along with IRNet International Academic Center, China. Like last year, this year also it was hosted in association with Kunming University of Science and Technology, Kunming, China. Due to COVID-19, it could not be possible to invite international speakers from other countries; however, it was academically very rich by online presentation.

Over the last four years, ICMIR has created a platform for the researchers and industry practitioners in the areas of mechatronics and intelligent robotics. The proceedings covers research work from the areas of mechatronics and intelligent robotic such as manufacturing, underwater robots, humanoid, biomedical applications, unmanned aerial vehicles, mobile/legged robot, and space applications. This volume also offers a comprehensive overview of the potential risks, challenges, and opportunities in the dynamically changing technologies as well as insights to assess and manage them effectively.

This year, it was not possible to have a large gathering, but we could arrange an online meeting by Zoom which was attended by more than 180 participants, and 71 papers were shortlisted and published in this proceedings. The papers covered in this proceedings are the result of the efforts of the researchers working in various domains of mechatronics and intelligent robotics. We are thankful to the authors and paper contributors of this volume.

We are thankful to Editor-in-Chief and the anonymous review committee members of Springer series on *Advances in Intelligent Systems and Computing*, for their support to bring out the proceedings of 2020 International Conference on Mechatronics and Intelligent Robotics. It is noteworthy to mention here that this was really a big boost for us to continue this conference series.

We are thankful to the experts and reviewers who have worked for this volume despite of the veil of their anonymity.

We shall not forget to inform you that the next edition of the conference, i.e., 5th International Conference on Mechatronics and Intelligent Robotics (ICMIR2021), will be held at Kunming, China, in association with Kunming University of Science and Technology, Kunming, China, during May 2–24, 2021, and we shall be updating you regarding further details of the conference from time to time.

Kunming, China — Prof. Zhengtao Yu
Bhubaneswar, India — Prof. Srikanta Patnaik
Montclair, USA — Prof. John Wang
Kolkata, India — Dr. Nilanjan Dey

Contents

About the Editors

Prof. Zhengtao Yu, Ph.D., is the dean of the Faculty of Information Engineering and Automation, the director of the Provincial Key Lab of Intelligent Information Processing, and the Chief Professor of KUST innovation team of Intelligent Information Processing. He is graduated from Beijing Institute of Technology with Ph.D. on Computer Application Technologies in 2005. He visited Purdue University as a visiting scholar in 2009. He has over 100 publications in international journals and conferences and holds numerous patents and copyrights. He had won numerous awards including the first prize and the third prize of Yunnan Province Science and Technology Progress Award and the second prize of Yunnan Province Natural Science Award. Prof. Yu is specialized in natural language processing, information retrieval, and information extraction. His research interests include: Chinese corpus and Southeast Asia minority languages corpus construction, linguistic information processing, question-answering system, cross-language information retrieval, multi-lingual machine translation, and cross-language public opinion analysis.

Dr. Srikanta Patnaik is a professor in the Department of Computer Science and Engineering, SOA University, Bhubaneswar, India. He has received his Ph.D. (Engineering) on Computational Intelligence from Jadavpur University, India, in 1999 and supervised 21 Ph.D. theses and more than 50 M.Tech. theses in the area of machine intelligence, soft computing applications, and re-engineering. Dr. Patnaik has published more than 100 research papers in international journals and conference proceedings. He is author of 2 text books and edited 52 books and few invited book chapters, published by leading international publisher like Springer-Verlag, Kluwer Academic, etc. Dr. Patnaik was Principal Investigator of TAPTEC project "Building Cognition for Intelligent Robot" sponsored by All India Council for Technical Education, New Delhi and Major Research Project "Machine Learning and Perception using Cognition Methods" sponsored by University Grant Commission. He is Editors-in-Chief of *International Journal of Information and Communication Technology* and *International Journal of Computational Vision and Robotics* published from Inderscience Publishing House, England, and also Editors-in-Chief of Book Series on "Modeling and Optimization in Science and Technology" published from Springer, Germany.

John Wang, Ph.D., works in the Department of Information Management and Business Analytics at Montclair State University, USA. Having received a scholarship award, he came to the USA and completed his Ph.D. in operations research from Temple University. Due to his extraordinary contributions, Prof. Wang has been honored with two special range adjustments in 2006 and 2009, respectively. He has published over 100 refereed papers and more than ten books. He has also developed several computer software programs based on his research findings. He is Editor-in-Chief of *International Journal of Information Systems and Supply Chain Management* and *International Journal of Applied Management Science*. Also, he is Editor of *Encyclopedia of Business Analytics and Optimization* (five-volume), *Data Warehousing and Mining: Concepts, Methodologies, Tools, and Applications* (six-volume), and Editor of the *Encyclopedia of Data Warehousing and Mining*, 1st (two-volume) and 2nd (four-volume). His long-term research goal is on the synergy of operations research, data mining, and cybernetics.

Nilanjan Dey is an Associate Professor in the Department of Computer Science and Engineering, JIS University, Kolkata, India. He is a visiting fellow of the University of Reading, UK. He is an Adjunct Professor of Ton Duc Thang University, Ho Chi Minh City, Vietnam. Previously, he held an honorary position of Visiting Scientist at Global Biomedical Technologies Inc., CA, USA (2012–2015). He was awarded his Ph.D. from Jadavpur University in 2015. He has authored/edited more than 90 books with Elsevier, Wiley, CRC Press, and Springer, and published more than 300 papers. He is the Editor-in-Chief of the *International Journal of Ambient Computing and Intelligence* (IGI Global), Associated Editor of *IEEE Access*, and *International Journal of Information Technology* (Springer). He is the Series Co-editor of Springer Tracts in Nature-Inspired Computing (Springer), Series Co-Editor of Advances in Ubiquitous Sensing Applications for Healthcare (Elsevier), Series Editor of Computational Intelligence in Engineering Problem Solving and Intelligent Signal Processing and Data Analysis (CRC). His main research interests include medical imaging, machine learning, computer aided diagnosis, data mining, etc. He is the Indian Ambassador of the International Federation for Information Processing—Young ICT Group and Senior member of IEEE.

Mechatronics

Simulation of Off-Wheel Problem of Triangle Tracked Vehicle Based on RecurDyn

Tian Wu, Xiaojun Xu, Dongwei Li, and Teng'an Zou

Abstract The off-wheel problem of the track is very common during the traveling of a tracked vehicle. In order to improve the reliability of triangular tracked vehicles, the track (HM) submodule of the multi-body dynamics simulation software RecurDyn was used to establish a virtual prototype model for the traveling mechanism of a triangular tracked vehicle. The final results showed that appropriately increasing the preload of the triangular track tensioning device and adding the torsion spring device in the balance suspension can effectively prevent the off-wheel problem of the track. The analysis results can provide reference for the design and optimization of triangle tracked vehicle walking system.

Keywords Triangle tracked vehicle · Off wheel · RecurDyn · Dynamics · Simulation

1 Introduction

Whether the mission objective of the ground unmanned platform is the fire support, the equipment transportation or the accompanying support, the first prerequisite to achieve the mission objective is to the implementation site. So, the reliability of walking mechanism is one of its most important basic performances [1]. However, the off-wheel problem of the track is an unavoidable problem during the use of tracked machinery. The triangular track mechanism has such advantages of low bottom pressure ratio and high trafficability and flexibility. The triangular track is used to replace tires and necessarily travels at a higher speed than traditional tracked

---

T. Wu · X. Xu · T. Zou (✉)
College of intelligent science, National University of Defense Technology, Changsha 410073, China
e-mail: tenganzou@163.com

D. Li
Xinxiang North Vehicle Meter Co., Ltd., Xinxiang 453000, China

Z. Yu et al. (eds.), *Advancements in Mechatronics and Intelligent Robotics*, Advances in Intelligent Systems and Computing 1220,
https://doi.org/10.1007/978-981-16-1843-7_1

vehicles. Therefore, the off-wheel problem of the triangular track is particularly prominent [2].

Empirically, the off-wheel problem is mainly caused by two reasons:

1. Insufficient track guidance due to too loose track for the tension provided by the tensioning mechanism is insufficient. The tensioning device can provide sufficient tension at rest. However, affected by ground excitation and when the displacement generated by the load-bearing wheel reaches a certain value, loose track and off-wheel problems will also be caused.
2. Loose track and insufficient meshing of track and sprocket due to lengthened track after long-term use.

2 Structure and Relevant Technical Parameters of Triangular Track

The 3D model of triangle track is shown in Fig. 1. The triangular track is mainly composed of the support frame, drive wheel assembly, guide wheel, load-bearing wheel, non-reactive suspension, tensioning wheel, tensioning device and track. The load-bearing wheel and the non-reactive suspension constitute a swinging load-bearing wheel, which is connected to the support frame through a pin shaft. The triangular track studied herein is configured with two sets of swinging load-bearing wheel and has a rigid tensioning device. The main parameters are shown in Table 1.

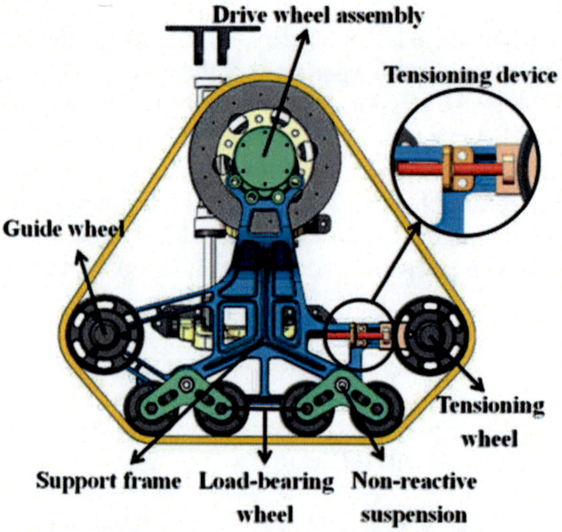

Fig. 1 Three-dimensional model of triangle track

Table 1 Main performance parameters

Platform full load mass (kg)	3100
Body size (mm)	5900 × 2300 × 1400
Track width (mm)	2100
Average velocity (km/h)	60
Drive wheel pitch diameter (mm)	380
Load-bearing wheel diameter (mm)	140
Guide wheel and tensioning wheel diameter (mm)	200
The numbers of track shoes	51
Track shoe length (mm)	3200

3 Simulation Test

3.1 Virtual Prototype Modeling

In order to carry out a dynamic simulation test, a dynamic model should be established in RecurDyn. First of all, the three-dimensional (3D) model was simplified to remove the parts that are not involved in the movement, such as nuts; the position of mass center was consistent with that of the 3D model without affecting the results of the simulation test as far as possible [3, 4]. The virtual prototype of the simplified triangular track and the vehicle is shown in Fig. 2. The reason is that, in the real vehicle test, the rigid tensioning could not provide the compensation for the tension; and the off-wheel phenomenon was serious. The tension spring stiffness was 100, the initial pretension was 1500 N, and the damping was 3, without limit.

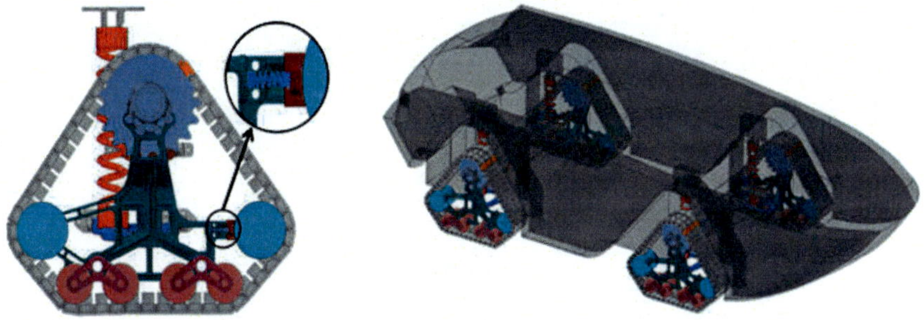

Fig. 2 Virtual prototype of the simplified triangular track and the vehicle

3.2 Simulation Test on off-Wheel Problem of Track

The contact settings between the track and the ground are shown in Fig. 3. The drive wheels were controlled by speed to make the vehicle move in a straight line first and then turn around. The function expression for the left drive wheel is step (time, 0, 0, 2, 3 * pi) and that for the right one is step(time, 0, 0, 2, 3 * pi) − step (time, 3, 0, 6, 6 * pi). The simulation duration was 7 s.

The process of the off-wheel problem of track triangular track is shown in Fig. 4. The changing curves for the spring tension and rotation angle of non-reactive suspension are shown in Fig. 5. At the simulation time of 5.2 s, the state of the right triangular track is shown in Fig. 4a. The drive wheel is in counterclockwise rotation with the right half of the track as the tight side and the left half as the loose side. It

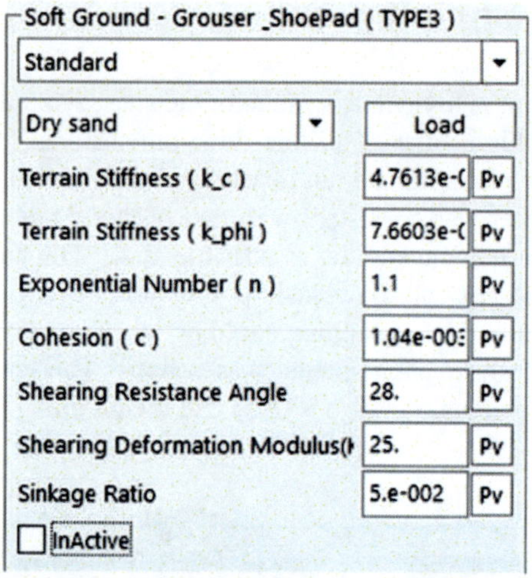

Fig. 3 Contact settings between the track and the ground

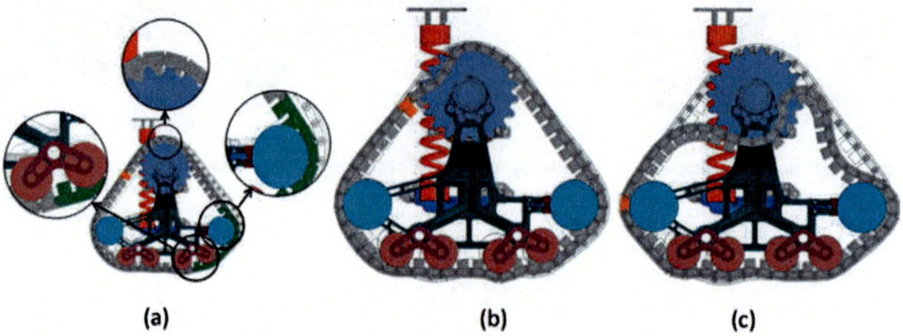

Fig. 4 Process of the off-wheel

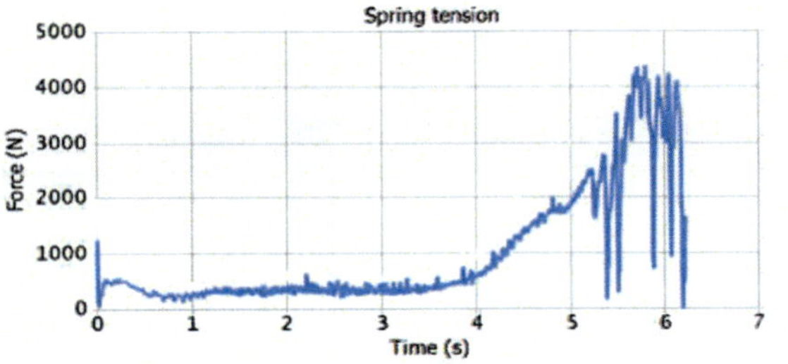

(a) Force curve of the tension spring

(b) Rotation angle curve of non-reactive suspension

Fig. 5 Changing curves for the spring tension and rotation angle of non-reactive suspension

could be obviously observed that the tension spring was compressed that the tension rose sharply to 4000 N and the swing angle of the right swinging load-bearing wheel reached 15°, which resulted in the loose track. This is consistent with the afore-mentioned first reason. The top track showed the tendency of the off-wheel problem and came off completely at the simulation time of 6.2 s, as shown in Fig. 4b, c.

3.3 *Simulation Test on Vehicle S Bend*

In order to avoid the off-wheel problem of the track, first of all, increasing the stiffness of the tension spring and the pretension may be considered. However, the tension from this measure will persist and aggravate the aging and relaxation of the track. Therefore, a limit device should be added to ensure that there will not be too much tension when the vehicle is at rest and avoid the failure of the tensioning device due to a large impact in complex road conditions [5]. Next, inhibiting the displacement of the load-bearing wheel, i.e., inhibiting the rotation angle of the non-reactive suspension, may be considered to ensure that the load-bearing wheel and track contact position will not change too much as far as possible and provide a certain track tension. Moreover, the tension will not be generated when the non-reactive suspension is in a horizontal state without imposing a great impact on the service life of the track. The model change location is shown in Fig. 6.

Fig. 6 Model change location

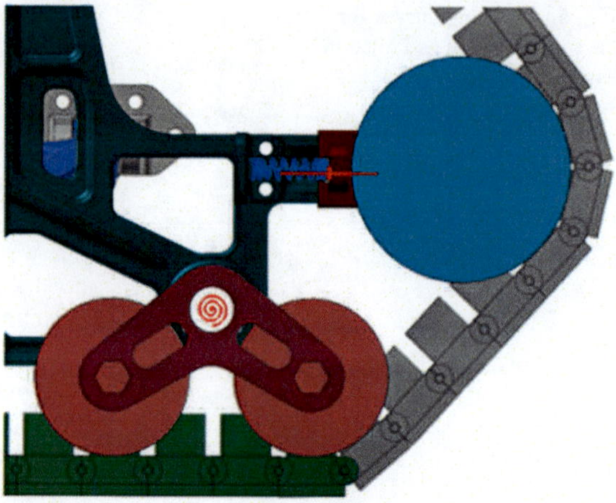

A torsional spring was added at the non-reactive suspension and a limit contact at the tensioning position.

After the change was completed, the simulation test was carried out on S bend. Except for the drive wheel function, other parameters were consistent with those in the previous simulation environment. The function for the right drive wheel is step(time, 0, 0, 2, −3 * pi) + step(time, 17, 0, 20, −3 * pi) − step(time, 30, 0, 32, −3 * pi); and function for the left one is step(time, 0, 0, 2, 3 * pi) + step(time, 7, 0, 10, 3 * pi) − step(time, 17, 0, 20, 3 * pi). The simulation duration was 40 s. The traveling path is as shown in Fig. 7.

Fig. 7 Traveling path

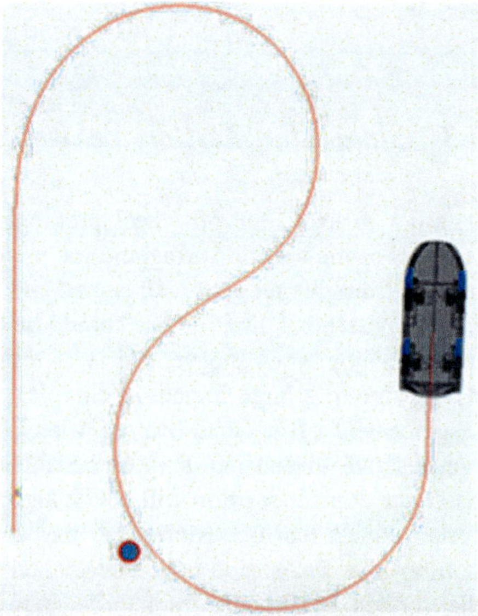

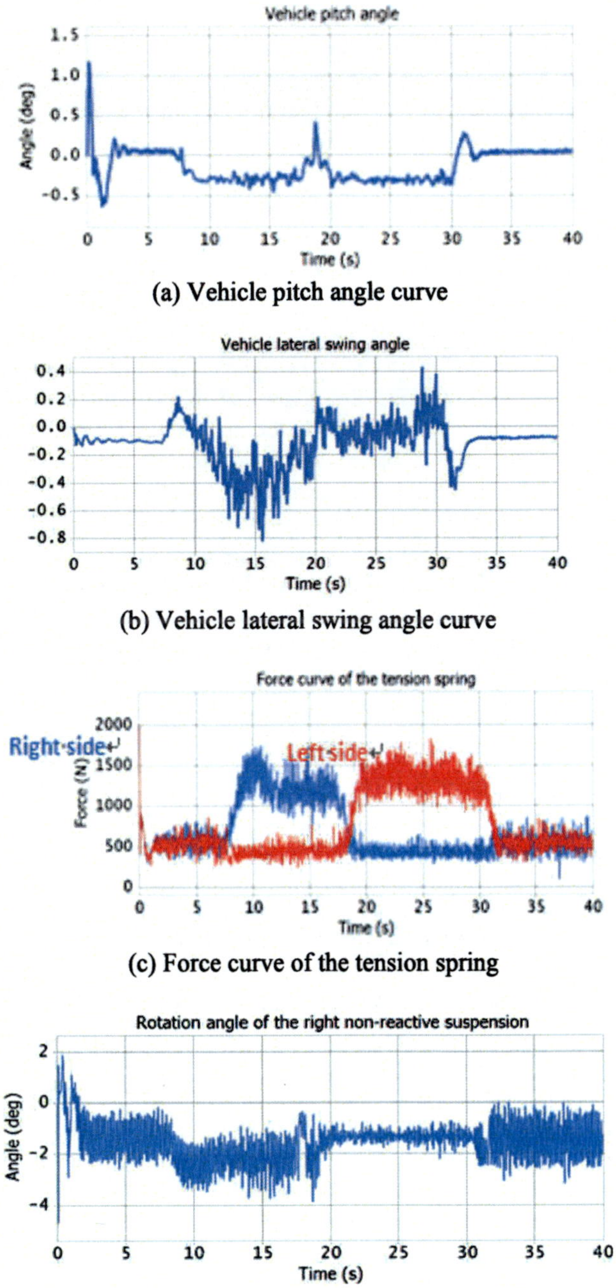

(a) Vehicle pitch angle curve

(b) Vehicle lateral swing angle curve

(c) Force curve of the tension spring

(d) Therotation anglecurve of the right non-reactive suspension

Fig. 8 Changing curves for the vehicle attitude, the force on the tension spring and the rotation angle of the right non-reactive suspension

The changing curves for the vehicle attitude, the force on the tension spring and the rotation angle of the right non-reactive suspension are shown in Fig. 8. In terms of the simulation time, 0–7 s was the straight driving stage; 7–30 s was the S bend driving stage; and 30–40 s was the straight driving stage again.

The pitch angle of the vehicle is shown in Fig. 8a. The maximum pitch angle was 1.2° and was caused by the impact of its own gravity at the moment of landing. Afterward, the pitch angle gradually stabilized at about 0.1° and increased to 0.4° at the 7–30 s turning stage. The overall range of change was not wide.

The lateral swing angle of the vehicle body is shown in Fig. 8b. It ranged from −0.8° to 0.2°. All peaks appeared in the 7–30 s turning stage. The overall change was reasonable.

The force on the tension spring is shown in Fig. 8c. The initial pretension was 1500 N. In the 0–7 s straight driving stage, the maximum force on the tension spring on both sides was around 800 N. In the 7–20 s turning-right stage, the force on the right spring increased to about 1700 N and that on the left spring decreased slightly. In the 20-30 s, turning-left stage, the force on the spring on both sides was opposite due to the addition of the limit device and did not exceed 2000 N.

The change curves for the rotation angle of the right non-reactive suspension are shown in Fig. 8d. Since the deformation of the tension spring decreased and a torsional spring was added at the suspension rotation, the overall rotation angle of the suspension was smaller than 4°. The contact position between the load-bearing wheel and the track showed no significant change.

Throughout the whole process, the track was in a good tensioning state without showing any tendency of the off-wheel problem.

4 Conclusion

In this paper, the fundamental causes of the off-wheel problem of the triangular track were analyzed. A virtual prototype of a triangular tracked vehicle was established based on the multi-body dynamics software RecurDyn. On the basis of the original mechanism, the rigid tensioning device was changed to an elastic tensioning device. However, it was found in the simulation test that the tension provided by the tensioning device was insufficient, which caused too large rotation angle of the swinging load-bearing wheel, finally resulting in the off-wheel problem of the track. The off-wheel problem of track disappeared, and the stability under the transport condition was improved after improvement. This method provided a reference for the design and optimization of the driving system of triangular tracked vehicles.

References

1. Jihong Tian (2018) Technology development trend of the ground unmanned system [J]. Foreign Tank 9:32–39
2. Yao-Yi Q, Liang H, Wen-Guang L et al (2013) Dynamical model of idler and simulation of wheel's falling-off for triangular tracked [J]. J Mech Electr Eng 30(2):134–137
3. Chen Y, Zhang Y, Du F et al (2019) Dynamic modeling and simulation of 4 × 4 wheeled vehicle [J]. J Phys: Conf Ser 1303:012060
4. Schramm D, Hiller M, Bardini R (2018) Vehicle dynamics: modeling and simulation [M]. Springer Publishing Company, Incorporated
5. Qing-Guo L, Dong-Ya SI, Zheng-Bo G et al (2011) Dynamic simulation of a tracked vehicle based on RecurDyn [J]. Veh Power Technol

Brief Design Requirements of Screen Printing Stencil

Fang Chen, Kaikai Han, Kangwei Chang, Shixun Luan, Wenbo Dou, Li Ma, and Yingjie Ding

Abstract The stencil used in screen printing is one of the key tools used in the welding process of SMT. The quality of screen printing directly affects the quality of PCB assembly. This paper mainly introduces the frame design, stretching design, mark point design, aperture design, and thickness design. At the same time, this paper focuses on the SMT device opening size and shape requirements.

Keywords Stencil · SMT · Quality · Design

1 Introduction

With the rapid development of electronic manufacturing, surface mount devices have been widely used in circuit board. Solder paste printing is the first process of reflow soldering for SMDs, it is to place the stencil on the solder paste printing machine, and the stencil and PCB need direct contact; the solder paste is evenly scraped into the mesh of the stencil by the scraper on the solder paste printing machine; when the stencil is detached from the PCB, the solder paste will fall off from the mesh to the corresponding pad pattern of the PCB in the shape of mask pattern, and then the solder paste is printed on the PCB. The quality of solder paste printing will directly affect the one-time qualification rate of SMT component welding. The printing quality of solder paste depends on the correct shape and position of the stencil. Therefore, this paper will discuss the design requirements of stencil.

F. Chen (✉) · K. Han · K. Chang · S. Luan · W. Dou · L. Ma · Y. Ding
Shanghai Aerospace Control Technology Institute, Shanghai 201109, China
e-mail: zdhchenfang@163.com

Z. Yu et al. (eds.), *Advancements in Mechatronics and Intelligent Robotics*, Advances in Intelligent Systems and Computing 1220,
https://doi.org/10.1007/978-981-16-1843-7_2

2 Stencil Design

Stencil is the key mold for solder paste printing, which is composed of frame, stainless steel mesh, and mask pattern [1]. The following will be a detailed introduction to some key parameters of stencil.

2.1 *Frame Design of Circuit Board Stencil*

The stencil frame is usually made of aluminum alloy. The stencil size depends on the type of solder paste printing machine, for example, the frame size of DEK Horizon03i4 solder paste printing machine is 736 mm × 736 mm, the frame profile specification is (thickness) 40 mm × (width) 40 mm, and the warpage between the corners of the stencil frame is less than 0.3 mm.

2.2 *Design of Stencil Tension Net*

The plate of stencil is stainless steel plate. Generally, glue and aluminum tape are used to stretch the stencil, glue shall be applied to the joint of aluminum alloy frame and stainless steel mesh plate, and scrape a layer of protective paint on the joint of aluminum alloy frame and adhesive. At the same time, in order to ensure that the stencil has enough tension [2] (35–50 N/cm) and good flatness, the space between the stainless steel mesh plate and the inner side of the mesh frame is 25–50 mm.

2.3 *Mark Point Design of Stencil*

The automatic solder paste printing machine needs to be automatically aligned through the mark point. Therefore, it is necessary to design the mark point when making stencil. The semi-automatic solder paste printing machine is generally through manual alignment, and there is no need to design mark point on the stencil. The mark point of stencil is generally opened in 1:1 mode according to the size and shape of mark points in the Gerber file of PCB. At the corresponding coordinates (including the diagonal), at least two mark points are opened diagonally for the whole stencil. Generally, the mark points are etched from the back of the stencil, such as DEK Horizon03i4, the mark points need to be half-engraved from the back of the stencil (that is, the side in contact with the printed circuit board) with the color set to black.

2.4 *Aperture Design of Stencil*

According to IPC-7525B stencil design guidelines, in order to ensure the solder paste can be released smoothly from the mesh opening to the PCB pad, the following principles should be followed in the aperture design of the stencil.

Aspect ratio and area ratio [3]

Both area ratio and aspect ratio are illustrated in Fig. 1, using the following formulas.

$$\text{Aspect Ratio} = \frac{\text{Width of Aperture}}{\text{Thickness of Stencil}} = \frac{W}{T} > 1.5 \tag{1}$$

$$\text{Area Ratio} = \frac{\text{Area of Aperture}}{\text{Area of Aperture Walls}} = \frac{L \times W}{2 \times (L + W) \times T} > 0.66 \tag{2}$$

The printing surface is the top of the stencil, and the lower opening of the stencil should be 0.01–0.02 mm wider than the upper opening. The opening of the stencil should be processed into an inverted cone.

When making stencil, PCB board should be placed in the middle.

The independent opening size of the stencil should not be too large, and the pad width should not be greater than 2 mm; if the pad size is greater than 2 mm, a 0.4 mm bridge is often needed in the middle of the pad opening.

2.5 *Thickness Design of Stencil*

The thickness of stencil determines the amount of solder paste printed on PCB. Too much or too little solder paste will lead to reflow defects. Too much solder paste will easily lead to bridging, while too little solder paste will easily lead to defects such as empty welding and virtual welding. The thickness of stencil should meet the requirement of the smallest pitch device and the smallest chip element. The thickness range of stencil is generally 0.12–0.2 mm, and the corresponding relationship between pitch and stencil thickness is shown in Table 1.

Fig. 1 Cross-sectional view of a stencil

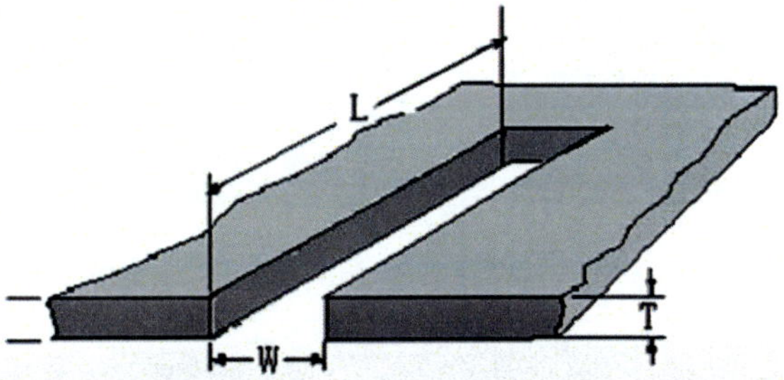

Table 1 Relationship between pitch and stencil thickness

Package	Pitch (mm)	Stencil thickness range (mm)
IC, QFP SO, PR	0.5–1.27	0.13–0.18
	0.4	0.1–0.12
BGA	1.25–1.27	0.15–0.2
	0.8–1	0.12–0.15
CCGA	1.0–1.27	0.15–0.18

3 Requirements for Opening Size and Shape of Stencil

Several aperture geometries are effective in reducing the occurrence of solder balls [3]. All these designs are aimed at reducing excess solder paste [4] trapped under the SMD.

3.1 Chip Components-Resistors, Inductors, Capacitors

The most popular designs for 0603, 0805, and 1206 package are shown in Fig. 2. The aperture geometries shall follow the following rules: the apertures internally tangent $B(B = 1/3\ X - R)$, X to plus 0.1 mm, $Y = Y1$, $A = 1/3Y$, $R = 1/2A$, round corners.

The most popular designs for 1210 package and above are shown in Fig. 3. Except for the inner side, the other three sides of the square pad need to be reamed. The aperture geometries shall follow the following rules: $X' = 0.85 * X$, $Y' = Y$, $X1' = 0.6 * X$, $Y1' = 0.4 * Y$, $D' = D - 0.2$ mm, further expansion of the outer three sides $X11' = 0.1 * X$, $Y11' = 0.1 * Y$.

The most popular designs for tantalum capacitor are shown in Fig. 4. The aperture geometries shall follow the following rules: $X1 = X$, $Y1 = Y$, $D1 = D$, $H = 0.25$ mm, when $Y < 3$ mm, cancel the bridge at H.

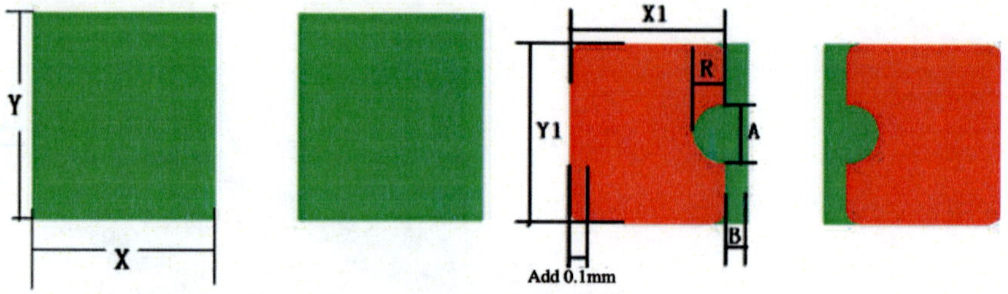

Fig. 2 0603, 0805, 1206 package

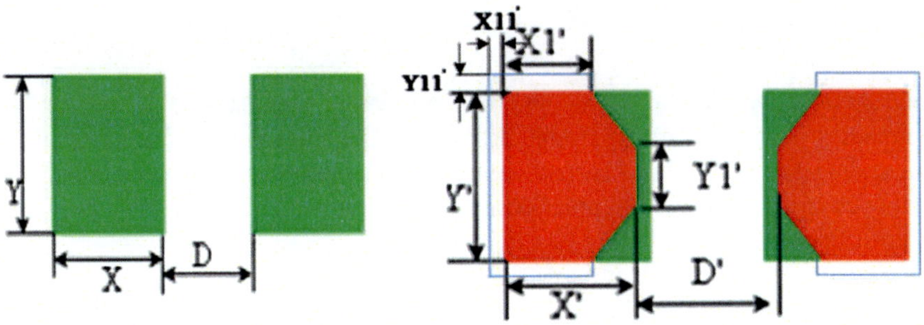

Fig. 3 1210 package

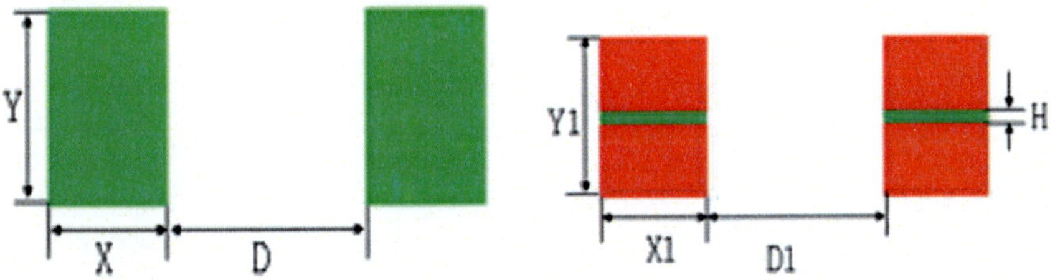

Fig. 4 Tantalum capacitor

3.2 *Power MOSFET*

The most popular designs are shown in Fig. 5. The aperture geometries shall follow the following rules: $X1 = 4/5 * X$, $Y1 = 4/5 * Y$, $X1' = X'$, $H = 0.3$ –0.4 mm, the apertures at pin are the same size as the pads, the apertures for the heat sink should be divided into 9, 16, or 25 equal parts, and the bridge H is 0.3–0.5 mm.

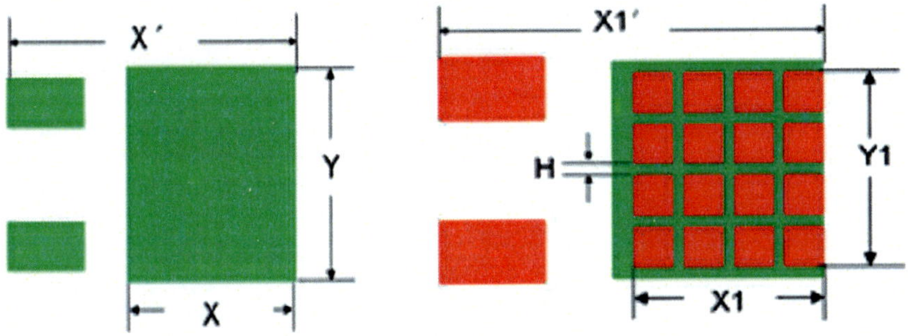

Fig. 5 Power MOSFET

3.3 QFP, SOP, QFN, IC

For devices with a grounding pad in the middle, the aperture of the middle grounding pad is 80% of the grounding pad, and the opening is inclined strip; if the grounding pad is less than 1.5 mm (regardless of length or width), the inclined strip is not required. The most popular designs are shown in Table 2.

3.4 BGA

The most popular designs are shown in Table 3.

Table 2 Aperture of QFP, SOP, QFN, and IC

Pitch	Aperture	Original pad	Open pad
0.4	$W1$ = 0.185 mm, $L1$ = L internally tangent 0.05 mm to plus 0.10 mm, round corners		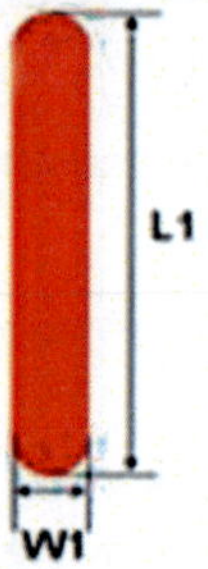
0.5	$W1$ = 0.23 mm, $L1$ = L internally tangent 0.1 to plus 0.10 mm, round corners		
0.65	$W1$ = 0.32 mm, $L1$ = L internally tangent 0.1 to plus 0.10 mm, round corners		
0.8	$W1$ = 0.42 mm, $L1$ = L internally tangent 0.1 to plus 0.15 mm, round corners		
1.0	$W1$ = W, $L1$ = L to plus 0.15 mm		
1.27	$W1$ = 1.1W, $L1$ = 1.1L		

Table 3 Aperture and thickness of BGA

Ball diameter	Pitch	Aperture	Thickness
0.15	0.25	0.13	0.05
0.20	0.30	0.18	0.07
0.25	0.4	0.23	0.10
0.30	0.50, 0.65, 0.75, 0.80	0.28	0.12
0.40	0.65, 0.75, 0.80	0.35	0.12
0.45	0.75, 0.80, 1.0	0.40	0.12
0.50	0.8, 1.0	0.45	0.13
0.60	1.0	0.55	0.15
0.75	1.27, 1.5	0.70	0.15

4 Conclusion

Good design of stencil can control the printing quality of solder paste. Due to the development of electronic components toward the direction of subminiaturization and precision assembly, higher requirements have been put forward for stencil printing and design.

References

1. Fei Xian (2007) Research of solder-paste printing process [J]. Printed Circ Inf 8:55–63
2. Cheng Qi (2007) A brief talk in electronics screen printing technique for PCB [J]. Printed Circ Inf 9:59–63
3. IPC-7525B 2011-October, Stencil design guidelines [S]
4. Zhang Y, Li H, Shen X (2018) A processing method for stencil opening of solder paste printing in microcircuit packaging [J]. Microelectronics 48(6):846–849

Review of Improved Collaborative Filtering Recommendation Algorithms

Lilin Pan and Jianfei Shao

Abstract In the Internet era with rapid data growth, it is difficult for users to quickly find the project information they need from the massive information on the Internet. As a technology based on data mining, the recommendation system effectively solves the problem of information overload. Collaborative filtering recommendation is a classic and effective recommendation algorithm. It focuses on the relationship between the user's interest preferences and items. The purpose is to recommend items that meet the user's interest preferences, such as product recommendation, audio recommendation, tourist attraction recommendation, etc. The field has a wide range of applications. Therefore, this article introduces the principle and related properties of collaborative filtering, as well as its improvement and application, and discusses future trends and directions for improvement.

Keywords Collaborative filtering · Recommendation system · Recommendation algorithm

1 Introduction

With the rapid development of Internet technology, the amount of information increases rapidly at an alarming rate. In the face of massive information, it is difficult for users to quickly acquire valuable parts, and the truly valuable information is inundated by a large amount of junk information, and when faced with a large amount of intricate information, users always feel at a loss. Therefore, it has become a research direction in the field of recommendation system to help users select needed information from a large amount of complex information and solve the problem of information overload for users. Nowadays, the recommendation system has been applied in many fields, such as commodity recommendation of

L. Pan (✉) · J. Shao
School of Information Engineering and Automation, Kunming University of Science and Technology, Kunming 650500, China
e-mail: 1156468319@qq.com

Z. Yu et al. (eds.), *Advancements in Mechatronics and Intelligent Robotics*, Advances in Intelligent Systems and Computing 1220,
https://doi.org/10.1007/978-981-16-1843-7_3

e-commerce, friend recommendation of social software, song recommendation of music player, scenic spot recommendation of travel software, and other fields based on user needs. The quality of the recommendation algorithm directly affects the accuracy of the recommendation content and then the user experience. As a traditional recommendation algorithm, collaborative filtering algorithm finds out the preferences of users through mining the historical behavior data of users, divides the groups of users with different preferences, and makes recommendations. At present, there are few studies on the improvement of collaborative filtering in the academic circle. Therefore, it is of great significance to systematically summarize the latest improvement of collaborative filtering and its application.

2 Overview of Traditional Collaborative Filtering Algorithms

Collaborative filtering recommendation algorithm is the earliest and more famous recommendation algorithm. The core idea is to calculate the similarity between users or projects through the user's historical score of the project, so as to personalized recommendation content for users. There are mainly two recommendation algorithms: memory-based collaborative filtering recommendation algorithm and model-based collaborative filtering recommendation algorithm. The former can be divided into user-based and item-based collaborative filtering recommendations.

2.1 User-Based Collaborative Filtering Recommendation Algorithm

The core idea of the user-based collaborative filtering algorithm is to first find out other users with similar interests and preferences as the target users, namely the nearest neighbor, using the score data of the nearest neighbor users to predict the score of the target users, analyze the score prediction results of all projects, and recommend the project with the highest probability to the target users.

2.2 Item-Based Collaborative Filtering Recommendation Algorithm

The core idea of the project-based collaborative filtering algorithm is to describe the degree of similarity between different projects through the project similarity matrix, and also to find out many other projects with the greatest similarity to the target project, the highly similar items that the users are not exposed to are recommended to the target users in the order of their degree of similarity with the target items.

Table 1 Comparison of collaborative filtering recommendation algorithms based on users and items

Classification	Target	Computational complexity	Recommend diversity	Applicable object
User-based collaborative filtering	User similarity	High	Good	Large recommendation system
Project-based collaborative filtering	Project similarity	Low	Poor	Small recommendation system

The basic step is to process the data generation user-project score matrix to calculate the project similarity to generate the nearest neighbor prediction score and the top-n recommendation [1]. The collaborative filtering recommendation algorithms based on users and items are compared as shown in Table 1.

2.3 *Model-Based Collaborative Filtering Recommendation Algorithm*

Model-based collaborative filtering recommendation is mainly about modeling, of all the user and project data, only part of the user and part of the data has historical scoring data, and the rest of the data has no scoring. In this way, it is necessary to predict the scoring relationship of the unknown part through the known scoring data, estimate some parameters in the model through the training data, and then use the model to predict the user's rating of the project [2]. The mainstream methods are divided into correlation algorithm, clustering algorithm, regression algorithm, Bayesian model, and so on.

3 Improvement and Application of Collaborative Filtering Algorithm

3.1 *The New Theory Is Introduced into Collaborative Filtering*

When the collaborative filtering algorithm faces the problem of data sparsity, its system performance will decline rapidly; therefore, many scholars study the decline of prediction accuracy under sparse data environment. Ji et al. used matrix decomposition to introduce collaborative filtering and improved the effectiveness and efficiency of the algorithm. A new collaborative clustering algorithm is proposed to ensure that each user or item in different clusters is closer to each other, and the original scoring matrix is decomposed into smaller matrices with lower dimensions. This clustering matrix maintains a high relationship with users and attempts to use potential analysis method to model the internal relationship between users or items in each submatrix [3]. After treatment, the accuracy of system prediction is improved effectively.

Transfer learning is introduced into collaborative filtering to facilitate learning in the target domain through the information of public users in the target domain and the auxiliary domain. Although there are a large number of public users in the target domain and the auxiliary domain, the data is still sparse considering that the historical record of public users' purchase or use of goods is still very small. Using the idea of transfer learning, Chen et al. introduced user rating information in the auxiliary field and proposed a concept based on transfer learning and union matrix, and they perform joint matrix decomposition of the user matrices in the target domain and the auxiliary domain and use public user information as the constraints of the two matrix decompositions [4]. The application of joint matrix effectively alleviates the problem of sparse public user data in migration learning.

Chai et al. introduced the dual attention mechanism into the migration learning collaborative filtering and used CNN to extract the features of the comment aggregation text corresponding to the user and the item [5]. This method can more effectively extract enough information, build neural networks in the target domain, auxiliary domain, and public domain separately and extract the feature values of different neural networks and shared domain feature values; through the transfer learning method, the knowledge of the auxiliary domain is applied to the target domain to improve the predictive ability of the system. Effectively alleviate data sparseness and cold start problems of new users.

3.2 *Data Dynamic Analysis Combined with Collaborative Filtering*

Using the user's interest information in a certain fixed period of time as evaluation data will lead to poor recommendation effect, so user interest migration has become the current research focus. The interest direction of a user will change over time, and the needs and hobbies of users in different periods are not the same, and traditional recommendation algorithms do not take this into consideration. Liu and Li clustered items based on user ratings to quickly find the nearest neighbors of the target item and introduced a time attenuation factor in the item similarity and predictive rating stages to objectively reflect user interests and improve recommendation accuracy [1].

After the user obtains the product experience, the evaluation information based on his own emotion becomes important modeling data. Lu et al. [6] took the transfer of user interest as a starting point and introduced sentiment analysis and sentiment forgetting into collaborative filtering. Through sentiment analysis and quantification of review information, user interest modeling is enriched, and recommendation accuracy is effectively improved.

3.3 Combination of Data Rule Mining and Collaborative Filtering

Considering the limited scope of data, it is not enough to simply study the attractiveness of a certain item to users. Therefore, the research on mining-related items has become a research hotspot. Items available for selection on a common thread may be related in a special relationship, that is, users will make different choices in different situations [7]. By studying the projects that have a deep correlation with a certain project, mining the user's interest preferences under multi-layer rules, and expanding the scope of data, the user's model establishment is better enriched and improved. Starting from the relevance of information, the scholars apply association rules to the recommendation algorithm, effectively solving the data sparse problem of collaborative filtering and alleviating the problem of poor recommendation effect during the initial recommendation of the system, providing a great help for the scalability of the algorithm.

4 Conclusion

After decades of development, the recommendation algorithm based on collaborative filtering has become an important factor in more and more fields with its efficient and accurate prediction of recommendation effects. With the integration and complementation of many technologies and recommendation algorithms, recommendation algorithms are becoming more and more mature and widely used in various fields. Although the collaborative filtering recommendation algorithm has achieved good research results in all aspects, there are still many challenges and new research topics. Internet technology has made the amount of data increasing. How to solve the storage problem caused by insufficient memory and how to improve the data processing speed are new problems faced by the collaborative filtering recommendation algorithm. The collaborative filtering recommendation algorithm can be combined with distributed computing technology, which can make the collaborative filtering recommendation system have a larger data storage space, and at the same time have a faster data calculation processing speed, and improve the effectiveness of the system. In the following improvements, while further optimizing data sparseness and cold start of new users, design and optimization of processing speed will also be a research focus.

References

1. Liu X, Li LJ (2020) Dynamic collaborative filtering algorithm based on item clustering and time decay. Comput Technol Dev 30(8):22–26 (2020)
2. Yang Q, Yang Y, YU C-J (2015) Survey of collaborative filtering recommender systems. Mod Comput 13:3–6
3. Ji P, Hu X-Y, Yang W-J, Liu X-L (2020) Co-clustering based matrix factorization for recommendation. J Hefei Univ (Comprehensive Edition) 37(5):10–18
4. Chen J-Y, Zhu Y-Q, Zhou G, Cui L-L, Wu S-M (2020) Collaborative filtering recommendation based on transfer learning and joint matrix decomposition. J Sichuan Univ (Natural Science Edition) 57(6):1096–1102
5. Chai Y-M, Yun W-L, Wang L-M, Liu Z (2020) A cross-domain recommendation model based on dual attention mechanism and transfer learning. Chin J Comput 43(10):1924–1942
6. Lu Z, Ma X, Wu W, Li Y (2020) A collaborative filtering recommendation strategy based on emotional analysis and forgotten rule. J Chongqing Normal Univ (Natural Science) 37(5): 103–108
7. Ji W, Wang H, Su G (2020) Review of recommendation methods based on association rules algorithm. Comput Eng Appl 56(22):33–41

A Preliminary Study on the Application of Computer Technology in Physical Education

Pengfei Qi

Abstract The application of modern information technology also promotes the development of school education, and how to better apply modern information technology in teaching has become a key project of modern research. This paper studies and integrates the practical application of information technology in the teaching process, finds out the problems in the application process of information technology, and constructs a new teaching model which combines computer technology and is suitable for physical education teaching, that named "Leader-independent physical education teaching model". In order to verify the feasibility of this teaching method, this paper carries out a comparative experiment on students' learning process of tennis knowledge. The experiment proves that this computer technology-based education model can enable students to master sports knowledge more quickly. The model can improve the way of physical education and promote the popularization and application of information technology in physical education.

Keywords Computer technology · Physical education · Guidance model · Classroom efficiency

1 Introduction

With the rapid development of computer Internet technology and multimedia technology, modern information technology has been widely used in various fields, especially in the global political, economic, cultural, and other fields, and has

P. Qi (✉)
National Institute of Standards and Technology, Boulder, CO 80305, USA
e-mail: author@boulder.nist.gov

P. Qi
Department of Physics, Colorado State University, Fort Collins, CO 80523, USA

P. Qi
Electrical Engineering Department, University of Colorado, Boulder, CO 80309, USA

Z. Yu et al. (eds.), *Advancements in Mechatronics and Intelligent Robotics*,
Advances in Intelligent Systems and Computing 1220,
https://doi.org/10.1007/978-981-16-1843-7_4

played a great role. Similarly, the application of modern information technology also promotes the development of school education, and how to better apply modern information technology in teaching has become a key project of modern research. In the 1990s, information technology education began to appear in the curriculum of primary and secondary schools and vocational schools in China. With the development demand, information technology education has now become one of the compulsory courses. With the development of modern information technology, there are more and more investments in new multimedia products such as computers and projectors, while fewer and fewer teachers use traditional chalk, blackboard, and other teaching methods. The main purpose of this study is to the actual situation of PE teaching, the necessity of information technology application, and carries on the detailed analysis and research, as well as how to use information technology to improve students' interest in learning, improve students' learning enthusiasm and ability to solve problems and improve the classroom efficiency, save the teaching resources put forward its own views and opinions.

2 Design of Instructional Model of Autonomous Physical Education Based on Computer Technology

The application of information technology in physical education can promote students' learning, but in modern physical education classroom, teachers are not skilled in the application of information technology. Therefore, in this study, based on the theory of constructivism, humanism, cognitive and basic guiding ideology of education information theory, combined with information technology and multimedia tools as well as the sports classroom situation, first of all for information technology in the teaching model of college teaching design, teaching according to teaching model can guide the university sports teaching to join the informationization teaching content correctly and provide theoretical guidance for colleges and universities sports teaching (Fig. 1).

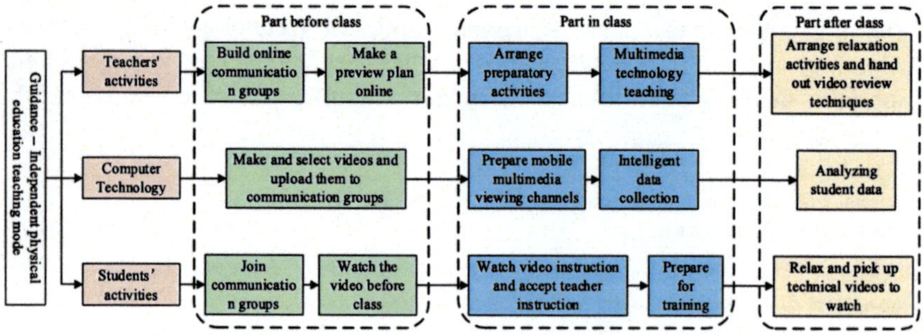

Fig. 1 Design drawing of guidance—independent physical education teaching mode based on computer technology

In this study, it is proposed that the network platform ("we media platform") and handheld devices should be used to assist physical education teaching. Using the modern information technology, based on the previous physical education, increase the network technology as a teaching assistant. The specific operation is to design and make sports technology teaching videos, set up online communication groups, and join the class students in the communication groups. Upload the short technical teaching videos to the we media platform. Send the link to the communication group, and students can click on and watch the video of technology teaching in the communication group by using mobile devices.

The main reasons for the design of this study are: first, the popularization of modern information technology and multimedia technology. Second, appearance plays an obvious role in the formation of motor skills. Thirdly, combining with the actual situation, physical education teaching, information technology and multimedia technology can be reasonably integrated together.

2.1 Part Before Class

Before class, the teacher will make a video of physical education skills and arrange students to preview before class, so as to preliminarily understand the basic body posture and movement track of physical education skills. The purpose is to improve students' interest in learning and let students develop the habit of active learning. In practical teaching, teachers should make preparations before class according to the different abilities of different students, design the teaching content and teaching plan suitable for students, plan the teaching key points and teaching difficulties of this class, and rehearsed the teaching process in their minds. On this basis, combining with the teaching content of the lesson, teachers will sports skills of body posture and movement trajectory into a small video, uploaded to the students' learning website, let the students in learning, learning this section before his students through the specified website, after open the browser can prepare in advance according to the arrangement of the teachers. The control group was prepared by routine jogging, static traction, etc.

2.2 Part in Class

In class, after organizing students for preparation activities, the physical education teacher will intersperse the special technical teaching with the display of special video technical actions, so that students can understand relevant skills by means of network video. And with the help of portable multimedia equipment, students can deepen the memory of relevant actions, so that students can master the relevant technology skillfully. In addition, the new teaching model is cross-applied in physical education. In traditional teaching, the part of the teaching content mainly includes the technology teaching, the students in pairs, the rest three contents, the main technical teaching by using the method of explanation and demonstration after the student practice tennis technology group, in the process of students' practice, teachers timely reminder of the students' mistakes, and correct action for demonstration.

2.3 Part After Class

After the class, according to students' classroom learning situation, arrange the task after class for students, using family (dormitory) or mobile video watching sports skills after class technology, and put forward their online learning problems in sports technology, through the interaction between teachers and students to improve students' physical education learning efficiency and achieve the teaching goal. At the same time, students can also timely put forward their doubts in network technology teaching and learning to physical education teachers in the form of WeChat group discussion, so as to make the relationship between teachers and students closer, make the learning atmosphere more relaxed and energetic. Control group in the traditional teaching, after—class stage, regular relaxation and cleaning work. The control group adopted the traditional teaching mode and had no contact with the content of information technology.

3 Application Analysis of Physical Education Teaching Model Based on Computer Technology

3.1 Experimental Subject

In order to ensure that the boot-independent application effect of physical education teaching mode, in this paper, some 60 college tennis learning experience students training experiment, respectively, on two groups of students before the start of the experiment for communication, the experimental group and control group in course before made a systematic study of tennis, and to the test of two groups of students' physical quality, guarantee two groups of students physical quality are about the same before the experiment, learning interest, learning attitude difference is small, and two groups of students by a teacher in the learning guidance, ensure that the experimental group and control group of learning content, learning time, exercise load and the amount of exercise are consistent.

In this paper, an interview was conducted in a college, and 60 students without tennis learning experience were randomly selected for the experiment. The 60 students were evenly divided into two groups: an experimental group and a control group.

3.2 Experimental Procedure

Students in the experimental group were taught tennis with multimedia technology, while students in the control group were taught tennis with traditional teaching mode. The learning content was basic theoretical knowledge of tennis, methods, and techniques of playing tennis, etc., except for different teaching methods, and other conditions were basically the same. Finally, the students in the experimental group and the control group were asked to analyze the evaluation results according to different forms of content evaluation, so as to verify the application effect of information technology in physical education teaching in the form of data. The statistical table of teaching content and class schedule is shown in Table 1.

In the process of using information technology in teaching, students have two class hours of tennis technique teaching every week, 16 class hours in total. In the tennis, technology teaching arranged 10 h, and in the tennis, theory knowledge study arranged 6 h. The basic tennis skills mainly include forehand and backhand. These two tennis skills are difficult for students. Through information technology teaching, students can not only display three-dimensional movements, but also better evaluate the learning effect of students.

3.3 Experimental Method

In the process of using information technology to teach tennis to students in the experimental group, the multimedia teaching materials for tennis were first constructed on the basis of consulting professional teachers and experts, then the multimedia classroom and computer laboratory were used for teaching, and finally

Table 1 Statistical table of teaching content and class schedule

Content of courses	Schedules
Basic tennis skills (12 credit hours) (01 grip, 02 prepare position, 03 forehand service, 04 backhand stroke, 05 volley, 06 return service)	Grip method (1 credit hour)
	Posture preparation (1 credit hour)
	Forehand service technique (3 credit hours)
	Backhand stroke (3 credit hours)
	Volley technique (2 credit hours)
	Receiving skills (2 credit hours)
Theoretical knowledge of tennis (4 credit hours) (Origin and development history of tennis, basic technical theory of tennis, relevant information of tennis match, tennis referee method)	4 h (1 credit hour for each module)

the link of independent learning was added after class. After eight weeks of study, the experimental group and the control group students the basic theoretical knowledge and basic technical performance compare, through the analysis of the two kinds of students learning, the traditional teaching mode and the information technology teaching model different effect to improve data support, validation in information technology help students learn sports knowledge and technology.

According to the requirements of the experiment, students' scores before learning were tested, respectively. The specific test indicators and methods are shown Table 2.

3.4 Experimental Result

In this study, the tennis skill test adopts the way of physical education teachers' technical evaluation. According to the contents and requirements of tennis ball teaching, the teacher evaluates the students' forward and backhand hitting skills. Each student has three forehand strokes and three backhand strokes. Finally, professional physical education teachers will evaluate them and take the average of the students' three scores. The full mark of this evaluation is 10 points. In this paper,

Table 2 Statistical table of test indicators and methods

Test index	Measuring method	Score
Forehand, backhand	Physical education teachers test students' forehand and backhand strokes according to the basic technical requirements of tennis. Each student has three forehand strokes and three backhand strokes, respectively, and the physical education teacher grades and evaluates the average scores of the three strokes	Technical evaluation adopts the 10-point system
Basic technical theory of tennis, tennis history, tennis match knowledge	Physical education teachers compile test papers of basic theoretical knowledge according to the teaching content and organize students to take the test of basic theoretical knowledge of tennis	The theory test is on a 100-point scale
Interest	Through the relevant questionnaire survey, the interest of students to carry out quantitative research	The learning attitude adopts the 5-level classification commonly used in sports psychology, which is reflected in the numerical value: 5-point scale

Fig. 2 Statistical table of skill test scores of the first two groups of students

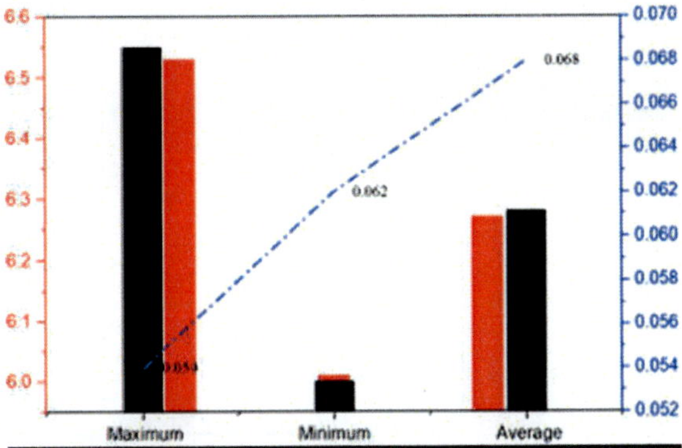

Exam skills	Maximum	Minimum	Average
experimental group	6.53	6.01	6.27
control group	6.55	6	6.28
T	0.054	0.062	0.068

the scores of each student are counted (see Annex 1 for detailed results), and the scores in Annex 1 are sorted out to calculate the average value and standard difference required in this paper. In order to ensure the authenticity and validity of the experimental results, the independent sample T-test was used in this paper to evaluate and score the students' scores. The results are shown in Fig. 2.

By analyzing the students in different periods in the same group skills, test results show that the experimental group and control group skills test scores of maximum, minimum, and average did not show significant differences, and differences in detection of the proceeds of the *P* value is greater than 0.05, shows that before the trial group and control group in the students skills test scores of maximum, minimum and average without showed evident differences, satisfy the requirement of experiment.

The comprehensive results show that there is no obvious difference in tennis skill level between the two groups of students, so the experimental results are scientific and reliable.

Then, observe the tennis level of students after the course, as shown in Fig. 3.

The horizontal analysis shows that the maximum, minimum, and average scores of the students in the experimental group are 7.11 points, 7.01 points, and 7.06 points, respectively. The maximum and minimum scores of the students in the control group were 6.72, 6.45, and 6.58, respectively. The *P* values obtained by difference detection were all less than 0.05. By means of two groups of students in different periods of analysis showed that the experimental group and control group skills test scores of maximum, minimum, and average showed significant differences, and differences in detection of the proceeds of the *P* value is less than 0.05, the differences of average detection even less than 0.01, shows that the experimental group and the control group after experimental skills students test scores of maximum, minimum, and average showed more significant difference.

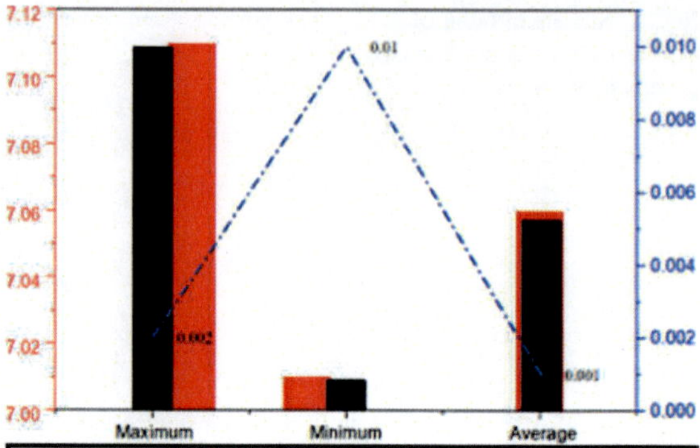

Fig. 3 Statistical table of skill test scores of students in the two groups after the experiment

Exam skills	Maximum	Minimum	Average
experimental group	7.11	7.01	7.06
control group	6.72	6.45	6.58
T	0.002	0.01	0.001

As can be seen from the experimental results, after the experiment, the tennis skill level of the two groups of students is significantly different, the experimental group students tennis level test scores is higher than the control group students, the experimental group students basic skills of the maximum, minimum and average value of the data have significant changes compared with previous experiment.

4 Conclusion

After learning, the basic technical level of students in the experimental group and the control group has been improved to some extent. However, according to the data analysis results, the application of information technology is very helpful to students' learning, and the level of students in the experimental group has been improved to a greater extent. This shows that both information technology and traditional teaching methods can improve students' tennis skills, but relatively speaking, using information technology in tennis teaching is more conducive to improving students' technical level.

References

1. Zhang Y, Han Y, Zhou Y (2014) The study on network examinational database based on ASP technology [J]. Phys Procedia 7:125–129
2. Psacharopoulos G, Patrinos HA (2013) Secondary vocational education and earnings in Latin America [J]. J Vocat Educ Training:453–459
3. Zhu X (2016) Moral education and values education in curriculum reform in China [J]. Front Educ China 12:152–158

4. Kostencka A, Szark-Eckardt M (2016) The estimation of educational needs of physical education teachers in the light of the new educational program basis [J]. Hum Mov 11(2): 33–35
5. Buchta K, Lisicki T (2011) Undergraduate studies in physical education in students' opinion [J]. Pol J Sport Tourism 18(2):99–102
6. Hall ND, Fishburne GJ (2017) Mental imagery research in physical education [J]. J Imagery Res Sport Phys Act 5(1):201–203
7. Nowak-Zaleska A (2019) Candidates for the first year of studies at the university level institution of physical education and their physical activity [J]. Baltic J Health Phys Act 3(3) :82–85
8. Jurak G, Kovač M (2019) Frequency and characteristics of excuses given by students attending special sports classes of secondary school to avoid participating in physical education class [J]. Slovenian J Public Health 50(2):52–58
9. Ivanna B, Eugene P (2015) Efficiency of inclusive physical education lessons for school children with minor deviations in health [J]. Adv Rehabil 28(3):361–367

Intelligent Systems

Design of Archives Management System for Teaching

Gongping Chen, Hong Wang, Wenyu Yang, and Shuhao Yu

Abstract In order to improve the management of a school's teaching archive information, this paper designs a management system. Because nowadays many schools still adopt manual management for teaching, and it is very low. In this work, we use Java language as development language and adopt MySQL as the database platform and SSM framework. The wet server employs Tomcat 8. We design the function module and the database in accordance with the actual requirements. The system function includes teaching file upload, teaching file management, security and user management, etc. Then, we code it and test the system. The test results show that the system can improve the archives management for teaching efficiency greatly.

Keywords Java · MySQL · SSM · jQuery · Teaching archives

1 Introduction

Since the development of archives management in history, archives have been kept in the form of physical records. After the birth of papermaking, the carrier of archives has become thin and thin, and paper archives have been kept for a long time and have been extended to modern times [1]. In an era of information, we will produce large amounts of information. If we have not a effective management system to help us to process these information, we have to pay much time on these information. At present, many schools still adopt manual management to deal with these teaching archives. These work waste a lot of manpower, and it is low efficiency [2].

G. Chen · H. Wang
College of Information and Electronic Engineering, Lu'an Vocational Technical College, Lu'an 237158, China

W. Yang · S. Yu (✉)
College of Electronic and Information Engineering, West Anhui University, Lu'an 237012, China

Z. Yu et al. (eds.), *Advancements in Mechatronics and Intelligent Robotics*,
Advances in Intelligent Systems and Computing 1220,
https://doi.org/10.1007/978-981-16-1843-7_5

Therefore, this work designs and develops a management system to deal with these teaching information for colleges or universities. The design objectives of this system include safety, reliability, stability and effectiveness. These objectives of teaching archives management system are very necessary [3]. The design can meet the needs of teachers in colleges or universities and help both teachers and administrators to convenient use system. By designing and implementing the system, it can manage and handle the teaching archives more reasonable and more scientific [4].

2 Requirement Analysis of System and Database Design

2.1 Requirement Analysis

The business requirements Nowadays, each university has more and more information, including the teaching plans of the courses taught by the teachers. The academic affairs office of the school wants to better manage the teaching information of each teacher. It is very complicated and troublesome to record the teacher's teaching information according to the previous paper materials, and it is easy to lose the information. Therefore, it is necessary to develop this platform to better manage teaching archives [5].

The user needs First of all, the overall design of the management system should be clear and concise. The page layout and color appropriate can be clear at a glance. It should quickly find the required function. The users have a good sense of experience. The information of teaching files displayed on the page should be accurate and detailed. Such as course time, teacher name, course c category, etc., should also be accurate and detailed. The font should be appropriate, and the size should be appropriate. After the administrator logs in, it is easy to find the module that needs to enter.

The functional requirements After entering the platform, teacher users can view and query the teaching files, submit the content of individual teaching files and submit the information after the relevant information is confirmed to be correct.

The teaching resource module displays the teaching file information of each teacher, such as name, teacher, category, credit and so on. Administrator logs in into the management interface. To be able to manage the data in teaching resources, teachers need to add their own courses after the administrator review. It can appear in the curriculum resources.

2.2 System Functional Structure Design

The system is divided into two modules, teacher user module and administrator module. Then, the two modules can be divided into a number of functional modules.

The user module is mainly composed of four functional modules as follows: login, personal information management, teaching course management, and check the status of course audit. The administrator module mainly consists of three modules: user management, course management and course audit management. The functional structure diagram is shown in Fig. 1.

2.3 Database Design

In the system, we design seven tables that conclude courseinstance, courseterm, department, nature, professional, teach_resourse and users. The following table structure is designed for these seven tables.

Courseinstance: store course name, course author, course time, process status (Fig. 2).

Courseterm: the name of the semester, the beginning of the semester and the end of the semester (Fig. 3).

Department: name of the depository (Fig. 4).

Nature: store course nature names (Fig. 5).

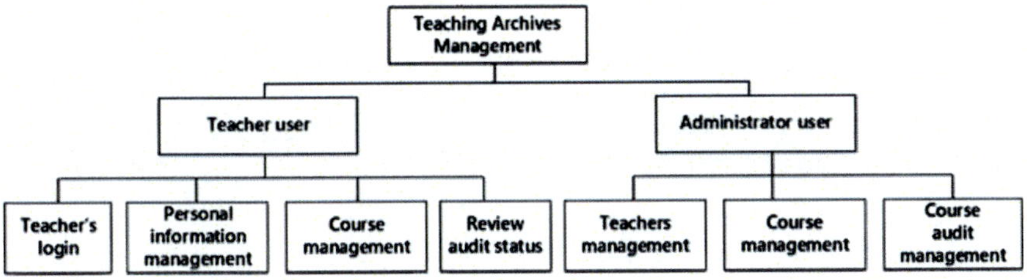

Fig. 1 System structure diagram

id	int	11	0	☑	☐	1
course_name	varchar	255	0	☐	☐	
submit_by	varchar	255	0	☐	☐	
submit_time	datetime	0	0	☐	☐	
fromstates	varchar	255	0	☐	☐	
other1	varchar	255	0	☐	☐	
other2	varchar	255	0	☐	☐	
other3	varchar	255	0	☐	☐	

Fig. 2 Design of the courseinstance table

id	int	10	0	☑	☐	1
term_title	varchar	255	0	☐	☐	
begin_time	datetime	0	0	☐	☐	
end_time	datetime	0	0	☐	☐	

Fig. 3 Design of the courseterm table

id	int	11	0	☑	☐	1
title	varchar	255	0	☐	☐	
other1	varchar	255	0	☐	☐	
other2	varchar	255	0	☐	☐	
other3	varchar	255	0	☐	☐	

Fig. 4 Design of the department table

id	int	11	0	☑	☐	1
title	varchar	255	0	☐	☐	
other1	varchar	255	0	☐	☐	
other2	varchar	255	0	☐	☐	
other3	varchar	255	0	☐	☐	

Fig. 5 Design of the nature table

id	int	11	0	☑	☐	1
title	varchar	255	0	☐	☐	
other1	varchar	255	0	☐	☐	
other2	varchar	255	0	☐	☐	
other3	varchar	255	0	☐	☐	

Fig. 6 Design of the professional table

Professional: store the title of the teacher (Fig. 6).

Teach_resourse: store course name, term, teacher, course type, course nature id, class hours, credits, syllabus, papers (Fig. 7).

Users: store teachers' name, password, account number, email, phone, id number, information, user type, school id and professional title id (Fig. 8).

id	int	11	0	☑	☐	1
couser	varchar	255	0	☐	☐	
trem	int	10	0	☐	☐	
teacher	int	10	0	☐	☐	
coursetype	int	10	0	☐	☐	
nature_id	int	10	0	☐	☐	
class_hour	varchar	255	0	☐	☐	
class_course	varchar	255	0	☐	☐	
syllabus	varchar	255	0	☐	☐	
testpaper	varchar	255	0	☐	☐	
other1	varchar	255	0	☐	☐	

Fig. 7 Design of teach_resourse table

id	int	11	0	☑	☐	1
name	varchar	20	0	☐	☐	
password	varchar	20	0	☐	☐	
account	varchar	20	0	☐	☐	
email	varchar	200	0	☐	☐	
telno	varchar	20	0	☐	☐	
cert_no	varchar	20	0	☐	☐	
message	varchar	20	0	☐	☐	
user_type	int	11	0	☐	☐	
dep_id	int	10	0	☐	☐	
pro_id	int	10	0	☐	☐	

Fig. 8 Design of the users table

3 System Function Realization

3.1 Login Module Design

The user login interface of the management system is only one. But it can be logged in with two identities: One is the teacher users, and the other is the administrator [6]. The teacher users and administrators have different authorities. The two different types of users are stored in the backend database. When a user login, the background identifies what type of user the logged-in user is. After successful login, the platform interface is displayed according to the judgment of user type [7]. When you open the management system, it will pop up the login interface. You can input name and password, and then, click login. If login is successfull, it will jump to the user interface. If your login fails, it will reminder you to login again. The login interface is shown in Fig. 9.

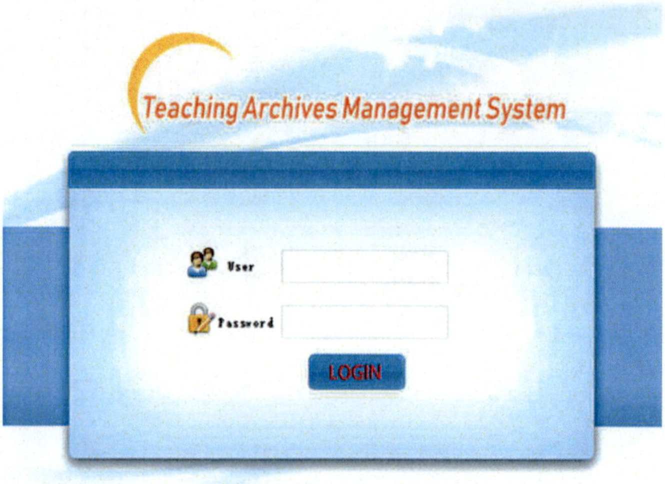

Fig. 9 Login interface of the teaching archives management system

3.2 Registration Module Design

Teaching archives management system is an internal software of a school, so the platform can set the teachers' initial password [8]. The super administrator can import the users' data, and the teachers need not to register. The teacher's information such as id and password can be added to the database. The user type can be set teacher user. Certainly, the teacher can log in the system and modify their account password.

4 System Test and Operation Result Analysis

After the coding of the whole system, we need to test the function. The administrator logs in the system and imports the teachers' account information. Then, the teachers log in the system and add the course information. The following are the test steps:

The teachers log in and input the correct account number and password. It can verify whether the login information is correct. If they input the wrong user name or password, the system will pop up the error message. After the users successful login, they click the menu of each module on the home page to test whether the page jumps smoothly and correctly and check whether the page information is loaded correctly.

The teachers test whether the course information submitted successful. They can add the course information and submit and check whether the course is in the state of auditing in the course auditing navigation bar. If displayed, it represents a successful addition.

The users can modify their personal information. For example, they can change the login password, after successful modification, exit the platform and log in again. Verify that the password is correct. After a series of information is modified, verify whether the data is changed successfully. The administrator logs in the system and enters accurate information and tests whether it can jump to the administrator interface. If the login information is incorrectly entered, the system will prompt an error message.

Administrator function module testing includes adding, modifying and deleting, etc. He can test whether data can be added correctly, modify data, and go to the database to query data. The Administrator also have the right to audit the course information submitted by teachers. The administrator can let it pass and then checks to see if the course information is displayed correctly in the teaching resource interface. The administrator also can check the teaching resources interface to see if there is no course information displayed.

5 Conclusion

This work completed the most basic part of a teaching archive management platform: the front page, background logic and data interaction. The management system realizes the functions of user login, submission of individual teaching files, administrator management of teaching files, review of submitted teaching files and so on. In the development of the front page of time, we use HTML, CSS, Javascript and jquery framework. The background code is developed based on the framework of SSM. The database we adopt is MySQL. The user can access the system through a browser. The controller of the background layer accepts the request. The Dao layer and database data interacted, and the service layer handles the request final processing of the data back to the user. This is the development process of the whole project. Because of the MVC pattern, the process is very clear. We just need to focus on the code and implement the function. And for some business log, performance detection module programming can be handed over to the IOC container.

In the future, we will design the system more intelligent and do some data mining and analysis.

Acknowledgements This research was financially supported by the top academic aid project for academic (professional) talents in colleges and universities of Anhui Province (gxbjZD74), the key project of natural science research in colleges and universities of Anhui Province (KJ2019A1065) and the quality engineering project of Anhui University (2018mooc340, 2019xfzx04).

References

1. Bo J (2010) Design and implementation of the archives management system based on SOA. Office Informatization
2. Cui SM (2013) Design and implementation of the web service-based archives management system of universities. Adv Mater Res:950–953
3. Zuo Y, Zhu W (2017) Aviation logistics information system based on SSM framework
4. Jun L, Jinshan D (2006) Research of lightweight web application based on spring MVC and iBATIS frameworks. Comput Appl 26(4):840–843
5. Lu J, Zhou W, Liu X (2017) Design and implementation of university teaching files management system based on J2EE lightweight framework. Sci Technol Square 000(004):189–192
6. Ward PL (2000) The management of information from archives (2nd ed). Lib Manage 21 (9):501–508
7. McGovern J et al (2003) Java web services architecture. Morgan Kaufmann Publishers Inc
8. Zhang D, Wei Z, Yang Y (2013) Research on lightweight MVC framework based on spring MVC and Mybatis. In: Proceedings of the 2013 sixth international symposium on computational intelligence and design, vol 01

An Algorithm for Distinguishing Between Speech and Music

Qing Jie Zheng and Hua Long

Abstract In the classification algorithm of speech music, most of the algorithms are to extract feature parameters and use the classifier to distinguish speech and music. On this basis, a speech music classification algorithm is proposed. The algorithm first preprocesses the audio signals before classification, calculates the MEL cepstrum coefficient of audio, then calculates the MEL parameter similarity matrix, then calculates the correlation matrix, and finally judges speech or music by threshold classification. The test results show that compared with other audio classification algorithms, this algorithm does not need a classifier, the classification accuracy is high, and the algorithm is simple.

Keywords Classification · Autocorrelation · Threshold · Classifier

1 Introduction

Speech and music are two important types of data in data analysis. The classification of speech and music signals is of great significance in various fields such as audio retrieval, beat tracking, and speech recognition. This article mainly provides a new algorithm to distinguish speech from music.

At present, there are various methods for distinguishing such classification problems. Such as literature [1] Lu et al. The feature of extracting audio is used to distinguish voice and music with hidden Markov. Literature [2] Wang et al. proposed a classifier based on the morphological mixed Gaussian density hidden Markov model, which is applied to the classification of speech, music, and mixed sound. Literature [3] Hu and others, based on the MLER classification method, first calculate the optimized low energy rate as a feature, use the primary classifier to obtain the preliminary classification results, finally use the context correlation of the audio category, use the context classifier to modify the initial classification, and get

Q. J. Zheng · H. Long (✉)
Faculty of Information Engineering and Automation, Kunming University of Science and Technology, Kunming 650000, Yunnan, China

Z. Yu et al. (eds.), *Advancements in Mechatronics and Intelligent Robotics*,
Advances in Intelligent Systems and Computing 1220,
https://doi.org/10.1007/978-981-16-1843-7_6

the final classification result. Literature [4] Wan et al. used echo-frequency time-frequency analysis to calculate the average energy spectrum and the optimized short-time low-energy ratio of fixed-length segments, and then, the classifier determined the type and revised the segmentation results according to the content continuity. The literature [5] Birajdar et al. proposed a feature extraction method for speech/music classification based on generalized Gaussian distribution descriptors. The visual representation of the IIR-CQT spectrogram provides superior time resolution at high frequencies and provides better spectral resolution at low frequencies compared to traditional short-time Fourier transform analysis that provides uniform frequency resolution.

This paper proposes another algorithm based on the above algorithm. First, the audio is classified as follows. After the audio has been preprocessed, it will be processed through the Mer triangle filter to obtain the MFCC parameters, and the MFCC parameters will be used to obtain the beat spectrum using the autocorrelation similarity matrix. Finally, the correlation threshold is used to determine the type of audio.

2 Audio-Related Features

The analysis of the Mel frequency cepstrum coefficient is based on the mechanism of human hearing characteristics, that is, the frequency spectrum of speech is analyzed according to the results of human hearing experiments. Because the level of the sound heard by the human ear is not linearly proportional to the frequency of the sound, the Mel frequency scale is more in line with the auditory characteristics of the human ear. The value of the Meier frequency scale generally corresponds to the logarithmic distribution relationship of the actual frequency, and the specific relationship with the actual frequency can be expressed by the following formula:

$$\mathrm{Mel}\{x\} = 2595 \times \log_{10}\left(1 + \frac{x}{700}\right) \tag{1}$$

In the formula, Mel is the perceived frequency in Mel; x is the actual frequency in Hz.

The similarity between the two vectors is calculated by calculating the cosine of the angle between the two vectors. The cosine of the 0° angle is 1, while the cosine of any other angle is not greater than 1; the minimum value is −1. The specific formula is expressed as follows:

$$d = \sqrt{[(x_2 - x_1)^2 + (y_2 - y_1)^2]} \tag{2}$$

where a and b are two feature vectors; $\cos(\theta)$ is the calculated cosine value.

3 Algorithm Steps

The steps of the algorithm of speech music classification according to the relevant features of audio are as follows:

Extract MFCC parameters. MATLAB provides the basic steps, as shown in Fig. 1.

Based on the literature [6], using the cosine similarity to calculate the similarity between the characteristic parameters, a similarity matrix can be obtained. The beat of the voice signal has no periodic rule, and the music signal will form a peak periodically.

Use the autocorrelation of the similar matrix to get the beat spectrum. Based on the literature [6] about the characteristics of the beat. It can reflect the periodic changes of rhythm (Fig. 2).

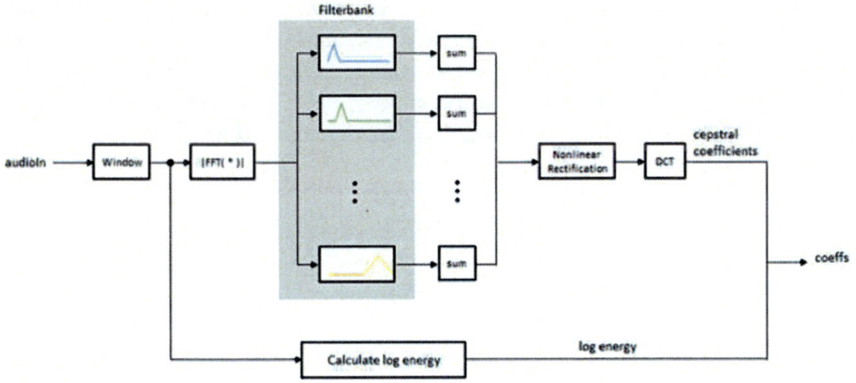

Fig. 1 Basic steps

Fig. 2 Beat spectrum

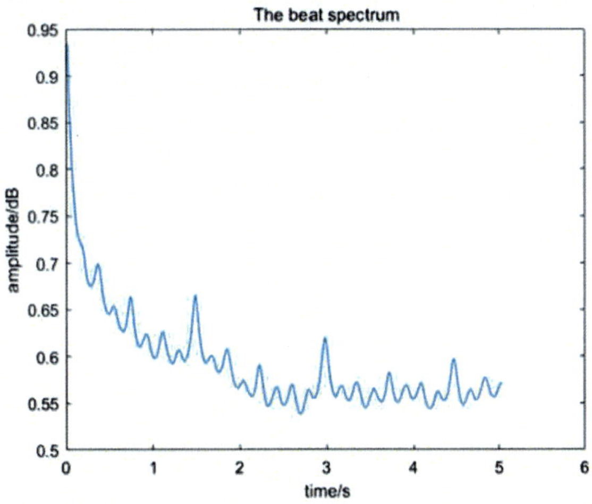

Normalized The purpose of normalization is to limit the obtained data to a certain interval, so that the data of each audio is easy to be processed later.

Count and calculate the threshold to determine the audio category. The threshold is set based on the accuracy of the overall judgment. When the accuracy of voice or music drops too much, the threshold is adjusted.

4 Experimental Result

The purpose of the experiment is to verify that the algorithm does not use a classifier to distinguish between audio and music.

The source of sample data for the experiment is radio stations, Mandarin learning network, MIR data, etc. This data is used to set the threshold.

The source of the experimental test data is the Mandarin learning network, music downloaded from major music apps of various radio stations, and so on. There are a total of 436 samples, including 198 speech and 238 voices. The sampling frequency of each sample is 16 kHz, the precision is 16 bit, mono, and the time is 10 s.

Experiment 1: It tested different results of 32-dimensional MFCC parameters and 13-dimensional MFCC parameters.

The experimental results are shown in Table 1.

Experiment 2: The 13-dimensional MFCC parameters and the 13-dimensional MFCC parameters were tested.

The experimental results are shown in Table 2.

Table 1 Experiment 1 result

Audio data set	Total	Correct sample	Error sample number	Correct identification rate/%
Speech	238	225	13	94.54
Music	198	193	6	97.47

Table 2 Experiment 2 result

Audio data set	Total	Correct sample	Error sample number	Correct identification rate/%
Speech	238	219	19	92.02
Music	198	196	2	98.99

5 Summary

It can be seen that the result of Experiment 1 is better than the result of Experiment 2 overall. The music distinction in Experiment 2 is close to 100% correct, but the accuracy of voice distinction is a bit low. It can be seen from both experiments that the accuracy rate of recognizing music will be about 2 percentage points higher than that of recognizing speech. After the experiment, a comparison of GTCC was made, and it was found that the accuracy of MFCC is relatively high. Compared with other algorithms, the accuracy of this algorithm needs to be improved. The next goal is to enlarge the data set and adjust the parameters.

Acknowledgements For the completion of my thesis, here, I would like to thank Longhua's academic guidance, Shao Yubin and Du Qingzhi, for their suggestions and my partners for their help.

References

1. Lu J et al (2002) Automatic audio classification based on hidden Markov model [J]. J Softw 13 (08):1593–1597
2. Wang C, Wu Y (2007) Automatic audio classification based on EMGD_HMM [J]. Voice Technol 31(11):52–54, 60
3. Hu Y, Wu J et al (2008) Voice/music classification method based on MLER [J]. J Tsinghua Univ (Sci and Tech) 48(S1):720–724
4. Wang Y, Zhou R et al (2013) Fast and accurate automatic music/voice segmentation method [J]. J Tsinghua Univ (Sci and Tech) 53(06):878–882
5. Birajdar GK, Patil MD (2019) Speech and music classification using spectrogram based statistical descriptors and extreme learning machine [J]. Multimedia Tools Appl 78:15141–15168
6. Guo W, Yu F (2015) Speech—music signal separation based on improved time—frequency ratio [J]. Comput Eng 41(3):287–291

Research on Multiple Overlapping Speakers Number Recognition Based on X-Vector

Lin-Pu Zhang, Hua Long, and Lin Duo

Abstract Based on the speaker recognition system, this paper studies how to recognize the number of speakers in the case that multiple speakers speak overlappingly. In order to improve the performance of the speaker recognition system, this paper is based on the x-vector system, combined with the convolutional neural network, focusing on the problems of the x-vector system, and explore effective solutions. Acoustic features are extracted by convolution neural network. MFCC is often used as input feature in the field of speech technology, but this empirical feature has some problems. In this paper, the most primitive acoustic parameter, spectrogram, is used as input feature, which contains more primitive speaker information. Meanwhile, using the mechanism of local perception and weight sharing of convolutional neural network (CNN), the spectrogram is automatically optimized, and dimensionality reduction is completed, thus avoiding the loss of information caused by empirical feature calculation. The experimental results show that the proposed scheme is reasonable and effective: extracting and optimizing features directly from the spectrogram by CNN.

Keywords Speaker recognition · Spectrogram · CNN

1 Introduction

Speaker recognition (speaker recognition, SR), also known as voiceprint recognition, is one of the core categories of audio classification and has high research and application value [1]. Voiceprint recognition technology has experienced a long time of development and evolution. Since the late 1980s, the research on speaker

---

L.-P. Zhang (✉) · H. Long · L. Duo
Faculty of Information Engineering and Automation, Kunming University of Science and Technology, Kunming 650500, Yunnan, China

H. Long · L. Duo
Yunnan Key Laboratory of Computer Technology Application, Kunming 650500, Yunnan, China

Z. Yu et al. (eds.), *Advancements in Mechatronics and Intelligent Robotics*,
Advances in Intelligent Systems and Computing 1220,
https://doi.org/10.1007/978-981-16-1843-7_7

recognition model has focused on pattern matching. At the same time, many speaker-related technologies are emerging, such as hidden Markov model (HMM) [2], artificial neural network (ANN) [3], dynamic time warping (DTW) [4], vector quantization (VQ) [5] and so on. In 2000, Reynolds applied GMM-UBM model to text independent speaker recognition, reducing the dependence of GMM model on speaker training dataset [6]. Around 2010, researchers proposed joint factor analysis (JFA) model [7] and i-vector model, which made a great breakthrough in speaker recognition.

Based on the speaker recognition system, this paper studies how to recognize the number of speakers in the case of multiple overlapping speakers. Different speakers have different frequency, volume and ending changes, etc., the above information can reflect the most original voice pattern characteristics of the speaker, and the statistics of these information can be more intuitive to judge the identity of the speaker. But when the speech items are superposed together, the frequencies may interfere with each other, and some speech features are weakened, so the feature extraction ability of the model has higher requirements.

In the previous studies, most of the problems are for speech separation or speech enhancement when multiple speakers overlap at the same time. In speech enhancement, the goal is to separate the speech from the noise. At this time, other speakers except the main speaker are treated as noise and removed. The goal of speech separation task is to separate the speech of each speaker and reduce the interference between each speaker as much as possible. However, the focus of speech separation is to extract clean audio, and its results serve the speech recognition task later. Therefore, in this process, the characteristics of the speaker may be destroyed, which will affect the accuracy of speaker recognition.

2 Approach

2.1 X-Vector

X-vector system is a speaker recognition system based on DNN. DNN is trained to extract the speaker's voiceprint features, and the extracted speaker embedding is called x-vector. As shown in Fig. 1, the whole system can be divided into two modules, which are the frame-level feature extraction module and sentence-level feature extraction module. The loss function adopts multiple types of cross entropy loss function:

$$E = -\sum_{n}^{N} \sum_{k=1}^{K} d_{nk} \ln\left(P\left(\mathrm{spkr}_k | x_{1:T}^{(n)}\right)\right) \tag{1}$$

In the above formula, suppose there are k individuals in N training statements, where $P(spkr_k|x_{1:T}^{(n)})$ means the probability of speaker K for a given T frame input

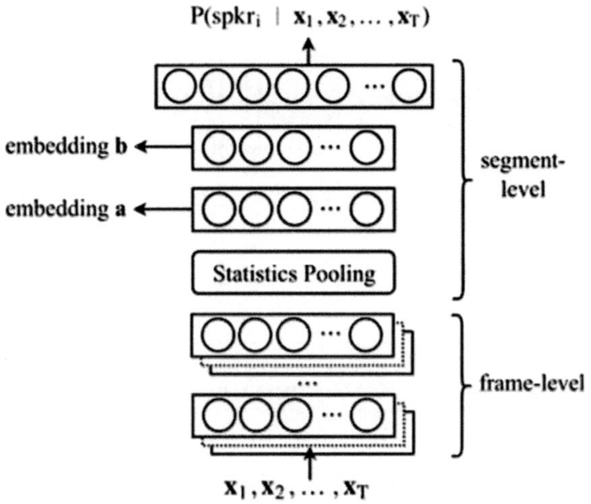

Fig. 1 X-vector speaker recognition system framework

$x_1^{(n)}, x_2^{(n)}, \ldots, x_r^{(n)}$, when sentence n belongs to speaker K, parameter d_{nk} is 1, otherwise 0.

2.2 *Data Preprocessing*

The dataset used in this paper is airshell-1, with a total time of 178 h per month. The audio sampling rate is 16 kHz. There are 400 speakers from different accent areas in China. Each speaker has hundreds of audio utterances, which is about 3–11 s long.

In order to facilitate processing, the audio is divided into 3 s. In order to prevent data waste due to the fact that the audio cannot be truncated to an integer of 3S, the method in this paper is to intercept the last 3 s of this audio segment if the remaining segment is greater than 1.5 s after the completion of the interception, so this segment includes the last remaining segment and the data of the previous segment, which will overlap partially, as shown in Fig. 2.

Audio Mixing In the mixing process, random segments of random speakers are selected. Take three speakers as an example: first, randomly select three from hundreds of speakers, and then randomly select a segmented 3S audio segment from their audio dataset, and then mix the three audio clips. A total of 50,000 audio streams were generated.

The input of the DNN is formed by stacking the 24-dimensional log filterbank energy features through a MFCC process extracted from a given frame, together with its context, 30 frames to the left and 10 frames to the right. The dimension of the training target vectors is 4, which is the same as the maximum number of speakers in one utterance.

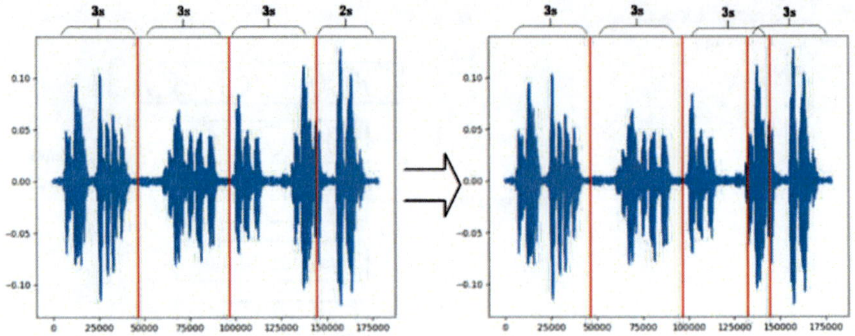

Fig. 2 Process of audio utterances separation

2.3 *Improving of Feature Extraction*

Because the acoustic feature MFCC is an empirical feature designed by human beings, there are some problems. In the process of MFCC extraction, the linear frequency is converted to Mel frequency through Mel filter bank, and the information of different frequency points is mixed together to a certain extent, which makes the key information difficult to distinguish. Using the advantages of convolutional neural network in the image field for reference, the spectrum is directly used as the input feature, and CNN is used to automatically optimize it, which can effectively replace the artificial filter bank designed according to experience to process the spectrum, and at the same time, it can reduce the data dimension, reduce the network parameters, and effectively improve the training speed of the network, and enhance the generalization ability of the model (Fig. 3).

CNN Network Settings In the frame-level feature extraction module, 3 * 3 small convolution kernels are used in all convolution layers conv1 ~ conv5 for convolution operation, with a step size of 1. This small convolution kernels are easy to capture some details of speech signals. In the first three pooled layers, pool1 ~ pool3, the features are dimensionally reduced in the frequency dimension, and in the fourth, fifth pool4 and pool5, the features are dimensionally reduced in the time dimension, all pooling operations adopt maximum pooling. The size of pooling window is 2 * 2 and the step size is 2.

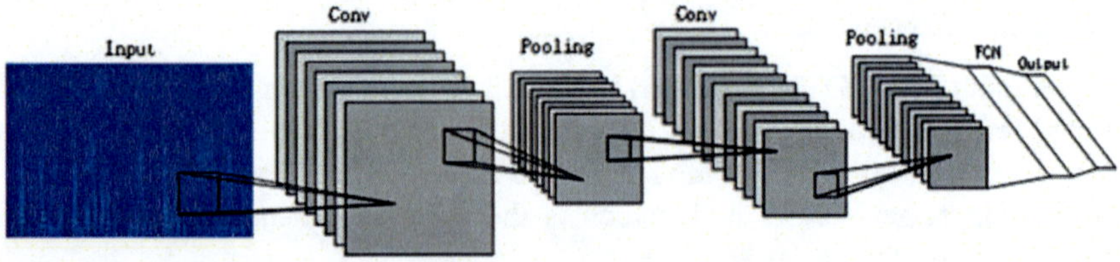

Fig. 3 Spectral features of CNN processing

Table 1 Network parameters of CNN

Model	Layers	Kernel size	Number of channels
CNN1	conv1 ~ conv5	3 * 3	32
CNN2	conv1, conv5	3 * 3	32
	conv2, conv3, conv4	3 * 3	64
CNN3	conv1, conv5	3 * 3	64
	conv2, conv3, conv4	3 * 3	128

Table 2 Results of different CNN systems on the test set

Model	Input feature	Network structure	EER (%)
Baseline	MFCC	TDNN	5.63
CNN1	Spectrogram	CNN	6.03
CNN2	Spectrogram	CNN	5.76
CNN3	Spectrogram	CNN	5.09

3 Experimental Results

In the experiment in this section, we use librosa in Python to extract the acoustic features of MFCC and spectrum and build several CNN-based x-vector speaker recognition systems. According to the spectrum characteristics, CNN is used for automatic optimization, and three network structures with different channel numbers are designed as shown in Table 1. The impact CNN with different structures on system performance is compared, and the test results are shown in Table 2.

4 Conclusions

According to Table 2, when spectrum is used as input feature, CNN may not be better than baseline system, so it is necessary to design CNN structure reasonably. The test results of CNN1 and CNN2 are worse than those of baseline. The reason is that the network parameters are set unreasonably. Too few channels will reduce the feature extraction ability of CNN. In the network structure CNN1 ~ CNN3, the number of channels in each volume layer is increasing, which enhances the feature extraction ability of the model. From the result, EER is also decreasing. When spectrum is used as input feature, model CNN3 is 0.54% lower than baseline model, 0.67% lower than CNN2, 0.94% lower than CNN1. CNN can extract local features in different spaces through different convolution kernels. Because the voice frame is continuous and the information correlation between adjacent frames is relatively large, CNN's pooling operation can aggregate the features in the region and effectively select the most representative features.

References

1. Kim J, Yang H, Kim S et al (2017) Histogram equalization using a reduced feature set of background speakers' utterances for speaker recognition. Front Inf Technol Electron Eng 18 (5):738–750
2. Sambur R, Rabiner R (2013) A speaker-independent digit-recognition system. Bell Labs Tech J 54(1):81–102
3. Wang WC, Xu J, Yan YH (2019) Identity vector extraction using shared mixture of PLDA for short-time speaker recognition. Chin J Electron 2:357–363
4. Chelali Z, Djeradi A (2017) Text dependant speaker recognition using MFCC, LPC and DWT. Int J Speech Technol 20(3):1–16
5. Iyer N, Ofoegbu O, Yantorno E et al (2007) Speaker distinguishing distances: a comparative study. Int J Speech Technol 10(3):95–107
6. Chandra E, Manikandan K, Kalaivani M (2014) A study on speaker recognition system and pattern classification techniques. Int J Inno Res Electr Electron Instrum Control Eng 2(2):963–967
7. Rastogi R, Pal A, Chandra S (2018) Multiclass universum SVM. Neurocomputing 332:151–165

Discussion on Production Technology and Testing Technology of 1553B Bus Cable Net for Satellite

Kaikai Han, Li Ma, Haihui Qiu, Fan Guo, and Fang Chen

Abstract With the development of the military industry, more and more aerospace electronic products have adopted new cable assemblies to ensure signal transmission, the quality of the signal cable assembly, and it is a key factor affecting the signal transmission performance. This article mainly introduces the 1553B bus cable network and the basic principles of bus transmission; taking common connectors as an example, it analyzes the technical difficulties in the assembly process of the bus cable network components and the causes of quality problems in the manufacturing process; in the bus cable network based on the principle of networking, the main test items, methods, and test indicators after assembly are introduced, which can effectively detect the structure or process defects of the cable and each connector, providing a powerful product for the subsequent production of highly reliable products theoretical basis.

Keywords MIL-STD-1553B · Assembly process · Testing technology

1 Introduction

The 1553B bus is the abbreviation of MIL-STD-1553 bus [1]. Before the 1960s, aircraft onboard electronic systems did not have a standard universal data channel, and the connection between various electronic equipment units often required a large number of cables [1, 2]. With the continuous complexity of airborne electronic systems, the cables used in this communication method take up a lot of space [3, 4]. In order to solve these problems, standard signal multiplexing the development of the system is imminent. In 1973, the US took the lead in developing the multiplex 1553B data bus and gradually replaced sensors, computers, indicators, and other data transmission equipment, greatly reducing the quality of the aircraft.

K. Han (✉) · L. Ma · H. Qiu · F. Guo · F. Chen
Shanghai Aerospace Control Technology Institute, Shanghai 201109, China
e-mail: hankk163@163.com

Z. Yu et al. (eds.), *Advancements in Mechatronics and Intelligent Robotics*,
Advances in Intelligent Systems and Computing 1220,
https://doi.org/10.1007/978-981-16-1843-7_8

With the introduction of advanced foreign technologies, China has also the corresponding military standard which was promulgated in 1987 [5].

2 Composition of 1553B Bus Cable

2.1 1553B Cable Structure

The use frequency of 1553B bus cable is about 1 MHz [6], and commonly used communication cables generally include coaxial cables, shielded twisted pair cables, and composite cables [7, 8]. The 1553B bus cable is a shielded twisted pair symmetrical cable. It is made of cross-linked ethylene for insulation and sheath. The cable core is formed by twisting two symmetrical insulated cores. A shielding layer is braided on the outside, and finally, the sheath is extruded. Structure shown in Fig. 1.

2.2 Composition of 1553B Bus Cable

The transmission of 1553B bus cable network has high reliability, can be used in harsh environments, and can meet the requirements of signal transmission in harsh environments such as aviation and aerospace. The bus signal transmitted by 1553B is a low-voltage differential signal. In order to realize the function of the bus signal transmission medium, the bus cable network also needs to be equipped with a coupler, a terminator and a connector, so that it can be assembled into a terminal

Fig. 1 1553B cable structure

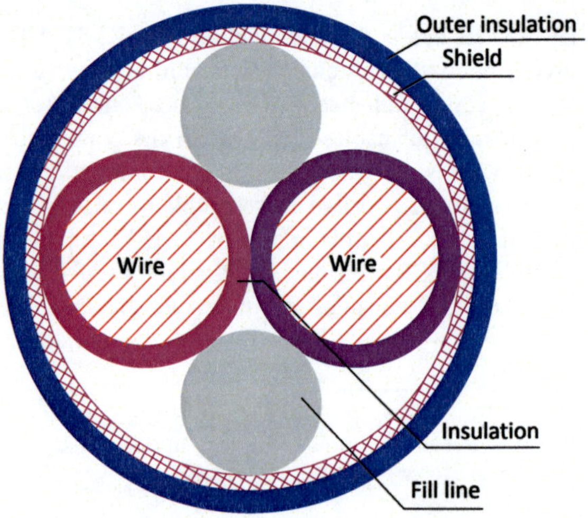

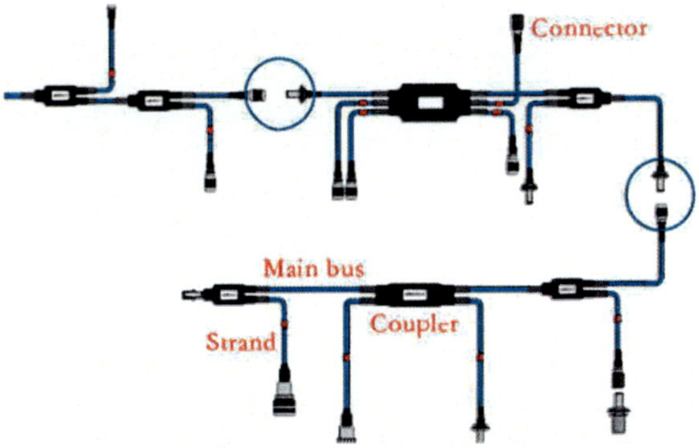

Fig. 2 Composition of 1553B bus cable

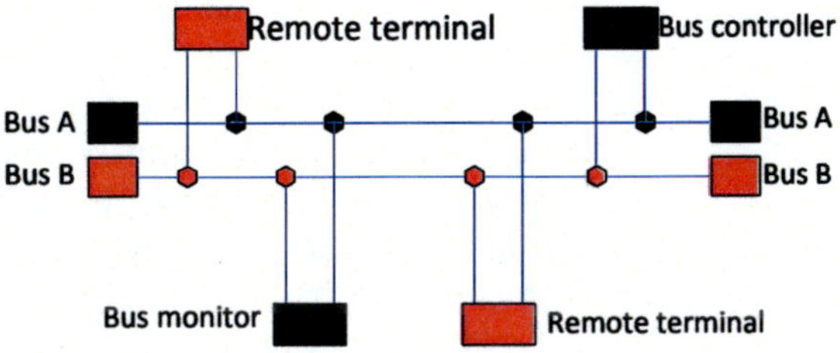

Fig. 3 Double-redundant 1553B bus

connection with a stub to form a signal transmission The bus cable network, the composition of the bus cable network, is shown in Fig. 2.

In order to improve the reliability of the 1553B bus cable network, the design of the 1553B bus uses the dual redundancy structure of bus A and bus B, so that the system can have more redundancy, as shown in Fig. 3. The structure of bus A and bus B is exactly the same, the size is the same, just to avoid misoperation when the plug is docked, the plug of one bus is all the head hole, and the plug of the other bus is all the head pin. In addition, the 1553B bus is also designed to support the interconnection with the low-frequency connector also improves the convenience of using the bus [6, 7].

3 1553B Bus Cable Network Assembly Process

The connector assembly is the core component of the 1553B bus cable network. A single connector consists of a shell, an intermediate contact piece, and a shielding sheath. The intermediate contact piece is divided into two layers of core wires inside and outside, and the two layers are isolated by an insulating medium. There are two solder rings on the inner and outer layers of the middle contact piece, and the twisted pair and the middle contact piece are connected by the solder ring in a "blow welding" manner. The DK-621 type connector structure is shown in Fig. 4.

Fig. 4 DK-621 type connector structure diagram

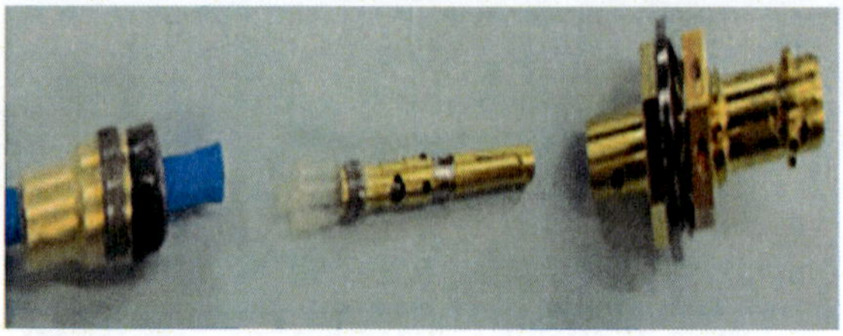

Since the twisted pair is connected to the intermediate contact piece by "blowing welding" with a hot air gun, the contact position between the stripped end of the twisted pair and the welding ring must be very precise. In order not to damage the core wire of the insulated wire during the operation, the end of the core wire of the insulated wire can be tinned. When tinning, make sure that the core wire is 0.5–1.0 mm away from the root of the insulation layer without tinning and then use scissors to cut the white and blue lines to the required size. During the blow welding, the wire and the intermediate contact assembly should be placed horizontally into the blow welding fixture. After the position is adjusted, the thermal welding will begin until the solder ring is completely melted; the housing should be installed after the thermal welding of the wire, and the intermediate contact is completed. After the shell is installed in place, push the solder ring onto the shell and heat it together. Finally, after the solder ring is cooled, the heat shrinkable sleeve is sleeved on the solder ring and protected by blowing.

Regarding the effect of connector blow welding, from the perspective of the failures in the assembly process, there are mainly two modes. One is the open circuit of the cable core wire. The main reason is that the signal line and the connector are in the process of cable end processing. The size is not completely matched, the core wire cannot form a good connection with the connector, and the molten solder does not form a reliable IMC layer with the wire end after enameling; another failure mode is caused by the heat shrinkable tube being heated and softened for a long-time short circuit. Through microscope observation and X-ray inspection of the faulty connector, it was found that too high blowing temperature of the solder ring and too long blowing time can easily cause the heat shrinkable tube to melt and short circuit. Therefore, optimizing the connector assembly process is the key to avoiding the above problems.

4 1553B Bus Cable Network Installation Quality Test

In order to ensure the reliability of the quality of the 1553B bus cable network installation, inspection and testing are required after the product is completed. The inspection items mainly involve the appearance inspection of the 1553B bus cable network product, static and dynamic testing of the product, and other content.

4.1 Bus Cable Net Appearance Inspection

After the assembly of the 1553B bus cable network is completed, the product needs to be subjected to routine appearance inspection. The main appearance inspection items include:

(1) The number and length of the bus branches, the model specifications and location of the coupler, the model specifications and corresponding locations of the electrical connector, the model specifications and weight of the terminal load, etc. should all comply with the drawing requirements;
(2) No connection parts or any components of the bus cable can be contaminated, scratched, cracked, deformed, without shielding layer exposed, without deformation of the cable seat, core, pin, etc. The cable label and connector code should be correct and clear. The cable is free from hard bending and cracking of the insulation layer.

4.2 1553B Bus Cable Network Static Performance Test

The 1553B bus cable network static performance test content mainly includes two parts: point-by-point test items and single-point test items. Among them, the point-by-point test items mainly include: insulation resistance between the bus and the branch line, shielding layer connectivity, and branch resistance; the single-point test items mainly include: bus parallel resistance, bus cable resistance hot spot value, bus cable resistance cold spot value, and insulation resistance between cable shield and internal cable. For the bus cable network containing the bus multiplexer, the on-off test of the electromagnetic relay inside the multiplexer is also required.

In the static performance test of the bus cable network, the single reflection and short-circuit method can directly measure the complex reflection coefficient, so that the resistance between the high and low signal lines of each stub will change. Therefore, the static test can accurately reflect the cable and each the structure or process defects of the connector, especially when the cable has insulation dielectric defects or conductor disconnection and virtual connection, the complex reflection coefficient will change suddenly at the defect, so that the resistance between the high and low signal lines will change significantly, coupling the schematic diagram and static test method of the device and multiplexer are shown in Fig. 5, and the static performance test indicators of the bus cable network are shown in Table 1.

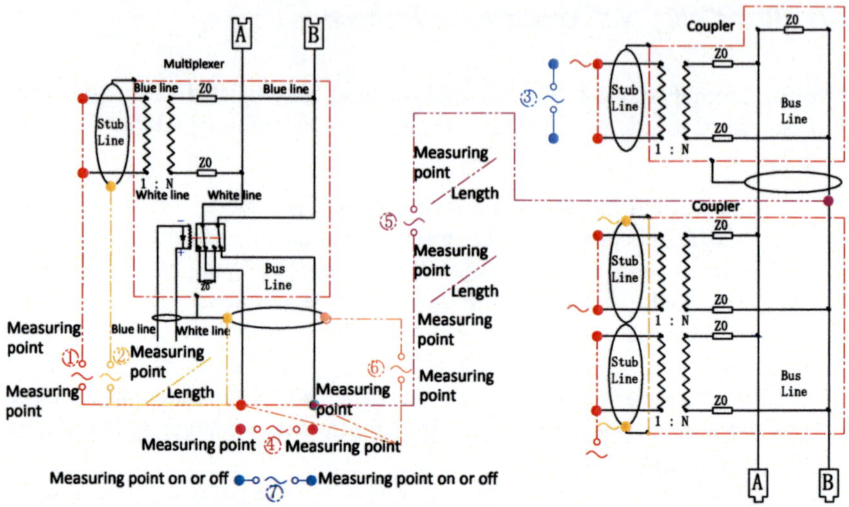

1,Insulation resistance between bus and branch 2,Shield layer connectivity 3,Branch resistance 4,Bus parallel resistance 5,Bus cable resistance hot (cold) point value 6,Insulation resistance between cable shield and internal cable 7,Multiplexer Break test

Fig. 5 Schematic diagram of bus cable network device

Table 1 Bus cable network static performance test index

Phase	No Dephasec
Insulation resistance between bus and branch	Ri > 1000 MΩ Under 250 VDC
Shield connectivity	Sc < 15 mΩ/m
Branch resistance	RSTUB < 5Ω
Bus parallel resistance	16.7Ω < RBUS < 23.4Ω
Bus cable resistance hot spot value	RB < 109 mΩ/m
Bus cable resistance cold spot value	RB < 109 mΩ/m
Insulation resistance between cable shield and internal cable	Ri > 1000 MΩ Under 500 VDC
Bus multiplexer performance test	28 ± 1.5 V 50 ± 10 mA

4.3 1553B Bus Cable Network Dynamic Performance Test

The 1553B bus cable network needs to be tested for dynamic performance after the networking is completed. During the test, the tester uses each bus branch as the sending end and the other branch ends as the receiving end for communication

Table 2 Bus cable network dynamic performance test index

Phase	No Dephasec
Vpp	1.0–14.0
Offset	<200 mv
Vmin	>0.66 V
Rise	<350 ns
Fall	<350 ns
Zcross	<0.125 μS
RtZero	<250 mV

traversal testing. Both when sending and receiving messages for dynamic waveform measurement, the main items of dynamic performance test are as follows:

(1) Peak-to-peak value: Signal transmission amplitude voltage value;
(2) Offset: The maximum offset of the signal transmission voltage;
(3) Waveform distortion: Minimum value of signal transmission voltage;
(4) Rise and fall time: Rise and fall time of signal transmission;
(5) Zero crossing offset: Maximum deviation of signal zero crossing;
(6) Waveform symmetry: Tail drag voltage for signal transmission.

The indexes of the dynamic performance test are shown in Table 2.

5 Conclusion

This article mainly introduces the basic principles of the 1553B bus cable network and its transmission, analyzes the technical difficulties and the causes of quality problems in the production of the bus cable network, and introduces the assembly connection based on the principle of the bus cable network. After the main test items, methods and indicators, the production practice found that this method is an effective means of detecting defects in the production process. Through the appearance, static and dynamic performance testing of the product, the reliability of the quality of the 1553B bus cable network assembly can be fully ensured, which provides a theoretical basis and engineering reference for the subsequent product assembly and testing.

References

1. Yulong W, Shouhong S, Zhang W (2012) Research on assembly process of 1553B cable assembly in military electronic products [J]. Electron Process Technol 000(4):219–221
2. Zhang Y, Zhang H, Ruimin P et al (2014) Test method of cable impedance for 1553B data bus [J]. Opt Fiber Cable Their Appl Technol (03):14–17
3. Liu Y (2012) On 1553B bus and its application development [J]. Friends Sci 8:167–169

4. Jian O (2012) 1553B data bus comprehensive test system [J]. Ordnance Ind Autom 1:45–48
5. Zhang K, Wu L, Zhang S et al (2018) Research on quality control method of assembly process of 1553B bus cable network used in spacecraft [J]. Qual Reliab 197(05):9–12
6. Zhou L, An J (2014) Research on wireless MIL-STD-1553B bus based on infrared technology [J]. J Netw 9(10)
7. Zhang Z, Yuan C, He Z et al (2018) 1553B data bus assembly technology [J]. Opt Fiber Cable Their Appl Technol (03):13–16
8. Huang W, Wang Z, Wang X (2017) Brief introduction of 1553B bus technology and its application [C], 7

Feasibility Analysis of Venture Capital Committee in the Innovative Design of Class Management in Colleges and Universities

Liu Xianming, Yu Qianqian, Yang Yong, Zheng Shishi, Xiong Yazhou, and Xu Xiaofang

Abstract The innovative design of class management in colleges and universities aims to provide class management tools that are flexible, scientific and systematic. This management mode, incorporating the initiative "mass entrepreneurship and innovation" which responds to the call of our country, keeps up with the pace of development of the times. In this paper, the innovative design is analyzed by literature research and case analysis to explain its practical application and confirm its practical feasibility. Taking class of 2010 majoring in business administration of Hubei Polytechnic University as an example, this paper can draw the conclusion that this management mode can serve the needs of the times concerning class management, contributing a lot to the development of individuals, countries and society.

Keywords Venture capital committee · Innovation and entrepreneurship · Higher education · Class management · Feasibility analysis

1 Introduction

The purpose of innovation and entrepreneurship (I&E) education is to enhance college students' consciousness, spirit and ability in this respect, and to foster high-quality talents for building our country into an innovative one. Since the official introduction of the policy and documents on I&E education in 2010, China has been attaching greater importance to this education and policy support has been given it as well, making significant effect eventually. However, few scholars have studied how to combine class management with I&E education so as to promote the development of the latter [1, 2].

L. Xianming (✉) · Y. Yong · Z. Shishi · X. Yazhou · X. Xiaofang
Hubei Polytechnic University, Huangshi 435003, Hubei, China

Y. Qianqian
Nanjing Audit University, Nanjing 211899, Jiangsu, China

Z. Yu et al. (eds.), *Advancements in Mechatronics and Intelligent Robotics*,
Advances in Intelligent Systems and Computing 1220,
https://doi.org/10.1007/978-981-16-1843-7_9

Since 2006, the author has deeply studied the class management mode, taking the venture capital committee as an important part of class organization, which not only helps to manage classes in school, but also produces greater economic value after graduation [3, 4]. This paper designed some charts, such as the class organization chart, job description, operation flow chart of class fee-generating and class fee-generating business project canvas, forming a complete set of class management tools.

2 Major Pain Points of Class Management in Colleges and Universities

2.1 *Different Opinions About Head Teacher's Impact on Class*

For a long time, the class organization in colleges and universities lacks flexibility, that is, the head teacher or counselor takes the lead and the class cadres are responsible for class affairs, so as to coordinate the class management. The main problems include head teachers' overstepping authority, management neglect and ability deficiency [5].

2.2 *Weak Initiative of Class Cadre*

The class cadres have weak initiative, which is shown in active competition in the early stage but shirking responsibility in the later stage, taking no initiative, lacking the corresponding power. Furthermore, they found their personalities may make greater differences than the position instead, as well as do thankless task [6].

2.3 *A General Lack of Class Fee*

At present, the funds for class activities in colleges and universities are generally paid by all students in equal amounts. Therefore, economizing on expense has become the only way to develop necessary class activities.

2.4 *Little Contact with Classes After Graduation*

After graduation, classes are in a state of unorganized management, and the contact among students is reduced, with their losing contact with the head teacher or school, but only class reunions after graduation for many years may bring students together again.

3 Pain Points Analysis of Class Management in Colleges and Universities

3.1 Urgent Need of System Construction Innovation

With the advancement of the I&E education and that of the science and technology, the class management system in colleges and universities is in urgent need of innovation. Firstly, the traditional class organization structure can no longer meet the needs of students' personal growth and development. Secondly, the nonstandard job functions result in unclear responsibilities and powers of class cadres [7].

3.2 Difference in Value Proposition

There are three main reasons for the weak initiative of class cadres: first, after the election of class cadres, some students may think that the elected class cadres can hardly play a substantial role in serving the class, which causes the devaluation of self-worth for elected ones; second, other students may consider that class cadres are nothing but busybodies, so class cadres lose all initiative eventually; third, class cadres think that "the less trouble the better," thus becoming absentee leaders.

3.3 Foundation

There is a single source of class fee in colleges and universities, that is, students pay their own expenses. The financial resources of college students, especially junior students, mainly depend on their parents.

3.4 Repulsion and Practical Difficulties

Due to the differences in family and geographical environment as well as growth experience, students' personalities vary obviously from person to person, thus having different values and making different friends. After graduation, they only keep in touch with their goods friends, and do not care whether there is a class reunion or not.

4 Innovative Design of Class Management in Colleges and Universities

4.1 Description of Class Management Tools

During the school years, the purpose of class management is to achieve class positioning, that is, to build an excellent class. The organization chart of the class treats the class as an enterprise for management. The following three departments are the league branch, the venture capital committee and the class committee (Fig. 1). The league branch is managed by the secretary, and the members of this department are publicity commissary and organization commissary. The evaluation of its contribution to class honor is 20%, which is mainly responsible for the work of class activities and honor evaluation. The venture capital committee is managed by the president and includes the student's life commissary, female commissary and disciplinary commissary as its members. This department's contribution to the class fee-generating is 30%. It is mainly responsible for the external communication, internship training, and the formulation and implementation of the class fee-generating plan for the class. The class committee is managed by the monitor, and it is including the recreation and sports commissary and studies commissary. This department is assessed that it is academic contribution to class is 50%, and it is responsible for daily management work such as in-school learning and safety protection of the class.

Each department has its own organizational chart to clarify its position and responsibilities. In addition, the head teacher, department head and its members all have job description to clarify the job conditions and responsibilities.

The innovation of class management structure lies in the venture capital committee. The principle of venture capital committee is to take the interests of all parties into consideration and encourage all members to participate in activities so as to make the long-term development of the class. While at school, the venture capital committee provides students with a platform of practice and drives them to participate in various activities, which not only helps students earn living expenses

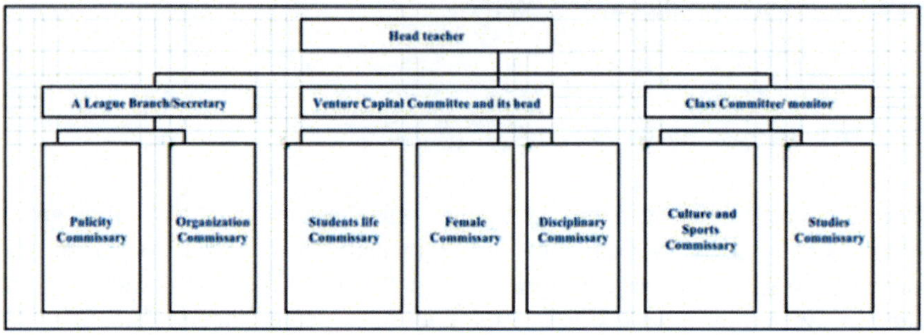

Fig. 1 Class organization chart

and gain practical experience, but also further promotes students' learning ability and learning effect. For each committee member, it is to give them rights of planning activities and making decisions, no longer just to provide a single service for the class. After graduation, the venture capital committee will become a formal management organization, responsible for the organization and coordination of post-graduation class league construction, employment and entrepreneurship, priority investment projects and other related activities.

See Fig. 2 operation flow chart of class fee-generating. These three departments are jointly responsible for the operation of generating profits of class fee. One of the operation processes about the business of class fee-generating is that the department obtains part-time job and venture capital information, sets up practical training team, organizes skill training, and organizes class students to practice in the internship base of the university. Another part of its operation process is to organize students to take part-time jobs in social practice bases, outsource business, employment and entrepreneurship, etc. The operation flow chart stipulates that departments should make contributions to increase class funds, students' extracurricular practice, internship experience, the rate of employment and entrepreneurship and other matters, and provides a scientific, reasonable and logical thinking in terms of management of class fee-generating, development of students' comprehensive ability and promotion of class cohesion.

See Fig. 3 for the project canvas of class fee-generating business. The business is divided into promotion, marketing outsourcing and talent customization. There are two different paying parties respectively distributing at the bottom of the three

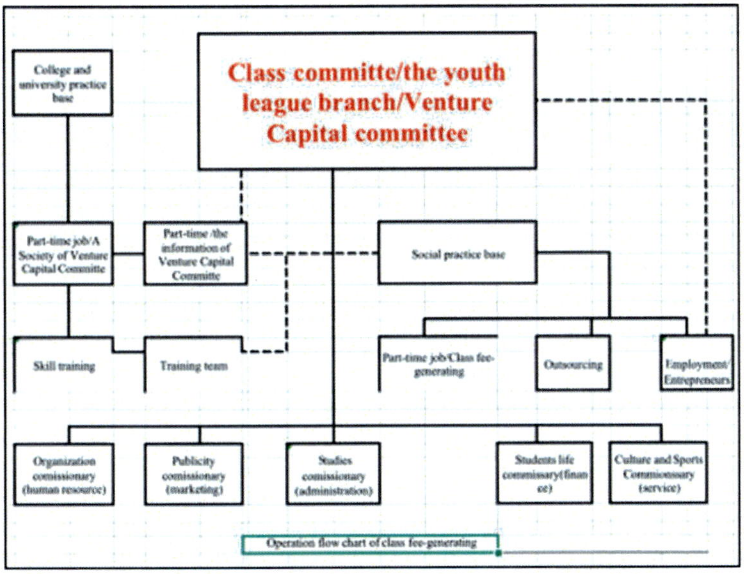

Fig. 2 Operation flow chart of class fee-generating

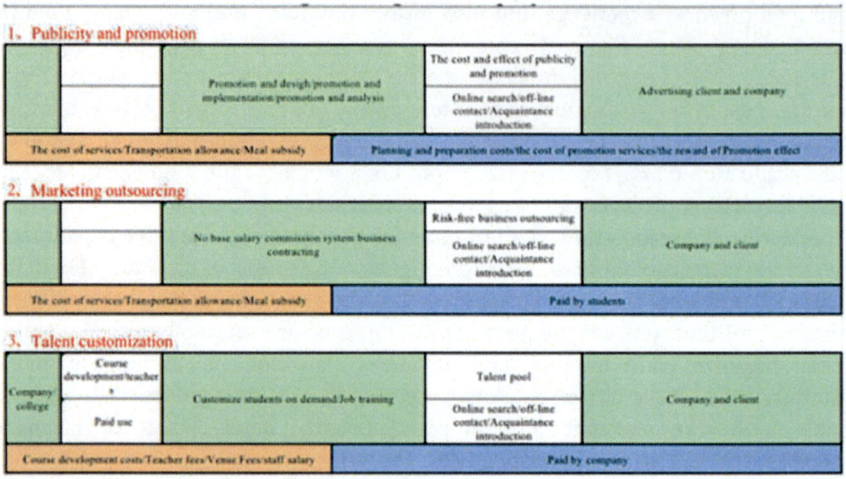

Fig. 3 Class fee-generating business project canvas

businesses table of the canvas. The left and right ends are channel partners and the businesses they undertake, and the blank space means that channel partners are not needed. The middle section introduces the specific transactions of each business. The canvas defines the types of channel partners required by the business and the businesses they should undertake, determines the pay parties of each business and can assist with the clear overall planning to develop the class fee-generating business.

4.2 The Feasibility Analysis of Innovative Design

This paper takes the class of 2010 business administration of Hubei Polytechnic University (hereinafter called "class M") as an example to analyze the feasibility of class management innovation design. First of all, in university, the well-organized cooperation of class M according to the organizational class structure played a significant role in the construction of class discipline and academic atmosphere and economic construction. The youth League branch is responsible for the class activities and the honor evaluation work builds a sound class collective consciousness for all the students in class. During the four years in the university, class M has won various honorary titles such as excellent class, excellent league branch and the class of excellent discipline. Some daily works such as organizing students' study in university and strengthening the safety protection were taken by class committee, which inspired all the students to learn by founding reading groups, appointing representatives and other ways. In addition, the class degree rate reached 100% upon graduation, which was closely related to the management and service of the class committee. Venture capital committee was responsible for the external liaison of the class, practice and training, class income generating scheme formulation and implementation, namely the development of class economic construction. In university, the committee provided students with part-time job information and relevant liaison work, and students paid a certain proportion of class fees as the class activity funds.

In 2016, all the students of class M jointly established the Charity Union as the formal management organization after graduation. A council would be established in the union and shall be responsible for the overall planning of the work of the union. The Charity Union is mainly responsible for the logistical support of class members' return to school, payment of membership dues, donation to Alma mater, organization of alliance activities, provision of project resources, organization of investment projects, etc.

5 Conclusions

This set of university class management tools proposed in this paper cannot be listed one by one because of the large number of forms. The form is simple, but complete and logical, which has been proven feasible by years of practice. After graduation, the class set up a formal organization for common development and progress, which meets the urgent needs of the current national I&E education and can effectively improve the educational, social and economic benefits. However, although the management tools are effective, there is no denying that users, using methods, and whether formality or not also play a vital role. Therefore, how to ensure that class managers can effectively implement this management mode is still a problem to be explored.

Acknowledgements The study is supported by the college students' innovation and entrepreneurship training projects of Hubei province Grant No. S202010920070.

References

1. Wei Y (2016) Study on the present situation and countermeasures of class construction management in colleges and universities [J]. Educ Modern (37):242–244
2. Li B, Yao T, Wenyan J (2018) A survey of class characteristics and management operation in colleges and universities [J]. Public Sci (11):139
3. Zhong L, Xu Y (2017) Research on the status quo, problems and countermeasures of head teachers' work in colleges and universities [J]. J Shandong Inst Sci Technol Shandong Shangye Zhiye Jishu Xueyuan Xuebao (05):26–28, 41
4. Li E (2019) Research on the status quo of student cadres in college and university [J]. Technol Econ Guide 31:102–103
5. Liao A, Shang S, Tian Y (2010) Current situation and management countermeasures of college and university student class construction [J]. Zhongguo Xiaowai Jiaoy (16):04, 05

6. Wu C (2019) A brief discussion on the status quo and improvement strategies of class construction management in colleges and universities [J]. China J Multimedia Netw Teach (The ten-day issue) (02):51, 52
7. Xiong Y, Zhang S, Lan J, Chen F (2019) Efficiency evaluation of regional economic development of mining and metallurgy city based on DEA model. Int J Appl Decis Sci 12(3): 242–256

A Survey and Study on Satisfaction and Influencing Factors Using Ant Credit Pay via Intelligent Processing

Jin Wu

Abstract With the rapid development of Internet finance, the various platforms of the network consumer credit gradually appeared in people's daily consumption, and "Ant Credit Pay" products of Alibaba company are widely popular among college students because of their low threshold and wide range of use. In order to understand the specific situation of the use of "Ant Credit Pay" by the college students' consumer group, the satisfaction of "Ant Credit Pay" and the factors of influence satisfaction via intelligent process, the study takes Hubei Polytechnic university students as a sample, using SPSS software descriptive analysis, chi-square test and Pearson correlation coefficient to analyze data. Through analysis and research, "Ant Credit Pay" needs to continue to improve the risk prevention system, so that consumers need to optimize the web interface, strengthen publicity efforts, increase consumer recognition, reduce the use threshold and transaction costs, simplify the transaction process and repayment process, so that consumers can use more easily. At the same time, the college students should treat the online credit products in a rational and objective attitude and establish the correct consumption values.

Keywords Influencing factors · Ant Credit Pay · Satisfaction theory · Intelligent processing

1 Introduction

With the development of science and technology, the Internet is gradually integrated into people's life. Around 2010, the Internet officially entered the financial sector. Up to now, China's Internet finance business model has become diversified but also developed a different operating mechanism. During the development of Internet finance, the consumption mode of consumption first and payment later came into being gradually. Ant Credit Pay, was an online credit product developed

J. Wu (✉)
School of Economics and Management, Hubei Polytechnic University,
Huangshi 435003, Hubei, China

Z. Yu et al. (eds.), *Advancements in Mechatronics and Intelligent Robotics*,
Advances in Intelligent Systems and Computing 1220,
https://doi.org/10.1007/978-981-16-1843-7_10

by Ant Financial, an Internet finance giant. As a product of Internet finance, it is essentially a credit product, providing credit service for users through the Internet. When consumers open ant chant function, the system will be based on the working conditions of consumers, as well as the financial situation. Ant Credit Pay judges the credit grade of the consumer from many directions and then determines the credit limit. As the low-interest, convenient and safe credit model, it is popular with students and has become a trend.

In this paper, a series of investigations and studies on the use, recognition and satisfaction of credit pay were carried out among college students of Hubei Polytechnic Technology through a sampling survey, after the basic information was obtained; it was used to study the satisfaction degree of university students to Ant Credit Pay and its influencing factors. Through the comparison of survey data, we can preliminarily analyze the degree of satisfaction of contemporary college students to credit pay and the change of their living conditions because of credit pay; furthermore, we can analyze the factors that influence the satisfaction degree of using Ant Credit Pay. Finally, the paper gives some suggestions on the improvement of the product and the future development of ant chant.

2 Literature Review

In recent years, most of the experts' research on consumer credit products in China has focused on the prospect of the development of online consumer credit platform in the era of big data, and the changes in consumer economy after the emergence of online credit products to the consumer and the traditional financial industry have caused certain impact and change. Li and Yang [1] analyzed the opportunities, challenges and risk types of Internet finance development in the big data era, and in order to avoid such risks, we need to improve credit reporting with big data, the mode of supervising Internet finance and the system of customer management. Wang [2] from the perspective of the development of Internet finance, a comprehensive study of Internet finance model, and this new financial model on the stability of the traditional financial sector and social economy can constitute a positive impact. Yu [3] used the data of consumer behavior of college students and Internet consumer credit to model the relationship between the two through AISAS; this paper puts forward some suggestions on how to better adapt the Internet consumer credit products to the rational consumption of college students and college students through online lending platform. Du et al. [4] believe that in today's social development, the continuous improvement of the network law, as well as the contemporary young people's consumption concept changes and the continuous development of online lending platform, the network credit platform has huge development space in the future.

By analyzing the factors that influence the satisfaction degree of the research object to the Ant Credit Pay, we could improve it, or make the Ant Credit Pay cover more college students. Not only can obtain the acceptance degree of this mainstream group of university students, but also can make the risk of Ant Credit Pay itself to the minimum, so that both sides can reach a win-win situation [5].

3 Data Analysis Method and Theoretical Basis

3.1 Chi-Square Test Principle

Chi-square test is a commonly used method in statistical inference of classified data. Its principle is to compare the theoretical and observed values of statistical samples, and the deviation between them represents the magnitude of chi-square value and chi-square value reflects the agreement between the theoretical and observed values. There is a positive correlation between chi-square value and the deviation between them, that is, the larger the deviation is, the smaller the deviation is, and the chi-square value will approach to zero when the two are in complete agreement, which indicates that the theoretical value is in complete agreement with the observed value.

The basic idea of analysis is to observe the close relationship between samples according to the numerical characteristics of things. And the close relationship is measured by the distance between the samples. Once the distance between the samples is defined, the close samples can be classified into one category.

3.2 Principle of Correlation Analysis

Correlation analysis is a method used to test correlation, which is usually used in the study of two or more random variables. In general, there is dependence among random variables. The change of one variable will cause the change of another variable. Although the change of value is not fixed, it all follows a certain law and fluctuates in a certain interval. In this paper, we use the Pearson Correlation Coefficient method (PCCs), and PCCs is that if we move y to c + dy and x move to $a + bx$, where a, b, c, and d are constants, which do not change the correlation between the two variables, x and y.

PCCs is usually calculated by the following formula (1):

$$\gamma = \frac{\sum_{i=1}^{n}(X_i - \bar{X})(Y_i - \bar{Y})}{\sqrt{\sum_{i=1}^{n}(X_i - \bar{X})^2}\sqrt{\sum_{i=1}^{n}(Y_i - \bar{Y})^2}} \quad (1)$$

Where the symbols of $\gamma, X_i, \bar{X}, Y_i, \bar{Y}$ represent the PCCs, sample value, the sample average value of X, Y, respectively.

3.3 Satisfaction Principle

Customer satisfaction is the degree to which an expectation of a customer in a service industry matches a customer's experience. In Layman's terms, it is an index of how customers feel about a product when they use it versus how much they expect them to do before they use it [6].

The RATER index is a universal and authoritative way to measure the quality of enterprises or products and services. The RATER index directly affects and even determines the degree of satisfaction of users. The RATER index comes from the initial letters of five English words: word "R" represents the word "reliability," word "A" represents the word "assurance," word "T" represents the word "tangible," word "E" represents empathy, and word "R" represents the word "responsiveness."

3.4 Establishing the Satisfaction Index System

The satisfaction index system of using "Ant Credit Pay" for college students is shown in Table 1 below.

4 Analysis on the Satisfaction and Influencing Factors

4.1 Reliability Analysis and Validity Analysis of Data

Reliability analysis. Using the alpha reliability coefficient method, this paper analyzes the five aspects indicators of "Ant Credit Pay" and obtains the alpha values of these indicators. Table 2 is a summary of the alpha values of these indicators.

As can be seen from Table 2 above, the alpha values of these indicators are all above 0.9, indicating that the reliability of the questionnaire is very high.

Validity analysis. Using KMO sample measure and Bartlett sphere test method, the validity of five aspects of "Ant Credit Pay" was analyzed, and KMO values of these five indicators were obtained. Table 3 is a summary table of KMO values and Bartlett sphere test of these indicators.

It can be seen from Table 3 above that KMO values of the above five indicators are all above 0.7, and the *P* values of Bartlett sphericity test of the five indicators are all 0. Therefore, the sample data can be used for factor analysis.

Table 1 Satisfaction index system of using "Ant Credit Pay" for college students

First-class indicator	Second-class indicator	Indicator interpretation
Basic service	Information transparency	The level of disclosure to users
	Customer service	The level of back end customer service provided to users
	Scope of use	The specific scope of use of Ant Credit Pay
	Repayment time	The minimum repayment time allowed for users' interest free
	Installment interest	The interest rate fixed for users' amortization
	Rate credit system	A system for determining the user's available quota
Using risk	Pay for security	The security of Ant Credit Pay's password when it is used
	Personal privacy security	Ant Credit Pay's privacy to users
Product image	Web interface	The design of Ant Credit Pay's web interface
	Overall image of the company	The image of Alibaba company that owns Ant Credit Pay
	Promotion efforts	Ant Credit Pay's promotion to consumers
Convenience	Using threshold	The conditions of using Ant Credit Pay
	Transaction cost	The cost of using Ant Credit Pay
	Transaction process	The specific operation process of consumption;
	Ease of payment and repayment	The convenience of consumption in payment and repayment
Recognition	Influenced by others	Influenced by the people to use Ant Credit Pay
	Brand recognition	Approve the brand of Alibaba company to use Ant Credit Pay
	Media outreach	The influence of various media propaganda on my use

Table 2 Summary table of alpha values of all indicators of questionnaire data

Indicators	Normalized alpha value	The number of secondary indicators
Basic service	0.977	6
Using risk	0.992	2
Product image	0.988	3
Convenience	0.991	4
Recognition	0.934	3

Table 3 Summary table of KMO value and Bartlett sphere test of questionnaire data

Indicators	KMO value	The corresponding *p* value of Bartlett sphere test	The number of secondary indicators
Basic service	0.893	0.000	6
Using risk	0.938	0.000	2
Product image	0.780	0.000	3
Convenience	0.853	0.000	4
Recognition	0.768	0.000	3

4.2 Analysis of College Students' Satisfaction with Using "Ant Credit Pay"

Correlation coefficient matrix. Pearson method is used for correlation analysis, and the correlation coefficient matrix of 12 indicators is calculated.

And all the correlation coefficients are all greater than 0.85, and the correlation coefficients are very high, indicating that the 12 variables present a strong linear relationship, and common factors can be extracted from them. Therefore, it is necessary to perform factor analysis on the 12 indicators.

Bartlett sphericity test and KMO test. The specific test results are shown in Table 2.

It can be seen that in Bartlett's sphericity test, the observed value of the test statistic is 765.125, and *P* value is close to 0, which means the test statistic is very significant. The null hypothesis is rejected, and the correlation coefficient matrix is considered unlikely to be a unit matrix, so it is suitable for factor analysis.

The KMO test has a value greater than 0.7. According to the KMO metric given by Kaiser, it is suitable for factor analysis.

4.3 Analysis of College Students' Satisfaction

This questionnaire investigates the satisfaction of Ant Credit Pay from the basic business, risk, product image, convenience and recognition, whose measurement results are shown respectively in Tables 4, 5, 6, 7, 8. And the comprehensive measurement result is shown in Table 9. The Minnesota Satisfaction Scale is used, in which very unsatisfied, dissatisfied, general, satisfied and very satisfied correspond to $1 \sim 5$ of the order of magnitude respectively. The higher the order of magnitude, the higher the degree of satisfaction.

Table 4 College students' satisfaction with basic business

	Very dissatisfied (%)	Not very satisfied (%)	Generally (%)	Satisfied (%)	Very satisfied (%)	Average score
Information transparency	0.00	1.09	32.61	46.74	19.57	3.85
Customer service	0.00	4.35	35.87	39.13	20.65	3.76
Scope of use	0.00	3.26	21.74	45.65	29.35	4.01
Repayment time	1.09	4.35	25.00	48.91	20.65	3.84
Installment interest	7.61	13.04	34.78	27.17	20.65	3.34
Rate credit system	1.09	2.17	23.91	41.30	31.52	4.00
Subtotal	1.63	4.71	28.99	41.49	23.19	3.80

Table 5 College students' satisfaction with using risk

	Very dissatisfied (%)	Not very satisfied (%)	Generally (%)	Satisfied (%)	Very satisfied (%)	Average score
Pay for security	0.00	1.09	21.74	50.00	27.17	4.03
Personal privacy security	1.09	0.00	22.83	52.17	23.91	3.98
Subtotal	0.54	0.54	22.28	51.09	25.54	4.01

Table 6 College students' satisfaction with product image

	Very dissatisfied (%)	Not very satisfied (%)	Generally (%)	Satisfied (%)	Very satisfied (%)	Average score
Web interface	1.09	1.09	23.91	51.09	22.83	3.93
Overall image of the company	0.00	1.09	19.57	52.17	27.17	4.05
Promotion efforts	0.00	1.09	23.91	48.91	26.09	4.00
Subtotal	0.36	1.09	22.46	50.72	25.36	4.00

Table 7 College students' satisfaction with convenience

	Very dissatisfied (%)	Not very satisfied (%)	Generally (%)	Satisfied (%)	Very satisfied (%)	Average score
Using threshold	1.09	2.17	19.57	52.17	25.00	3.98
Transaction cost	0.00	3.26	19.57	53.26	23.91	3.98
Transaction process	0.00	2.17	17.39	52.17	28.26	4.07
Ease of payment and repayment	0.00	2.17	18.48	51.09	28.26	4.05
Subtotal	0.27	2.45	18.75	52.17	26.36	4.02

Table 8 College students' satisfaction with recognition

	Very dissatisfied (%)	Not very satisfied (%)	Generally (%)	Satisfied (%)	Very satisfied (%)	Average score
Influenced by others	2.17	5.43	21.74	46.74	23.91	3.85
Brand recognition	1.09	7.61	21.74	42.39	27.17	3.87
Media outreach	6.52	11.96	29.35	30.45	21.74	3.49
Subtotal	3.26	8.33	24.28	39.86	24.28	3.74

Table 9 Comprehensive analysis of college students' satisfaction

	Very dissatisfied (%)	Not very satisfied (%)	Generally (%)	Satisfied (%)	Very satisfied (%)	Average score
Basic service	1.63	4.71	28.99	41.49	23.19	3.80
Using risk	0.54	0.54	22.28	51.09	25.54	4.01
Product image	0.36	1.09	22.46	50.72	25.36	4.00
Convenience	0.27	2.45	18.75	52.17	26.36	4.02
Subtotal	3.26	8.33	24.28	39.86	24.28	3.74

4.4 Analysis of Influencing Factors

Correlation analysis between satisfaction and basic business. This questionnaire is set up as follows: basic business 1 refers to the degree of information disclosure; basic business 2 refers to customer consulting service; basic business 3 represents

Table 10 PCCs between the satisfaction degree and the factors of "basic business"

Basic business	1	2	3	4	5	6
Relevance	0.777	0.745	0.730	0.700	0.579	0.767
Significant bilateral	0.000	0.000	0.000	0.000	0.000	0.000
Number of cases	168	168	168	168	168	168

the scope of application; basic business 4 represents repayment time; basic business 5 represents installment interest rate; basic business 6 represents credit system.

PCCs was made between the satisfaction degree of college students and the factors of "basic business." The results are shown in Table 10 below.

As can be seen from Table 10 above, except for basic business 5, the correlation coefficient between other five basic businesses and satisfaction is between 0.6 and 0.8, which belongs to strong correlation. It shows that the better the basic business is, the higher the satisfaction will be.

Correlation analysis between satisfaction and using risk. The correlation analysis process of the two is similar to the above, and the result is that they are negative correlation.

Correlation analysis between satisfaction and product image. The correlation analysis process of the two is similar to the above, and the result is that they are positive correlation.

Correlation analysis between satisfaction and convenience. The correlation analysis process of the two is similar to the above, and the result is that they are positive correlation.

5 Some Suggestions

Based on the above survey and analysis results and some conclusions on the current situation of the use of "Ant Credit Pay" of college students of Hubei Polytechnic College, we put forward some constructive opinions on the consumer groups (especially the college students) and the operators of the "Ant Credit Pay."

5.1 Individual Consumption Needs to Be Tailored

Ant CCredit Pay is a consumer credit product. This kind of consumption mode is only temporary consumption in advance, which only slows down the payment time, and the amount consumed is still paid by yourself. Therefore, we need to consume rationally, consider our own repayment ability before consumption and do not blindly compare with consumption and blindly pursue the brand.

5.2 *Pay Attention to Self-credit and Enhance Self-protection Consciousness*

"Ant Credit Pay" has been put into the credit investigation mechanism of the central bank, which is a regular credit platform, so users should pay attention to punctuality and trustworthiness. Bad lending behaviors, such as deliberately avoiding repayment, long overdue time and long-term non-response to the platform, will be recorded by the credit reference system, which will have an impact on the reputation of users.

5.3 *Appropriately Reduce the Withdrawal Procedure Rate and Expand the Use of "Ant Credit Pay" in Daily Life*

Some items of "Ant Credit Pay" still need to be improved. Some students think that the scope of application of "Ant Credit Pay" is narrow, because many businesses do not accept the use of "Ant Credit Pay," because of the high withdrawal procedure rate and large loss.

References

1. Li S, Yang F (2019) Research on the development of internet finance in the era of big data. Mod Bus 35:131–133
2. Wang D (2015) The study of internet finance's development in China. Jilin University, Master's Thesis
3. Yu T (2019) Research on the influence of Internet consumption credit on college students' consumption behavior. Hainan University, Master's Thesis
4. Du C, Wang Z, Tu G, Peng Y (2018) Analysis on the development status and prospect of overdraft network credit consumption platform in universities—taking "Ant credit pay" as an example. Market Weekly 10:153–154
5. Xiong Y, Zhang S, Lan J, Chen F (2019) Efficiency evaluation of regional economic development of mining and metallurgy city based on DEA model. Int J Appl Decis Sci 12 (3):242–256
6. Xiong Y, Jie Z, Jie L (2019) Performance evaluation of food cold chain logistics enterprise based on the AHP and entropy. Int J Inf Syst Supply Chain Manage 12:57–67

A Review on Main Optimization Method of ALBERT in Natural Language Processing

Xiaowei Zhang, Jianfei Shao, Changli Hu, Rongjian Wei, and Rong Pu

Abstract This paper mainly sorts out the related research and optimization route of the ALBERT [1] model released by Google and introduces the main application scenarios and model combination directions of ALBERT. Experiments by many researchers show that: The current research direction of this model is mainly focused on different application scenarios and combined with other improved models, only a few researchers will improve and optimize the structure of the model itself. Through literature research, this paper also looks forward to the possible future development direction of ALBERT.

Keywords ALBERT · LSTM · BERT · GRU

1 Introduction

ALBERT is called A Lite Bert for Self-Supervised Learning for Language, and it is an improved model from BERT [2] proposed by Google in September 2019. This new model is not only a simple upgrade on the original model but also a new parameter sharing mechanism. Moreover, the model solves the problem of OOM in video memory or memory caused by increasing the scale of the model, and long-term training may cause model degradation. Because in this model, two methods of reducing memory are proposed, and at the same time, it improves the speed of training and breaks through the limitations of hardware to a certain extent. Compared with the original BERT model, both ALBERT and BERT use the same encoder result of transformer [3], and the activation function also uses GELU.

X. Zhang (✉) · J. Shao · C. Hu · R. Wei · R. Pu
Faculty of Information Engineering and Automation, Kunming University of Science and Technology, Yunnan, Kunming 650000, China

Z. Yu et al. (eds.), *Advancements in Mechatronics and Intelligent Robotics*, Advances in Intelligent Systems and Computing 1220,
https://doi.org/10.1007/978-981-16-1843-7_11

2 Recent Development

In the field of natural language processing before this, most researchers used recurrent neural networks such as LSTM [4]. However, LSTM is currently obsolete in the field of natural language processing, for example, in the list (Table 1) of Stanford University's Reading Comprehension Data Set (SQuAD2.0). Among the ten models occupying the top five positions, five model algorithms have the participation of the ALBERT model.

At present, the ALBERT model is widely used in natural language processing under the transformation of some researchers and in combination with other models. Through literature research and analysis, this article summarizes the various improvements of the ALBERT model and the combination with other models. Based on this, this article hopes to clarify the researcher's optimization and improvement mode of ALBERT in natural language processing and the future application development direction of this model.

Table 1 SQuAD2.0. (2020.10.29)

Rank 1	Model	EM	F1
	Human performance *Stanford university*	86.831	89.452
1-Apr 06, 2020	SA-Net on **AlBERT** *QIANXI*	90.724	93.011
2-May 05, 2020	SA-Net-V2 *QIANXI*	90.679	92.948
2-Apr 05, 2020	Retro-reader *Shanghai Jiao Tong University*	90.578	92.978
3-Jul 31, 2020	ATRLP + PV *Hithink RoyalFlush*	90.442	92.877
3-May 04, 2020	ELECTRA + **ALBERT** + EntitySpanFocus *SRCB_DML*	90.442	92.839
4-Jun 21, 2020	ELECTRA + **ALBERT** + EntitySpanFocus *SRCB_DML*	90.420	92.799
4-Sep 11, 2020	EntitySpanFocus + AT *RICOH_SRCB_DML*	90.454	92.748
4-Mar 12, 2020	**ALBERT** + DAAF + Verifier *PINGAN Omni-Sinitic*	90.386	92.777
5-Jan 10, 2020	Retro-Reader on **ALBERT** *Shanghai Jiao Tong University*	90.115	92.580

3 Solve the Problem of Insufficient Text Word Vector Learning by Classifying the Output Features of ALBERT

The ALBERT dynamic word vector is different from other traditional dictionary mapping methods. In actual use, the network structure needs to be re-planned. Zhou et al. [5] proposed to encode the text based on word level and add the classification label CLS and the end label SEP, respectively, in the first position, and then in order to learn more sequence features in the text data, they proposed the ALBERT-RNN model (Fig. 1).

The model uses the ALBERT network to relearn the fixed word vector after dictionary mapping and then sends it to the RNN network and extracts the sequence features in the sentence tags. Then the CLS and SEP tags are fused in the feature fusion part. Finally, the fused features are input to the classification network. From the experimental results, it can be seen that by processing the same data set, compared to the traditional algorithm models such as KNN and SVM, the ALBERT-RNN model performs better, but the accuracy value is slightly reduced, and the original model is positive overfitting of samples.

4 Solve the Problem of Gradient Explosion and Disappearance in the Process of Long Text Training by Combining the Bi-LSTM Model

Many researchers have achieved some results in the identification of rumors on Weibo. And now, some researchers apply the improved BERT or ALBERT model to this text processing scenario. However, the ALBERT model has the problem of gradient explosion and gradient disappearance in the training process of long text. Therefore, Sun et al. [6] solved this problem by combining the Bi-LSTM model improved by the LSTEM model. Compared with LSTM, Bi-LSTM can better capture the dependence of bidirectional semantics. By combining ALBERT and Bi-LSTM, this model (Fig. 2) will train the pre-processed sentence vectors through

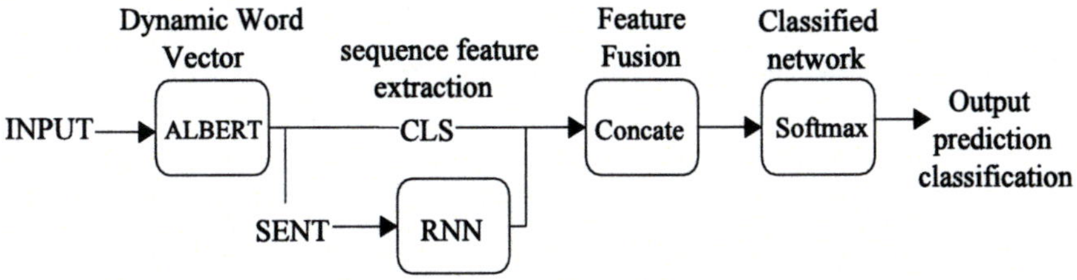

Fig. 1 ALBERT-RNN model

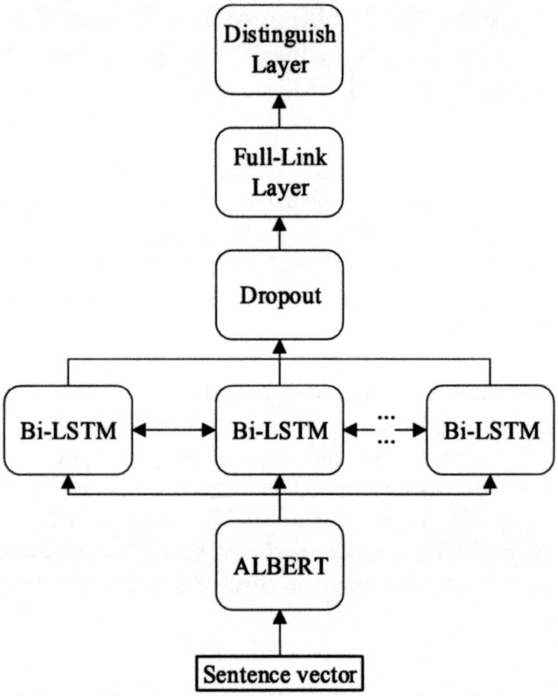

Fig. 2 Bi-LSTM model

the ALBERT model for word training and then perform feature extraction on the neural network layer after Bi-LSTM, and add the dropout layer in the middle. This effectively prevents the overfitting problem and finally enters the fully connected layer for dimensionality reduction and outputs the corresponding label in the discrimination layer. It is not difficult to see through experiments: Compared with the original ALBERT model, the accuracy value, recall rate and F1 value are increased by 0.161%, 0.202% and 0.162% respectively, although the increase is not obvious. But this model has been able to identify Weibo rumors well.

5 By Combining CRF Model to Better Recognize Long Texts

The language model mentioned in this paper 2 is one-way to a certain extent and cannot fully understand the meaning of the entire sentence. The longer the sentence, the lower the agreement between the recognition result and the meaning of the original sentence. The CRF model can multi-classify the components in the sentence, extract the components predicted to be specific tags and form an entity. Gao et al. [7] proposed a model that combines ALBERT, Bi-LSTM and CRF. The model inputs the vector into the Bi-LSTM feature extraction network and uses CRF

Table 2 Compare precision

Model	Accuracy (%)
ALBERT	85.69
Bert + Bi-LSTM + CRF	88.53
Albert + Bi-LSTM + CRF	89.23

to fuse the context information in the long text to obtain the feature. It represents the vector and then uses CRF to learn the constraint relationship between the classified tags to predict the generated entity tags. Through the experimental results (Table 2), it is obvious that the accuracy of this model in solving long text recognition is higher than that of the other two control models.

6 By Combining the BiGRU Model to Maximize the Preservation of the Semantic Association of Word Vectors in Long Texts

GRU is a variant after optimization and transformation based on the LSTM model, but because the GRU [8] model is not a two-way propagation model, it cannot make full use of the word sense association between contextual word vectors in long texts. BiGRU [9] uses a two-way propagation model, which can solve this problem to a certain extent. Wen et al. [10] proposed an ALBERT-BiGRU model, which not only can replace the traditional Word2vec to train static word momentum but also improve the representation ability of word vectors in long texts. Experimental results show that the model has a significant effect on the word meaning classification of long text word vectors. But when the pre-training data reaches a certain level, the accuracy of the model begins to decline.

7 Conclusion

ALBETRT, as a natural language processing model improved on the basis of BERT, has been favored by researchers in many application scenarios. Researchers can make up for the deficiency of Albert by combining different pre training models, the shortcomings of ALBERT can be com-pensated for. Several optimization and improvement routes of the model and future research directions can include the following points:

1. Optimization of input data set
2. Optimization of pre-training targets
3. Choose different language training models to combine according to different application scenarios
4. The ALBERT model structure is optimized to achieve smaller memory and even break through the limitations of hardware.

References

1. Chi PH, Chung PH, Wu TH et al (2020) Audio ALBERT: a lite BERT for self-supervised learning of audio representation
2. Devlin J, Chang MW, Lee K et al (2018) BERT: pre-training of deep bidirectional transformers for language understanding
3. Correia GM, Niculae V, Martins AFT (2019) Adaptively sparse transformers
4. Hochreiter S, Jü Schmidhuber RA (1997) Long short-term memory. Neural Computation
5. Zhining Z, Binjun W, Yiming Z et al (2020) Spam filtering model based on ALBERT dynamic word vector. Netinfo Security (9)
6. Weichao S, Tao C (2020) Research on Microblog rumor recognition method based on ALBERT-Bi LSTM model. Comput era 8:21–26
7. Xuepan G, Chun D, Jinliang W (2020) Research on method of military named entity recognition based on BiLSTM-CRF. Comput era 5:19–22
8. Cho K, Van Merrienboer B, Gulcehre C et al (2014) Learning phrase representations using RNN encoder-decoder for statistical machine translation. Comput ence
9. Jat S, Khandelwal S, Talukdar P (2018) Improving distantly supervised relation extraction using word and entity based attention
10. Chaodong W, Cheng Z, Junwei R et al (2020) Patent text classification based on ALBERT and bidirectional gated recurrent unit. J Comput Appl (8)

Study on the Influencing Factors of Customer Loyalty in Large Smart Home Furnishing Stores

Jin Wu

Abstract This study mainly studies the following aspects: (1) it discusses the theory of experience marketing, customer satisfaction, and customer loyalty and puts forward the main reasons affecting customer loyalty. (2) Based on the development situation of Ikea store in Wuhan, some hypothesized factors affecting customer loyalty are analyzed. (3) Conduct questionnaire survey, collect data, and prove research hypotheses. Through the above research work, the conclusions are as follows: (1) The influencing factors of customer loyalty in Ikea store in Wuhan are formed in many aspects and the influence of experience marketing and customer satisfaction is significantly and positively correlated. Customer satisfaction is the mediating variable between experience marketing and customer loyalty.

Keywords Smart home · Experience marketing · Customer loyalty · Customer satisfaction

1 Introduction

With the rapid development of science and technology, it has more and more influence on our life. In recent years, people also put forward the concept of smart home. Smart home is literally a home product that can be connected with technology; it is a home-based platform, using the integrated cabling technology, communications technology, and so on to develop smart home products for home use and to enhance home security, convenience, comfort as the goal of environmental protection and energy-saving living environment with a sense of fashion.

In today's competitive market, improving customer loyalty can effectively promote the economic development of enterprises. In recent years, along with the economic growth of our country, people's purchasing power has gradually increased. Now many first- and second-tier cities have also gradually increased their

J. Wu (✉)
School of Economics and Management, Hubei Polytechnic University, Huangshi 435003, Hubei, China

Z. Yu et al. (eds.), *Advancements in Mechatronics and Intelligent Robotics*, Advances in Intelligent Systems and Computing 1220,
https://doi.org/10.1007/978-981-16-1843-7_12

home furnishing stores. The home furnishing market has undergone a long period of development, and it has already seen its first scale, Ikea's sales model is very human, which is an important factor in Ikea's success. In China's home furnishing market, no single home furnishing company has a share of 1% of the Chinese market, while Ikea has a share of 25%. Although there are many well-known domestic enterprises in China, such as Red Star Macalline and Easyhome, their enterprises are still mainly based on OEM, and the real formation of their own brand awareness is also the beginning of the last few years, but compared to Ikea's various aspects, China's home furnishing industry has fallen far behind. Retention of relatively loyal customers and cultivation of new customer loyalty are the basis for enterprises to obtain greater profits and play an important role in the sustainable development of enterprises.

An organization's long-term profitability and sustained reputation depend on customer loyalty, which can continuously create great value for customers and bring long-term benefits to the organization [1]. Not all loyal customers are profitable, and not all profitable customers are loyal, but loyal customers must bring profits [2]. Customer loyalty is influenced by perceived value, trust, and habit and customer satisfaction. Customer satisfaction plays an important intermediary role in the relationship between perceived value and trust and loyalty [3]. Experiential marketing is a kind of marketing method that allows enterprises and customers to create costs through interaction and intervention in the whole experience process, and enterprises will own products and services through the experience of this platform for customers to create costs to hold profit [4]. Although there were differences in academic research conclusions, he basically agreed that customer experience had a strong influence on the composition of customer piety; customer experience can influence customer piety through customer satisfaction, customer cost, customer switching costs, etc. [5]. Enterprises need to have a loyal customer base in order to have a certain position in the peer [6].

In theory, there are many references on customer loyalty, but there are few on the basis of the research on the influencing factors of Ikea customer loyalty, from which specific aspects can achieve a stable customer base, stable customer base can reduce the marketing costs of enterprises and create good word-of-mouth, so in doing research on this area is powerful and feasible.

2 Related Theory of Influencing Factors of Customer Loyalty

2.1 Experience Marketing

Experiential marketing refers to the consumer in the consumption process, customers can be integrated into it, have the opportunity to experience goods and the greatest degree of all-round understanding of goods performance and its characteristics, more intuitive feeling, direct contrast, to maximize the benefits of selling a product. Experiential marketing brings people a real and reliable experience, no longer a situation where they feel so clueless about a product they do not understand that they have no choice, in some ways, it can also guide consumers in the direction of shopping, but also can better improve the practicality and applicability of goods, is that consumers feel worthy of trust.

2.2 *Customer Satisfaction*

From the consumer's point of view, customer satisfaction depends on the consumer's point of view, refers to the consumer, the product brings him is satisfied with the joy, or dissatisfied with the disappointment. When the service provider believes that the quality of the product or service, etc., is not meeting his or her expectations, the brand loses the trust of the customer, even if for some specific reason the customer still chooses to believe, but the consumer will still have a grudge against the brand, or even dissatisfied.

2.3 *Measurement of Customer Loyalty*

The measurement direction of customer loyalty is obvious, customer loyalty is passive, many times is activated by the enterprise's brand advantage, solidified down, not that consumers bought this thing or come once immediately loyal, the generation of consumer loyalty, is passive, is the consumer after weighing the pros and cons of their own hearts after the rational choice. So its measurement has to do with three things: (1) The frequency with which customers comes to the store and the rate at which they buy similar products. (2) The type, quantity, and proportion of customers' purchases. (3) Customer tolerance of product accidents.

3 Establishment of Influencing Factor Model of Customer Loyalty

3.1 *Selection of Modeling Dimension*

Dimension of experiential marketing. Experience marketing is to abandon the traditional marketing, starting from the demand point of customers, in order to create the interests of enterprises and consumers. In the experience marketing, the enterprise presents its enterprise idea and the product to the consumer from each aspect very well. Experience marketing is mainly divided into five aspects: They are action, thinking, emotion, relevance, and sensory marketing.

Dimension of customer loyalty. Customer loyalty is mainly reflected in the evaluation of consumer behavior, focusing on the analysis of consumer style and repeated consumption motivation, consumers are also divided into short-term customers and long-term customers, so the two dimensions of behavior and attitude.

Dimension of customer satisfaction. From the buyer's point of view, satisfaction should be lower in industries with homogeneous supply and heterogeneous demand and higher when the heterogeneity/homogeneity of demand matches supply. So we use the dimension of customer perceived value to study customer satisfaction and make an explanation to the influence of customer loyalty.

3.2 Research Hypothesis

Relationship between experiential marketing and customer loyalty. Experiential marketing is an important factor that affects consumer loyalty, therefore, based on this, the following hypothesis is proposed:

H1a: experiential marketing has a significant positive impact on customer loyalty;
H1b: Sensory marketing has a significant positive impact on customer loyalty;
H1c: Affective marketing has a significant positive impact on customer loyalty;
H1d: Thinking marketing has a significant positive impact on customer loyalty;
H1e: Behavioral marketing has a significant positive impact on customer loyalty;
H1f: Affiliate marketing has a significant positive impact on customer loyalty.

Relationship between experiential marketing and customer satisfaction. What the consumer feels at the time of purchase can promote consumer satisfaction. It is proved that consumer perception may promote the formation of customer loyalty. It is believed that customer *loyalty* can only be generated by providing superior value to customers. The following assumptions about satisfaction are made:

H2a: experiential marketing has significant positive impact on customer satisfaction;
H2b: Sensory marketing has significant positive impact on customer satisfaction;
H2c: Affective marketing has significant positive impact on customer satisfaction;
H2d: Thinking marketing has significant positive impact on customer satisfaction;
H2e: Behavioral marketing has significant positive impact on customer satisfaction;
H2f: contextual marketing has significant positive impact on customer satisfaction.

The role of consumer satisfaction in experiential marketing and customer loyalty.

H3: Consumer satisfaction is the intermediary between the two sides.

Relationship between customer satisfaction and customer loyalty.

H4: Consumer satisfaction has a significant positive effect on consumer loyalty.

3.3 *Description of Research Model*

In this paper, Ikea's experience marketing is set as a antecedent variable, which includes five aspects of sense organ, emotion, thinking, behavior and association, customer. Loyalty is an independent variable to study the factors that influence Ikea's customer loyalty.

4 Data Analysis on Influencing Factors of Customer Loyalty

4.1 *Reliability Test*

In this paper, the reliability of the collected questionnaires is analyzed, and the specific results are shown in Table 1.

The result of Table 1 shows that the Cronbacha coefficient and split-half reliability of the questionnaire are above 0.8, which shows that the questionnaire has good reliability. The content of the questionnaire was tested by relevant experts, and the validity of the questionnaire was good.

4.2 *Correlation Analysis*

SPSS22.0 software was used to analyze the related variables of large-scale Home Shopping Center. The data were sorted as Table 2. From the below Table 2, (1) there were significant positive correlation between the two of the five dimensions in the table; (2) the five dimensions of experiential marketing have significant positive correlation with customer loyalty and customer satisfaction; (3) customer loyalty has significant positive correlation with customer satisfaction, and the correlation is high. In view of the results of the correlation analysis of the three variables, it is suitable for regression analysis among the three variables.

4.3 *Regression Analysis*

Regression analysis of experience marketing to customer loyalty. SPSS22.0 software is used to analyze the linear regression between experience marketing and

Table 1 Reliability test of preliminary test questionnaire

Cronbacha coefficient α	Split-half reliability
0.930	0.860

Table 2 Summary table of KMO value and bartlett sphere test of questionnaire data

	Sensory experience	Emotional experience	Thinking experience	Action experience	Connected experience	Customer loyalty	Customer satisfaction
Sensory experience	1						
Emotional experience	0.708**	1					
Thinking experience	0.693**	0.542**	1				
Action experience	0.703**	0.631**	0.611**	1			
Connected experience	0.484**	0.386**	0.537**	0.561**	1		
Customer loyalty	0.689**	0.487**	0.614**	0.655**	0.611**	1	
Customer satisfaction	0.721**	0.604**	0.602**	0.686**	0.527**	0.779**	1

Remarks: *** denotes $p < 0.001$; ** denotes $p < 0.01$; * denotes $p < 0.05$

customer loyalty of large-scale home stores. The results are summarized in Table 3 of multiple regression analysis. From the above table, we can see that the multiple correlation coefficient between experience marketing and customer loyalty is 0.744, and the square of multiple correlation coefficient is 0.553, which means that experience marketing can explain 55.3% variation of customer loyalty. From the above table, it can be seen that experience marketing has a significant impact on customer loyalty. The standardized regression coefficient of experience marketing is 0.744, and the direction is positive, indicating that experience marketing can significantly positively predict customer loyalty. The results show that the hypothesis H1 holds.

Regression analysis of five dimensions of experience marketing on customer loyalty in large home stores. From Table 4 below, we can see that the sense, association, and action experience of large-scale home furnishing stores have an impact on customer loyalty, and they are significantly positive. The standardized regression coefficients of the three are 0.391, 0.214, and 0.275, respectively, and the direction is positive, indicating that sensory experience, mobile experience, and related experience can significantly positively predict customer loyalty. Therefore, we assume that H1a, H1e, and H1d are valid; in addition, the influence of emotional experience and thinking experience on customer loyalty is not significant, so we assume that H1b and H1c are not tenable.

According to the above data results, the standardized regression equation of experience marketing and customer loyalty of large home stores is as follows:

Table 3 Summary table of multiple regression analysis of experience marketing on customer loyalty

Predictive variable	B	Standard error	β	t value
Intercept	0.401	0.296		1.356
Experience marketing	0.931	0.081	0.744	11.508***
$R = 0.744$ $R2 = 0.553$ adjusted $R2 = 0.549$ $F = 132.438$***				

Remarks: *** denotes $p < 0.001$; ** denotes $p < 0.01$; * denotes $p < 0.05$

Table 4 Summary table of multiple regression analysis of five dimensions of experience marketing on customer loyalty

Predictive variable	B	Standard error	β	t value
Intercept	0.403	0.298		1.353
Sensory experience	0.443	0.123	0.391	3.589**
Emotional experience	−0.097	0.095	−0.093	−1.018
Thinking experience	0.100	0.080	0.115	1.248
Action experience	0.202	0.092	0.214	2.195*
Connected experience	0.341	0.097	0.275	3.502**
$R = 0.776$ $R2 = 0.603$ adjusted $R2 = 0.583$ $F = 31.259$***				

Remarks: *** denotes $p < 0.001$; ** denotes $p < 0.01$; * denotes $p < 0.05$

Customer loyalty = 0.391 * sensory experience + 0.214 * action experience + 0.115 * thinking experience + 0.275 * related experience − 0.093 * emotional experience

Regression analysis of experiential marketing on customer satisfaction. From Table 5 below, we can see that the multiple correlation coefficient between experience marketing and customer satisfaction of large home stores is 0.767, and the square of multiple correlation coefficient is 0.588, which means that experience marketing can explain 58.8% variation of customer loyalty. From the above table, experience marketing has significant influence on customer satisfaction. The standardized regression coefficient of experience marketing is 0.767, and the direction is positive, indicating experience marketing can significantly positively predict customer satisfaction. The results show that the hypothesis H2 holds.

Regression analysis of five dimensions of experience marketing on customer satisfaction in large home stores. It can be seen from Table 6 below that sensory experience and action experience of large-scale home stores have significant impact on customer satisfaction. The standardized regression coefficients of the two dimensions are 0.357 and 0.251, respectively, with positive directions, indicating that both sensory experience and action experience can significantly positively predict customer satisfaction. In addition, the influence of emotional experience,

Table 5 Regression analysis of experiential marketing on customer satisfaction

Predictive variable	B	Standard error	β	t value
Intercept	0.625	0.261		1.356
Experience marketing	0.882	0.071	0.767	11.508***
$R = 0.767$ $R2 = 0.588$ adjusted $R2 = 0.584$ $F = 152.824$***				

Remarks: *** denotes $p < 0.001$; ** denotes $p < 0.01$; * denotes $p < 0.05$

Table 6 Summary table of multiple regression analysis of five dimensions of experience marketing on customer satisfaction

Predictive variable	B	Standard error	β	t value
Intercept	0.521	0.273		1.908
Sensory experience	0.372	0.113	0.357	3.293**
Emotional experience	0.096	0.087	0.100	1.105
Thinking experience	0.059	0.073	0.075	0.814
Action experience	0.218	0.084	0.251	2.584*
Connected experience	0.153	0.089	0.134	1.718
$R = 0.779$ $R2 = 0.606$ adjusted $R2 = 0.587$ $F = 31.718$***				

Remarks: *** denotes $p < 0.001$; ** denotes $p < 0.01$; * denotes $p < 0.05$

Table 7 Comprehensive analysis of college students' satisfaction

Predictive variable	B	Standard error	β	t value
Intercept	0.056	0.266		0.212
Experience marketing	0.445	0.111	0.355	4.018***
Customer satisfaction	0.551	0.096	0.507	5.730***
$R = 0.812$ $R2 = 0.659$ adjusted $R2 = 0.652$ $F = 102.333$***				

thinking experience, and related experience on customer satisfaction is positive, but not significant. This result shows that emotional experience, thinking experience, and related experience have no significant predictive effect on customer satisfaction. Therefore, according to the above results, it is concluded that H2A and H2E are tenable, while H2B, H2C, and H2D are not.

According to the above data results, the standardized regression equation between experience marketing and customer satisfaction of large home stores is as follows:

Customer satisfaction = 0.357 * sensory experience + 0.075 * thinking experience + 0.134 * related experience + 0.100 * emotional experience + 0.251 * action experience

The role of customer satisfaction. It can be seen from Table 7 below that experience marketing has a significant positive predictive effect on customer loyalty. After adding customer satisfaction into the regression model, we can still draw the above conclusion. The standardized regression coefficients of the two are 0.355 and 0.507, respectively, and the direction is positive. Combined with Tables 3, 4, 5, it can be concluded that experiential marketing has an impact on customer satisfaction, and there is a significant positive impact. Through the above results, we can know that: customer satisfaction is a bridge between customer loyalty and experiential marketing and plays an important role in both. Therefore, H3 is assumed to hold.

5 Summary

This part analyzes the results of the questionnaire through SPSS22.0 and verifies the research hypotheses through the results of the analysis. The main methods used are: correlation analysis and regression analysis. The analysis results show that the reliability of the questionnaire is good and the data have strong persuasion.

References

1. Koo D (2003) Inter-relationships among store images, store satisfaction, and store loyalty among Korea discount retail patrons. Asia Pacific J Mark Logistics 15(4):42–71
2. Reinartz W, Kumar V (2002) The mismanagement of customer loyalty. Harvard Bus Rev 80(7):86–94
3. Yu T (2019) Research on the influence of internet consumption credit on college students' consumption behavior. Hainan University, Master's Thesis
4. Lin HH, Wang YS (2006) An examination of the determinants of customer loyalty in mobile commerce contexts. Inf Manage 43(3):271–282
5. Xiong Y, Zhang S, Lan J, Chen F (2019) Efficiency evaluation of regional economic development of mining and metallurgy city based on DEA model. Int J Appl Decis Sci 12(3): 242–256
6. Xiong Y, Jie Z, Jie L (2019) Performance evaluation of food cold chain logistics enterprise based on the ahp and entropy. Int J Inf Syst Supply Chain Manag 12:57–67

A Detection Algorithm of Lung Nodule Based on Faster R-CNN

Xianggui Xian

Abstract At present, with the deterioration of environmental pollution and air quality, the number of lung cancer patients has increased year by year. Lung cancer has become the highest incidence rate and mortality rate in the world. The early symptoms of lung cancer were pulmonary nodules on CT images. Therefore, improving the detection efficiency of pulmonary nodules can detect lung cancer in advance and strive for the best treatment opportunity. In order to realize the effective and accurate identification of pulmonary nodules in CT images and reduce the phenomenon of missed detection and false detection caused by the difference of doctors' level, this paper proposes a lung nodule detection algorithm based on the improved Faster R-CNN (Ren et al. in IEEE Trans Pattern Anal Mach Intell 39 (6):1137–1149 [1]). This algorithm has a good effect on the detection of pulmonary nodules. The experimental results on LUNA16 dataset show that the accuracy of the algorithm is 92.10%, the sensitivity is 94.21%, and AP is 85.30%.

Keywords Image measurement · Pulmonary nodules · Faster R-CNN

1 Introduction

Lung cancer is one of the highest incidence rate and mortality rate in the world, and it is a serious threat to human life and health. Among them, in 2017, the number of lung cancer cases in China reached 800 thousand, and the number of deaths due to lung cancer reached 700 thousand. The early symptoms of lung cancer are basically no special symptoms or its symptoms are difficult to find, which leads to most patients often go to the hospital for examination in the late stage of lung cancer with

X. Xian (✉)
Faculty of Information Engineering and Automation, Kunming University of Science and Technology, Kunming 650500, China

Z. Yu et al. (eds.), *Advancements in Mechatronics and Intelligent Robotics*,
Advances in Intelligent Systems and Computing 1220,
https://doi.org/10.1007/978-981-16-1843-7_13

obvious symptoms. However, the best time to treat lung cancer has been missed. Therefore, it is very important to improve the early detection of lung cancer (Fig. 1).

Pulmonary nodule is the early manifestation of lung cancer. In low-dose chest CT images, the diameter of pulmonary nodules is 3–30. The pulmonary nodules have the characteristics of complex shape, uncertain distribution, easy adhesion of other tissues, and no specificity in clinical, so it is difficult to make accurate judgment in clinical practice. Therefore, accurate detection of pulmonary nodules has become a key and difficult point in medical CT image research.

The early detection method of pulmonary nodules is to set different thresholds according to different tissues to achieve the purpose of segmenting nodules [2]. This method is easy to be affected by factors such as CT image quality, and it is difficult to achieve accurate segmentation. Then, researchers manually constructed the geometry, texture, and location of pulmonary nodules and used classifiers to detect pulmonary nodules [3, 4]. However, the absorption degree of the organs in medical images is reflected by the gray level, so it is very difficult to extract the features [5]. Therefore, it is difficult to complete the feature selection, artificial construction signs, classifier selection, and training workload [6, 7]. However, it is not easy to confuse the results of CT images with other planar features if the depth information of CT images is not fully used.

In view of the above problems in the detection of pulmonary nodules, combined with the *excellent* performance of deep learning neural network in the field of medical detection, we propose a lung nodule detection algorithm based on improved Faster R-CNN.

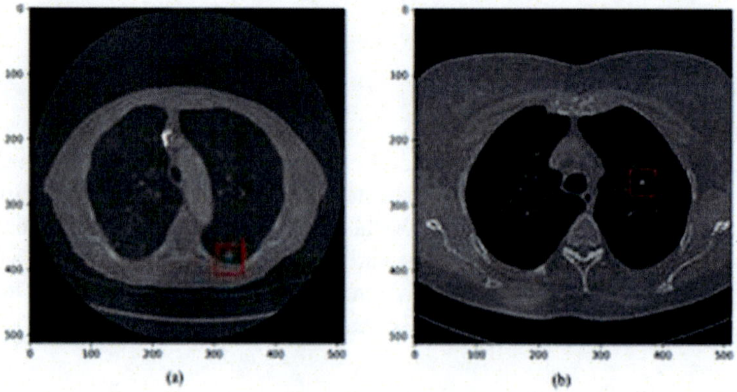

Fig. 1 Pulmonary nodule target

2 Deep Learning Detection Framework and Faster R-CNN

In recent years, deep learning target detection algorithms emerge in endlessly. Faster R-CNN is a prominent representative. Faster R-CNN is an improvement of faster R-CNN [8]. Faster R-CNN uses selective search algorithm to select candidate frames. This step takes a long time and becomes the bottleneck of faster R-CNN to improve real-time performance. Faster R-CNN replaces the selective search algorithm with regional recommendation network (RPN), which improves the selection time of candidate frames without affecting the detection accuracy. The overall framework of faster R-CNN network is as follows.

2.1 *Feature Extraction Network*

As a method of image target detection, faster R-CNN uses Conv + Relu + pooling layer to extract feature map of image, and vgg16 is used to extract feature map. The feature map is shared for subsequent RPN layer and full connection layer.

2.2 *RPN*

On the feature map extracted from vgg16, RPN performs further convolution through 3 * 3 convolution layer and then goes through two parallel 1 * 1 convolution branches. One branch combined with setting anchor parameter predicts the probability of selecting candidate frames that may contain targets, while the other branch generates the probability of whether the candidate frames are targets. Among them, anchor is the essence of RPN, and anchor is actually a group of rectangular boxes generated based on each feature point of feature map.

2.3 *ROI Pooling*

Faster R-CNN innovatively proposed region of interest pooling (ROI), which is a pooling method similar to spatial pyramid pooling, which contains only one pyramid layer, that is, ROI pooling only generates one pyramid scale.

2.4 Classifier

The classifier uses the proposal feature maps obtained in the above steps to calculate the category and probability vector of each proposal through the full connect layer and softmax; at the same time, the position offset bbox of each proposal is obtained by back calculation with bounding box expression, which is used to return more accurate target boxes.

3 Pulmonary Nodule Detection Network

Faster R-CNN performs very well in Pascal VOC and coco natural image datasets. However, due to the difference between CT image and natural image set, the effect of faster R-CNN directly used in pulmonary nodule image detection is not good. The main reason is that the lung nodule image contains less features such as color and texture, and the size of pulmonary nodule target is different from that of natural image set.

For these two reasons, faster R-CNN is not ideal in the detection of pulmonary nodules. According to the characteristics of CT images of pulmonary nodules, this paper improves the original faster R-CNN and proposes the detection algorithm in this paper.

3.1 Improvement of Feature Extraction Network

Vgg16 network is a very effective feature extraction network with five layers. The original faster R-CNN uses the fifth-layer output of vgg16. Because the feature information of pulmonary nodule image is less than that of natural image, only using the fifth-layer information of vgg16 as output will result in the loss of shallow information. Therefore, this algorithm uses the third- and fifth-layer information of vgg16. The third-layer feature map is fused with the fifth-layer feature map with the same resolution as the fifth-layer feature map through the expansion network. The fused feature map is used as the output of feature extraction network.

3.2 Modify Anchors

K-means [9] clustering algorithm is a simple and classic clustering algorithm based on distance. It uses distance as the evaluation index of similarity, that is, the closer the distance is, the greater the similarity is. The algorithm considers that the class cluster is composed of objects close to each other, so the final goal is to get a compact and independent cluster. In this paper, we use k-means clustering to get nine anchors which are consistent with the dataset of pulmonary nodules. These anchors are a combination of size (2,8,16) and aspect ratio (1:1,1:2,1:4).

4 Experimental Data and Result Verification

4.1 Dataset

This paper uses the LUNA16 dataset, which is the largest subset of lung image database Consortium (LIDC-IDRI) [10] provided by the National Cancer Research Association. LIDC-IDRI deleted CT images with slice thickness greater than 3 mm and pulmonary nodules less than 3 mm, leaving the LUNA16 dataset. In LIDC-IDRI dataset, a total of 36378 nodules was identified in 888 slices. However, in LUNA16 dataset, nodules less than 3 mm in diameter and larger than 3 mm in diameter and non-nodules were not taken as test samples. After four authoritative experts jointly labeled, a total of 1186 nodules was generated as the target nodes to be detected, and the notes of nodal coordinates and maximum diameter were provided.

4.2 Evaluation Criterion

In order to show the accuracy of pulmonary nodule detection, we select the following three indicators to comprehensively evaluate the performance of the algorithm. The three standard indexes were sensitivity (SEN), accuracy (ACC), and average accuracy (AP).

4.3 Experimental Result

In this study, 700 candidate pulmonary nodules were selected as test samples. Among them, 396 algorithms can be tested correctly. Compared with the algorithm [2] and algorithm [3], the results are shown in Table 1.

Table 1 Comparison of test results

Algorithm	Detection effect/%		
	SEN	ACC	AP
Algorithm [2]	85.00	68.97	–
Algorithm [3]	92.37	73.24	–
This paper	94.21	92.10	85.30

Through data comparison, the improved faster R-CNN detection algorithm proposed in this paper is higher than the algorithm [2] and algorithm [3] in terms of ACC and SEN, and the algorithm can obtain the location information of pulmonary nodules more accurately. Therefore, this algorithm can effectively improve the detection accuracy and reduce the misdiagnosis rate of pulmonary nodules. The algorithm not only depends on the two-dimensional information presented in CT images to fully understand the internal structure of pulmonary nodules, but also uses the spatial information of pulmonary nodules to obtain the overall structure, which is fully different from other tissues, so as to obtain better detection results. The experimental results show that it is a very effective method to detect pulmonary nodules by using the planar and spatial features of pulmonary nodules through deep learning network.

5 Conclusion

In view of the high false positive rate of pulmonary nodules, it is difficult to distinguish blood vessels from pulmonary nodules. In this paper, an efficient detection network of pulmonary nodules is designed. The improved faster CNN network model, which combines deep features and shallow **features**, is used to make full use of the feature information contained in CT images, and good detection results are achieved. However, the recognition rate of this algorithm for the small size of adhesive nodules needs to be improved, and further research is needed.

References

1. Ren S, He K, Girshick R et al (2017) Faster R-CNN: towards real-time object detection with region proposal networks. IEEE Trans Pattern Anal Mach Intell 39(6):1137–1149
2. Shi Yu, Xiaoguang Zeng, Hong Zhang et al (2006) Research on automatic detection algorithm of pulmonary nodules. Acta Instrumenta Sinica 27(6):2265–2267
3. Chen S, Chao Y, Zou L (2016) Detection of solitary pulmonary nodules based on geometric features. J Biomed Eng 33(4): 680–685
4. Kan Chen, Bin Li, Lianfang Tian (2012) Pulmonary nodule detection algorithm based on local threshold and cluster center iteration. Comput Sci 39(2):302–304
5. Mengmeng L, Tao Z, Feifei Z et al (2018) Convolutional neural network and its application in medical image analysis. J Biomed Eng 36(6):977–985
6. Tian JX, Liu G, Gu S et al (2018) Research and challenge of deep learning methods for medical image analysis. Acta automatica Sinica 44(3):401–424
7. Zhiyong X, Xiang Z (2019) Application of medical image analysis based on machine learning in drug development and precision medicine. Chinese J Bioeng 39(2):90–100
8. Girshick R (2015) Faster R-CNN[J]. Computer ence
9. Huang Z (1998) Extensions to the k-Means algorithm for clustering large datasets with categorical values. Data Min Knowl Disc 2(3):283–304

10. Lii SGA, Mclennan G, Bidaut L et al (2011) The lung image database consortium(LIDC) and image database resource in-itiative (IDRI): a completed reference data based of lung nodules on CT scans. Med Phys 38(2):915–931

Information Extraction from Contract Based on BERT-BiLSTM-CRF

Xiang Hu and Wenwei Su

Abstract The entire process of traditional power grid company contract management requires professional legal personnel to communicate back and forth with the contract initiating unit and rely on manpower to conduct special review of professional contract review content. In order to reduce the dependence on manual labor, a neural network-based sequence labeling is proposed, information extraction model. This model uses bidirectional encoder representation from transformers (BERT) to obtain the vectorized representation of words, uses BERT as the sentence encoding part, and then uses the BiLSTM+CRF model to classify the characters, and uses the contract corpus to do migration learning makes BERT better adapt to the semantic expression of the contract field. The experimental results show that the extraction effect of this model on contract information is much higher than that of the traditional CRF, and it has a certain improvement compared with other models.

Keywords Bidirectional long short-term memory network · BERT · Transfer learning

1 Introduction

With the development of the Internet, natural language processing (NLP), machine learning, and other artificial intelligence technologies can be used in multiple contract review links, such as discovery, analysis, and compliance review. Compliance review and contract management have become the most important legal application scenarios for the combination of artificial intelligence and production business. This is also an area where innovations continue to emerge.

X. Hu (✉)
Faculty of Information Engineering and Automation, Kunming University of Science and Technology, Kunming 650500, China

X. Hu · W. Su
Electric Power Research Institute of Yunnan Power Grid Co. Ltd, Kunming 650217, China

Z. Yu et al. (eds.), *Advancements in Mechatronics and Intelligent Robotics*,
Advances in Intelligent Systems and Computing 1220,
https://doi.org/10.1007/978-981-16-1843-7_14

Therefore, how to automatically extract structured legal person and company information from the contract is a problem. An important research topic, this topic can be transformed into a sequence labeling task, that is, the label corresponds to each word in the contract. Sequence annotation plays a very important role in the field of natural language understanding. According to the realization principle, it can be divided into three categories: rule-based methods, statistical machine learning methods, and deep learning methods. The rule-based method needs to obtain frequently occurring text pattern rules from a large amount of manually annotated data and then use the rule template to extract the corresponding information. Although this method can obtain higher extraction accuracy, it has poor portability and is not suitable for large-scale data.

Most methods based on statistical machine learning rely on conditional probability models. Commonly used models include conditional random field (CRF) model, hidden Markov model (HMM), maximum entropy Markov model (MEMM), and so on. These models have been applied to sequence labeling tasks in many fields. For example, Li [1] analyzed the internal and external characteristics of product aliases and used CRF to realize the extraction of aliases in ancient local chronicles; He et al. [2] used character-level CRF to construct a medical entity recognition model, which avoided the use of word segmentation. Accumulation of errors is in entity boundary recognition. Although this method of using statistical machine learning has achieved certain results in sequence labeling tasks in some fields, it requires manual extraction of features, which requires relatively high requirements for knowledge in this field. With the rapid development of deep learning, many neural network models have also been applied to the task of sequence labeling. Since bidirectional long short-term memory (BiLSTM) can simultaneously obtain context information and store memory on sequence problems, the BiLSTM model is widely used in sequence labeling tasks and has become the mainstream model for sequence labeling using deep learning. For example, Zhai et al. [3] use BiLSTM to vectorize the input text, and then uses CRF to mark it, which improves the accuracy and recall rate of knowledge map entity extraction. Based on word vectors, Zhang [4] used the BiLSTM-CRF model to identify product attributes and automatic extraction of attribute evaluations from Chinese online shopping reviews, and achieved good results on three products: camera, menswear, and child safety seats. Gridach [5] uses both the character-level vector representation of words and word vectors and through the BiLSTM-CRF model has achieved good results in the recognition of biomedical named entities. In recent years, some contracts have been introduced the attention mechanism based on the structure of the BiLSTM-CRF model. For example, Yang et al. [6] first used BiLSTM to obtain the contextual representation of the word, then used the attention mechanism to obtain the contextual representation of the word in the full text range, and finally used CRF to give the final annotation result. This model has achieved better results in the identification of chemical drug entities. Compared with the methods based on machine learning and rules, these methods based on deep learning do not need to extract features manually, reduce the dependence on domain knowledge, and have good scalability and generalization capabilities.

2 Related Research

According to the research results released by AMiner, about 85.62% of contract come from universities, scientific research institutions or enterprises [7]. For contract from the same university or company, most of their corporate information will have the same template, and they can use a rule-based method for information extraction; but for contract from different universities companies or enterprises, their corporate information is different in form. Similarly, it is difficult to extract with uniform rules. Therefore, most of the existing researches use machine learning-based methods to mark the contract sequence and extract information from it. For example, Tang et al. [8] proposed a tree structured conditional random fields (TCRF), which can use the hierarchical information modeling of the contract information structure, and achieved a comparative advantage in the contract information labeling. Better results are based on rules. Gollapalli et al. [9] adopted a semi-supervised method of feature labeling, using dictionary features and neighboring features to improve the F1 value of the labeling results. Gu et al. [10] proposed a Markov logic factor graph (Markov logic factor graph, MagicFG) model, which specializes in e-mail extraction and other tasks, and has also achieved good extraction results.

The above several methods use traditional machine learning methods. Although good results have been achieved on the sequence labeling task of contract information, there are still some shortcomings, such as the inability to capture long-distance context information and the need for manual use of experience in specific areas. There are limitations to the characteristics, versatility, and scalability of knowledge formulation. Compared with traditional machine learning methods, deep learning using neural networks has great advantages. Neural network can automatically mine hidden features from the text, reduce the dependence on artificial features, and have better versatility. BiLSTM neural network can capture the long-distance information of the context, and it has great advantages in processing sequence labeling tasks.

In summary, since most of the existing methods of information extraction use machine learning-based methods, which rely heavily on artificial features and cannot capture long-distance context information, and this paper proposes a new method from the perspective of deep learning. Considering the superiority of the BiLSTM-CRF framework in dealing with sequence problems, this paper introduces the BERT model on the basis of the BiLSTM-CRF model to generate the vectorized representation of words, and proposes a BERT-BiLSTM-CRF model. The model uses neural networks to automatically mine text hidden features, reducing the dependence on artificial features and specific domain knowledge, and can fully capture long-distance context information, and by introducing a BERT model with powerful ideographic capabilities, it can generate richer semantics the word vector of, realizes the sequence tagging task of contract information end to end. The experimental results prove that the model proposed in this paper has achieved good results in the task of labeling contract information.

3 Algorithm Model

3.1 Contract-Based Language Model (BERT)

BERT is stacked with a transformer structure. The transformer structure is shown in Fig. 1.

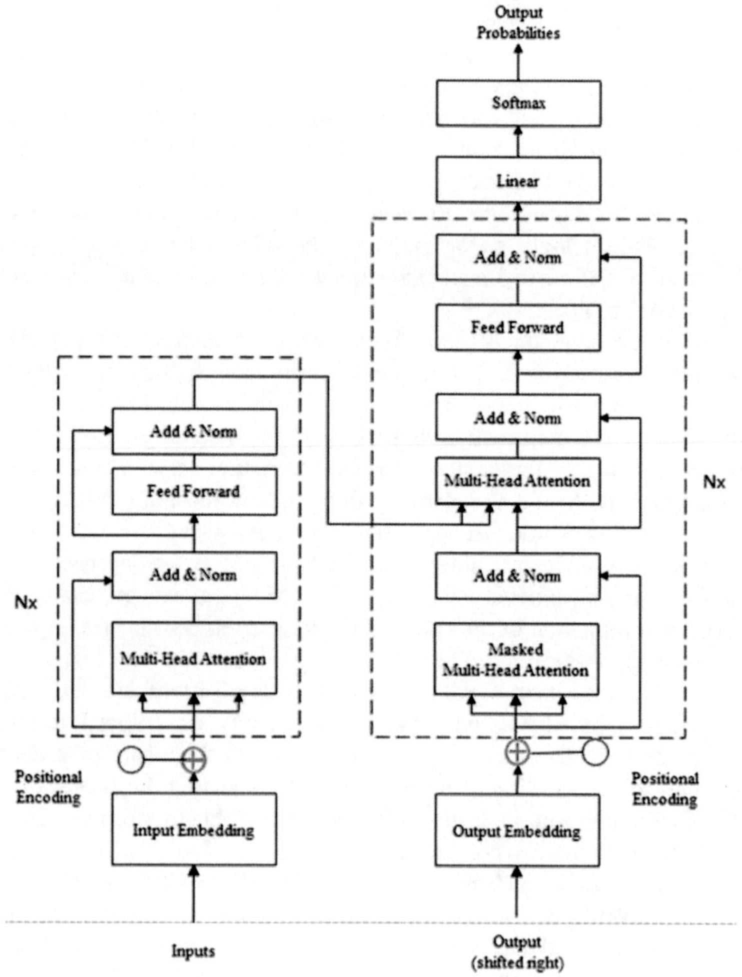

Fig. 1 Transformer structure diagram

3.2 BiLSTM-CRF Method Based on Migration Optimization

This model uses a large number of contract corpus to tune BERT in advance, so that BERT can become a pretrained language model in the contract scenario, and then uses BERT as the sentence encoding part, and finally uses the BiLSTM+CRF model to classify characters and determine the contract entity information.

The structure of BiLSTM+CRF is shown in the figure above. For a sequence input, BiLSTM+CRF first corresponds to each input character to a word vector pretrained on a large corpus. Then the word vector sequence is bidirectionally coded by LSTM, and the BiLSTM word code is obtained by concatenation. BiLSTM word encoding is used as the input of the top-level CRF, and the result identified by BIO is obtained, and finally connected to the loss layer (using cross entropy as the loss function), and then the gradient is calculated by the BP algorithm to optimize the network parameters.

The specific architecture is shown in Fig. 2.

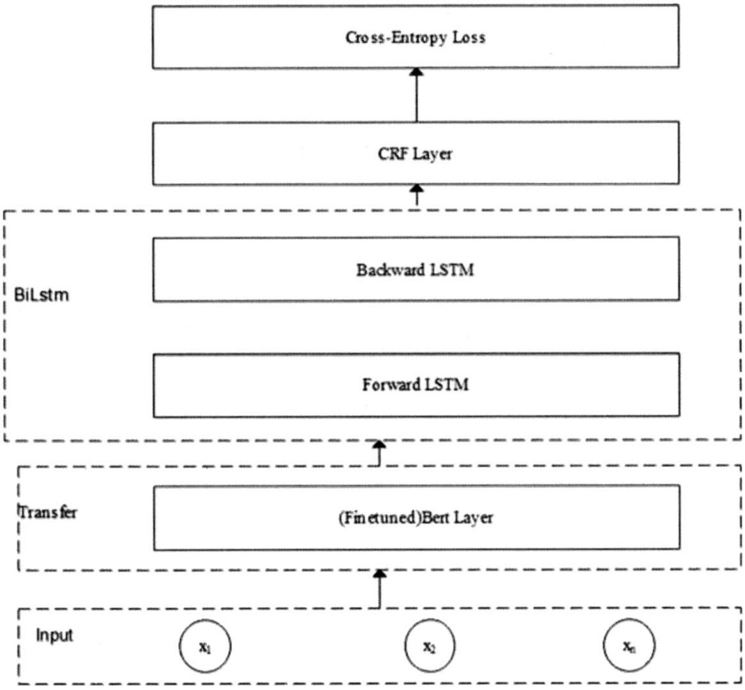

Fig. 2 Specific architecture diagram of the model

4 Test Results and Analysis

All the experiments in this article use pytorch to build the model, which was completed in the Ubuntu 16.04 environment, and used a GeForce GTX 1050Ti GPU for acceleration. Three models of BILSTM-CRF, CNN-BiLSTM-CRF, and BERT-BiLSTM-CRF were used in the experiment (Table 1).

Table 2 shows the prediction results of each entity obtained by using BERT-BiLSTM-CRF in this paper. It can be seen from Table 2 that the labeling effect of the BERT-BiLSTM-CRF model used in this paper is good overall, especially on some labels with obvious characteristics.

5 Concluding Remarks

This article studies mainstream algorithms for the small sample size and finds that the latest BERT language model based on fine-tune can be attempted to be modified. We combine the current power grid samples and the latest algorithm to carry out algorithm innovation.

Self-developed migration algorithm based on pretrained language model (BERT) in the contract field, which meets the technical requirements of smart audit of power grid contracts.

Table 1 Experimental comparison results

Frame model	Accuracy	Recall rate	F1 value
BiLSTM-CRF	0.9063	0.9145	0.9103
CNN-BiLSTM-CRF	0.9326	0.9281	0.9303
BERT-BiLSTM-CRF	0.9584	0.9639	0.9612

Table 2 BERT-BiLSTM-CRF labels the results of each entity

Entity	Accuracy	Recall rate	F1
Full text_basic information of the parties_Bank account_Party A	0.9444	0.9444	0.9444
Full text_basic information of the parties_Address_Party B	0.8125	0.9286	0.8667
Text_Price_Currency	1.0	0.9706	0.9851
Body_price_Total Contract_Lowercase	0.9643	1.0	0.9818
Text_Price_Total Contract_Uppercase	0.9167	1.0	0.9565
Full text_Basic information of the parties_Bank account_Party B	0.9474	0.9474	0.9474

References

1. Li N (2018) J Chinese Inf Process 32(11):41–48
2. He B, Guan Y (2019) Character-based CRF for medical entity recognition. Intell Comput Appl 9(2):130–134
3. Zhai S, Duan H, Li Z (2019) Knowledge graph entity extraction based on BILSTM_CRF. Comput Appl Soft 36(5):269–274, 280
4. Zhang S (2019) Commodity attributes extracting in Chinese shopping reviews based on Bi-LSTM and CRF. Comput Modernization 282(2):97–101
5. Gridach M (2017) Characterlevel neural network for biomedical named entity recognition. J Biomed Inform 70:8591
6. Yang P, Yang Z, Luo L, Lin H, Wang J (2018) An attention-based approach for chemical compound and drug named entity recognition. J Comput Res Dev 55(7):1548–1556
7. Yuan S, Tang J, Gu X (2018) A surveyon scholar profiling techniques in the open internet. J Comput Res Dev 55(9):1903–1919
8. Tang J, Yao L, Zhang D (2010) A combination approach to Web user profiling. ACM Trans Knowl Discov Data 5(1):1–44
9. Gollapalli SD, Qi Y, Mitra P (2014) Extracting researcher metadata with labeled features. Proc of SIAM Int Conf Data Mining, pp 740–748. SIAM, Philadelphia, PA
10. Gu X, Yang H, Tang J (2016) Web user profiling using data redundancy. Proc of IEEE ACM international conference on advances in social networks analysis and mining. Piscataway, pp 358–365. IEEE Press, NJ

Research on Linear Detection Technology for Massive MIMO in Wireless Communications

Hongfei Pu, Jianfei Shao, Rongjian Wei, Rong Pu, and Peng Guo

Abstract Massive multiple-input multiple-out (massive MIMO) transmission technology is a key technology of next-generation mobile communication 5G and one of the leading research topics in the field of wireless communication. This article first introduces the research background and current status of traditional MIMO and massive MIMO technology and points out the significance of the research. Then the massive MIMO model was established to provide theoretical support for the following simulation experiments. Finally, the research on the massive MIMO linear detection algorithm is studied, mainly on the three types of linear detection: maximum combining ratio (MRC), zero forcing (ZF), and minimum mean square error (MMSE). Through experiments on these three algorithms, and then analyze their advantages and disadvantages.

Keywords Massive MIMO · Linear inspection technology · Maximal ratio combining · Zero forcing · Minimum mean square error

1 Introduction

Massive MIMO technology refers to a MIMO system with hundreds or thousands of antennas. The number of antennas in a traditional MIMO system is generally 4, 8, or 16, while the number of antennas in a massive MIMO system can reach hundreds of thousands. The antennas of the ordinary MIMO system are only arranged in the horizontal direction, but the antenna positions of the massive MIMO system are always arranged in the horizontal and vertical directions, thereby forming a plane. Compared with ordinary networks, the advantage of MIMO networks is that they can increase the capacity of the system without occupying too much spectrum and transmit power. Massive MIMO brings the following benefits: (1) The channel capacity is greatly increased. (2) The beam becomes narrower, the

H. Pu (✉) · J. Shao · R. Wei · R. Pu · P. Guo
School of Information Engineering and Automation, Kunming University of Science and Technology, Kunming, China

Z. Yu et al. (eds.), *Advancements in Mechatronics and Intelligent Robotics*,
Advances in Intelligent Systems and Computing 1220,
https://doi.org/10.1007/978-981-16-1843-7_15

energy efficiency becomes higher, and the noise interference in the MIMO system is smaller. (3) With the improvement of resolution, massive MIMO system technology meets the simultaneous communication of multiple users and base stations by mining dimensional resources.

2 Model of Massive MIMO Uplink System

The following figure shows a model of a multi-user uplink massive MIMO system. Each user's transmitting end uses one antenna. In the following figure, there are eight users at the transmitting end. Channel H is the sum of different transmission paths, and then noise is superimposed. At the base station, we use hundreds of antennas to receive. The number of antennas is much greater than the number of transmitting antennas [1] (Fig. 1).

The data is sent out through each transmitting end; the receiving end receives the data detection and then demodulates to get the output signal. In order to reduce mutual interference of data and separate data efficiently, the distance between antennas should be more than half a wavelength in order to transmit signals correctly [2].

Suppose that there are m antennas at the transmitting end and n antennas at the receiving end. The number of receiving antennas n is greater than m. In addition, the influence of noise on the signal cannot be ignored in communication, and it is affected by Gaussian white noise in the channel. We use N to represent noise. At a certain time t, it is used to represent the signal vector at the transmitting end, and the first antenna at the receiving end at time t is expressed as: $y_1(t) = h_{11}(t)x_1(t) + h_{12}(t)x_2(t) + \cdots + h_{1n}(t)x_n(t) + n_1$. Then you can get:

$$y_n(t) = h_{m1}(t)x_1(t) + h_{m2}(t)x_2(t) + \cdots + h_{mn}(t)x_n(t) + n_n.$$

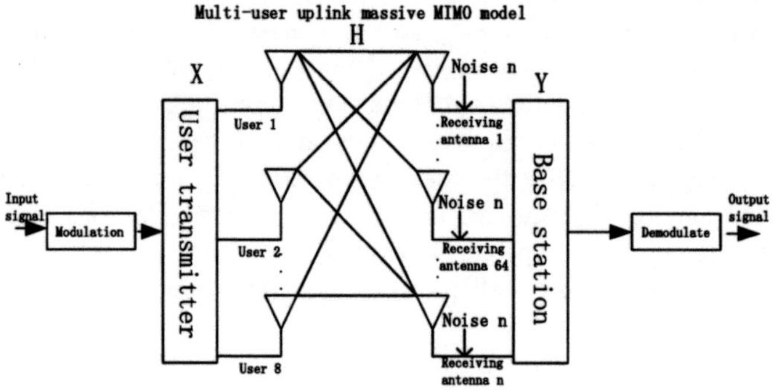

Fig. 1 Multi-user and multi-receiving antenna model in uplink

Then we will express $y_1(t), y_2(t), y_3(t)\ldots y_n(t)$ in matrix form:

$$\begin{bmatrix} y_1(t) \\ y_2(t) \\ \vdots \\ y_n(t) \end{bmatrix} = \begin{bmatrix} h_{11} & h_{12} & \cdots & h_{1n} \\ h_{21} & h_{22} & \cdots & h_{2n} \\ \vdots & \vdots & \ddots & \vdots \\ h_{m1} & h_{m2} & \cdots & h_{mn} \end{bmatrix} \times \begin{bmatrix} x_1(t) \\ x_2(t) \\ \vdots \\ x_n(t) \end{bmatrix} + \begin{bmatrix} n_1 \\ n_2 \\ \vdots \\ n_n \end{bmatrix} \tag{1}$$

We express $X(t) = [X_1(t), X_2(t)\ldots, X_n(t)]^T$ as a vector at the sending end at a certain time, and then express $Y(t) = [Y_1(t), Y_2(t)\ldots, Y_n(t)]^T$ as a vector at the receiving end at the corresponding time. Use $H = [h_1, h_2\ldots, h_n]$ to denote the channels of the *m*th receiving antenna and the *n*-th receiving antenna. In particular, h_n denotes the column vector and uses $h_n = [h_{1n}, h_{2n}\ldots, h_{nm}]^T$ to indicate which column of the vector. It obeys a complex Gaussian distribution with zero mean and variance 1. *N* ideal additive complex Gaussian vector. Then the entire MIMO system model can be written as:

$$Y = HX + N \tag{2}$$

3 Linear Detection Algorithm for Massive MIMO System

3.1 Linear Detection Algorithm

In general, the principle of linear detection technology is to calculate the filter matrix using the linearity criterion at the receiving end, let the signal at the receiving end pass through the filter matrix, and let the filter matrix remove the interference of interference signals and noise, leaving only the detection signal. At the receiving end, if we want to get the transmitted signal, we need to multiply the received signal by the filter matrix, and secondly, pass the judgment on the signal passing through the filter matrix [2]. *G* is the filter matrix, *X* is the vector passing through the filter matrix *G*, denoted by *X*, and *Y* is the signal detected by the receiving end. Available formulas. Then we use Quant() to represent the decision and *x* to represent the received signal after the decision [3].

$$\tilde{X} = GY \tag{3}$$

$$\hat{X} = Quant(\tilde{X}) \tag{4}$$

Linear detection requires that the receiving days of the system be greater than or equal to the number of transmitting days. Linear detection algorithms mainly include zero forcing and minimum mean square error [4].

3.1.1 Maximum Ratio Combining Algorithm (MRC)

The maximum ratio combining algorithm (MRC) is the optimal algorithm in diversity merging. Compared with selective merging and equal gain merging, it can obtain better communication indicators. The advantage of the MRC algorithm is that the implementation is relatively simple, but the MRC algorithm has low error resistance performance and only meets the special channel.

When a channel has Nt subchannels, let $H = [h_1, h_2, \cdots h_{N_t}]$ $G = [g_1, g_2, \cdots g_{N_t}]$ then have:

$$G = H^H y$$

$$\hat{i} = \arg msx|G| \tag{5}$$

$$\hat{s} = Q(G)$$

Q is the constellation point decision operation, S is the modulation symbol statistics, and i is the emission estimation [5].

3.1.2 Zero-Forcing Algorithm (ZF)

The principle of the zero-forcing algorithm (ZF) is to zero the interference caused by the channel matrix H, so zero forcing comes from this. Because the zero-forcing algorithm divides the channel into several M channels with noise n signals, the noise is boosted by G_{ZF}. The simplest detection algorithm in MIMO system is it G_{ZF} is a simple m by m matrix, and the zero-forcing matrix is given by the following formula:

$$G_{ZF} = \sqrt{\frac{M_T}{E_s}} H^+ \tag{6}$$

The detection value of the signal is:

$$\hat{X}_{ZF} = G_{ZF} Y = H^{\&} Y = X + H^{\&} N \tag{7}$$

H is a channel matrix with full columns, and the filter matrix G is its pseudo-inverse matrix. Its formula is as follows:

$$G_{ZF} = H^{\&} = (H^H H)^{-1} H^H \tag{8}$$

In summary, although the zero-forcing algorithm is simple, G will amplify the noise and increase the bit error rate, so the anti-noise performance of ZF is very poor.

3.1.3 Minimum Mean Square Error Detection Algorithm (MMSE)

The minimum mean square error (MMSE) eliminates the interference of multi-stream interference and noise because of the compromise and reduces the transmission error [6]. The minimum mean square error (MMSE) is converged to the zero-forcing algorithm (ZF), where the minimum mean square error (MMSE) performance is better than the zero-forcing receiver.

For matched filters, it can be obtained by using orthogonality [7]:

$$G_{\text{MMSE}} = (H^H H + \delta^2 I_{n_t})^{-1} H^H = H^H (H^H H + \delta^2 I_{n_t})^{-1} \tag{9}$$

The output signal is

$$\hat{X}_{\text{MMSE}} = G_{\text{MMSE}} Y \tag{10}$$

From the above formula:

$$\hat{X}_{\text{MMSE}} = G_{\text{MMSE}} Y = (H^H H + \delta^2 I_{n_t})^{-1} H^H Y \tag{11}$$

3.2 *Simulation*

In the six pictures above, we can clearly see that no matter what kind of modulation or signal-to-noise ratio, as the number of antennas increases, the bit error rate will be greatly reduced, and the performance of the three algorithms will be improved. This verifies that the increase in the number of antennas in the MIMO system increases the system's anti-interference ability. In particular, we also see that the performance of the MMSE algorithm is better than that of the ZF algorithm at 2 dB, but in a massive MIMO system, there is not much difference between the performance of the ZF algorithm and the MMSE algorithm. In addition, we can see in the above figure that the bit error rate of the MRC algorithm is much higher than that of the ZF algorithm and the MMSE algorithm. It can be seen that the MRC algorithm is not suitable for massive MIMO systems. Moreover, it can be seen from the ZF algorithm and the MMSE algorithm that when the number of receiving antennas in the uplink is much larger than the user antennas, we can also get better performance if we use the linear algorithm [8] (Figs. 2, 3, 4).

Fig. 2 BPSK (2 dB)

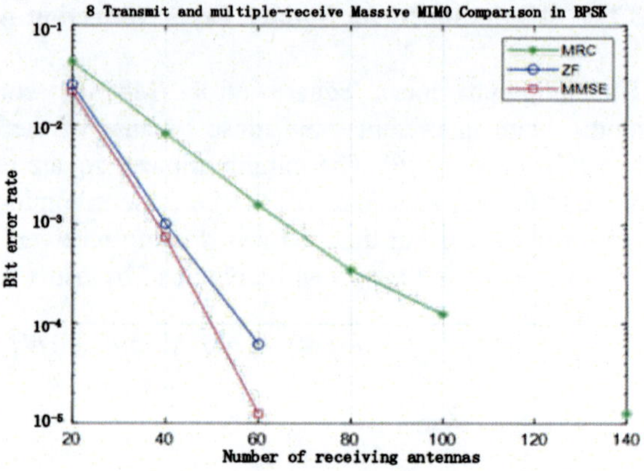

Fig. 3 QPSK (2 dB)

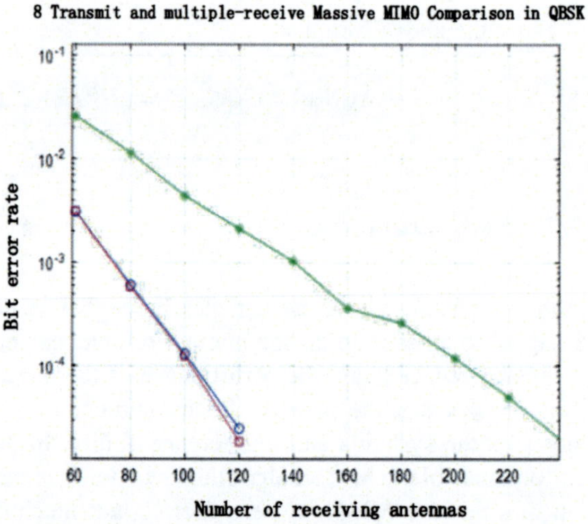

Fig. 4 16QAM (2 dB)

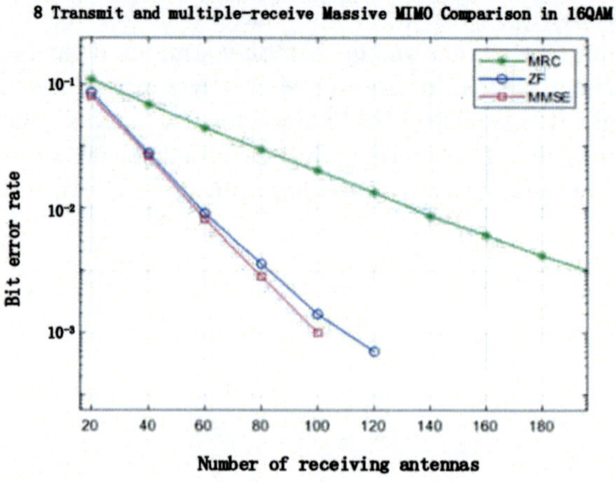

4 Conclusion

In massive MIMO, when the number of antennas in the base station is greater than the number of users, it is approximately considered that each column in the channel is orthogonal to each other, and the MRC algorithm can significantly reduce the interference between users at this time. In the channel processing, the noise is not amplified, so when the number of antennas is greater than the number of users, good results can be achieved. But compared with ZF and MMSE algorithms, the performance of MRC algorithm is a bit worse.

For the ZF algorithm, the ZF algorithm resets the interference in the channel to zero by multiplying the channel pseudo-inverse matrix on the left side of the received signal, but the noise will be amplified when multiplying, and the performance will deteriorate, although the complexity of the ZF algorithm is relatively low, but its performance is very poor, especially at low signal-to-noise ratio. So in order to improve performance, the influence of noise should be taken into consideration.

Fig. 5 BPSK

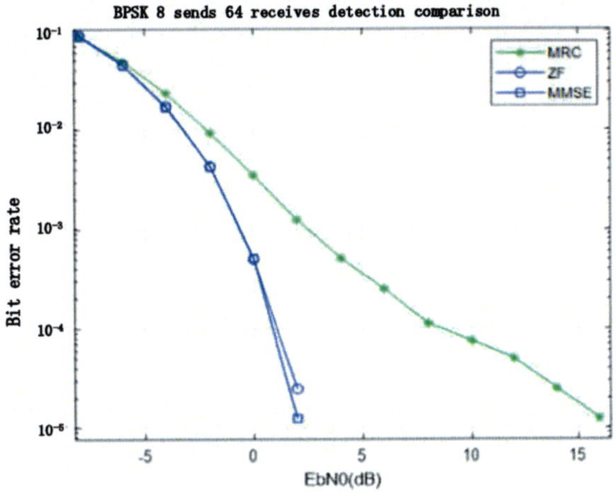

Fig. 6 QPSK

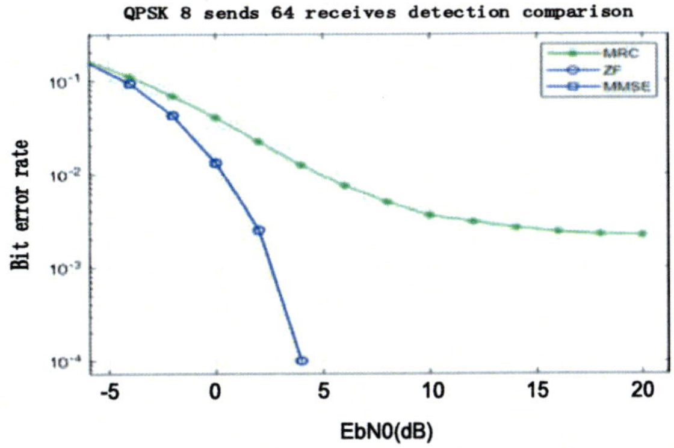

Fig. 7 16QAM

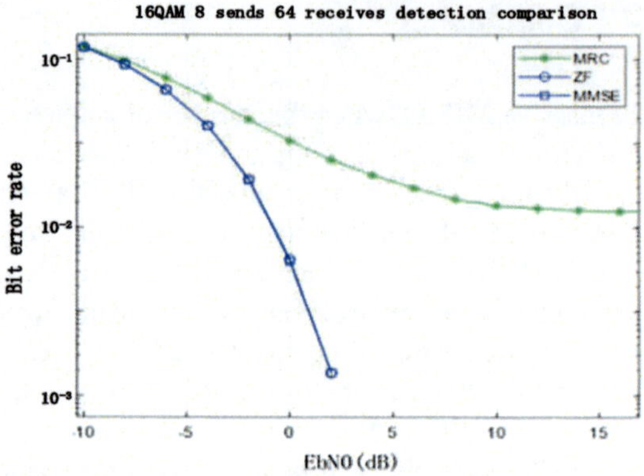

While experimenting, we can clearly see that with the increase of the modulation number, such as BPSK modulation to 16QAM modulation, as shown in Figs. 5, 6, and 7, we can see that the bit error rate of the massive MIMO system will vary with the modulation. The increase in the number of digits in the base number proves that the effectiveness and reliability of the communication system is a contradiction.

References

1. Yin B, Wang H, He Y (2018) Signal detection method for massive MIMO uplink. Commun Technol 51(317)(05):7–12
2. Zhao Cheng H, Haiyan C, Shiyi X et al (2017) Study on low complexity of MMSE linear detection algorithm for multi-cell Massive MIMO system. Wirel Internet Technol
3. Zhang Y (2009) Research on low-complexity massive MIMO detection algorithm. Xidian University
4. Liu L, Peng G, Wei S (2019) Linear massive MIMO detection algorithm. Massive MIMO detection algorithm and VLSI architecture
5. Wu Y (2015) Research on MIMO signal detection algorithm under V-BLAST coding. Jiangxi University of Science and Technology
6. Zhang Y (2015) Research and simulation of massive MIMO detection technology. University of Electronic Science and Technology of China
7. Wei L (2017) Research on signal detection algorithm of MIMO system. Nanjing University of Posts and Telecommunications
8. Wang X, Yang L (2015) Research on massive MIMO signal detection algorithm. J Nanjing University of Posts and Telecommunications: Natural Science Edition 03:56–60

Research on Intelligent Reconfiguration and Recognition Technology of Mobile Environment for Substation Operation

Cai Bin

Abstract Firstly, this paper studies the key technologies of scanning modeling for dynamic complex environment based on mobile terminal and carries out handheld scanning modeling for typical power grid equipment in power grid to realize the functions of dynamic complex environment mapping. Secondly, the device recognition technology based on mobile terminal is studied to realize the device object detection and power device recognition based on the physical similarity. Thirdly, the positioning and navigation technology based on mobile terminal in dynamic and complex environment is studied to realize the functions of operation positioning and navigation in substation operation environment. Finally, the prototype system of intelligent reconfiguration and recognition of mobile environment are developed, and the overall integration scheme of the prototype system is studied. The software and hardware integration of intelligent mobile device, depth of field modeling and scanning terminal, intelligent reconfiguration and recognition of mobile environment prototype system are realized. Through the research of this paper, the data collection and value mining of three-dimensional model of substation are realized, and the intelligent level of on-site inspection is improved.

Keywords Mobile side · Intelligent reconfiguration · Substation operation

1 Introduction

With the continuous development of industrial automation and intelligence, the role of environmental modeling and cognition as the core technology has become increasingly prominent [1]. In order to solve the contradiction between the scale growth of power grid and the synchronous growth of operation and inspection personnel allocation, change the traditional operation and inspection mode with a large amount of human investment, traditional information acquisition mode and single source [2], and improve the utilization rate of advanced means and data such

C. Bin (✉)
State Grid Xianyang Power Supply Company, Xianyang, China

Z. Yu et al. (eds.), *Advancements in Mechatronics and Intelligent Robotics*,
Advances in Intelligent Systems and Computing 1220,
https://doi.org/10.1007/978-981-16-1843-7_16

as robots [3], UAVs and individual equipment, it is necessary to study the intelligent reconstruction and identification technology of mobile side environment for substation operation. Under the traditional power inspection operation mode, due to the complicated operation of power equipment, the component logic is not visible, especially the highly integrated module and more complex logical relationship of intelligent station, the operation and inspection personnel cannot complete the task independently, and have a high dependence on the manufacturer. In the operation process, it is prone to misoperation [4], misentering the interval and other dangerous situations, resulting in casualties. At present, with the development of IT technology, environment scanning modeling and object recognition technology have been applied in consumer electronics, industry, medical treatment, robot, and other fields [5]. For example, Boeing has established the bar platform, which uses a large number of fixed cameras to realize the perception of factory production line environment and the positioning of people and objects. Siemens is developing this technology on mobile terminals for its application to guidance systems of complex equipment maintenance. Through the research and development of the basic common key technology, it will provide new automation and intelligent solutions for power grid field operation, improve operation mode, improve operation efficiency, and realize active safety protection.

In order to realize the autonomous modeling, cognitive understanding and positioning navigation of the work environment, this paper intends to take the portable terminal equipment as the information collection, processing and interaction center, and take the task as the guidance, to study the low-cost self fast scanning and three-dimensional modeling technology of the work environment, to study the terminal side efficient and accurate power equipment identification technology, and to study the efficient and accurate three-dimensional model matching and positioning. In addition, a prototype system is developed and implemented in the scene of substation operation. Intelligent reconfiguration, cognition, positioning, and navigation technology of complex dynamic environment based on depth vision is one of the core key technologies of automation and intelligent development in the industrial field. Through the research of key technologies, it will lay the foundation for the follow-up research and development of intelligent operation terminal equipment and automatic operation robot equipment in the power industry with independent intellectual property rights, and the research and development of extended business application support system foundation.

2 Object Scanning and Modeling

Object scanning and modeling are based on KinectFusion, as shown in Fig. 1. The depth image is converted into three-dimensional point cloud, and the normal vector of each point is calculated; second, the point cloud with normal vector and the point cloud projected from the model according to the pose of the previous frame are calculated by ICP. According to the calculated pose, the point cloud of the current

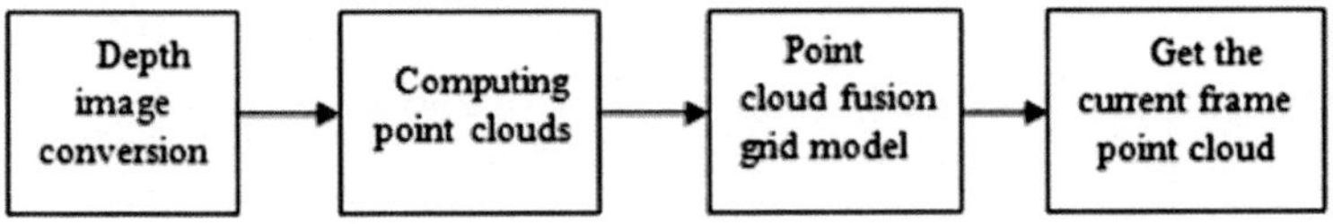

Fig. 1 Object scanning and modeling flowchart

frame is fused into the grid model; according to the pose of the current frame, the point cloud of the current frame is projected from the model by ray projection algorithm, and the normal vector is calculated to register the input image of the next frame.

3 Modeling and Scanning Modeling of Dynamic Complex Environment

The modeling of dynamic complex environment consists of two parts: creating 3D point cloud map of environment and creating two maps on the plane of environment. Three-dimensional point cloud map is created by using rgb-d sensor, and plane grid map is created by using laser sensor. When creating a planar grid map, we need to use the zed camera sensor to build a synchronous map to build a topological environment profile map and align the topological map with the planar grid map. The plane grid map is used for users to view and calculate the path planning according to the grid map. The topology map is used to locate the user in the environment. The topology map is used to get the user's position in the current scene, and the user's position information is placed in the grid map for display. Three-dimensional point cloud map is created by using rtab mapping, and plane grid map is created by using laser sensor and gmapping in ROS.

In this paper, the basic module of scanning modeling for dynamic complex environment based on mobile terminal is designed and developed, which is mainly divided into object scanning and modeling, dynamic complex environment modeling and scanning modeling. This paper mainly studies the key technology of object device scanning modeling based on rgb-d sensor and selects Microsoft Kinect sensor to build the handheld and fixed base scanning modeling experimental platform. Using Microsoft kinectfusion SDK as the secondary development platform, the handheld object scanning and modeling software are developed, and the handheld scanning modeling of typical power grid equipment in power grid is carried out. In view of indoor environment, the design and implementation method of 3D map's establishment, positioning, and guidance function is studied, as well as the implementation method of dynamic complex environment mapping, positioning, and guidance software.

4 Weak Supervision Instance Segmentation of Substation Equipment Based on RGB-T Self Labeling

Infrared and color cameras are common devices in power equipment monitoring applications. The thermal conductivity of the metal materials of power equipment is obviously different from that of the environment, so their infrared images can be used to guide the generation of mask labeling of related RGB images. However, there is a big difference in resolution between RGB image and infrared image. The infrared image with smaller resolution has a great loss of information compared with the RGB image with larger resolution. Therefore, simply using the mask label generated by the infrared image foreground segmentation guidance to train the segmentation model will lead to the weak generalization ability of the model because the training sample is too simple.

Therefore, a new automatic annotation of masks of samples (AAMs) based on RGB-T image is proposed. Compared with many weak supervised semantic or instance segmentation methods based on rough annotation training model, this method is simple but effective and can generate more accurate mask annotations when image category labels are used. In addition, because of the independence of this method, it is compatible with any existing split network structure. This project further proposes a progressive optimized model (POM), which first optimizes the segmentation boundary of each object by using the fully connected conditional random field and the loss function constrained to the boundary and solves the problem that the generated annotation is noisy and the spatial details of depth CNN are lost. Then, the model uses self-learning technology to train the segmentation network from simple to complex to solve the problem that the training sample is too simple.

5 Model Compression and Acceleration for Edge Devices

In order to deploy instance segmentation model on resource constrained devices, this paper studies the compression and acceleration technology of instance segmentation model, so that the model can be deployed on resource constrained devices with guaranteed performance. Compression and acceleration process of the model are shown in Fig. 2. Firstly, the backbone network of the instance segmentation model is replaced by an efficient and concise network to get a new small instance segmentation model. Then, the new model is trained by simulation and supervision based on knowledge distillation technology. Finally, the model is stored quantitatively.

The specific implementation scheme is as follows: First, we use mobilenetv3 to replace the backbone network of the instance segmentation model taking mask r-cnn as an example to get the newly created small model mobilenetv3 mask r-cnn, then we use the trained mask r-cnn to train mobilenetv3 mask r-cnn for knowledge

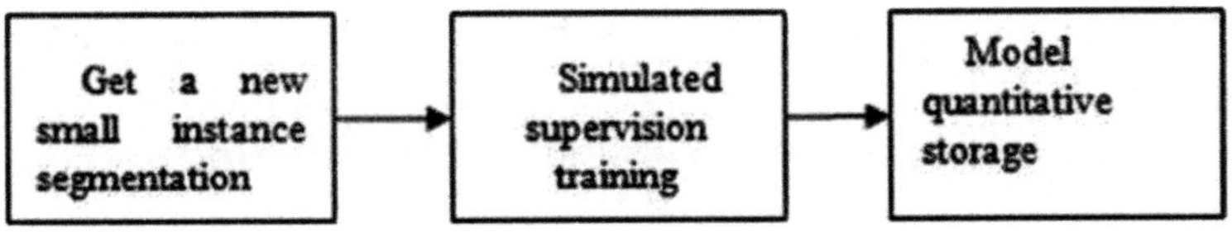

Fig. 2 Flowchart of model compression and acceleration

distillation, and finally use the trained mobilenetv3 mask r-cnn for knowledge distillation training R-cnn quantifies the model to get the final instance segmentation model to be deployed.

Model post training quantization is a common model quantization technology, which usually results in little performance loss of the model, so it can be performed on floating-point model. The weights and activation values of the trained model are usually distributed in a relatively small range, and the neural network is robust to noise, so the noise error caused by quantifying the model to a small range will not have much impact on the performance of the model.

The weak supervised instance segmentation method based on RGB-T fusion is mainly divided into two modules: AAMs and POM. AAMs are mainly used to provide the mask label required by POM training.

The dataset used in this paper is composed of paired RGB-T images of power equipment, and the RGB images are cropped to have a similar field of view and scaled to the same size as the infrared thermal images to get a simple RGB image (low resolution, only containing a single target). Here, the original RGB image is defined as a complex RGB image (high resolution, including one or more targets), and all samples are only RGB images. There are category labels.

Mask samples are labeled automatically. In the case of using only category labels, the semantic label of each foreground pixel is obtained. Firstly, the infrared thermal image is mapped to the corresponding simple RGB image coordinate system by using image registration technology. Then, the pre- and background coarse segmentation prior markers are obtained by saliency detection. Finally, on the basis of coarse segmentation prior marker, the foreground fine segmentation result of infrared thermal image is obtained by using grabcut algorithm, and the target mask label of RGB image is obtained.

The case segmentation model is optimized step by step. Firstly, a case segmentation model based on boundary optimization is proposed, which is built on the pre trained mask r-cnn, and uses the fully connected conditional random field and the loss function constrained to the boundary to optimize the segmentation. The case segmentation model based on boundary optimization is applied to simple RGB image data set. Then the model based on self-learning is optimized. The mask of complex RGB image is predicted by using the model trained from simple RGB image dataset, and the mask label of complex RGB image is obtained by denoising with category label. The complex RGB image data set is constructed to train the final instance segmentation model.

6 Research and Development of Environment Intelligent Reconfiguration and Recognition Prototype System for Mobile Side Software and Hardware Platform

The main functions include: (1) depth vision 3D scanning modeling function to realize indoor and outdoor environment modeling in complex and noisy environment. (2) The large-scale model stitching function can realize segmented scanning, sub map stitching, topology creation and optimization, synchronous object scanning and other functions. (3) The semantic marking function of substation operation environment realizes the semantic marking of regional structure of substation operation environment and supports the subsequent model feature recognition. (4) The function of equipment model library is to build the feature recognition library of power grid equipment model by using deep learning technology. (5) Intelligent recognition function realizes online deep learning model compression based on mobile terminal and efficient identification of power grid equipment. (6) The indoor and outdoor integrated positioning function realizes the fast decimeter level accuracy positioning of operators in indoor and outdoor environment. (7) High precision indoor and outdoor operation navigation is based on the scene point cloud map to realize the indoor and outdoor operation path planning and real-time navigation. (8) The model shows the interactive function and realizes the job site feature recognition, model display, and job guidance based on the point of interest. (9) Mobile environment intelligent reconfiguration and recognition prototype system.

7 Conclusion

In this paper, firstly, the 3D model construction technology of dynamic scanning based on depth vision can quickly and low-cost establish the 3D model of the operation site, which has the ability of large-scale promotion and application; secondly, the 3D model is the digital reproduction of the operation site, which can help the operators quickly grasp the inspection points by integrating the equipment information and inspection operation experience and knowledge; thirdly, it can help the operators to quickly locate the work site by using the portable equipment, and improve the efficiency of operation. Based on the reconstruction and positioning technology of the field environment, it can determine the location of the operators and the status of the equipment, prevent accidents and effectively reduce accidental losses. The research and development and verification of smart environment reconfiguration and identification prototype for substation operation mainly include two aspects: the research and development of basic module of smart environment reconfiguration and identification based on the software and hardware platform of mobile side, and the research and development and technical verification of smart environment reconfiguration and identification prototype system of mobile side.

References

1. Liao KJ (2011) Browser/Server platform solution to the financial software of management type. Comput Eng Sci 78–91
2. Perez-Luque MJ, Little TDCA (2003) Temporal reference frame work for multimedia schronization. IEEE J Select Aieas Conllnunlcat 14(l):36–37
3. Bertino E, IEEE M, Ferrari E (2008) Temporal synchronization models for multimedia data. IEEE Trans Knowl Data Eng 10(4):612–631
4. Research on streaming media playback and codec based on smart phone, pp 12–15. Nanjing university of posts and telecommunications, Nanjing
5. Li S (2012) Design and implementation of streaming media system based on Android, pp 21–14. Jilin university, Changchun

Development and Application of Power Grid Disaster Intelligent Perception and Emergency Command System

Xu Min, Xu Xiyuan, Yang Jianhua, Yang Jian, Peng Lin, Tang Shiyang, Li Nige, Hou Zhansheng, He Zhimin, Wang Gang, Wang He, Bao Xingchuan, Yu Hai, Zhu Liang, and Zhang Zehao

Abstract Research and development of multi-information intelligent acquisition module of power grid disaster loss, power grid disaster loss information fusion analysis and prediction module, and multi-person information real-time interaction module. Based on these three modules, a prototype system of power grid disaster intelligent perception and emergency command is developed for the application scenarios of emergency drill, training, and command operation. Develop the integrated interface of the prototype system, realize the software and hardware integration of intelligent mobile terminal, background equipment monitoring system and electric power emergency command platform, and realize the field application verification in the company headquarters and typical electric power emergency drill, rescue and command site of provincial company, realize the intelligent perception of power grid disaster and auxiliary decision-making and visual command under the condition of missing information in the emergency site, so as to realize the real-time monitoring of power grid disaster. It can effectively improve the efficiency of emergency command in disaster judgment, operation command, and material dispatching.

Keywords Multi-information · Intelligent collection · Real-time interaction

---

X. Min (✉) · X. Xiyuan · P. Lin · T. Shiyang · H. Zhansheng · H. Zhimin · W. Gang · W. He · B. Xingchuan · Y. Hai · Z. Liang · Z. Zehao
Global Energy Interconnection Research Institute, Nanjing, Jiangsu 210003, China

X. Min · X. Xiyuan · L. Nige
State Grid Key Laboratory of Information & Network Security, Nanjing 210003, China

Y. Jianhua · Y. Jian
State Grid ZheJiang Electric Power Co. LTD, Zhejiang Hangzhou 310000, China

Z. Yu et al. (eds.), *Advancements in Mechatronics and Intelligent Robotics*,
Advances in Intelligent Systems and Computing 1220,
https://doi.org/10.1007/978-981-16-1843-7_17

1 Introduction

In recent years, a variety of natural disasters occur frequently and have a growing trend. Typhoon, heavy rainfall, geological earthquake, rain, snow, freezing, and other disasters have brought huge losses to the power grid, seriously affecting the safe and stable operation of the power grid and the normal production and operation of the company. State Grid Corporation of China (SGCC) started the construction of emergency command information system in 2008, which has played an active role in the emergency response of SGCC in recent years. The system focuses on the management of emergency plans, emergency teams, and other emergency information and has insufficient support ability in the field information collection, information interaction, information fusion, situation analysis and other emergency command and auxiliary decision-making. The specific performance is as follows: (1) The efficiency of emergency site information collection is not high, and the traditional way of disaster damage information collection mainly relies on manual site investigation and reporting. When major emergencies come, there are some problems, such as difficult data acquisition, slow data acquisition, large limitations and few data sources, which affect the scientific decision-making of emergency response. It is unable to collect the damaged data of the equipment in the disaster site quickly and comprehensively, and the efficiency of data collection needs to be improved [1]; (2) after the natural disasters such as typhoon, heavy rainfall, geological earthquake, snow, and freezing, the information communication between the on-site rescue personnel and the on-site rescue headquarters and the emergency command center is mainly by means of telephone or short message. The flexibility is insufficient, and it is difficult to upload and release the on-site pictures, audio and video disaster information; (3) there is a lack of fusion analysis technology and disaster loss prediction model, the analysis of disaster data and fault data of each specialty is relatively independent, and there is a lack of emergency data fusion analysis technology for cross-specialty comprehensive analysis of disaster data in various fields. In addition, the emergency command center lacks of real-time data mining for the disaster scene, and the analysis is mainly based on the current situation statistics of power grid damage, so the degree of disaster damage prediction is not high, which affects the rapid and accurate decision-making of emergency command.

2 Electric Power Disaster Emergency Situation Map and 3D Visual Command Technology

Through the integration of 3D visualization technology and emergency command system, the situation map and 3D visualization technology of power disaster emergency are studied. Through the data visualization function of the emergency command system, combined with the GIS module, the current situation of the emergency and the loss of power equipment are displayed. According to the development of disaster and the loss or recovery of equipment, the visualization data is updated dynamically [2].

On this basis, the short-term fuzzy dynamic power grid loss prediction function studied in the study predicts the next development of disaster and loss situation, and displays on another interface based on data visualization function, which does not overlap with the current disaster loss situation.

Based on the short-term loss prediction results of a disaster, this paper studies the development trend indicators of emergencies, studies the scientific, intuitive and easy to understand development trend indicators for the progress of emergencies and the loss and repair of power equipment of different emergencies, and finally identifies the development of emergencies through the development trend indicators.

The indicators of emergency development situation at least include: disaster development situation (typhoon intensity development situation, urban development situation covered by typhoon road strength, regional temperature change situation of ice and snow disaster, rainfall development situation, etc.), power grid loss development situation (loss prediction, personnel input prediction, repair quantity prediction, etc.).

The visualization of disaster situation based on GIS, the visualization of power equipment loss prediction results, and the development trend indicators of emergencies constitute the power disaster emergency situation map. Based on the scientific prediction results, the development of emergencies in the future is visually displayed to the emergency commanders through visualization technology, and the corresponding data indicators are summarized at the same time, which is of great help to the literal processing of emergency situation [3] (Fig. 1).

3 Research and Development of Power Grid Disaster Intelligent Perception and Emergency Command Prototype System

Based on the multi information collection technology of power grid disaster loss, remote information real-time interaction technology, and disaster loss information analysis and prediction technology, the prototype system of power grid disaster intelligent perception and emergency command is developed. The system is divided into terminal and master station service functions, and the main functions include: (1) intelligent identification function of equipment, using "UAV-based line disaster investigation and identification technology," realizing the use of deep learning technology to build the feature recognition library of power grid equipment model [4], using UAV automatic detection system to identify the power disaster scene, realizing the automatic identification, and exploration of the disaster scene. (2) The field data acquisition function uses the "multi-information rapid acquisition and

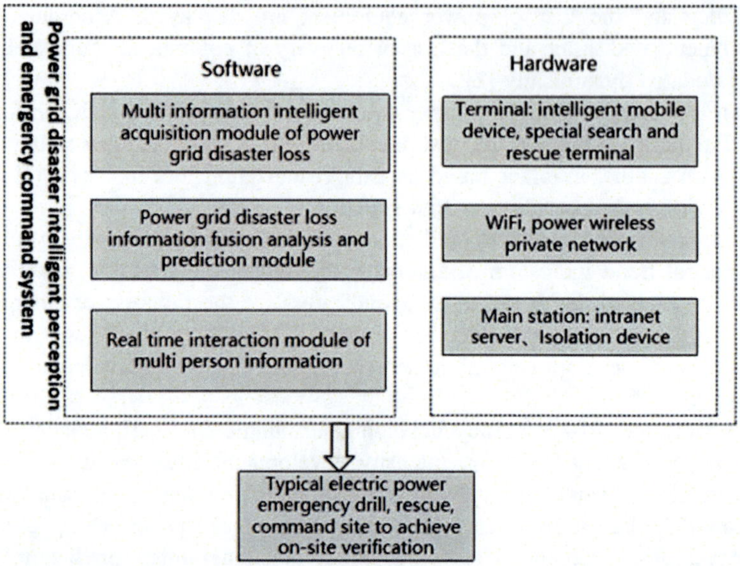

Fig. 1 System development plan

automatic acquisition technology of typical power equipment disaster damage based on the Internet of things," and the integrated field power equipment intelligent identification acquisition based on three-dimensional point cloud, two-dimensional image and radio frequency RFID assistance. Through the integration with the background intranet PMS system and monitoring system, the field equipment account, equipment drawings, historical maintenance and operation are realized Maintenance records, real-time load operation status and other data are pushed in real time to assist field operators to carry out diagnosis and analysis of damaged equipment in the emergency site. (3) The function of disaster loss information fusion analysis and prediction uses "power grid disaster loss information fusion analysis and prediction technology" to realize the information fusion of power grid emergency command comprehensive database and disaster loss model database and fuzzy dynamic power grid loss prediction. (4) Multi-person information real-time interaction technology uses "information real-time interaction technology between emergency scene and command center" to realize the real-time interaction between disaster damage scene terminals and multi person real-time interaction between scene and remote command center, build efficient multi-type information submission method, and improve the information interaction ability among emergency departments and points [5] (Fig. 2).

The prototype system can be divided into user layer, application layer, service interface layer, and database service layer. (1) The user layer mainly includes intelligent mobile devices, networking devices, Intranet servers, and other hardware devices. (2) The application layer mainly includes business pages and platform

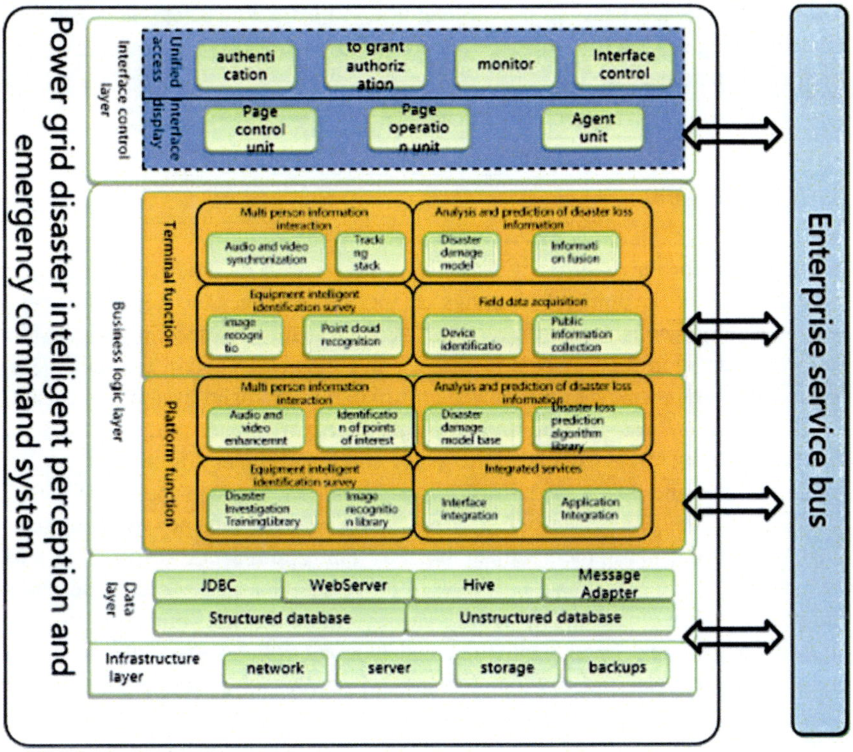

Fig. 2 Functional framework of power grid disaster intelligent perception and emergency command prototype system

presentation framework. The platform presentation layer provides various page modules to form business pages. The business application module develops the business logic in this layer according to the business characteristics. (3) The service interface layer includes the communication client running in the browser and the service access point running in the server. The service interaction layer provides the functions of user request data format, transmission, packet and unpacking, URL processing and distribution. The platform supports restful and Blazeds service invocation and provides Web service support for inter system integration. The service call data inside the platform is serialized in JSON format. The server uses dispachservlet in spring MVC to dispatch requests. The business logic layer consists of the business logic module of the business application system, the standard logic module provided by the platform and the public service agent. Platform services expose public service interfaces in the business system running environment. These interfaces can be referenced by business logic and can also be exposed to client calls through service interaction layer (to solve the cross-domain problem of browser HTTP requests). (4) In the database service layer, MySQL database is used to analyze and back up the structured and unstructured data, and cstor distributed storage is used to manage the data, so as to ensure the efficiency and security of the data.

4 Conclusions

In this paper, for the emergency disposal of power grid disaster emergencies, we make full use of the Internet of things technology, conquer the key technologies such as multiple collection, information fusion, real-time interaction and situation prediction, develop the power grid disaster intelligent perception and emergency command system, innovate the power grid disaster command system, and realize the rapid, automatic collection and real-time access of power grid disaster scene information to all levels of power emergency command center, through information fusion analysis and power grid disaster loss prediction, improve the efficiency and accuracy of emergency decision-making, improve the efficiency of emergency disposal. Support joint emergency command, joint optimization dispatching analysis and other comprehensive business, improve the emergency command ability and power grid safe operation level. (1) Through rapid disaster perception, accurate disaster analysis and trend prediction, and efficient collaborative interaction, the prototype system can improve the efficiency of power emergency disposal, reduce the outage time of power grid, and reduce the company's economic losses. (2) The prototype system will form a power grid disaster intelligent perception and emergency command platform, which can realize the transformation of results through industrial units and produce economic benefits. (3) Through the operation and use of the system, it can shorten the repair recovery time, highlight the social responsibility of central enterprises, and produce social benefits.

Acknowledgements This work was financially supported by the science and technology project to State Grid Corporation "Research on the key technologies of power grid disaster intelligent perception and emergency command (5700-202019185A-0-0-00)."

References

1. Kang L, Shang Y, Sun L, Li Z, Zhang M (2019) Deep perception of power failure of substation area in smart grid. IOP Conf Series Earth Environ Sci 371(4):052016
2. Yuan H, Liu J, Liu Y (2018) Design of external information fused power dispatch data platform. J Electr Power Sci Technol 28(3):102–106
3. Guowei Z, Su, Y, Guoqing Z, Pengyue F, Boyan J (2020) Smart firefighting construction in China: status, problems, and reflections. Fire Mater 44(4):479–486
4. Giordani I, Archetti F (2017) Models and architectures for emergency management. J Ambient Intell Human Comput 8(1):1–8
5. Journal of Burn Care & Research (2013) Proceedings of the American Burn Association. J Burn Care Res 34(1):S1–S63

Research on State Perception of Power Grid Equipment Based on Intelligent Analysis of Multi-Spectral Data

Hai Yu, Gang Wang, Lin Peng, He Wang, Zhansheng Hou, and Zhimin He

Abstract State grid puts forward the development strategy of "three types and two networks," fully applies modern information technologies such as "big, cloud, things, mobile, and intelligence," and strives to create a ubiquitous power Internet of things with comprehensive status perception, efficient information processing, and convenient and flexible applications, intelligent analysis and calculation of big data based on advanced sensing is one of the key technologies for the construction of ubiquitous power Internet of things. The research of this project can greatly improve the coverage of the power grid equipment status perception dimension, improves the intelligent analysis ability of multi-spectral time series data, effectively enhances the real-time performance of massive data input and effective feedback of the computing system, and improves the efficiency of power grid equipment status perception and fault warning. Benefits, forming a monitoring and early warning system for power grid equipment is based on intelligent analysis of multi-spectral data [1].

Keywords Multi-spectrum · State awareness · Intelligent analysis · Fault warning

1 Introduction

As the country's electricity demand continues to grow, the scale of grid construction continues to expand, and the power load is increasing. Electrical equipment has been exposed to high-pressure, high-*temperature*, and high-load working conditions outdoors for a long time. Once it is damaged or malfunctions, it may

H. Yu (✉)
Southeast University, Nanjing 210096, Jiangsu, China
e-mail: yuhai@geiri.sgcc.com.cn

H. Yu · G. Wang · L. Peng · H. Wang · Z. Hou · Z. He
Global Energy Interconnection Research Institute Co, Ltd, Nanjing 210000, Jiangsu, China

H. Yu · G. Wang · L. Peng · H. Wang · Z. Hou · Z. He
State Grid Key Laboratory of Information & Network Security, Nanjing 210000, China

Z. Yu et al. (eds.), *Advancements in Mechatronics and Intelligent Robotics*,
Advances in Intelligent Systems and Computing 1220,
https://doi.org/10.1007/978-981-16-1843-7_18

cause huge economic losses. More importantly, the status of the equipment after a power failure, such as operating voltage and temperature, does not match the dynamic conditions in operation, which will affect the accuracy of the diagnosis. According to statistics, in 2019 alone, there were as many as 6 power equipment accidents and more than 30 power safety incidents that caused a direct economic loss of more than 1 million yuan due to equipment failures across the country. Since most electrical equipment is damaged suddenly, the traditional cycle-based planned maintenance method can no longer meet the current development needs of the power grid. At the same time, the number and complexity of various types of electrical equipment continue to increase. New types of electrical equipment are closed in structure and insulated from materials. Many traditional detection and maintenance methods are no longer applicable [2, 3].

2 Feature Distribution and Time-Domain Correlation of PoWer Equipment Status Under Different Spectrums

Relying on the power equipment fault sample library of three-light fusion imaging, analyze the characteristic distribution patterns of different types of power equipment states in the three different spectrums of ultraviolet, visible and infrared, and train to establish infrared, ultraviolet, and visible image input and faults Convolutional neural network model between categories to realize fault diagnosis of power equipment.

Study the characteristic distribution of different power equipment under different spectrums. The types of faults produced by different power equipment due to their own structural differences are generally quite different, and even the same type of power equipment will have different types of faults due to its own environment and working conditions, and thus behave in different spectrums. Based on the data in the power equipment fault sample library, the spectral feature distribution of different faults of different equipment is analyzed. Different types of power equipment faults have different characteristics [4].

Study the time-domain correlation of power equipment characteristics in different spectrum bands. Relying on the data in the power equipment failure sample library, according to practical experience and multi-spectral image data, each equipment failure is divided into four stages: early warning stage, mild failure stage, moderate failure stage, and severe failure stage. Because the same type of power equipment failure may exhibit different characteristics at different stages, these characteristics may have a certain time-domain relationship, and even the characteristics of different power equipment may also have a certain time-domain correlation. By dividing the fault categories into early warning, mild, moderate, and severe, four states analyze the relationship between the characteristics of power equipment at different stages.

3 Online Monitoring and Fault Warning of Electrical Equipment Status of Stations and Transmission Lines

In response to the demand for online monitoring and fault warning of power grid equipment status in stations and stations, based on the research of three-light fusion intelligent imaging technology, three-light fusion time series data intelligent analysis technology, and event detection-oriented deep learning lightweight technology, online monitoring, and fault early warning system application.

In the field station application scenario, on the one hand, the three-light fusion intelligent imaging device can be fixed at a location that is convenient for monitoring power equipment. On the other hand, for the location where it is inconvenient to install fixed monitoring equipment, the three-light fusion intelligent imaging device in this project can be combined with the market. The existing inspection robots are combined to complete the collection of multi-spectral images and multi-spectral time series data of key electrical equipment at the stations, and return them to the monitoring platform in real time. In the transmission line scenario, consider combining the three-light fusion intelligent imaging device in this project with the existing UAV intelligent inspection equipment on the market to complete the collection of multi-spectral images and multi-spectral time series data of key electrical equipment of the transmission line. After the data collection is completed, the three-light fusion time series data intelligent analysis technology and the event detection-oriented deep learning lightweight technology can be used to analyze the collected data. According to the current status of the key electrical equipment of the stations and transmission lines displayed by the analysis results, as well as the fault category, the monitoring personnel can take corresponding equipment maintenance and troubleshooting measures in time [5].

4 Establish a Sample Library of Power Equipment Faults Based on Three-Light Fusion Imaging

In order to realize the state detection of power equipment based on deep learning, a labeled power equipment fault sample library is essential. Establish a sample library of electrical equipment failures based on multi-spectral imaging technology and proceed in sequence according to the following three steps.

Analyze the types of national power equipment and corresponding types of faults. To obtain a sample library of power equipment failures, we must first clarify what types of current power equipment in the country, as well as the application scenarios and application environments of various types of power equipment, and the general differences in the types of faults caused by different power equipment due to their own structural differences larger, even the same type of power equipment will have different types of faults due to its own environment and working conditions, and these faults will show different characteristics in different bands.

In view of the power equipment fault types obtained from the research of equipment types and fault types in the previous step, a safe and reliable fault simulation experiment is designed. Considering that the duration of different fault types may vary greatly, it should be considered in the design stage of the fault simulation experiment plan, try to fully acquire the multi-spectral image data for the full time period of the corresponding fault, and determine the corresponding multi-spectral image data collection time segment length for different fault types.

According to the designed fault simulation experiment, take multi-spectral image data. According to the failure simulation experiments designed for various failures in the previous step, through cooperation with relevant units, the corresponding shooting experiment environment is established, and the multi-spectral image collection of equipment failure characteristics is completed under the premise of ensuring safety and reliability [6].

5 Comprehensive Evaluation Method of Multi-Spectral Data Analysis Model

The regression problem of time series data is common in multi-spectral data analysis tasks. The main evaluation indicators are as follows:

(a) Root Mean Square Error (referred to as RMSE). It is a common evaluation index that represents the sample standard deviation of the difference between the predicted value and the true value and is mainly used to evaluate the magnitude of the prediction error. Usually at different times, a value is used to express the predictive ability, which is related to the numerical range, so it can only be used to compare the predictive ability between different models. The calculation formula of RMSE is as follows, where represents the real data, represents the predicted value, and is the predicted time period.

$$\mathrm{RMSE} = \sqrt{\frac{1}{m}\sum_{i=1}^{m}(y_i - \hat{y}_i)^2} \tag{1}$$

(b) Mean square error (MSE for short). Loss function commonly used in linear regression problems. MSE has done a square calculation on the RMSE numerical result, and its calculation formula is as follows:

$$\mathrm{MSE} = \frac{1}{m}\sum_{i=1}^{m}(y_i - \hat{y}_i)^2 = \mathrm{RMSE}^2 \tag{2}$$

(c) Mean absolute error (referred to as MAE). MAE is a measure of the difference between two continuous variables and is the average vertical distance between each point and the identified point. MAE has clear mathematical meaning and strong interpretability. The calculation formula of MAE is as follows:

$$\mathrm{MAE} = \frac{1}{m}\sum_{i=1}^{m} |y_i - \hat{y}_i| \tag{3}$$

(d) Mean absolute percentage error (referred to as MAPE). MAPE is a measure of prediction accuracy and has nothing to do with the specific size of the data set, so it can be compared on a variety of data sets. The ratio of the difference between the predicted value and the true value to the true value is used to identify the ability to predict. The calculation formula of MAPE is as follows:

$$\mathrm{MAPE} = \frac{100\%}{m}\sum_{i=1}^{m} \frac{|y_i - \hat{y}_i|}{y_i} \tag{4}$$

The above four evaluation indicators are selected to evaluate the pros and cons of the network model and measure the predictive ability. Among them, MSE is used as the loss function during model training to update the model parameters. The remaining three evaluation indicators are used for result analysis and comparison between models.

For the recognition and classification problems in multi-spectral data analysis tasks, the common evaluation indicators are accuracy (also known as precision) and recall (recall). According to the statistical learning method, the labels of the binary classification problem are divided into positive and negative categories. The prediction results of the classifier model have four situations: predict the positive class as the total number of positive classes TP (True Positives), predict the positive class as the total number of negative classes FN (False Positives), and predict the negative class as the total number of positive classes FP (False Positives), predict the negative class as the total number of negative classes TN (True Negatives). The definition formulas for accuracy and recall are as follows:

$$\text{precision} = \frac{\mathrm{TP}}{\mathrm{TP} + \mathrm{FP}} \tag{5}$$

$$\text{recall} = \frac{\mathrm{TP}}{\mathrm{TP} + \mathrm{FN}} \tag{6}$$

In addition to accuracy and recall, there is an important evaluation criterion called accuracy. It reflects the discriminative ability of the classifier model for the entire sample data. The definition formula of the correct rate is as follows:

$$\text{accuracy} = \frac{\mathrm{TP} + \mathrm{TN}}{\mathrm{TP} + \mathrm{FN} + \mathrm{FP} + \mathrm{TN}} \tag{7}$$

6 Conclusions

In this paper, the infrared/ultraviolet sensing technology is studied. The high-speed and high-resolution camera is used to capture the operating status of the large-scale core equipment of the valve chamber of the converter valve. The location of the typical fault occurrence area is clarified, and the viewing point of the monitoring equipment is optimized. According to the infrared diagnostic technology and standard of typical power equipment, the fault intelligent analysis and detection and recognition technology based on infrared/ultraviolet image are studied for the typical fault causes of valve hall equipment and its imaging characteristics in infrared/ultraviolet image. The overall architecture of the system is designed. The module includes fault image region segmentation, fault feature extraction, and abnormal state determination.

Acknowledgements Thanks for the support of the science and technology project to State Grid Corporation "Research and application of power grid equipment state perception based on intelligent analysis of multi-spectral data" (SGTYHT/19-JS-215).

References

1. Lee YH, Medioni G (2016) RGB-D camera based wearable navigation system for the visually impaired. Comput Vis Image Underst 149:3–20
2. Rublee E, Rabaud V, Konolige K, Bradski G (2011) ORB: an efficient alternative to SIFT or SURF. In: Proceedings of IEEE international conference on computer vision. 2011 International Conference on Computer Vision, pp 2564–2571
3. Hoersting H, Bilitchenko L, Dodds Z (2009) Visual loop-closing with image profiles. In: Proceedings of the 2009 ACM symposium on applied computing, pp 1166–1170
4. Chrysanthakopoulos G, Shani G (2010) Augmenting appearance-based localization and navigation using belief update. In: Proceedings of the 9th international conference on autonomous agents and multiagent systems, pp 559–566
5. Labbe M, Michaud F (2011) Memory management for real-time appearance-based loop closure detection. In: IEEE/RSJ international conference on intelligent robots and systems. IEEE, Piscataway, USA, pp 1271–1276
6. Badino H, Yamamoto A, Kanade T (2014) Visual odometry by multi-frame feature integration. In: IEEE international conference on computer vision workshops. IEEE, pp 222–229

Elementary Analysis on the Testing Method of the Contact Fixity of the Crimp-Oriented Electrical Connector

Haijian Shi, Penghui Ding, Kangwei Chang, Menglong Wang, and Jianyon Shi

Abstract Crimping-type connectors are widely used in the aerospace industry due to their high reliability. At present, all manufacturers have clear requirements for crimping-type electrical connector contacts, crimping pliers selection, and wire gauge matching relationship, but they have not mentioned the reliability test after the contacts which are installed in the connector. In this paper, by researching the detection method of the contact fixity of the contact into the connector, it solves the problem that the contact is installed in the connector and further improves the reliability and detectability of the contact into the connector.

Keywords Crimp-type connector · Contact piece · Fixity detection · Reliable

1 Introduction

In recent years, many people in the industry have conducted research on process technologies such as the crimping method of the crimping connector, the matching relationship between the wiring harness and the crimping contact, and the contact pull-off detection method, but few of the contacts are: The assembly reliability is studied [1, 2]. This paper studies the testing method for the reliability of the crimp connector contacts. By comparing the holding force of the contacts of different specifications, it is J599, D-Sub, J6W crimping-type connection and provides a reference basis for the installation and detection of the device contact piece.

H. Shi (✉) · P. Ding · K. Chang · M. Wang · J. Shi
Shanghai Aerospace Control Technology Institute, Shanghai 201109, China

Z. Yu et al. (eds.), *Advancements in Mechatronics and Intelligent Robotics*,
Advances in Intelligent Systems and Computing 1220,
https://doi.org/10.1007/978-981-16-1843-7_19

2 Fixed Form of Electrical Connector Contact

2.1 Contact Loading

Hold the plug-in tool, put the wire into the tool insertion end along the tool gap, and the plug-in tool should be pushed to the terminal boss. Hold the plugging tool with the wire and slowly push it vertically into the sealing body at the end of the electrical connector. When the terminal is plugged into place, the operator feels an obvious stop and can hear the "click" sound, then move the tool along with the wire slipped out [3, 4].

2.2 Contact Locking Mechanism

Take the J6W crimp connector as an example. During the mating process of the contact, as shown in Fig. 1, there is a cylindrical boss in the middle of the needle contact. The width of the boss is about 0.8 mm, and there is a circle behind the boss. Bit slot, after the limit slot, is the mark slot. The socket of the connector body has a retaining clip, which looks like a horn-shaped reed from the cross-sectional view, as shown in the black solid part in Fig. 2. Wrapped by the needle feed tool, the boss of the contact piece gradually expands the reed of the holding clip until the reed completely fits the inner wall of the socket. When the boss completely slides over the reed, the contact is pushed by the needle feed tool. When the boss is pressed to the inner end surface of the limit slot of the connector jack, the reed recovers its deformation from the tightly attached state, and after deformation, it withstands the limit slot behind the boss. At this time, the pin contact is inserted in place and the reed is fully opened. Hold the pin to prevent retraction.

At present, the commonly used crimping-type electrical connectors include J599 crimping circular electrical connectors, SAC circular composite material

Fig. 1 Needle contact

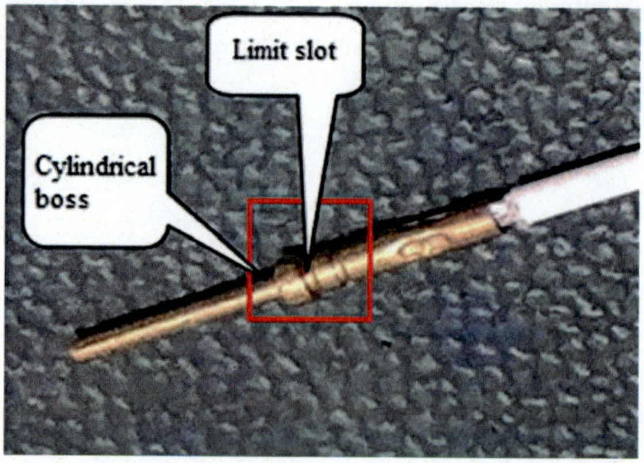

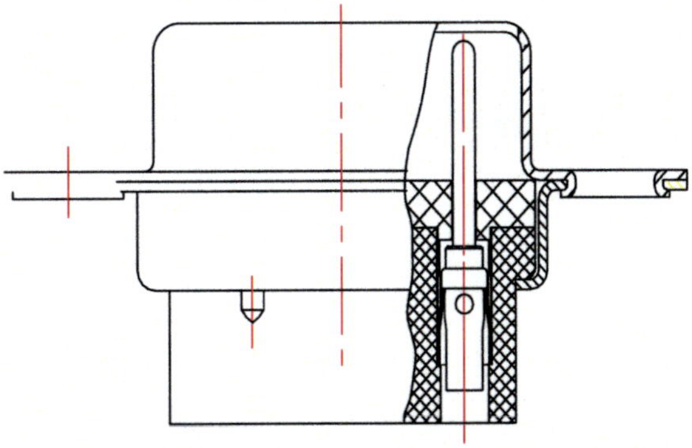

Fig. 2 Contact insertion process

connectors, D-Sub-crimping electrical connectors, J6W crimping electrical connectors, etc., as shown; they all adopt similar contact structure. The contact is fixed in the insulating assembly by holding clips or elastic claws in the housing. The insulating assembly is fixed by riveting the housing assembly.

2.3 Failure Analysis

Under normal circumstances, the needle contact piece with the wire is fed into the main socket of the electrical connector, and the pin is inserted in place. Due to the matching gap of the mechanical size, the inserted needle contact piece can still move up and down. According to the official data of the 693 factory, the movement amount of the J6W standard density crimping pin contact is 0.2–0.4 mm.

When in a critical state, there is a gap between the cylindrical boss of the needle contact and the inner end surface of the socket of the electrical connector body, and the reed of the retaining clip in the socket of the connector body just catches on the cylindrical boss of the pin. When the reed is deformed, the closer the reed is to the rear end of the boss, the greater the amount of deformation of the reed, and the greater the pressure of the retaining clamp on the boss at this time.

Therefore, when the needle feeding is only sent to the critical position of the boss just caught by the retaining clip and then the needle feeding tool is withdrawn, the critical locking state is caused, as shown in Fig. 3. At this time, through conventional visual inspection, it is difficult to judge whether the contact piece is in place.

Especially when the contact harness supports the product, as shown in Fig. 4, after the harness is formed, the wire and the needle contact form a whole, and the formed section of the wire forms the resilience generated after the wire harness is deformed, that is, along the horizontal direction. The component force Ft.

The greater the insertion force Fs received by Ft, the greater the wire deformation and the greater Ft. At this time, the holding force of the pin is equal to the

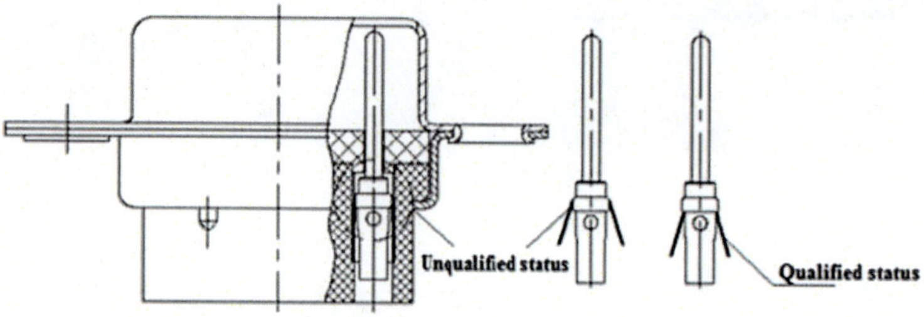

Fig. 3 Critical state

Fig. 4 Connector plug and close diagram

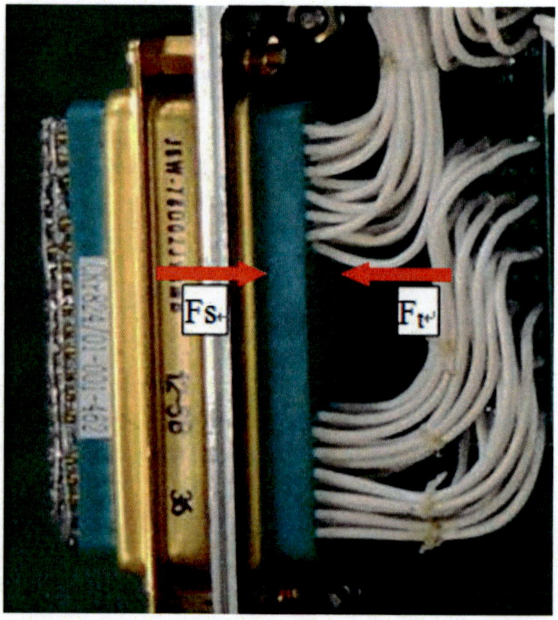

combined force of the frictional force Fn caused by the pressure of the reed and the elastic force Ft of the horizontal section after the wire is formed. The combined force of the two is close to the insertion force of the jack, but is slightly smaller than the insertion force. The greater the deformation of the wire forming section during insertion and removal, the deeper the insertion, the greater the Ft.

Therefore, in the critically locked state, even if there is an insertion phenomenon in the middle, it is difficult to find whether the contact piece is inserted in place at the initial stage. With the increase of the number of insertions and removals, the contacts that are not completely locked are gradually separated, resulting in that the contacts are eventually not inserted.

2.4 Existing Detection Methods

At present, the method for detecting the fixity of the contact piece of the crimp-type connector is that in the stage of contact piece installation, the operator feels the obvious stop after the terminal is inserted in place, distinguishes the "click" sound during installation, and the contact piece is installed. After that, the operator gently pulls the root of the wire harness to ensure that the wire harness does not exit.

2.5 There is a Problem

Due to certain differences in all contacts, retaining clips, and insulators, different people have different resolutions for the sense of stop and "click," especially the operation method of lightly pulling the wire harness cannot be quantified.

Different operators have different feelings for light pulling, and there is a risk that they cannot effectively judge whether the contact is fixed in place.

3 Detection Method of Contact Fixity

3.1 Selection of Fixed Detection Power

In response to this problem, this paper proposes to test the retention of the installed contacts. After the connector contact piece is installed, a certain horizontal thrust is applied at a certain speed from the mating surface of the contact piece, so as to solve the problem that the contact piece is installed in the shell and cannot detect whether it is completely installed in place. At the same time, through the range of holding force that the contacts of different specifications bear, the process of quantitative detection and control of the contacts is realized.

After investigation, the range of the holding force that the contact can withstand is currently in accordance with NASA standards [4], and the holding force requirements are shown in Table 1.

Table 1 Standards specifies holding force requirements

Serial number	Contact specifications	Retentivity N(P)
1	22, 22D, 2M	17.8～ 26.7 (4～6)
2	20	22.2～ 31.1 (5～7)
3	16	35.6～44.5 (8～10)
4	12	44.5～53.4 (10～12)

Fig. 5 Test tool with needle

3.2 Detection Process

By using the manual holding force inspection tool, the contact piece is tested for the holding force. The retention test tools are shown in Fig. 5. The main operation method is [5, 6].

According to the specifications of the contact, adjust the corresponding holding force value through the push–pull force meter.

From the mating surface of the connector, push the test head of the force measuring tool to the head end of the connector contact piece, and apply force at a uniform speed to make the test tool reach the predetermined position.

Observe the contact, and it is not allowed to leave the contact from the normal position. If this happens, reinstall the contact in place for testing.

The corresponding pairing relationship of the retention force test tool is shown in Table 2.

3.3 Detection Stage

For stand-alone products, priority should be given to the retention test after the connector harness is installed and before the board is soldered.

For cable harness products, the connector should be tested first before the tail cover is installed and the cable is covered.

Table 2 Selection of retaining force tools

Serial number	Retention tool	Retentivity N(P)
1	HT250-1	6.6 ~ 13.3 (1.5 ~ 3)
2	HT250-2	14.2 ~ 24.4 (3.2 ~ 5.5)
3	HT250-3	17.7 ~ 35.5 (4.0 ~ 8.0)
4	HT250-4	31.1 ~ 80.0 (7.0 ~ 18.0)
5	HT250-5	75.6 ~ 111.2 (17.0 ~ 25.0)
6	HT250-6	106.7 ~ 177.9 (24.0 ~ 40.0)

4 Usage

4.1 Verification Method

The J6W crimp-type connector is selected for the verification of the contact fixity detection method. The 22# contact body of the J6W-type connector is selected for crimping. In order to achieve the verification effect, during the verification process, the two conditions of contact locking and pseudo locking are simulated. The test scenarios are as follows:

Contact 5# is not inserted in place artificially, and the contact is kept in a pseudo-locked state in its natural state;

The contact piece 1#–4# is normally installed in place, and the contact piece remains locked in its natural state. The state of the contact is shown in Fig. 6.

In the two scenarios, refer to the method in Chap. 3 and use the retention test tool to test and verify the samples respectively. Through verification, contact 5# has the phenomenon of needle withdrawal. After contact 5# is withdrawn, after contact 5# is reinstalled in place normally, the samples are tested with the retention force test tool, and there is no needle withdrawal phenomenon.

4.2 Use Effect

The method of detecting the fixity of the contact has been implemented in some products, and the test is performed after the connector is installed with the contact, and the situation that has not been installed in place has been actually screened to prevent the problematic product from flowing into the subsequent process links.

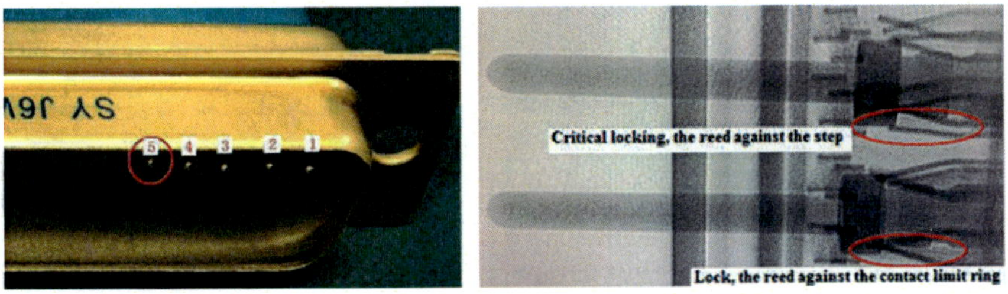

Fig. 6 False lock state

5 Concluding Remarks

In this paper, through the research and analysis of the crimping process and contact locking mechanism of the pit crimping-type electrical connector, improve the detection method after the contact is installed. This solves the problem that it is impossible to judge whether the contact piece is in place after the contact piece is installed in the connector housing. At the same time, it provides a reference for the fixation detection method of similar crimp-type connectors after the contacts are installed.

References

1. Zhao G (2014) Connector compression technology. Wirel Internet Technol 10(1):165
2. Xu Y (2009) JY Type connector compression technology. Electron Technol 04(2):225–229
3. Li X (2003) Pratical electronic assemble technology. Electron Technol 24(2):89–90
4. NASA-STD 8739.4 with change 6
5. Q/Jc483—2015 J6W series rectangular electrical connector application guide for spacecraft
6. GJB 5020–2001 Crimping connection technical requirements

Research on Flexible Interference Suppression Technology for Large Solar Panel of Spacecraft

Peihua Fu, Jiachen Zhang, Yunhong Huang, Chenghua Ding, Yongjie Zhang, and Wanliang Zhao

Abstract The solar panel of large spacecraft has the characteristics of large size and large flexibility. In order to improve tracking control accuracy for large flexible solar panel and reduce the flexible interference to spacecraft attitude, this paper presents a flexible interference suppression technology based on filter. The spectrum characteristics of Butterworth filter, notch filter and periodic interference filter are discussed in detail, and their advantages and disadvantages are analyzed, respectively. As well as a speed closed-loop control system based on filter is designed to suppress the flexible interference. Simulation results are presented to illustrate the effectiveness of the flexible interference suppression technology.

Keywords Large flexible solar panel · Filter · Interference suppression · Spacecraft

1 Introduction

With the development of apace technology, more and more large spacecrafts are used to perform various tasks like space station and large satellite which all have large flexible solar panels. During the working period of large flexible solar panels, the flexible interference may affect the stability of the spacecraft if there is no action being taken to suppress the interference. Examples indicated the importance of carrying out the research on flexible interference suppression of large solar panels could be found by the Hubble space telescope and OGO-III satellite. The observation accuracy of Hubble was degraded, and the attitude of OGO-III became uncontrolled because of the flexible interference.

P. Fu (✉) · J. Zhang · Y. Huang · C. Ding · Y. Zhang · W. Zhao
Shanghai Aerospace Control Technology Institute, Shanghai 201109, China
e-mail: fupeihua@126.com

P. Fu · C. Ding · Y. Zhang · W. Zhao
Shanghai Engineer Research Center of Inertia, Shanghai 201109, China

Z. Yu et al. (eds.), *Advancements in Mechatronics and Intelligent Robotics*, Advances in Intelligent Systems and Computing 1220,
https://doi.org/10.1007/978-981-16-1843-7_20

Recent years, researchers around the world have taken some deep research on the flexible interference suppression [1, 2]. The engineers in NASA used the input shaping method to suppress the flexible interference during the 90s. Hu Jianheng and Doherty proposed a robust input shaping method based on solar panel [3, 4]. The sliding mode control method was also adopted by Hu to suppress the vibration of the flexible solar panels [5]. Besides, one kind of active suppression technology by adding piezoelectric ceramics on the solar panels was also presented by some researchers [6]. For most of these technology, they more or less exist some problems. The sliding mode control has the drawback of chattering phenomenon. The input shaping method must undergo the stages of starting, acceleration, inertance and deceleration which may bring interference to the spacecraft. The method of adding piezoelectric ceramics is not an easy work and even not applicable for some kinds of rolling solar panels in spite of its effectiveness for suppressing flexible interference.

In this paper, three kinds of filters—four orders low-pass Butterworth filter, notch filter and periodical interference filter—are presented with the research target of flexible solar panels. The spectral characteristics of these three filters are analyzed, respectively, and one kind of flexible interference suppression technology base on notch filter is proposed. The effectiveness of the proposed approach was verified by simulation.

2 Dynamic Model of Solar Panel System

During the modeling of solar panel system, there are many aspects that need to be considered like transmission clearance, friction and flexibility [7]. In this paper, the flexibility is taken into attention. The physical transmission model of solar panel is presented in Fig. 1.

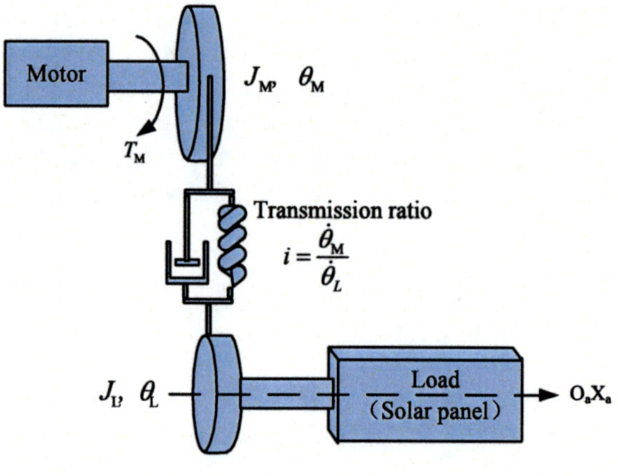

Fig. 1 Physical transmission model of solar panel

From Fig. 1, we get the transmission dynamic model:

$$\begin{cases} J_M\ddot{\theta}_M = T_M - \tau \\ J_L\ddot{\theta}_L = i\tau - T_f \end{cases} \tag{1}$$

where T_M is electromagnetic torque, J_M, θ_M are the moment of inertia and mechanical angle, respectively, from the motor side, J_L, θ_L are the moment of inertia and mechanical angle, respectively, from the load side, τ is the elastic force and damping force between gears, and T_f is the friction of the load.

Taking the flexibility of solar panel into account, the dynamic model is:

$$\begin{cases} J_M\ddot{\theta}_M = T_M - \tau \\ J_L\ddot{\theta}_L = i\tau - T_f - \boldsymbol{B}_{\mathrm{rot1}}\ddot{\boldsymbol{q}}_1 - \boldsymbol{B}_{\mathrm{rot2}}\ddot{\boldsymbol{q}}_2 \\ \ddot{\boldsymbol{q}}_1 + 2\boldsymbol{\varepsilon}_1\boldsymbol{\Lambda}_1\dot{\boldsymbol{q}}_1 + \boldsymbol{\Lambda}_1^2\boldsymbol{q}_1 + \boldsymbol{B}_{\mathrm{rot1}}^{\mathrm{T}}\ddot{\theta}_L = 0 \\ \ddot{\boldsymbol{q}}_2 + 2\boldsymbol{\varepsilon}_2\boldsymbol{\Lambda}_2\dot{\boldsymbol{q}}_2 + \boldsymbol{\Lambda}_2^2\boldsymbol{q}_2 + \boldsymbol{B}_{\mathrm{rot2}}^{\mathrm{T}}\ddot{\theta}_L = 0 \end{cases} \tag{2}$$

where $\boldsymbol{B}_{\mathrm{rot1}} = -\boldsymbol{B}_{\mathrm{rot2}}$, $\boldsymbol{q}_1$ and $\boldsymbol{q}_2$ are the modal coordinates of solar panel 1 and 2, $\boldsymbol{B}_{\mathrm{rot1}}$ and $\boldsymbol{B}_{\mathrm{rot2}}$ are the rotating coupling coefficients on solar panel 1 and 2, ε_1, ε_2 are the damping matrixes on solar panel 1 and 2, and Λ_1, Λ_2 are the stiffness matrixes on solar panel 1 and 2.

From (2), we can get the flexible coupling torque:

$$\boldsymbol{T}_{\mathrm{cou}} = -\boldsymbol{B}_{\mathrm{rot1}}\ddot{\boldsymbol{q}}_1 - \boldsymbol{B}_{\mathrm{rot1}}\ddot{\boldsymbol{q}}_2 \tag{3}$$

The polarities of the flexible coupling torque on two solar panels are the same, and method to decrease this coupling torque could improve the stability and control accuracy of the solar panels.

3 Design of Interference Suppression Filters

There are different filters that can be used to suppress flexible interference. For a system, a high bandwidth is the character that we expect. But too high a bandwidth may magnify the noise. In addition, the phase delay caused by filters is another factor that should not be neglected. In the following, three filters are presented, and their spectral characteristics are analyzed.

3.1 Butterworth Filter

Butterworth filter is a low-pass filter that could suppress the high-frequency interference. But the phase delay caused may decrease the stability margin.

The advantages and disadvantages of Butterworth filter are as follows: (a) effective in suppressing the high-frequency interference, (b) not sensitive to the change of modal frequency, (c) phase delay.

The transfer function of a four-order Butterworth filter is designed as:

$$G_{\mathrm{Bufferworth}}(s) = \frac{\omega_c^4\left(\frac{s^2}{\omega_o^2} + \frac{1.4s}{\omega_o} + 1\right)}{\left(s^2 + 0.7654\omega_c s + \omega_c^2\right)\left(s^2 + 1.8476\omega_c s + \omega_c^2\right)} \tag{4}$$

With proper parameters, the bode diagram of above four-order Butterworth filter is shown in Fig. 2.

From Fig. 2, we get that in order to effectively suppress the interference in high frequency, the phase delay caused by four-order Butterworth filter is very serious and even may result in instability of the system.

3.2 *Notch Filter*

Notch filter is one kind of filter that can not only suppress the interference on specific frequency but also can suppress the vibration modal on high-frequency points to some extent. Another merit of the notch filter is its little phase delay than Butterworth filter which could enable the system to have a better stability margin. The transfer function of notch filter is denoted in (5).

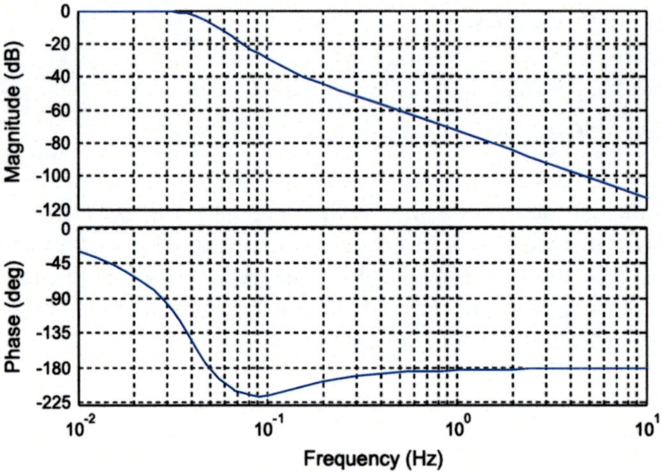

Fig. 2 Bode diagram of four-order Butterworth filter

$$G_{\text{notch}}(s) = \frac{(\frac{s}{\omega_z})^2 + 2\zeta_z(\frac{s}{\omega_z}) + 1}{(\frac{s}{\omega_p})^2 + 2\zeta_p(\frac{s}{\omega_p}) + 1} \; (\omega_z > \omega_p) \tag{5}$$

With proper parameters, the Bode diagram of notch filter is depicted in Fig. 3.

Figure 3 shows that the notch filter can suppress the flexible interference in 10^{-1} Hz dramatically while the phase delay is very little. In addition, by choosing proper parameters, the interference in high frequency can also be suppressed with even no phase delay.

3.3 *Periodic Interference Filter*

The characters of periodic interference filter are similar to that of notch filter which can be seen in Fig. 4. The periodic interference filter is mainly used to suppress periodic interference. In practical application, the periodic interference filter and notch filter are sometimes replaceable with each other as long as proper parameters are adopted. The transfer function of periodic interference suppression filter is shown in (6).

$$G_{\text{rejection}}(s) = \frac{20s^4 + 30s^3 + 10s^2 + 3.2s + 1}{50s^4 + 0.01s^2 + 1} \tag{6}$$

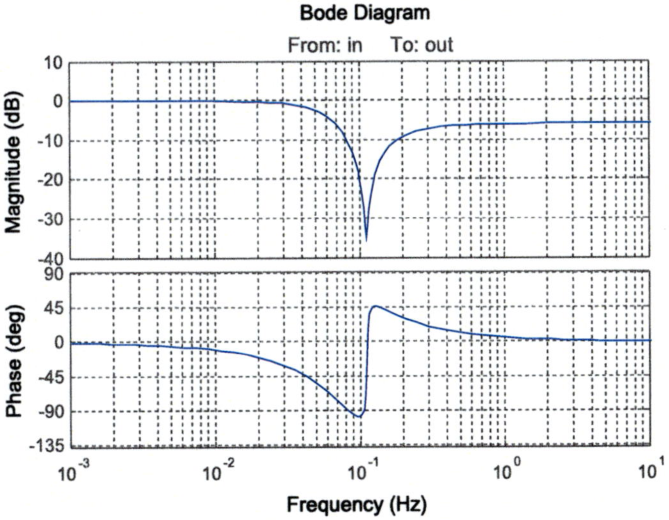

Fig. 3 Bode diagram of notch filter

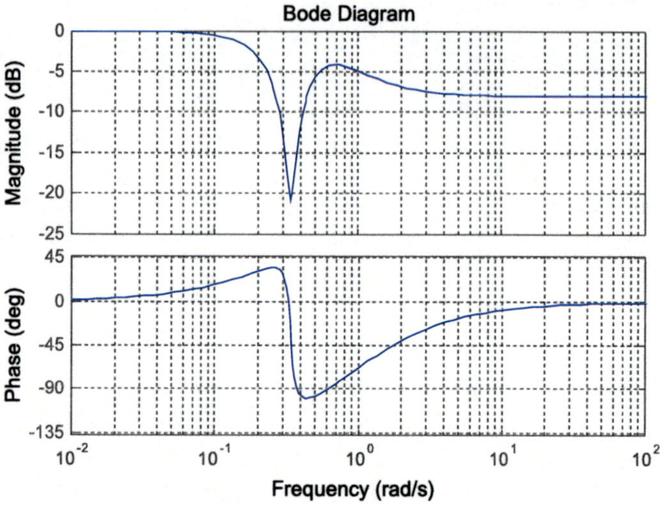

Fig. 4 Bode diagram of periodic interference suppression filter

4 Simulation

By the above analysis of Butterworth low-pass filter, notch filter and periodic interference filter, we get to know that the notch filter and periodic interference filter are more effective in suppressing specific frequency interference. Also the high-frequency interference can be suppressed to some extend with little phase delay. So, we take the notch filter as an example to suppress the flexible interference in this paper. Figure 5 is the system function diagram using filter method to suppress flexible interference. Figure 6 shows the Bode diagram of the system speed.

As shown in Fig. 6, after adding the filter, the gain in the principal modal points has been suppressed under −10 dB, and the biggest modal gain has been reduced from 15.6 dB to −14.98 dB. The phase margin of the system in the point of 0.0152 Hz is about 41°. So, by using notch filter, the flexible interference of solar panel can be suppressed effectively with satisfied performance.

Fig. 5 System function diagram

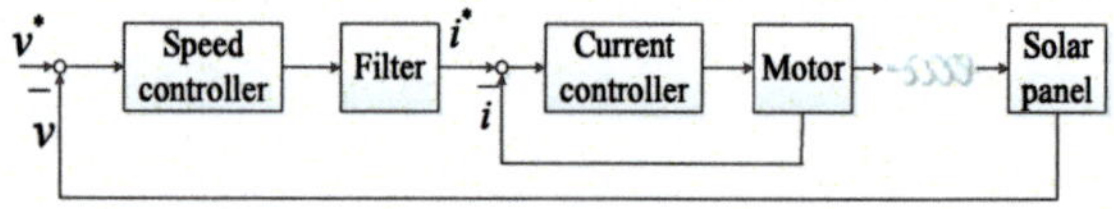

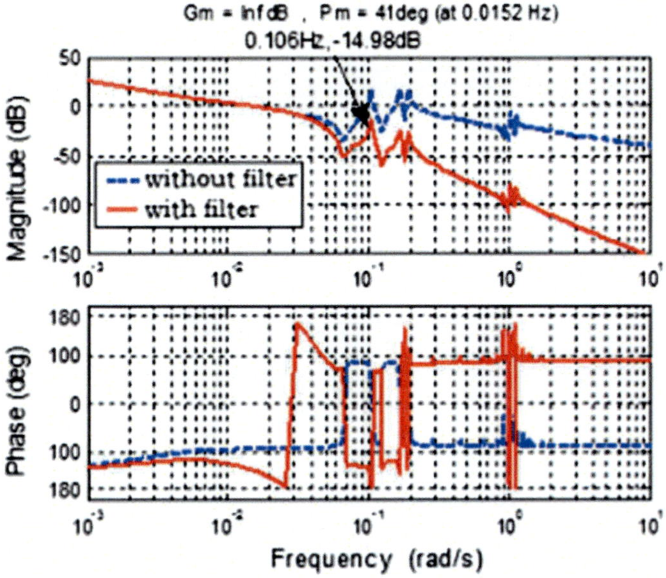

Fig. 6 Bode diagram of the system speed

5 Conclusion

In this paper, an approach to suppress flexible interference of large solar panels using filter is presented. Three kinds of filters—four orders low-pass Butterworth filter, notch filter and periodic interference filter—are introduced. By analyzing their spectrum characteristic, the notch filter has a better performance and is more proper to be used for suppressing the flexible interference of solar panels. Simulation is carried out to verify the effectiveness of proposed method, and the result is desirable.

References

1. Lu D, Liu Y (2013) Research on the control of flexible solar panel. Aerosp Control Appl 39(1): 27–33
2. Sun J, Huang TX, Zhu DF (2019) Dynamics modeling and vibration suppression of spacecraft based on macro fiber composites. Flight Control Detect 2(3):70–76
3. Hu J, Li Y, Shi G (2016) Input shaping robust control of space station solar panel. Shanghai Spaceflight 33(1):13–17
4. Doherty MJ, Tolson RH (1998) Input shaping to reduce solar array structural vibrations. NASA/CR-1998-208698
5. Hu YR, Ng A (2005) Active robust vibration control of flexible structures. J Sound Vibr 288(1–2):43–56

6. Fei J, Fang Y (2006) Active feedback vibration suppression of a flexible steel cantilever beam using smart materials. In: Proceeings of the first international conference on innovative computing, information and control, pp 286–292
7. Cao D, Liu M, Zhu D, Li Y (2020) Research progress and prospect of equivalent dynamics model of space deployable truss structure. Flight Control Detect 3(1):8–17

Intelligent Sensors and Automation

Research on Tax Policy Supporting for the Construction of China Pilot Free Trade Zone

Hongxia Rong

Abstract The establishment of pilot free trade zone is an important decision to promote the leap-forward development of China foreign economic and trade relations. The in-depth analysis of tax policy supporting for the construction of China pilot free trade zone is of great practical significance. This paper analyzes the theoretical basis of tax policy supporting the construction of China pilot free trade zone, summarizes the problems existing in tax policy supporting for the construction of China pilot free trade zone, and puts forward relevant suggestions for improving the tax policy of the existing pilot free trade zone.

Keywords Pilot free trade zone · Tax policy · Preferential tax policy

1 Introduction

China has signed Free Trade Agreement with a number of countries and planned to build Free Trade Area. Domestic pilot free trade areas have also been established. On August 22, 2013, the Shanghai Pilot Free Trade Zone was officially approved to be established. In May 2015, the State Council approved the establishment of Guangdong, Tianjin and Fujian Pilot Free-Trade Zones. In August 2016, seven provinces, including Liaoning, Hubei, Henan, Zhejiang, Shanxi, Sichuan and Chongqing, also set up pilot free trade zones, which symbolizes that China pilot free trade zone has entered a new stage of development. In 2016, the four major pilot free trade zones of Shanghai, Guangdong, Tianjin and Fujian achieved a total tax revenue of 409.055 billion yuan, far higher than the national tax growth rate of the same caliber. This paper focuses on the tax policy of the four major pilot free trade zones.

H. Rong (✉)
Harbin Finance University, No. 65 Electric Carbon Road, Xiangfang, Harbin, China

Z. Yu et al. (eds.), *Advancements in Mechatronics and Intelligent Robotics*,
Advances in Intelligent Systems and Computing 1220,
https://doi.org/10.1007/978-981-16-1843-7_21

2 Correlation Theory of Tax Policy Supporting for the Construction of China Pilot Free Trade Zone

2.1 Definition of Pilot Free Trade Zone

Pilot free trade zone is actually a trading area established by a country or region within its territory to impose special tariffs on it. In essence, there are no restrictions in the process of trading, manufacturing and processing activities in the region. The emergence of pilot free trade zone is gradually driven by regional economic integration, which has many successful examples in the global scope. The establishment of a free trade area within the territory of a country is to plan a special area within its territory where a country can freely implement relevant trade policies.

2.2 Function of Tax Policy in Pilot Free Trade Zone

As for the function of tax policy in pilot free trade zone, the most obvious one is to promote the rapid development of the regional economy. Its original intention was to drive the rapid growth of the economy in the port or surrounding areas. This is because reasonable and scientific tax policies can bring the relevant trade development and industry competition of a region to a new height, to a large extent, and promote the economic development of the region, while optimizing the industrial structure. In general, the scope of tax policy implementation is not a specific economic subject, so it also provides a fair competitive environment and opportunities for many enterprises. With the encouragement of scientific and reasonable tax preferential policies, we should guide capital, talents and other factors to enter the pilot free trade area, so as to form a certain cluster effect in the pilot free trade area, and lay a good foundation for the economic development of the region and its surrounding areas.

The impact of tax policy on economic and social development is twofold: if tax policy is compatible with economic development, it will promote economic and social development, and vice versa, it will hinder economic and social development. Therefore, in the construction of pilot free trade zone, we should arrange tax policies reasonably and play its positive role.

3 Game Analysis of Tax Policies Supporting for the Construction of China Pilot Free Trade Zone

From the point of view of the significance and function of the tax policy supporting for the construction of China Pilot Free Trade Zone, we can regard the tax policy support of the pilot free trade zone as a kind of regional public goods. Assume that the main body of tax policy support for participating in the pilot free trade area is two: the central government and local government, namely *X* and *Y*, the original earning is for the x_1 and y_1, if earning revenue from tax policy support is for x_2, y_2, Assuming that the support cost for a single party's decision is *C*, if both parties support together, the average shared support cost is for *c*/2. As long as $x_1 - c$, $y_1 - c$, $x_2 - c$ and $y_2 - c$ is greater than or equal to 0, the payment matrix faced by both parties of the game is shown in Fig. 1:

3.1 *Probability of Prisoner's Dilemma*

Consider that *X* and *Y* each make independent decisions and there is no opportunity to communicate with each other. In Fig. 1, if both *X* and *Y* adopt the strategies that are not supported by tax policies at the same time, then the benefit is obviously less than if one party coordinates while the other party does not, and this benefit is suboptimal for each party. Therefore, when each player adopts a non-supporting strategy, if *X* and *Y* both adopt fiscal policies that are not supported, the prisoner's dilemma will appear (x_1, y_1), but the probability of this occurrence is extremely low, so the prisoner's dilemma does not always appear.

3.2 *Behavioral Studies of Individual Subjects*

Consider the situation of non-zero-sum game and the mutual exchange and repeated game of information. It is a situation that is more in line with the reality of free trade zone. Non-zero-sum game contains factors of cooperation at the same time. It is in the common interest of participants to achieve a higher total utility. This is the basis for the existence and deepening of international economic cooperation relations in the process of the continuous development of regional economic cooperation integration. For each individual subject of *X* or *Y*, the optimal choice is to make the opposite decision after considering the other party's decision, so as to avoid the possibility of falling into the prisoner's dilemma. As long as one side chooses the strategy, the other side's optimal choice is established.

X \ Y	Tax policy does not support	tax policy support
Tax policy does not support	x_1; y_1	x_1+x_2; $y_1+(y_2-c)$
tax policy support	$x_1+(x_2-c)$; y_1+y_2	$x_1+(x_2-c/2)$; $y_1+(y_2-c/2)$

Fig. 1 Game analysis of tax policies supporting the construction of pilot free trade zones

The game will produce two different game equilibrium results for each player. For X, the result is $(x_1 + x_2)$ or the $x_1 + (x_2 - c)$, similarly, the result of Y in Fig. 1, you can see if X choose to support, the optimal choice of Y is not supported, because $y_1 + Y_2 \geq y_1 + (y_2 - c/2)$, in the same way, if Y choose to support, the optimal choice of X does not support it. Because the $X_1 + X_2 \geq X_1 + (x_2 - c/2)$, if one party adopts a policy to support the other party to take a lift is the optimal choice. This fully shows that one party who actively establishes a free trade zone may have to contribute to the establishment of a free trade zone. It does not choose free rides, and may even be willing to let the other party free rides, which reflects a certain degree of the first concession and cooperation.

3.3 The Tax Policy Coordination Game of the Pilot Free Trade Zone

In reality, the economy of both parties is highly complementary and dependent. One party participating in the game will not choose to be a free rider. Therefore, considering that both parties adopt fiscal and tax policies to support the game, the benefits of the single party will be different

$$\begin{aligned}[x_1 + (x_2 - c/2)] + [y_1 + (y_2 - c/2)] &= [x_1 + (x_2 - c)] + (y_1 + y_2) \\ &= (x_1 + x_2) + [(y_1 + y_2) - c] \\ &= [(x_1 + x_2) + (y_1 + y_2) - c].\end{aligned} \tag{1}$$

In a word, the tax policy support of the pilot free trade zone construction is a game. Although it is possible to fall into the prisoner's dilemma, there is also a free ride phenomenon. However, if both parties can consider the overall interests and long-term interests brought by the establishment of the pilot free trade zone, they will cooperate to support the construction of the pilot free trade zone.

4 Tax Policies and Existing Problems in China Pilot Free Trade Zone

4.1 Tax Policy of China Pilot Free Trade Zone

The common tax policies of China pilot free trade zone, (1) tax policies to encourage investment. The first category refers to the tax policies which involve the investment of non-monetary assets. These policies favorable object is mainly referred to some in the test zone registered enterprise or individual shareholders, the preferential benefit includes some because of non-monetary assets in the investment process, to the appreciation of the company's asset restructuring of, policy sets can be less than 5 years' time, value-added part of the income tax by installment. The second kind is mainly about the tax preference policy of equity incentives. The purpose of these policies is to attract talents and foreign capital, that is, through the form of equity, the enterprises in the experimental area will be rewarded in real terms (2) tax policies to encourage trade. The import value-added tax and consumption tax will be levied on the goods produced and processed by enterprises in the pilot free trade zone and sold to the mainland through the "second line" (domestic market demarcation line). According to the application of the enterprise, try out the policy of imposing tariffs on the domestic sold goods according to the corresponding imported materials or the actual inspection status. On the premise of strictly implementing the tax policy on import of goods, it is allowed to set up the bonded exhibition trading platform in specific areas [1].

Tax difference policy of China pilot free trade zone. The pilot free trade zone has different preferential tax policies, which mainly focus on the following aspects: (1) preferential tax policies for special economic and trade talents. Shenzhen deep before the seaport of modern service industry zone of high-end talent and a shortage of talent interim measures for the individual income tax subsidies (deep mansion [2012] no. 143) in the overseas work, in line with the former preferential industry direction of overseas talents and shortage of talents, its income from wages and salaries in the sea before pay personal income tax is the tax payable by more than 15% of the wages and salaries taxable income part, by the Shenzhen city people's government to give subsidies [2]. The above financial subsidies obtained by the applicant are exempt from individual income tax. Effective from Jan 1, 2013, and valid for 5 years. After the state has made explicit provisions on the fiscal and tax subsidies for the personal income tax of high-level talents and talents in short supply abroad, the provisions shall be implemented according to the state and the present interim measures shall be repealed at the same time [3]. The departments of Guangdong Province about working in Zhuhai Hengqin new area of Hong Kong, Macao residents interim measures for the administration of individual income tax burden difference subsidies (Guangdong wealth [2012] no. 93) on the citizens of Hong Kong, Macao, Zhuhai Hengqin employment in Hengqin actual pay personal income tax and the tax payable calculated by the Hong Kong and Macao region balance part taxes, will be fully subsidized by the government, the concrete shall be borne by the provincial government and all cities and counties in government 50%. The Fujian Pilot Free-Trade Zone also encourages young people from Taiwan to start a business and set up a company in Fujian. Those who pay taxes on normal business for more than 6 months can be granted a subsidy of up to 100 m^2 of the site rent, and the subsidy term shall not exceed 2 years (2) encourage preferential tax policies for relevant industries. The service industry is one of the driving forces for the development of the pilot free trade zone. On the basis of giving play to the advantages of the tertiary industry, the pilot free trade zone in Guangdong further optimizes the preferential tax system related to the industry. Shenzhen Qianhai and Zhuhai Hengqin new area impose enterprise income tax at a low rate of 15% on eligible encouraged industries. In Pingtan, Fujian Province, the same incentive policy is adopted to levy corporate income tax on target industries at a preferential rate. Tianjin Dongjiang to the registration of the pilot modern service taxpayers in the implementation of the value-added tax levy and refund tax preferences. In 2015, the Fujian Pilot Free-Trade Zone stipulated that financial equipment, ships and other bonded financial leasing businesses were allowed, and relevant lessees could pay the rent in installments and check with the customs on schedule and pay the value-added tax and customs duty of the financial leasing goods, (3) tax service policy. The Shanghai pilot free trade zone has adopted the e-invoice policy. In 2016, the accumulated audit and refund (exemption) tax of the paperless tax refund was nearly 26 billion yuan, the processing time of refund (exemption) was reduced by nearly 30%, and the annual average annual number of enterprise paper invoices and statements was reduced by more than 3.5 million. The Guangzhou Nansha development zone local taxation bureau took the lead in launching "independent tax declaration" nationwide, boldly and innovatively abolishing the "zero tax declaration" system, that is, taxpayers declare their own taxes when they have the duty to pay taxes, and establishing a new model of tax management in the pilot free trade zone. As a result, nearly 10,000 taxpayers in the region do not have to make "zero returns", greatly reducing the burden on taxpayers. Tianjin municipal tax bureau has piloted "one certificate, one code, one chapter, one vote" in the pilot free trade zone to complete the tax service one day, and the total time for the approval of new enterprises has been reduced from the past 9 working days to 1. At the same time, the comprehensive tax mode of "one window, one person, one machine and two systems" was launched, which realized the mutual allocation, business integration and process standardization of national and local tax system authorities. Zhuhai Hengqin new area duty land tax department offers tax services, in addition to Hong Kong and Macao enterprises including Cantonese, English, Portuguese, mandarin, etc. Multilingual barrier-free, tax consulting and reservation tax-related personalized services, was carried out joint local mutual recognition "had agreed" countries, further reduce the tax risk, strengthen the investment confidence.

4.2 Tax Policy Supports the Problems in the Construction of China Free Trade Pilot Area

The overall tax policy is not complete and lacks unified policy legislation. Tax policies in China free trade pilot area are not detailed, especially in the areas of circulation tax and customs duties. The development mode of tax policy in China pilot free trade zone is usually "first try" and "policy can be copied". At present, the current policies of the pilot free trade zone have been "pasted" on the mainland, and the mature policies of other pilot projects will be promoted one after another. Accordingly, the corresponding problems arose. The overall tax policy of the pilot free trade zone is not complete, which is easy to be exploited by some enterprises, which can shake the regional tax stability, and then have a negative impact on the national tax. The establishment of tax policies in the pilot free trade zone should also adapt to the development of industries in the region so as to produce expected policy effects. The Shanghai pilot free trade zone focuses on the offshore financial market because it has good resources for its regional positioning. To achieve the same policy effect, other regions need not only copy the preferential tax policies of the pilot free trade zones, but also the industries of the pilot free trade zones. However, due to limitations of location factors, it is difficult to realize industrial duplication. Therefore, the measures taken to integrate pilot projects with national promotion cannot adapt to the effect of tax policies in China pilot free trade zone. Therefore, it is necessary to take a long-term consideration and improve tax policies according to local conditions [4].

Policy projects and innovation needs deviate from each other. The pilot free trade zone is not the same as the bonded zone. The pilot free trade zone is established within the territory of a sovereign country or region, which is essentially a customs separation zone in the form of a free port. The bonded area, established within the territory of a country, is an economic area under special supervision by the customs. However, the actual situation is that the tax policies implemented by China pilot free trade zone still follow the relevant policies of the original bonded zone, such as the "export tax rebate policy", "bonded processing policy", "tax exemption and relief policy" and "tax refund policy". This makes the "internal customs", which is usually followed internationally, become "internal customs" in China. China tax policy imposes a 25% income tax on domestic companies and a 10% withholding tax on overseas companies' investment income in China. In contrast to Hong Kong, China, which is also a pilot free trade zone, only 16.5% of income tax is levied on domestic and overseas companies. According to the free trade area reforms, offshoring is the important development direction, should be in accordance with the general train of thought about "offshore financial development and foreign investment to promote the financial innovation" the theme of tax policy design and implementation of the current corporate income tax policy is obviously increased the test in the free trade zone abroad is restricted the domestic enterprise innovation cost: the presence of a high and new technology enterprise, the policy and innovation needs the phenomenon of deviating project.

Narrow coverage of tax preferential policies and low intensity of preferential policies. The preferential tax policy of China pilot free trade zone has the problem of narrow coverage. In China, pilot free trade zone, the preferential tax policy covers mainly income tax [5]. Specific performance is to reduce taxes and extend levy mainly. As far as corporate income tax is concerned, only part of the pilot free trade zone is currently levied at 15%, and most traditional enterprises in the pilot free trade zone do not enjoy the low tax rate. Moreover, for the high-tech industries in the pilot free trade zones, the differences in the preferential tax rates between them and those in other parts of the country are not significant. Even the pilot free trade zones like Shanghai and Tianjin do not have the policy of reducing and exempting the income tax for the high-tech industries. This will greatly affect the entry and expansion of technology-based industries. The project subsidiaries of the financial leasing enterprises or financial leasing companies registered in the free trade pilot area are included in the pilot scope of export tax rebate of financial leasing, which obviously has little preferential power. The pilot free trade zone still needs to consider other related tax preferences, such as tariff, stamp duty, urban construction maintenance tax, education fee surcharge and so on.

5 Implementation Path of Tax Policy Supporting China Free Trade Pilot Area

5.1 Unify the Policies and Laws of Various Districts and Improve the Existing Tax Policies

The unity of legislative level and policy content. The new policies and preferences concerning the central tax in the free trade pilot zones should be formulated by the central government in order to ensure that the preferential tax policies in the free trade pilot zones are treated fairly; and the preferential tax policies in the free trade pilot zones should abide by the "decision of the CPC Central Committee on comprehensively advancing several major issues governing the country according to law" and other laws [6]. Maintain the original way of formulating relevant policies after consultation with the central government, ensure the authority of the central government and local government, and ensure the healthy development of the free trade pilot area. In addition, China should unify the content of tax preferences among free trade pilot zones to ensure the consistency of tax preferential policies at all legislative levels. Only when the legislative level and policy content are unified, can China free trade pilot area develop smoothly.

5.2 Maintain the Existing Policy Advantages and Innovate Research on Tax Policy

On the premise of "conforming to the direction of tax reform and international practice, and without profit transfer and tax base erosion", local governments implementing free trade pilot zones can boldly draw lessons from mature experience at home and abroad, such as selective tariff collection, no capital gains tax, cancellation of interest advance tax and so on, and establish common standards. Local governments in implementing free trade pilot zones can boldly draw lessons from mature experience at home and abroad, such as selective tariff collection, no capital gains tax, and cancellation of interest withholding tax, and establish a tax policy system that conforms to the general standards and has its own advantages, which is conducive to encouraging the investment development of enterprises in the region and stimulating domestic sales, reducing tax costs and enhancing the competitiveness of enterprises. Of course, tax equity cannot be ignored. While trying out new tax policies, we should pay attention to following the principle of tax fairness. The state and local governments should also actively think about new countermeasures and actively deal with the uneven tax burden of different industries.

5.3 Improve Policy Coverage and Strengthen Policy Preferences

Improve policy coverage. At present, the export tax rebate policy is adopted in the circulation tax. It is only for the financial leasing enterprises registered in the free trade pilot area or the project subsidiaries set up by the financial leasing companies in the free trade pilot area to be included in the pilot scope of export tax rebate for financial leasing. It is suggested to broaden the types of export tax rebate enterprises and give certain incentive policies to export industries. The transformation of tax incentives will improve and broaden other tax preferences. In the aspect of income tax, if the preferential tax rate is not implemented, appropriate measures should be taken to implement the preferential tax policy in accordance with local conditions, and indirect preferential measures can be tried, that is, to reduce other expenses or defer tax payment. We can also consider expanding other tax preferential policies such as land value-added tax, real estate tax and deed tax while improving the coverage of current tax preferential policies [7].

5.4 Establish an Evaluation and Promotion Mechanism and Strengthen the Supervision of Preferential Tax Policies

The free trade pilot area should sum up the experience and achievements of reform and innovation in a timely manner. The Ministry of Commerce and local governments, together with relevant departments, shall conduct a comprehensive and special evaluation of the implementation of pilot policies for free trade pilot zones, entrust third-party agencies to conduct independent evaluation when necessary, and report the evaluation results to the State Council. The results of the pilot projects that are good and can be duplicated and promoted will be duplicated and promoted after the approval of the State Council. Those which have conditions will be further extended to other parts of the country. We should improve the tax preferential budget management system. Combined with China actual development set up a professional assessment and exit mechanism for tax preferences. Specifically, the Ministry of Finance can be specifically responsible for formulating and improving the relevant evaluation methods and systems, making a comprehensive assessment of the effects of some policies that have been implemented, and then making accurate predictions and judgments on the necessity and feasibility of their continued implementation. For some newly promulgated policies that have not yet been implemented, we should also assess the necessity and feasibility of their implementation, so as to accurately judge the possible effects of these preferential policies.

Acknowledgements Getting fund Project:Heilongjiang philosophy and Social Science Foundation Project (General project) – Based on the counterfactual model perspective of Heilongjiang free trade zone construction and regional economic development (project number: 20JYB038).

References

1. Wang W, Jin D (2017) Preliminary thoughts on tax service in China ASEAN Free Trade Area. Tax Res (2)
2. Yang F, Yan C (2015) Research on the implementation mechanism of China foreign direct investment under the “one belt and one way” strategy. Theor Discuss (5)
3. Chen Z, Zhou G (2017) Research on “going out” business of “maritime silk road”. Tax Res (2)
4. Chen L, Luo L (2014) Research on the policy effect of China foreign investment access barriers —also on the policy dividend of Shanghai free trade area reform. Econ Res (4)
5. Fan Z, Li J, Xue Q (2011) Export industry development and financial and financial support system. People’s Publishing House, Beijing
6. Research group of Wuhan Local Taxation Bureau (2016) Study on tax policy of China free trade area. Learn Pract (1)
7. Xue J, Guo X (2015) Evaluation and rational response to the tax policy in the development of China free trade area. Asia Pac Econ (5)

Geomagnetic Field Simulation in Hardware-in-the-Loop Simulation System for Geomagnetic Navigation

Zhi feng Lyu, Li guo Xu, Xiao hu Fan, Jian yong Wang, and Ning Liu

Abstract Geomagnetic navigation is a navigation technique by using geomagnetic feature, and it has wide application prospects. But the reliability of its technologies remains to be verified. This paper reports the construction of the hardware-in-the-loop simulation system for geomagnetic navigation in laboratory, which is a high creditability platform to verify and evaluate the geomagnetic navigation technologies. As the most important part of the system, the high-precision simulation of geomagnetic field is studied. The magnetic shielding device shields the external disturbing magnetic field and provides a stable environment for magnetic simulation. The solenoid coil is used as a magnetic field simulator. The high-resolution programmable current source provides accurate energy driven for magnetic simulation. Based on the above hardwares, a magnetic field simulation experimental system is built. The experimental results indicate that the geomagnetic field on the vehicle motion path can be accurately simulated. It demonstrates that the proposed design method can provide accurate geomagnetic field environment for the hardware-in-the-loop simulation system of geomagnetic navigation.

Keywords Geomagnetic navigation · Geomagnetic field simulation · Hardware-in-the-loop simulation · Magnetic shielding device · High-resolution programmable current source

1 Introduction

Geomagnetic field is the inherent resource of the earth. The geomagnetic vectors of different locations on the Earth's surface are different, and there is a one-to-one correspondence between geomagnetic field and locations. Thus, geomagnetic information can be used for navigation and location [1]. In recent years, great achievements have been made in the field of geomagnetic navigation, such as the

Z. f. Lyu (✉) · L. g. Xu · X. h. Fan · J. y. Wang · N. Liu
High-Tech Institute, Fan Gong-Ting South Street on the 12th, Qingzhou, Shandong, China
e-mail: lvzhifeng1213@163.com

Z. Yu et al. (eds.), *Advancements in Mechatronics and Intelligent Robotics*,
Advances in Intelligent Systems and Computing 1220,
https://doi.org/10.1007/978-981-16-1843-7_22

establishment and improvement of geomagnetic field model [2], reference map creation [3], magnetic sensor development [4], magnetic field distortion compensation [5] and improvement of geomagnetic matching algorithm [6]. However, as the most important terminal part of geomagnetic navigation, matching algorithm still relies on pure computer simulation. But in an actual complex system, it is difficult to take into account all the details based on pure computer simulation, so the credibility of the simulation results needs to be verified. If the actual test is carried out, the trial cost is relatively expensive, and the repeatability is poor. Hence, it is necessary to establish a low cost, strong applicability and high creditability platform to verify and evaluate the geomagnetic navigation technologies.

As an important branch of simulation technology, hardware-in-the-loop simulation has been widely used in automation engineering, aerospace, power system, manufacturing, robot design and other fields [7]. Hardware-in-the-loop simulation can generate realistic environment scene by the environment simulator, and some nonlinear or difficult-to-model key components can be introduced into the simulation loop, which can greatly improve the credibility of the simulation. Besides, by the hardware-in-the-loop simulation, the development cycle can be shortened, and the development cost can be reduced. Therefore, the hardware-in-the-loop simulation system is undoubtedly a better way for online real-time verification and evaluation of the functions and performances for geomagnetic navigation system. The purpose of this paper is to preliminarily explore the geomagnetic field simulation in hardware-in-the-loop simulation system for geomagnetic navigation, which can promote the engineering process of geomagnetic navigation.

2 Basic Principles

2.1 Geomagnetic Navigation Principle

Geomagnetic navigation is a navigation technique by using geomagnetic feature. Firstly, geomagnetic field in the matching area is measured and stored in navigation computer in advance. When the vehicle passes through the matching area, geomagnetic field is measured by the magnetic sensor and constitutes a measurement sequence. After the measurement sequence compensated, the measurement sequence and the reference sequence are matched by the matching algorithm, and the location result is output to correct the accumulated error of the inertial navigation system (INS). The geomagnetic navigation principle is shown in Fig. 1.

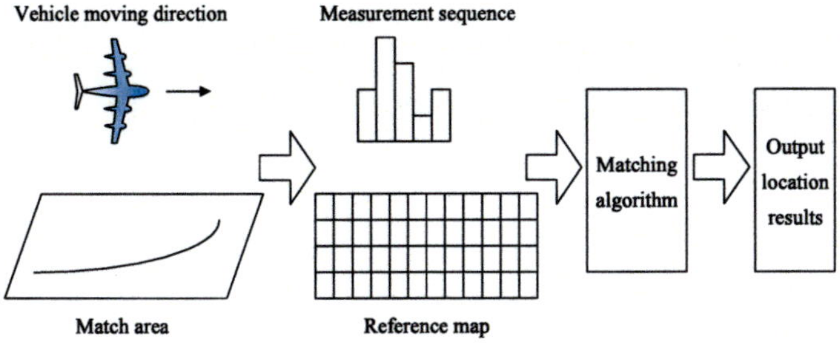

Fig. 1 Geomagnetic navigation principle

2.2 *Hardware-in-the-Loop Simulation for Geomagnetic Navigation*

As everyone knows, it is difficult to realize real motion of the vehicle in laboratory. In order to simulate the geomagnetic environment which the vehicle passes, geomagnetic field simulator is needed. The geomagnetic field simulator mainly reproduces the relative motion between the vehicle and the geomagnetic field. The motion of the vehicle relative to the geomagnetic field is converted to the motion of the geomagnetic field relative to the vehicle by the geomagnetic field simulator. That is, the vehicle does not move but the magnetic field changes in real time. And the magnetic sensor fixed on the vehicle measures the change of the geomagnetic field. The working process is as follows. Firstly, according to the motion parameters such as the moving speed, moving direction and initial position of the vehicle, the geomagnetic field values on the path are determined from the geomagnetic reference map. Then, according to the input-output relationship of the geomagnetic field simulator, the control instructions are determined. And the geomagnetic field on the path is generated by sequential control instructions, which can be measured by the magnetic sensor in real time. Finally, after the measurement sequence compensated, the measurement sequence and the reference sequence are matched by matching algorithm, and the location result is output to correct the accumulated error of the INS. The principle of the hardware-in-the-loop simulation system for geomagnetic navigation is shown in Fig. 2.

As can be seen from Fig. 2, in the hardware-in-the-loop simulation system, INS, magnetic sensor, simulation computer, geomagnetic measurement error compensation algorithm and geomagnetic matching algorithm can be consistent with the actual system. Besides, the vehicle can be a scaling model of the same material and the same structure as the actual vehicle to produce magnetic interference close to the real situation. As a consequence, hardware-in-the-loop simulation is more realistic and credible than pure computer simulation.

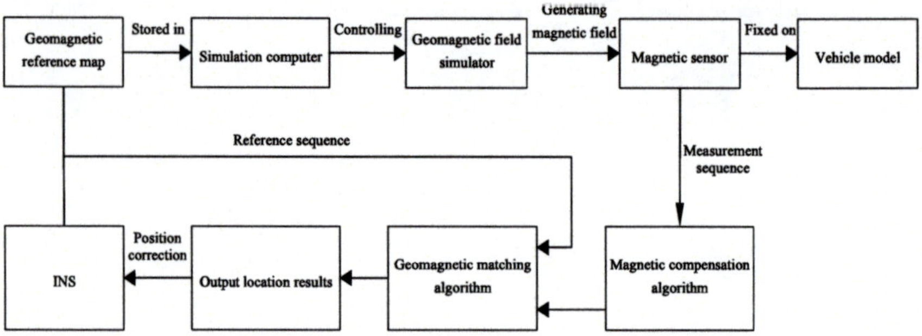

Fig. 2 Schematic of hardware-in-the-loop simulation system for geomagnetic navigation

3 Key Components of Geomagnetic Field Simulation System

Obviously, the most important part of hardware-in-the-loop simulation for geomagnetic navigation is to dynamically simulate the geomagnetic field on the vehicle motion path. So, the construction of geomagnetic field simulation system is the crux of the hardware-in-the-loop simulation. The key components of geomagnetic field simulation system are as follows.

3.1 Magnetic Shielding Device

Geomagnetic field is a weak magnetic field, and the magnetic field environment in the laboratory is relatively complicated, which can easily interfere with the magnetic field generated by the geomagnetic field simulator. Therefore, in order to simulate the required magnetic field accurately in laboratory, it is necessary to shield the interference magnetic field.

The cavity device composed of permalloy with high magnetic permeability can shield the external interference magnetic field well and form stable near-zero magnetic environment inside [8]. We built a small cylinder magnetic shielding device made up of five-layer permalloy. And the device has an excellent shielding effectiveness. So, if the geomagnetic field simulator is placed in the device, the simulated magnetic field will not be disturbed by the external interference.

3.2 Field Coil

According to Biot-Safar's law, the energized wire can generate magnetic field in the surrounding space, and the direction of the generated magnetic field and the direction of the current satisfy the right-hand rule. Accordingly, solenoid coils or Helmholtz coils can be used to produce magnetic field dynamically by controlling the change in current [9].

At present, most of the research on geomagnetic navigation takes the total geomagnetic field intensity as the matching characteristic variable. A solenoid coil can generate magnetic field intensity in a single direction. So, without consideration of the vector properties of the geomagnetic field, the solenoid coil can be used as a magnetic field simulator in the hardware-in-the-loop simulation system.

We fabricated a small solenoid coil and placed it in the magnetic shielding device. It can produce dynamic magnetic field driven by the programmable current source, which is controlled by timing sequence instructions of the simulation computer.

3.3 High-Resolution Programmable Current Source

Before simulating magnetic field, it is necessary to calibrate the input-output relationship of the magnetic field simulator. As we all know, when the coil structure parameters are fixed, the relationship between magnetic field and coil current meets the following formula:

$$B = kI \tag{1}$$

In formula (1), B represents magnetic field intensity, and I represents the coil current. k is a constant, and its value is related to the structural parameters of the coil.

From formula (1), we can see that magnetic field simulation accuracy is determined by the resolution of the current source. In order to achieve precise timing control, the current source needs program control. So, we fabricated a high-resolution programmable current source as the energy drive, whose resolution is 0.1 μA.

4 Experiments and Results

4.1 Experimental System

The experimental system consists of magnetic shielding device, a solenoid coil, a high-resolution programmable current source, a three-axis fluxgate magnetometer and a simulation computer. The experimental setup for the geomagnetic field simulation system is shown in Fig. 3.

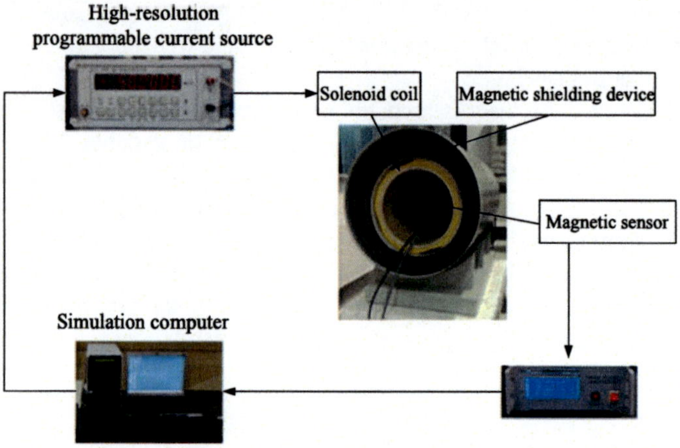

Fig. 3 Experimental setup diagram for the geomagnetic field simulation system

4.2 *Experiment and Results*

Prior to the simulation, we had a geomagnetic survey of a region and built up a high-resolution contour geomagnetic map. The map is 60 × 60 grids (1 grid = 200 m), as shown in Fig. 4. The blue line in the map is a planned path for the vehicle. The starting point is (25,40), and the direction of movement is from west to east at a speed of 200 m/s (i.e., 1 grid/second). The sampling frequency of the magnetic sensor is 1 Hz. Ten geomagnetic field values were continuously simulated and

Fig. 4 Geomagnetic reference map and planned path

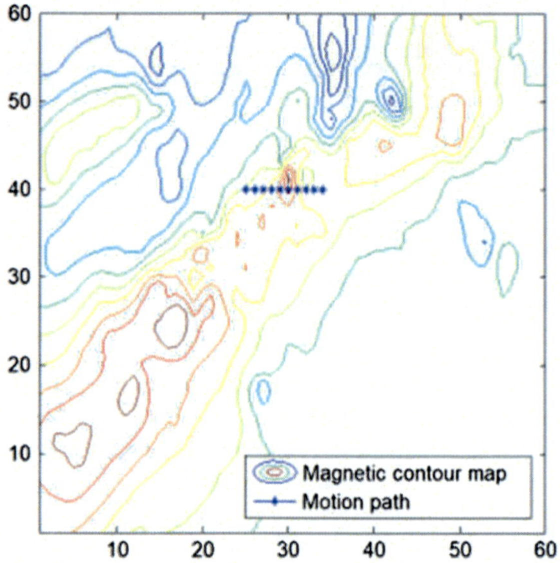

Table 1 Comparison between simulation data and reference data

Grid coordinates	Reference map data (nT)	Simulation data (nT)	Values of deviations (nT)
(25,40)	52483.9	52482.8	1.1
(26,40)	52543.3	52544.6	1.3
(27,40)	52614.6	52614.3	0.3
(28,40)	52718.3	52717.9	0.4
(29,40)	52887.0	52885.4	1.6
(30,40)	53136.0	53137.5	1.5
(31,40)	52784.2	52784.0	0.2
(32,40)	52470.7	52470.3	0.4
(33,40)	52495.4	52495.9	0.5
(34,40)	52582.7	52583.6	0.9

constantly measured by magnetic sensor. Measurement data and raw data are shown in Table 1. It can be seen from Table 1 that the maximum deviation between simulated magnetic field and expected magnetic field is only 1.6 nT, indicating that the geomagnetic field simulation system can accurately simulate the geomagnetic field value on the path. By analyzing the structure of entire system, we can find that the magnetic shielding device and high-resolution current source are the necessary hardware foundation for high-precision magnetic simulation.

5 Conclusion

In this work, a hardware-in-the-loop simulation system is proposed for online real-time verification and evaluation of the functions and performances for geomagnetic navigation. And the high-precision simulation of geomagnetic field is studied. The experimental results show that the combination of magnetic shielding device and high-resolution programmable current source can accurately simulate the geomagnetic field on the vehicle moving path. Limited to the experimental conditions, this paper only makes a preliminary exploration on the hardware-in-the-loop simulation system for geomagnetic navigation and only simulates the total geomagnetic field intensity. We believe, with sufficient experimental funds, a large-scale magnetic shielding device and three-dimensional Helmholtz coils can be constructed to generate three components of geomagnetic field vector, which will be a more realistic geomagnetic field for the hardware-in-the-loop simulation system.

References

1. Goldenberg F (Apr 2006) Geomagnetic navigation beyond the magnetic compass. In: Proceedings of IEEE/ION position, location, navigation symposium, pp 684–694
2. Navabi M, Barati M (Jun 2017) Mathematical modeling and simulation of the earth's magnetic field: a comparative study of the models on the spacecraft attitude control application. Appl Math Model 46:365–381
3. Zeng X et al (Apr 2015) Iterative Wiener filter for unstable linear transformations of potential field data. J Appl Geophys 115:100–109
4. Paul S et al (Mar 2018) Parametric design analysis of magnetic sensor based on model order reduction and reliability-based design optimization. IEEE Trans Magn 54(3) (Art. no. 8000204)
5. Liu Z et al (Jan 2017) Distortion magnetic field compensation of geomagnetic vector measurement system using a 3-D Helmholtz coil. IEEE Geosci Remote Sens Lett 14(1):48–51
6. Chen Z et al (Aug 2018) A new geomagnetic matching navigation method based on multidimensional vector elements of earth's magnetic field. IEEE Geosci Remote Sens Lett 15(8):1289–1293
7. Qi C et al (Jul 2018) Distortion compensation for a robotic hardware-in-the-loop contact simulator. IEEE Trans Contr Syst Tech 26(4):1170–1179
8. Dickerson S et al (Jun 2012) A high-performance magnetic shield with large length-to-diameter ratio. Rev Sci Instrum 83 (Art. no. 065108)
9. Batista DS et al (Feb 2018) Three-axial Helmholtz coil design and validation for aerospace applications. IEEE Trans Aerospace Electron Syst 54(1):392–403

Evaluation of Water Resource Utilization Efficiency Based on Super-Efficiency DEA: A Case of Hubei Province

Keer Li, Shiyu Zhang, Haiyun Gong, Xuejing Zhang, Ying Zhou, Yazhou Xiong, and Ruishan Chen

Abstract How to improve the utilization efficiency of water resources is the central issue of building a water-saving society. Realizing the sustainable development of water resources is a difficult problem facing the continuous development of the economy and society of Hubei Province. Therefore, this study chose the super-efficiency DEA model to analyze and study the water use efficiency of twelve cities in Hubei Province from 2014 to 2018. It was found that only Wuhan's super-efficiency increased every year, proving unreasonable in water utilization efficiency of other 11 cities. Based on the super-efficiency value, the study used tobit model to analyze the e the impact of the efficiency of water resources utilization from the three aspects including society, nature, and industry, and it was concluded that socioeconomic factors and water use efficiency were positively correlated, and the other two were negatively correlated. Finally, this study put forward four improvement measures based on the above results.

Keywords Super efficiency · Data envelopment analysis (DEA) · Water resource · Utilization efficiency · Tobit model

1 Introduction

Water resources are one of the necessary conditions for human survival and development, but the amount of global water resources available is limited. All human production and business activities are inseparable from water. Whether it is the three major industries or human daily activities, it promotes the continuous development of human civilization. From the perspective of regional distribution,

K. Li · S. Zhang · H. Gong · X. Zhang · Y. Zhou · Y. Xiong · R. Chen (✉)
School of Economics and Management, Hubei Polytechnic University, Huangshi 435003, Hubei, China

Z. Yu et al. (eds.), *Advancements in Mechatronics and Intelligent Robotics*, Advances in Intelligent Systems and Computing 1220,
https://doi.org/10.1007/978-981-16-1843-7_23

the distribution of water resources in China is not concentrated and unbalanced; from the perspective of water resources construction, the utilization of water resources is in an unsatisfactory state; from the perspective of the amount of available water resources per capita, the amount is decreasing year by year, and industrial and agricultural water The quantity cannot be guaranteed, and the supply and demand situation is tight. In recent years, the economy of Hubei Province has been developing at a rapid pace, the living conditions of residents have been continuously improved, and the people's demand for water has also increased. The harsh water environment in some areas has affected the water resources cycle. In order to solve the situation of water resource utilization efficiency in Hubei Province, this paper establishes a universally applicable water resource utilization efficiency evaluation index system, comprehensively uses super-efficiency DEA model and tobit model to rationally evaluate water resource utilization efficiency in Hubei Province, and scientifically allocates water resources rationally, promote a good ecological environment, meet the normal needs of human life, and take the socialist road of sustainable development.

As people's awareness of water resource utilization efficiency increases, more and more scholars in China have begun to conduct related research on water resource utilization efficiency [1–8]. From the perspective of the research progress of foreign scholars, Charnes et al. used DEA model in the early stage to collect relevant data of Chinese urban agglomerations as the research object, and calculated the industrial efficiency of Chinese cities [9]; subsequent research on water resources utilization efficiency starting to use DEA and SFA methods [10–12]. Stephent, Colenbrander and others conducted research and analysis on industrial water use efficiency, all of which indicated that water resource recycling technology is an important way to improve water efficiency [13, 14].

Through consulting and researching relevant information, many scholars at home and abroad have conducted research on water resource efficiency evaluation. Among these, the research on water resource efficiency in agriculture and industry is more extensive, and the research on overall water resources is relatively rare. Overall, it reflects the issue of water resource utilization efficiency. It is common to use traditional DEA model in water resources analysis and research, but when the efficiency value exceeds 1, the efficiency values of these cities cannot be compared, and effective improvement plans are proposed. It is found that domestic research on water use efficiency is more focused on agriculture and industry, and less involved in total water use efficiency. Therefore, this article uses super-efficiency DEA model and tobit model to analyze and research the comprehensive water resources utilization efficiency of major cities in Hubei Province.

2 Concept Definition and Related Theoretical Models

2.1 The Concept of Water Resources

Some people think that water resources include "air water" and "sea water", while some people oppose it. Based on the research content, this article decides to generalize the definition of water resources as a certain amount of water sources that can be used in human production and life.

2.2 The Concept of Water Resource Utilization Efficiency

The existing domestic and foreign scholars have different views on water resource utilization efficiency in the related topics of water resource utilization efficiency, most of which are in the two forms: unilateral water production method and total factor productivity:

2.3 Data Envelopment Analysis Model

As a new research field, Data Envelopment Analysis (DEA) can evaluate its relative effectiveness when there are multiple input-output decision-making units. The DEA model shows unique advantages in solving multiple inputs and multiple output indicators. The DEA method can be defined as a non-parametric statistical evaluation model. It does not consider the functional relationship between input and output and does not need to estimate the parameters in advance. It can effectively avoid subjective influencing factors, simplify the algorithm, and reduce errors [15, 16]. In addition, this method can analyze a variety of input and output indicators, and the resulting analysis results are optimized for each DMU, and specific optimization suggestions are proposed for the analysis results.

CCR model: The CCR model is the earliest DEA model. It assumes that the return to scale remains unchanged. Under the production technology conditions of the CCR model, the overall efficiency and effectiveness are determined by whether the scale of the decision-making unit and the technology are relatively effective at the same time; On the contrary, if the size of the decision-making unit is effective, but the technology is not effective, then DEA is also ineffective [17].

BCC model: BCC adds a convexity hypothesis to the theoretical basis of the CCR model, dividing technical efficiency (comprehensive efficiency) into two parts, one is pure technical efficiency, and the other is scale efficiency. This model can evaluate the relative technical effectiveness of each decision-making unit together, but does not evaluate the scale efficiency.

Super-efficiency DEA model: Anderson and Peterson jointly proposed the super-efficiency DEA model, which can be compared and sorted when the water efficiency is completely effective, that is, when the efficiency value is 1. This model can distinguish the efficiency value of each decision-making unit.

3 Evaluation of Water Resources Utilization Efficiency Based on Super-Efficiency DEA

3.1 Selection of Evaluation Indicators

By combining the actual situation in Hubei Province with the reference to the author's literature and considering the availability of data, this article selects some main factors, the total water consumption and total population of each city in Hubei Province and fixed asset investment are used as input indicators and are represented by X_1, X_2, and X_3; output indicators are selected as economic indicators GDP, represented by the letter Y_1. The total population here takes the number of permanent residents. Because of the lack of data in some areas, the five cities of Enshi, Shennongjia, Xiantao, Tianmen, and Qianjiang are excluded in this article. This article is based on the super-efficiency DEA model from the perspective of input. The data of water resource utilization are obtained from the "China Statistical Yearbook" and "The Bulletin of the State of Water Resources in Hubei Province".

3.2 Evaluation Result Analysis

This article assumes that under the condition of variable returns to scale, DEAP2.1 is used to measure the overall efficiency, pure technical efficiency and economic benefits of 12 cities in Hubei Province from 2014 to 2018, and DEA-SOLVER selects the input-oriented super-efficiency model. The operation calculated the super-efficiency value of water use in 12 cities in Hubei Province from 2014 to 2018. The results in 2014 are shown in Table 1. As space of this study is limited, the table of 2015–2018 will not be shown here.

Based on the above five-year super-efficiency calculations from 2014 to 2018, we can draw the following conclusions:

(1) Wuhan City, as the capital of Hubei Province and a second-tier economically developed city in China, has a water use efficiency above 1, which means DEA is relatively effective, and the efficiency value increases every year. As a railway transportation hub, Wuhan is also a university city with the largest number of students. Water resources sometimes face water shortages. For this reason, every drop of water has to be used.

Table 1 Water resources utilization efficiency and ranking in 2014

City	Comprehensive efficiency	Super efficiency	Scale return	Super-efficiency ranking
Wuhan	1.000	1.380	Constant	2
Huangshi	0.594	0.594	Increasing	6
Xiangyang	0.685	0.685	Increasing	3
Jingzhou	0.443	0.443	Increasing	9
Yichang	1.000	1.472	Constant	1
Huanggang	0.413	0.413	Increasing	12
Ezhou	0.677	0.677	Increasing	4
Shiyan	0.599	0.599	Increasing	5
Xiaogan	0.429	0.429	Increasing	11
Jingmen	0.568	0.568	Increasing	7
Xianning	0.466	0.466	Increasing	8
Suizhou	0.442	0.442	Increasing	10

(2) On the whole, the efficiency values of other cities in the past five years are uneven, and they are all lower than 1, which shows that there may still be unreasonable places in the use of water resources in the past few years. And the efficiency values of Yichang, Xiangyang, Shiyan are relatively high, but they are room for hints.

(3) The efficiency values of Huanggang, Xianning and Suizhou are relatively low, which have unreasonable input scales, resulting in ineffective overall efficiency. Therefore, they should increase the scale of investment to achieve a relatively effective state.

4 Evaluation of Water Resources Utilization Efficiency Based on Tobit Model

4.1 Tobit Model Introduction

Tobit model is often used to study the regression problem of dependent variable under certain constraints. This model can handle the problem of model construction with limited and truncated dependent variables. It contains two parts, one is to meet the constraints of equation modeling, and the other is to achieve Continuous variable modeling constraints. In 1958, James Tobin promoted the use of the standard tobit model of probit regression when studying the expenditure of household economic and daily necessities and later extended it to other research fields. The standard Tobit regression model is as follows (1):

$$\begin{aligned} y^* &= \beta' x_i + u_i \\ y_i^* &= y_i \quad \text{if } y_i > 0 \\ y_i^* &= 0 \quad \text{if } y_i \leq 0 \end{aligned} \tag{1}$$

As in (4), y_i^* is the latent dependent variable, which is observed when the latent variable is greater than 0, and the value is y_i. When the latent variable is less than or equal to 0, y_i is equal to zero. x_i is the independent variable vector, β' is the coefficient vector, u_i is the error term, which is independent and obeys the normal distribution. This article estimates that the dependent variable changes due to other factors by using the maximum likelihood method tobit model.

4.2 Explanation of Influencing Factors

Natural factors: The difference in water use efficiency is largely caused by different natural conditions, and the water use concepts and water use methods in different regions will be affected by the amount of water resources in different regions. Therefore, the index used in this paper to distinguish the natural endowment of water resources in different regions is the natural value of water resources per capita (cubic meter per person).

Socioeconomic factors: According to the economic development degree system of different provinces and cities, local governments will formulate relevant water resources utilization, development and protection policies, thereby affecting the efficiency of water utilization. The data used in this article to represent the economic development level of a region is the natural logarithm of per capita GDP.

Industrial structure factors: Agriculture and industry are industries that use the most water every year. In agriculture, the efficiency of water use will be affected by irrigation technology and facilities. In industry, the utilization efficiency of water resources is affected by the sewage discharge system and the degree of mechanization, so this paper selects the ratio of agricultural water to industrial water to represent the industrial structure.

And all the correlation coefficients are all greater than 0.85, and the correlation coefficients are very high, indicating that the 12 variables present a strong linear relationship, and common factors can be extracted from them. Therefore, it is necessary to perform factor analysis on the 12 indicators.

4.3 Empirical Result Analysis

The DEA value of the twelve cities in the data is the value of y, the per capita GDP (yuan per person) is X_1, the per capita water resources is X_2, the proportion of the primary and secondary industries are X_3 and X_4, respectively. The data comes from

Table 2 Descriptive statistics of related variables

Variable	Sample no.	Mean value	Max value	Min value	Standard deviation
X_1	60	57136.68	135136	25319	26211.53
X_2	60	1863.05	5158	315	1156.161
X_3	60	47.14417	119.58	17.46	19.69486
X_4	60	34.72	72.51	13.36	17.70029
Overall efficiency	60	0.641017	1	0.33	0.1899

Table 3 Regression results of panel tobit model

Variable	Regression coefficient	Standard deviation	*Z* value	*P* value
C	0.5026795	0.149449	3.360000	0.0010
X_1	5.91E–06	1.09E–06	5.410000	0.0000
X_2	−0.0000254	1.79E–05	−1.420000	0.1620
X_3	−0.0015593	0.001517	−1.030000	0.3080
X_4	−0.001803	0.001512	−1.190000	0.2480

the Hubei Statistical Yearbook and the Water Resources Bulletin. This paper selects panel data from 2014 to 2018 for five years for tobit model analysis. The descriptive characteristics of the relevant variables are shown in Table 2.

To analyze the above data from 2014 to 2018 through Stata14.0 software, and the results are shown in the following Table 3.

It can be seen from the analysis from Table 3.

(1) From the perspective of socioeconomic factors, per capita GDP and water use efficiency are positively correlated. This shows that the more developed the economy, the higher the efficiency of water use.
(2) There is a negative relationship between per capita water resources and water resource utilization efficiency. And it shows that abundant water resources may not necessarily improve water resource utilization efficiency.
(3) The proportion of agriculture and industry has a negative relationship with water use efficiency. Due to the low level of crop irrigation, cities dominated by agriculture are prone to wastewater; cities dominated by industry have high energy consumption and lack of recycling of water resources, resulting in low water use efficiency.

5 Some Suggestions

5.1 To Build and Improve Infrastructure

In the process of building infrastructure, we can repair existing infrastructure and improve flood control procedures. Especially for basic flood control equipment, the flood control capability can be improved through the improvement of existing equipment. The other is to widen the spillway of the reservoir, which can increase the discharge capacity of the reservoir.

5.2 To Implement the Concept of Harmony Between Man and Nature

The government can restrict residents' water consumption behavior by adjusting water price policies, thereby saving water and improving water resource utilization efficiency. In peacetime, the government can convey the concept of water-saving to residents through water-saving promotional videos. Implement relevant water resources management methods, do a good job in disseminating the basic situation of water resources to the public, and form a good social atmosphere for everyone to save water.

5.3 To Establish a Complete Information Management System

The template is designed so that author affiliations are not repeated each. Water conservancy departments can make full use of modern information technology to strengthen management. In the water resources information management system, we can use water consumption control, water resources development management, water source protection and other sections to refine management, and provide data background guarantee for the sustainable use of water resources through the Internet technology.

5.4 To Formulate a Reasonable Water Pricing Policy

A reasonable water price policy can control the use of water resources to a certain extent, thereby effectively using water resources. Not only can it help water users to establish water conservation awareness, but it can also help water users change their water use behavior and make water conservation awareness a conscious behavior.

Therefore, the government can improve water resource utilization efficiency through different water price measures in different industries.

Acknowledgements The study is supported by National college students' innovation and entrepreneurship training projects Grant No. 202010920004 and 201810920012X.

References

1. Chen G (2009) Research on regional differences of water use efficiency in China. Dalian University of Technology, Dalian
2. Liu Y, Du J (2011) Empirical analysis of agricultural water resources utilization efficiency in Hubei Province. China Rural Water Hydropower, 37–39
3. Lu L (2008) Research on the utilization efficiency of industrial water resources in Zhejiang Province. Zhejiang University, Zhejiang
4. Dong Y, Liao H (2011) Research on water resources utilization efficiency of western provincial capital cities based on DEA. Bull Soil Water Conserv 31:134–139
5. Liao H (2011) Research on water resources utilization efficiency of 12 western provinces based on DEA and malmquist index. Kunming University of Science and Technology, Kunming
6. Lu Q, Lu G, Fan Z, Mo L (2018) Research on Guangxi agricultural water resources utilization efficiency and its influencing factors—based on global super-efficient DEA and tobit model. Water Saving Irrig 08:54–58, 65
7. Shen X, Zhao M (2015) Study on water efficiency of Jiangsu Province based on super efficiency DEA. Water Conservancy Econ 33:9–13, 73
8. Qin Y (2018) Evaluation of water resources utilization efficiency and analysis of influencing factors based on DEA and tobit models in the economic zone of the northern slope of Tianshan Mountains. Xinjiang University, Xinjiang
9. Charnes A, Cooper WW (1989) Using DEA to evaluate the efficiency of economic performance by Chinese cities. Socio-Econ Plan Sci 23:325–344
10. Howell TA (2001) Enhancing water use efficiency in irrigated agriculture. Agron J 93:281–289
11. Condon AG, Richards RI, Rebetzke GJ et al (2002) Improving intrinsic water-use efficiency and crop yield. Crop Sci 42:122–131
12. Winter K, Aranda JE, Holtum JA et al (2005) Carbon isotope composition and water-use efficiency in plants with crassulacean acid metabolism. Funct Plant Biol 32:381–388
13. Stephent A (1998) Water use management and planning in the United States. Academy Press, San Diego
14. Colenbrander HJ (1986) Water in the Netherland. The Netherlands Organization for Alied Scientific Research, Hague
15. Xiong Y, Jie Z, Jie L (2019) Performance evaluation of food cold chain logistics enterprise based on the AHP and entropy. Int J Inf Syst Suly Chain Manag 12(2):57–67
16. Xiong Y, Zhang S, Lan J, Chen F (2019) Efficiency evaluation of regional economic development of mining and metallurgy city based on DEA model. Int J Alied Decis Sci 12 (3):242–256
17. Song C (2008) Data envelopment analysis method to evaluate the effect of technological progress. Technol Econ Manage Res, 15–16

Application of Bitter Gourd Leaf Disease Detection Based on Faster R-CNN

Zehua Liu, Xianzhen Yuan, Jianhong Weng, Yonghong Liao, and Liming Xie

Abstract In this paper, the purpose of this paper is to realize the automatic detection of many kinds of diseases in balsam pear leaves, and a target detection algorithm based on Faster R-CNN is proposed to detect the diseases of balsam pear leaves in natural environment. In this algorithm, the pre-trained ImageNet depth network model is used for migration learning, and the three convolution neural networks ZF-Net, VGG_CNN_M_1024, VGG-16 are used as the feature extraction networks of this experiment. Combined with the small size of bitter gourd leaf disease, the parameters of the original faster R-CNN are modified to increase the recommended size of the area during training. The results showed that the deep learning network model trained with VGG-16 as feature extraction network had the best performance. The average accuracy rates of healthy leaves, powdery mildew, gray spot, vine blight and gray spot were 0.899, 0.830, 0.819, 0.795, and the average mean precision (mean average precision, mAP, was 0.836) after increasing the size of candidate frame, the MAP value of the model was 0.999%, which was increased by 7%. This method can effectively realize the classification and location of balsam pear leaf diseases and has important research significance for the prevention of melon and fruit diseases.

Keywords Automatic detection · Faster R-CNN · Feature extraction network

1 Introduction

In recent years, deep learning has been greatly developed and promoted, and computer vision and artificial neural network algorithms have been greatly improved. Among them, face recognition, pedestrian detection, road obstacle detection, cell nucleus segmentation and other technologies are relatively mature. The research of disease recognition based on image processing has always been an

Z. Liu (✉) · X. Yuan · J. Weng · Y. Liao · L. Xie
Guangdong Industry Polytechnic, Guangzhou 510300, Guangdong, China
e-mail: 383285107@qq.com

Z. Yu et al. (eds.), *Advancements in Mechatronics and Intelligent Robotics*,
Advances in Intelligent Systems and Computing 1220,
https://doi.org/10.1007/978-981-16-1843-7_24

important topic. Jindan et al. [1] used a convolutional neural network to design a disease recognition model for strawberry powdery mildew, with a correct recognition rate of 98.61%, but it was not taken in a natural environment, and the recognized disease type was single; Amara, Bouaziz and Algergawy [2] combined the CNN model based on the LeNet framework extracts the characteristics of the color, shape and texture of the banana leaf and uses the stochastic gradient descent (SGD) algorithm to optimize the weights and deviations of the neural network. The loss function is minimized, and the result is not low. Recognition accuracy rate; Jun et al. [3] combined the 14 crop diseases of the Plant Village project and the improved AlexNet proposed a model that can recognize a variety of leaf diseases, with an accuracy of more than 90%, but the crop samples are all in a blank background shooting and did not test the generalization ability of the model in natural scenes.

The above methods all use the excellent learning ability of neural network and the flexibility and adaptability to the environment. This paper is based on Faster R-CNN to study the leaf diseases of balsam pear in the natural environment, and is committed to locating and identifying healthy leaves and diseased areas, including the location and classification of balsam pear powdery mildew, gray spot, spot disease and vine blight.

2 Faster R-CNN

After the precipitation of R-CNN and Fast R-CNN, Girshick [4] proposed a better-performing Faster R-CNN in 2015. The algorithm uses the region proposal network (RPN) instead of Fast. R-CNN's search selective (SS) algorithm, this approach not only reduces the time to calculate the region proposal but also improves the accuracy of target detection. The network structure of Faster R-CNN is complex, including feature extraction networks, also called conv layers, regional proposal networks (RPN), roI pooling layers and classification and recognition. As shown in Fig. 1.

3 Data Processing

3.1 Data Collection

The image data of balsam pear leaf diseases in this experiment was collected at Dongsheng Farm, Nansha District, Guangzhou City, Guangdong Province. The collection time was from June to July 2019, and the time period was from 10:00 to 13:00 and 15: 00–17:00, and the weather is sunny and sunny. Image acquisition equipment includes a miniature single-lens digital camera, model SONY

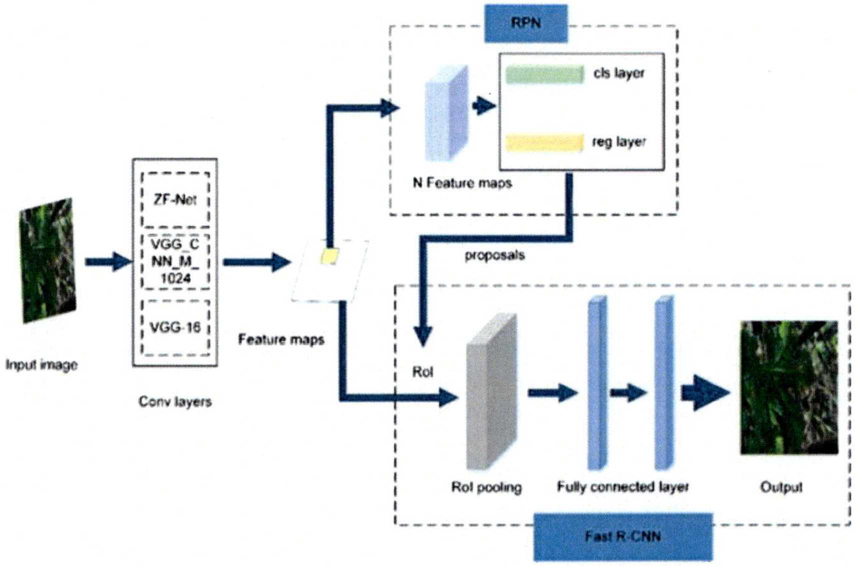

Fig. 1 Network structure of faster R-CNN

DSC-HX400, image resolution is 5184 × 3888, image type is JPEG format; Huawei Honor 9 mobile phone, image resolution is 2976 × 3968, image type is JPEG format. In order to prevent the image data from being too singular, the acquisition method is a top view, a side view and other random shooting angles.

3.2 Image Data Analysis and Annotation

In this experiment, 2146 images of balsam pear leaf were determined by field collection and manual screening as raw data for subsequent model training and testing. Among them, there were 491 images containing balsam pear powdery mildew and 343 images containing balsam pear blight. There are 570 sheets containing bitter gourd spot disease, 367 sheets containing bitter gourd gray spot disease and 510 sheets containing bitter gourd healthy leaves. In order to facilitate subsequent labeling and reduce the occupation of CPU and GPU during model training, the resolution of the original image needs to be compressed to 2048 × 2048. The image labeling adopts the manual labeling method, and a rectangular frame is used to label each diseased spot or healthy leaf, and the content of the rectangular frame is as simple as possible.

3.3 *Image Data Enhancement Technology and Data Set Classification*

When training a deep learning network, a large amount of data is often needed. If the sample size is insufficient, artificial image data enhancement can be appropriately performed. It is mentioned in the literature [5] that through the enhancement of image data, the overall learning performance of the model can be improved, the generalization ability of the model can be increased, and the ability to deal with changes in the real scene is more capable. In this experiment, in order to prevent the model from over-fitting and enhance the generalization ability of the model, the original image data was rotated 90°, 180°, 270° counterclockwise, and the horizontal mirror flip and vertical mirror flip were performed to the amplified data. There are 10,730 samples. Randomly select 80% (8584) images in the total sample as the training set, and randomly select 25% (2146 images) in the training set as the validation set, modify the model parameters, and randomly select 20% (2146 images) from the total sample) as a test set to evaluate the generalization ability of the final model.

4 Test Setup

4.1 *Test Platform*

The operating platform of this test is a desktop computer, the running environment is Ubuntu 18.04, the processor is Interl Core i7-3750, the main frequency is 3.7 GHz, 8G running memory, 1T mechanical hard drive, the graphics card is NVIDIA GeForce GTX 1080 Ti, the running video memory is 11 G, GPU accelerates computing and builds the Caffe deep learning framework under the Python programming language.

5 Parameter Setting

In this paper, the approximate joint training method is adopted, which reduces the time by about 25–50% compared with the alternate training, and the accuracy rate is similar. The remaining parameters are set as follows. The learning rate is updated every 700 iterations during training.

Parameter name	Parameter value	Remarks
RPN batch size	256	RPN batch size
Base learning rate	0.001	Basic learning rate

(continued)

(continued)

Parameter name	Parameter value	Remarks
Gamma	0.1	Learning rate decay
Momentum	0.9	Momentum
Iteration	32000	Number of iterations
Step size	11200	Stride
Iteration size	16	Number of input images per iteration

6 Results and Analysis

6.1 Experimental Design

This experiment carried out two sets of comparative experiments: (1) combine the original Faster R-CNN network with the feature extraction layer ZF-Net, VGG_CNN_M_1024, VGG-16, ZF-Net is a small network, VGG_CNN_M_1024 is a medium model of VGG16, VGG16 it is a large-scale network; (2) increase the number of anchor points. The RPN network is a branch network after the feature extraction network, which can realize the extraction of candidate frames. The original Faster R-CNN generates nine candidate regions of different scales (1282, 2562, 5122) and different aspect ratios (0.5, 1, 2) in the center of each sliding window, which makes it difficult to capture small diseases such as balsam pear leaves spots and other targets. Spots and other targets. Therefore, the candidate frames with sizes of 32 × 32 and 64 × 64 are added, and the aspect ratio remains unchanged to adapt to the target detection in this experiment.

6.2 Analysis of Model Results

In this section, different image feature extraction networks (ZF-Net, VGG_CNN_M_1024, VGG16) are used to detect the disease in the leaves of Momordica charantia and compare the performance of the trained models. The comparison results are shown in Table 1. As the number of model training iterations (Iterations) increases, the real-time overall loss function of the three feature extraction networks in their respective deep network models is compared. The specific comparisons are shown in Figs. 2 and 3. Figure 2 shows the overall loss function curve of three different feature extraction networks combined with the ImageNet model, and Fig. 3 shows the overall loss function curve after increasing the suggested size of the region on the original basis. It can be seen that using different feature extraction networks or using different size region suggestions, the loss of the function will oscillate to different degrees in the initial training stage,

Table 1 Comparison of model results

Model	Feature extraction network	Average precision (%)					Average precision mean accuracy (%)
		Healthy leaves	Powdery mildew	Gray spot	Vine blight	Spot disease	
Original faster R-CNN	ZF-Net	86.3	72.6	67.5	62.6	34.9	64.78
	VGG_CNN_M_1024	89	78	89	74	80	82
	VGG-16	87	78	73	75	87	80
Improved faster R-CNN	ZF-Net	86.3	72.6	67.5	62.6	34.9	64.78
	VGG_CNN_M_1024	84	80	83	85	73	81
	VGG-16	84	86	72	78	86	81.2

but as the number of iterations increases, the loss has a convergence trend that is learn the characteristics of the target.

According to the evaluation indicators of the model, this experiment made a detailed comparison of the average accuracy rate, the average accuracy rate average and the average detection time, and the comparison results are listed in Table 1. The original Faster R-CNN has not modified its network parameters. The improved Faster R-CNN has increased the size of the proposed production frame in the region by 16 × 16, 32 × 32 and 64 × 64 to adapt to the target of small lesions detection. Through comparison, it can be found that the Faster R-CNN deep learning network with VGG-16 as the feature extraction network shows good performance under the condition that the average detection time is very small. The average accuracy rate reaches 70%, which is better than ZF-Net and VGG_CNN_M_1024 are increased

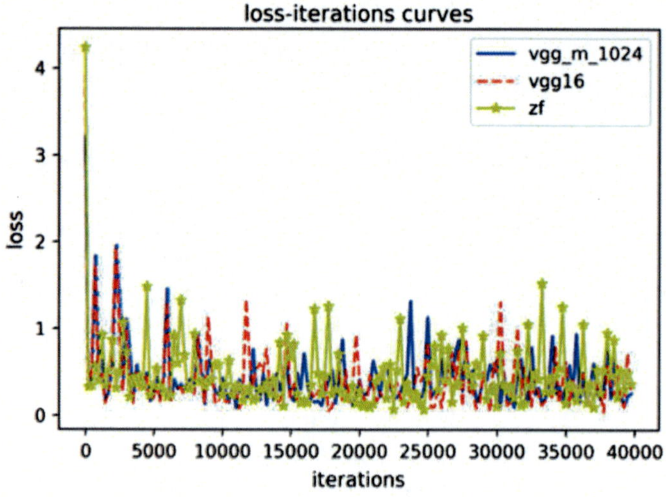

Fig. 2 Comparison of different feature extraction networks to overall loss

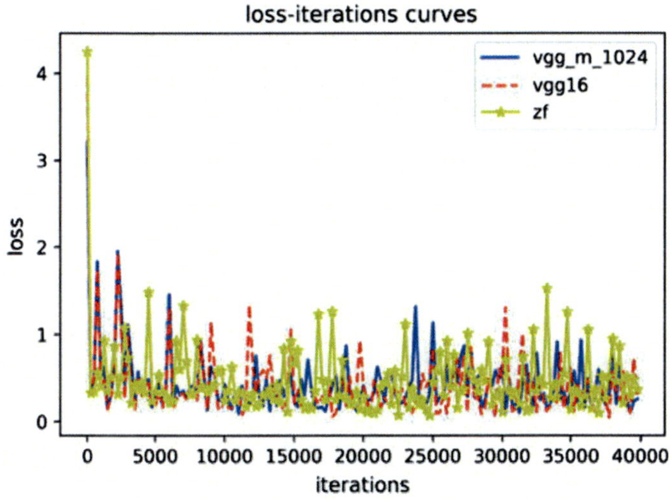

Fig. 3 Comparison of overall loss after adding regional recommendations

by 5 and 6%, respectively, and after increasing the recommended size of the area, although the detection accuracy of healthy leaves and vine blight has decreased, it is still within the error tolerance range, while gray spot disease and spot disease The detection accuracy has increased by 7 and 8%, respectively, which is better than the original model.

6.3 *Disease Identification Results*

This experiment uses the Faster R-CNN deep learning network model with ZF-Net, VGG_CNN_M_1024, and VGG-16 convolutional neural network as the feature extraction network. According to the actual size of the lesions, three scales of 16×16, 32×32, 64×64 area suggestion boxes are added, and the area suggestion network is used to generate the area suggestion boxes, which improves the recognition accuracy of the bitter gourd disease target. The recognition result of the trained model in a complex natural environment is shown in Table 1.

7 Conclusion

In this paper, healthy leaves and diseased parts of balsam pear are the research objects, and the results of the model established by the ZF-Net, VGG_CNN_M_1024 and VGG-16 feature extraction network are compared and combined with the characteristics of the small size of the disease of the balsam. The parameters of R-CNN are modified to increase the size of the region suggestion frame during training. The results show that the performance of the deep learning network model trained with VGG-16 as the feature extraction network is the best. After increasing the size of the candidate frame, the map value of the obtained model is 0.99, which is 7% higher than the original model. The model is in practical application. It performs well in the medium, can adapt to the complex natural environment and has good robustness. It has laid a research foundation for the subsequent target positioning of melon and fruit diseases and spraying pesticides on demand, reducing the use of chemical pesticides is of great significance for protecting farmland ecology.

Acknowledgements [Fund Project] Research project of Guangdong Provincial Department of Education: Design and Application of Crop Disease Identification System Based on Deep Learning(2020KQNCX147); Natural Science Research Project of Guangdong Industry Polytechnic: Research on Crop Disease and Pest Identification Algorithm Based on Deep Learning (KJ2019-015).

References

1. Jindan Y, Tao Y, Teng M, Chao Z, Qiucai S, Yufei P, Bozhang M, Yuqing D (2018) Strawberry leaf powdery mildew disease identification based on convolutional neural network. Jiangsu J Agric 34(03):527–532
2. Amara J, Bouaziz B, Algergawy AA (2017) A deep learning-based approach for banana leaf diseases classification. In: Datenbanksysteme Für Business, Technologie Und Web
3. Jun S, Wenjun T, Hanping M, Xiaohong W, Yong C, Long W (2017) Recognition of multiple plant leaf diseases based on improved convolutional neural network. Trans Chin Soc Agric Eng 33(19):209–215
4. Girshick R (2015) Fast R-CNN. Comput Sci
5. Kamilaris A, Prenafeta-Boldu FX (2018) Deep learning in agriculture: a survey. Comput Electron Agric 147:70–90

Study on the Use and Satisfaction of Short Video APP for College Students—Based on the Empirical Survey of Hubei College Students

Yan Liu and Fen Liu

Abstract Short video app is welcomed because of its fragmentation, rich content, accurate recommendation mechanism and other characteristics of rapid preemption of the market, by the vast number of college students welcome. In order to better understand the use of short video app by college students in Hubei Province, the study conducted a questionnaire survey of college students in Hubei Province. The results show that, from the basic use of the situation, Hubei college students on the short video APP understanding channel is recommended by friends around, generally install 1 short video app; the most used is shaking sound; the average daily spend 1–3 h on the short video app; often watch, sometimes like, rarely comment, rarely forwarded, never recorded, never published; Pay attention to funny short videos. From the motivation point of view, Hubei college students use short video app's main motivation is interpersonal communication, information and escape from reality. From the impact point of view, Hubei college students use short video app after three major impact, enrich themselves, enrich life and improve the mentality. The purpose of this study is to grasp the specific use of short video app by college students in Hubei Province, the motivation of use and the impact of use, as well as to improve the development of short video app.

Keywords College students · Short video APP · Use and satisfaction

1 Introduction

Short video app is a new thing in recent years. Since the short video app has begun to enter the public eye and win a lot of attention, all kinds of short video applications have entered the market quickly, such as fast hands, jitter, micro-vision and a series of other short video applications. As of December 2016, China has 500 million video users. Short video is considered to be the export of the Internet industry, and talents and funds are entering the market on a large scale. The

Y. Liu (✉) · F. Liu
School of Teachers, Hubei Polytechnic College, Huangshi 435003, Hubei, China

Z. Yu et al. (eds.), *Advancements in Mechatronics and Intelligent Robotics*, Advances in Intelligent Systems and Computing 1220,
https://doi.org/10.1007/978-981-16-1843-7_25

development of short video in China has formed a complete production and dissemination chain.

By March 2020, nearly 540 papers have been searched on CNKI with "short video app" as the subject word. Generally speaking, from 2015 to 2019, the research on short video app shows an increasing trend year by year. The research content mainly focuses on three aspects: the development status and Strategy Research of short video, communication research, business model and business value discussion. The academic research on short video app is very limited, most of them are case studies, and systematic research is less [1].

Young people are the main users of the Internet college students, as the most representative part of the youth group, have more free time than the general group and like new things. Undoubtedly, they have become the main users of short video app. By October 2020, 69 articles have been found on CNKI with the theme of "short video app and college students". There are only three related studies in 2020.

Therefore, based on the use and satisfaction theory, this paper will investigate Hubei college students' use behavior of short video app. This study can effectively fill in some gaps in the research content of short video app and college students, and broaden the depth and breadth of short video app research.

2 Research Questions and Methods

This paper investigates the application of short video app by college students through questionnaire, which fills in the vacancy of relevant research content to a certain extent, and enriches the theoretical support for the research of short video application hold. From the practical level, the questionnaire can intuitively understand the basic use of short video app for college students, effectively understand the internal needs of college students for short video app, provide reference strategies for the development of short video industry, and promote the development and improvement of short video app [2].

Specifically, this study mainly analyzes three issues: the current situation, motivation and influence of Hubei college students using short video app. In order to promote the better development of short video app and reveal the main needs of young people using short video app, this paper takes Hubei college students as an example to investigate the use of short video app, aiming to understand the basic use of short video app and the main needs of Hubei college students.

This topic is based on the use and satisfaction theory, using questionnaire survey method. The questionnaire consists of 15 questions, which are basically divided into three parts: the first part consists of six questions to understand and collect the basic demographic attributes of users; the second part consists of six questions to investigate the specific behavior of users using short video app; the third part consists of three questions to explore the main needs of users for using short video app.

In this study, a total of 400 questionnaires were distributed to Hubei college students, and 400 questionnaires were recovered, of which 323 were valid, and the effective rate was 80.75%. SPSS software analysis showed that the reliability of the questionnaire was 0.929. From the six dimensions of gender, age, grade, major category, college location and student origin, the sample distribution is relatively uniform.

3 Research Results

3.1 *The Basic Situation of Hubei College Students Using Short Video App*

3.1.1 Understanding Channels of Short Video App

Based on Chi square test, there were significant differences in the channels for Hubei college students to understand and use short video app.

$$x_2[\mathrm{DF} = 3] = 83.848, P = 0.000$$

Most college students in Hubei use short video app through the recommendation of their friends. Specifically, 37.5% of college students use the short video app through the recommendation of their friends; 37.2% of the college students use the short video app through media advertising; 17.3% of the college students understand and use the short video app through the mobile app market software ranking list; only 8.0% of the college students understand and use the short video app through other ways [3].

3.1.2 Number Selection of Short Video App

Based on Chi square test, there was a significant difference in the number of short video apps installed on mobile phones of college students in Hubei.

$$x_2[\mathrm{DF} = 3] = 159.935, P = 0.000$$

Most college students in Hubei have only one short video app on their mobile phones. Among them, 48.9% of college students installed one short video app on their mobile phones; 34.1% installed two short video apps on their mobile phones; 13.0% installed three short video apps on their mobile phones; only 4.0% of college students installed 4 or more short video apps on their mobile phones.

3.1.3 Popularity of Short Video APP

Based on Chi square test, there was a significant difference in the short video app that Hubei college students used most frequently.

$$x_2[\mathrm{DF} = 6] = 913.746, P = 0.000$$

College students in Hubei love to use TikTok more. 72.4% of Kwai students used most of the most vibrant sounds; 11.8% of college students used most of the most frequent use of fast hands; 7.4% of college students used most of the other short videos TikTok APP; 4.6% of college students used most of the micro-vision; 2.2% of the college students used most videos of pear, 1.2% of them used most seconds, and 0.3% of college students used most of the American pictures.

3.1.4 Usage Duration of Short Video App

Based on Chi square test, the average daily time spent on short video app of Hubei college students was significantly different.

$$x_2[\mathrm{DF} = 3] = 120.356, P = 0.000$$

Each week, Hubei college students who spent 1–3 h a day on short video app were the most. Among them, 38.1% of college students use short video app for less than 1 h every day; 42.1% of college students use short video app for 1–3 h; 12.4% of college students use short video app for 3–5 h; 7.4% of college students use short video app for more than 5 h every day.

3.1.5 Use Behavior of Short Video App

See Fig. 1.

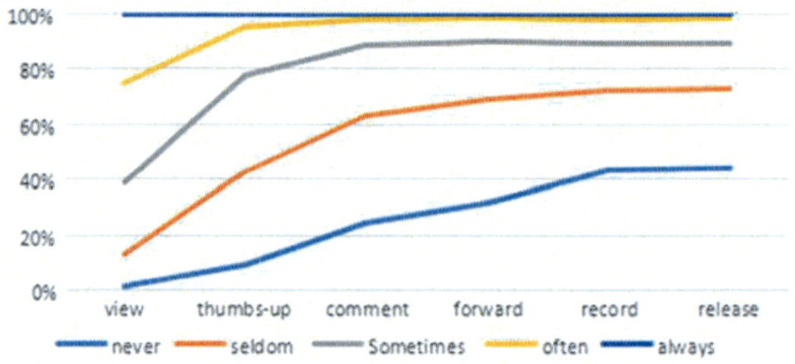

Fig. 1 Use behavior of Hubei college students' short video app

3.1.6 Usage Preference of Short Video App

See Fig. 2.

3.2 Motivation Analysis of Hubei College Students Using Short Video App

The motivation of Hubei college students to use short video app has 10 indicators. Through factor analysis, the 10 indicators of motivation can be divided into three major motivations.

$$\text{kmo} = 0.824, \text{DF} = 45, P = 0.000$$

The first is interpersonal communication, and its explanatory variation is 32.2%; the second is to obtain information, and the second is to obtain information, and the third is to evade reality, with the explanatory variation of 15.9%. The factor analysis explained that 69.6% of the students' motivation to use short video app (Fig. 3).

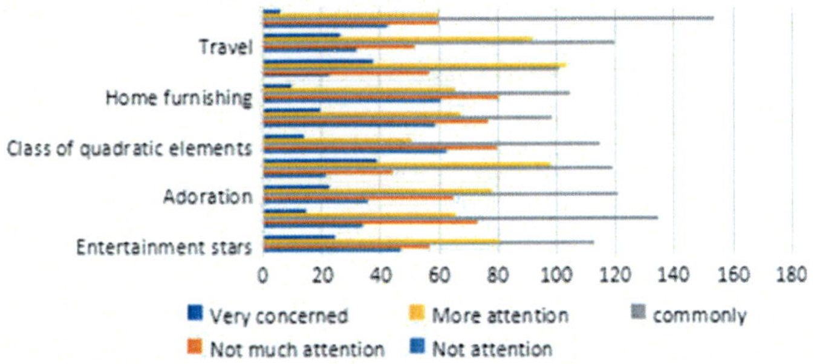

Fig. 2 Content preference of Hubei college students using short video app

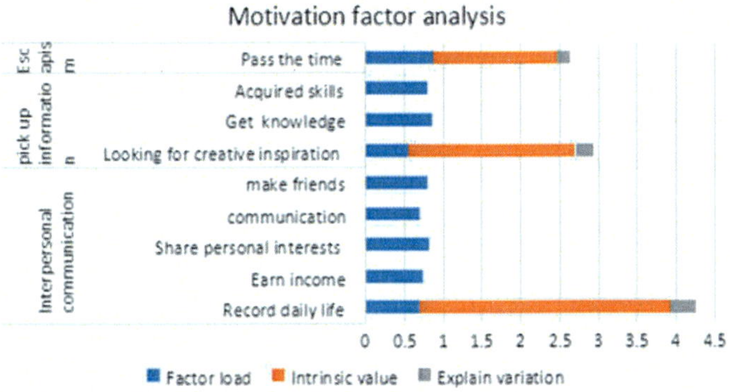

Fig. 3 Motivation factor analysis

Through the single factor ANOVA analysis of grade and motivation, we found that compared with the higher grade students, the lower grade students have stronger motivation to use short video app, which is mainly reflected in four aspects: looking for creative inspiration, acquiring some knowledge, acquiring some skills and sharing personal interests (*N*: number, *M*: mean value, SD: standard deviation) (Tables 1, 2, 3 and 4).

In conclusion, there are three motivations for Hubei college students to use short video app, which are interpersonal communication, information acquisition and escapism. Compared with the senior students, the junior students in Hubei tend to use short videos to find creative inspiration, acquire some knowledge, acquire some skills and share personal interests [4].

Table 1 Single factor ANOVA analysis of grade and creative inspiration

Motivation	Grade	*N*	*M*	SD	*F*	Post hoc test (Scheffe)
Looking for creative inspiration	Freshman (a)	64	3.16	0.895	5.866	b > c b > d
	Sophomore (b)	77	3.43	0.895		
	Junior (c)	80	2.94	0.998		
	Senior (d)	102	2.87	0.972		

Table 2 Single factor ANOVA analysis of grade and acquiring some knowledge

Motivation	Grade	*N*	*M*	SD	*F*	Post hoc test (Scheffe)
Get knowledge	Freshman (a)	64	3.64	0.743	5.551	a > d b > d
	Sophomore (b)	77	3.56	0.819		
	Junior (c)	80	3.30	0.848		
	Senior (d)	102	3.17	0.924		

Table 3 Univariate ANOVA analysis of grade and some skills acquired

Motivation	Grade	*N*	*M*	SD	*F*	Post hoc test (Scheffe)
Acquired skills	Freshman (a)	64	3.80	0.739	8.070	a > c a > d b > c b > d
	Sophomore (b)	77	3.77	0.841		
	Junior (c)	80	3.29	0.903		
	Senior (d)	102	3.32	0.903		

Table 4 Single factor ANOVA analysis of grade and sharing personal interests

Motivation	Grade	*N*	*M*	SD	*F*	Post hoc test (Scheffe)
Share personal interests	Freshman (a)	64	3.28	1.175	4.219	b > d
	Sophomore (b)	77	3.44	1.118		
	Junior (c)	80	2.98	1.031		
	Senior (d)	102	2.91	1.135		

3.3 *Analysis of the Impact of Short Video App on College Students in Hubei Province*

There are 16 indicators of the impact of short video app on college students in Hubei Province. Through factor analysis, the 16 indicators of impact can be divided into three major impacts (kmo = 0.942, DF = 120, $P = 0.000$): firstly, to enrich the self, the explained variation is 30.4%; secondly, to enrich life, the explained variation is 22.5%; thirdly, to improve the mentality, the explained variation is 16.9%. The factor analysis explained that the total variation of short video app use was 69.8% ($P < 0.05$, *N*: number, *M*: mean value, SD: standard deviation) (Tables 5, 6, 7 and 8).

Above all, Hubei college students' use of short video app has three major effects, which are enriching themselves, enriching life and improving mentality. From the perspective of gender, short videos have a greater impact on female students. From the perspective of college location, short video has a greater impact on students in Wuhan. From the perspective of grade, short video has a greater impact on lower grade students. From the perspective of major categories, short videos have a greater impact on students majoring in Humanities and social sciences. Generally speaking, the main impact of short video app on college students is to enrich their life.

Table 5 Independent sample *t*-test of gender and rich life

Impact of using short video app	Gender	*N*	*M*	SD	*T*
Enrich life	Male	152	−0.17	1.049	−2.978

Table 6 Independent sample *t*-test of college location and rich life

Impact of using short video app	College location	*N*	*M*	SD	*T*
Enrich life	Wuhan	188	0.10	0.988	2.136
	Other cities in Hubei Province	135	−0.14	1.003	

Table 7 Univariate ANOVA analysis of grade and rich life

Impact of using short video app	Grade	*N*	*M*	SD	*F*	Post hoc test (Scheffe)
Enrich life	Freshman (a)	64	0.43	0.839	6.990	a > c a > d
	Sophomore (b)	77	0.09	1.075		
	Junior (c)	80	−0.28	0.968		
	Senior (d)	102	−0.12	0.974		

Table 8 Single factor ANOVA analysis of major category and rich life

Impact of using short video app	Major	*N*	*M*	SD	*F*	Post hoc test (Scheffe)
Enrich life	Humanities and social sciences (a)	99	0.26	0.922	3.943	a > d
	Science and engineering (b)	172	−0.08	1.029		
	Agriculture, forestry and medicine (c)	14	0.03	0.905		
	Others (d)	38	−0.31	0.977		

4 Discussion and Conclusion

In the new media era, the short video industry has been a good development. Short video application has become a part of people's daily life. However, according to the results of the questionnaire, there are still some problems in the specific use of short video app among college students, which also leads to our further discussion.

First, college students are only "bystanders" in the use of short video app. According to the survey results, from the perspective of use behavior, most college students in Hubei often watch, sometimes like, rarely comment, rarely forward, never record and never release short videos. The results reflect the current situation of Hubei college students using short video app to a certain extent. That is to say, in the process of using short video app, Hubei college students are more likely to play the role of "bystander", with low participation, low platform activity and weak interaction. Compared with other roles, "bystander" does not need to invest too much energy, time and thinking, and has low degree of engagement and loyalty, which can reduce the cost of attention and have strong privacy. Therefore, compared with the participants, more Hubei college students are more willing to join the ranks of "bystanders".

Second, the demand of college students to use short video applications is diversified. Different users have different needs, such as meeting entertainment needs, learning needs, or social needs. This shows that the construction of short video platform should fully understand the internal needs of users, draw portraits for users, improve the algorithm mechanism, and accurately recommend the content that users like, so as to meet the needs of users as much as possible.

Third, college students are vulnerable to the influence of "environment". From the results of the questionnaire survey, Hubei college students of different gender, college location, grade and professional category are more or less affected after using the short video app. The rapid spread of short video app has both advantages and disadvantages. While it brings positive effects to users, it also brings negative effects. While enriching people's lives, it also presents a series of problems, such as short video addiction, cognitive limitations, vulgar content, serious homogenization, excessive entertainment, and so on.

In view of the problems exposed in the questionnaire on the use of short video app among college students in Hubei Province, we can feel that there are some problems in the short video industry, and the short video app is also facing many challenges in the rapid diffusion. Looking forward to the future, I believe that more and more scholars will participate in the research of short video, analyze the needs of users, improve the short video app according to the actual needs and satisfaction of users, give advice and suggestions for the rapid development of short video app, and help the development of short video app become better and better.

Acknowledgements The key project of Hubei Province's Mining and Metallurgical Culture and Economic and Social Development Research Center in the Middle Reaches of the Yangtze River, "Artisan Spiritual Connotation and Modern Value from the Perspective of Mining and Metallurgical Culture" (2019kyz01) Hubei Province Humanities and Social Sciences Key Research Base Public Cultural Center Key Project "Research on the Historical Value of Jingchu Culture and Its Contemporary Construction" (2019GKY03Z).

References

1. Feng H (2020) Research on the operation mode and guiding strategy of short video apps promoting college online culture education. Educ Teach Forum 53
2. Zhang X (2020) Research on pear video information communication strategy. Insight-News Media 9
3. Zhao W (2020) Research on short video transmission strategy under the background of new media. Mark Ind 51
4. Li J (2020) Tencent released a video clip app. Comput & Netw 22

Design and Implementation of OA Office System for Intelligent Teaching

Yu Guo

Abstract With the continuous improvement of software and hardware facilities in colleges and universities, the demand for office affairs is also growing. Therefore, the traditional office mode in colleges and universities can no longer meet the basic needs. It is very important to introduce the computer intelligent teaching office system into office management. The realization of information statistics summary, file sharing, resource sharing, and collaborative office work in the university can greatly improve the work efficiency, reduce unnecessary expenses, and enhance the soft power of colleges and universities. The system uses spring MVC technology and Java language development, MySQL as the background database.

Keywords Intelligent teaching · Intelligent OA system · Java

1 Introduction

In recent years, the application of Internet technology has reached a new height. All units are using the network for data processing. School office processing involves a lot of data management. The traditional office mode of paper filing and manual management is relatively common. In the era of rapid development of computer and network, with the continuous expansion of school running conditions and enrollment scale, the disadvantages of this traditional office mode relying on paper phone are more and more obvious, such as complex process, waste of manpower, and material resources. It is difficult to share information due to the poor timeliness, error prone, and isolation between departments. Therefore, the establishment of information sharing, intelligent teaching office, and quickly complete the timely processing of the school's notice become the necessary demand of intelligent teaching equipment.

Y. Guo (✉)
Faculty of Information Engineering and Automation, Kunming University of Science and Technology, Kunming 650500, China

Y. Guo
Electric Power Research Institute of Yunnan Power Grid Co. Ltd., Kunming 650217, China

Z. Yu et al. (eds.), *Advancements in Mechatronics and Intelligent Robotics*, Advances in Intelligent Systems and Computing 1220,
https://doi.org/10.1007/978-981-16-1843-7_26

To strengthen the communication between departments and improve the office efficiency, most colleges and universities have adopted OA office system. OA office management system has the advantages that traditional office mode cannot surpass: high efficiency, convenient communication, good confidentiality, high reliability, good authority, low cost, reducing complex, and complicated procedures, and more importantly, it improves the timeliness of information and makes the school's work arrangements increasingly perfect. At the same time, it can greatly promote the level of the school for various office management, which is an important milestone for the school to standardize and scientific intelligent teaching office management [1–3].

2 Research Status at Home and Abroad

There is no consensus on the definition of automatic office system at home and abroad, and the description of automatic office system is also different. American expert M. D. Zisman holds that automatic office system is actually a comprehensive technology, including computer information processing technology, information communication technology, and system engineering technology. It is mainly used to realize office business support [4]. Domestic-related people believe that China's automatic office system refers to the use of advanced technology to improve office efficiency of an application system [5]. Automatic office system began to appear in the mid-1970s, until now, a total of three stages of development, as follows. In the first stage, the automatic office system at that time was mainly based on the application of some office software by personal computer [6]. Using these office software to manage and maintain office data, after the emergence of management information system in the later stage, the institutional data is selected as the main processing object of the system, and the computer is used to store the data and effectively use the computer to realize the document writing[7].

3 System Requirement Analysis and Design

According to the visit to schools and colleges, it is found that the processing of different documents is the most commonly used business, including the approval and filling of various forms or use the network to transfer documents and communicate with each other, which is the basic feature of intelligent teaching OA office system.

This system is suitable for intelligent teaching office in colleges and universities and has complete functions. For example, for the exchange and use of the forum, the background can also get graphical representation according to the forum content, which is more intuitive; workflow based on jBPM, such as leave approval process, reimbursement approval process, activity application approval process, etc., in order to meet the requirements of the school to frequently count different

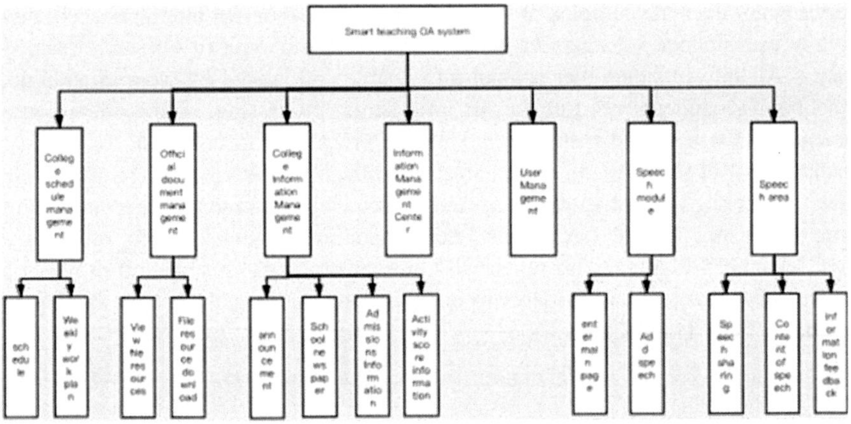

Fig. 1 Relationship model diagram

Fig. 2 Table summary

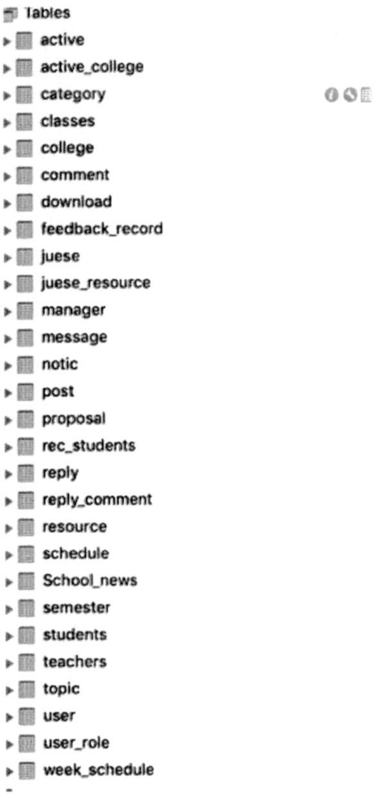

forms and reduce the complex process, it can generate different interfaces according to any user-defined information, so as to reduce the complexity of table maintenance. All information tables are put into a table and judged by type; in addition, teachers and students need document information processing, so the official document module is an indispensable module for processing information; for the user management of students and classrooms, the college's schedule management, work plan, and college information management, such as the release and coordination of announcements, school newspapers, enrollment information, speech provides a good channel for students to reflect all kinds of problems in time and provides a medium for teachers and students to communicate with each other for intelligent teaching (Figs. 1 and 2).

4 Detailed Design and Implementation of the System

4.1 Student Module: It Is Divided into Five Modules

4.1.1 The First Module: File Download Module

Synchronize the course uploaded by the teacher in the background.

4.1.2 The Second Module: Video Viewing Module

After watching all the course videos uploaded from the background, students can enter the video viewing module and click to view the uploaded videos.

4.1.3 The Fourth Module

Teaching resources download module: In the teacher module, teachers can upload teaching resources, and students can download their required course materials in this module.

4.1.4 The Fifth Module; Forum Management

Query posts, create new posts, and reply to them. Classroom and students can query existing posts, create new posts, and reply to others' posts, so as to realize the functions of posts.

4.2 *Teacher Module*

4.2.1 Schedule Management Module

Including schedule and weekly work plan.

Teachers can arrange and query their own schedule and weekly work plan.

4.2.2 Official Document Management Module (View File Resources, File Download Resources)

Including view file resources, file download resources, upload the information required by students.

4.2.3 College Information Management Module

See Figs. 3 and 4.

4.2.4 Speech Module

Enter the home page, add comments, etc. This module can only be issued by teachers with permission in the speech module.

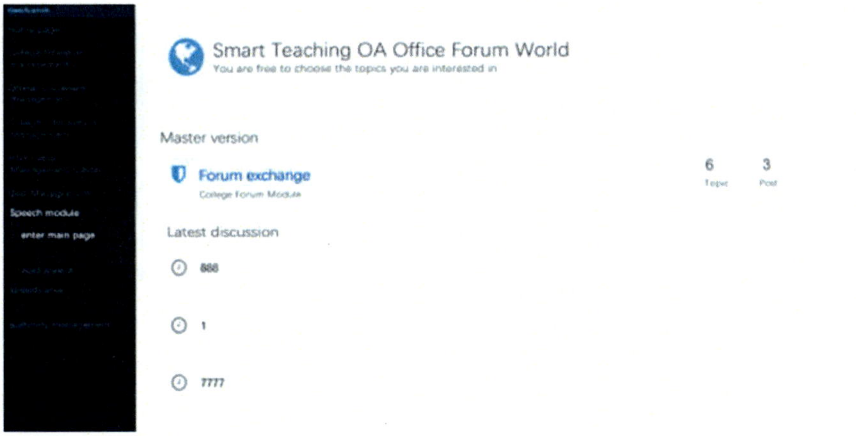

Fig. 3 Speech interface display

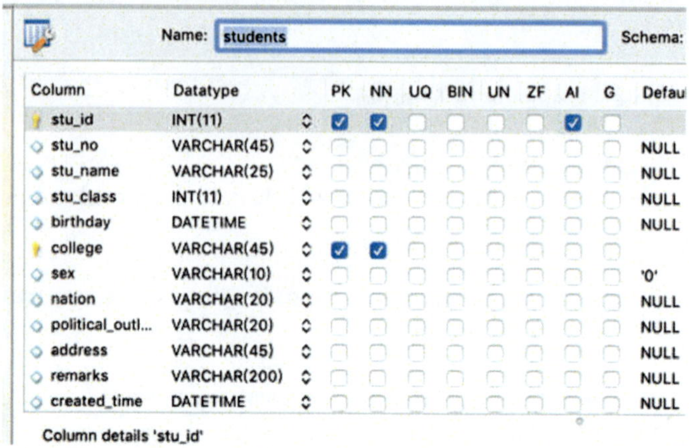

Fig. 4 Student table database

5 Test of Intelligent Teaching Office System

5.1 Test Purpose

After the development of the system, in order to better test whether the interface of the system is friendly, whether the module function is correct, whether the interface between different modules is correct, whether the system design and function meet the requirements analysis, it is necessary to test the developed system. Through the system test, we can find the system problems as soon as possible and modify them in time, improve the reliability of the system, and ensure the normal operation of the system.

5.2 Test Contents and Results

Generally speaking, white box test, black box test, and grey box test can be used to test the system. In order to verify the module function of the system, this paper uses the black box test method to test the function of the system module. Due to the space, this paper selects user management module test, forum module. No problems were found in all modules (Tables 1 and 2).

Table 1 Summary of user management module test

Test unit	Test page	Function description	Expected results
Administrator account management	Add administrator account	Add administrator account	Added successfully
	Administrator account modification	Modify the administrator account	Modification succeeded
	Administrator account view	Check the administrator account details	Find correct
	Administrator account deletion	Delete the administrator account	Deletion succeeded
Teacher account management	Add teacher account	Add teacher account	Added successfully
	Teacher account number modification	Modify the teacher's account number	Modification succeeded
	View teacher account	View the teacher account	Find correct
	Teacher account deletion	Delete the teacher account	Deletion succeeded
Student account management	Add student account	Add student account	Added successfully
	Student account number modification	Modify student account number	Modification succeeded
	Student account check	Check student account	Find correct
	Student account deletion	Delete student account	Deletion succeeded

Table 2 Forum module test summary

Test unit	Test page	Function description	Expected results
Forum	Topic addition	Start a topic	Added successfully
	Reply to topic	Reply to the topic and realize interaction	Reply successful
	Level 3 response	Multi-layer reply	Reply successful
	Message tips	You will be prompted when you receive a message reply	Prompt success

6 Summary and Prospect

6.1 Summary

At present, although many colleges and universities have already used the intelligent teaching network office system, its function is mainly a bridge between the management department and the colleges to convey official documents. The daily office work of each college is complicated, including not only the transmission of information and the sending of official documents, but also the statistics of various information forms, as well as the communication between teachers and students. Therefore, the school needs a set of office management system that allows teachers and students to share, so as to show the purpose of intelligent teaching. The purpose of the system development is to enable the office management-related information exchange, information statistics and summary, file sharing, resource sharing, document notice sending, approval process processing and other work on the Internet, improve office efficiency, reduce unnecessary expenses, and complete information automation processing. The main work of this paper is as follows:

(1) According to the business requirements of OA collaborative office system in colleges and universities, the demand analysis of the system is given from the perspective of feasibility analysis, and the main system function modules are given.
(2) According to the requirement analysis, the overall framework of the system is designed, and the database design is given.
(3) Starting from the system function module, the system is designed and implemented in detail.
(4) Some functions of the system are tested.

6.2 Outlook

The OA collaborative office system designed and implemented in this paper has basically completed the main function modules, but there are still some imperfections, which will be improved in the future work to make the intelligent teaching office system more perfect.

(1) The interface needs to continue to be beautified.
(2) The universality of the system needs to be strengthened.
(3) The system is sometimes slow to load data during operation.

References

1. Xiaowei L (2013) Research on Problems and optimization of enterprise information management system. Netw Secur Technol Appl 11:1–3
2. Junxing W (2015) Research status and development trend of OA system at home and abroad. Inf Comput (Theor Ed) 21:45–46
3. Xiaoni W (2016) Application and practice of mobile OA system in university informatization construction. Off Autom 21(15):52–53
4. Xin W (2019) Design and implementation of OA administrative office platform. Dalian Maritime University, Dalian
5. Honghong C (2018) Design and implementation of OA office information management system for Taiyuan Urban Investment Company. Dalian Maritime University, Dalian
6. Yingli T (2015) Design and implementation of university office automation system. Shijiazhuang Railway University, Shijiazhuang
7. Zhili Y (2014) Design and implementation of collaborative office system for a University of traditional Chinese medicine. Xiamen University, Xiamen
8. Jichen Z (2015) Design and implementation of OA system in universities. Dalian Maritime University, Dalian
9. Yantong C (2018) Design and implementation of office automation system for University of traditional Chinese medicine. University of Electronic Science and Technology, Chengdu

Research on Charging and Discharging Characteristics of Solar Water Storage System

Qiong Li, Xiaoqiao Huang, Yonghang Tai, and Wenfeng Gao

Abstract Intelligent control of the solar thermal utilization system requires the understanding of the dynamic characteristics of the instantaneous flow of water in the hot water storage tank, which has practical application value. The computer numerical simulation is used to simulate and analyze the instantaneous flow characteristics in the solar hot water storage tank in charging and discharging mode under the actual application parameters. The results show that the change of the thermocline is very different between the discharging mode and charging mode. The discharging mode thermocline is greatly affected by the incident cold water. In the area where the thermocline exists, the fluid movement appears meandering.

Keywords Solar water storage tank · Intelligent control · Thermal stratification

1 Introduction

With the gradual improvement of the efficiency of solar collector and energy storage device, solar energy is widely used in people's daily hot water, heating and ventilation. At present, the large-scale solar water heating system is gradually integrated with the building to realize automation, intelligence and multi-energy complimentary use. In the actual operation process of the system, due to the gap and time-varying of solar radiation, and the irregularity of hot water extraction, it is necessary to make in-depth research on the influence of these factors on the thermal stratification characteristics of the solar thermal storage tank under various operation modes. Combined with the actual operation of the solar water storage tank

Q. Li · X. Huang
Solar Energy Research Institute, Yunnan Normal University, Kunming 650500, Yunnan, China

X. Huang · Y. Tai (✉) · W. Gao
School of Physics and Electronic Information, Yunnan Normal University, Kunming 650500, Yunnan, China
e-mail: taiyonghang@126.com; zhchynnu@foxmail.com

Z. Yu et al. (eds.), *Advancements in Mechatronics and Intelligent Robotics*, Advances in Intelligent Systems and Computing 1220,
https://doi.org/10.1007/978-981-16-1843-7_27

under various working conditions of various proportions of in-depth study is to provide theoretical basis and simulation data for the intelligent control of solar water heating system.

The temperature stratification (thermal stratification) in the solar thermal storage tank is based on the following process: when the hot water from the collector enters the tank, because its density is often different from that of the water at the inlet height of the tank, it will accumulate and mix in the tank in the form of horizontal buoyant jet; when the temperature of the incoming jet is high, the formed jet will be positive buoyant jet; when the temperature of the incoming jet is high, the formed jet will be positive buoyant jet. On the contrary, when the temperature of the incoming jet is low (which will appear when the solar radiation becomes relatively small, such as 4–5 p.m., or when the sun is covered by clouds for a period of time), the formed jet will be a negative buoyancy jet. Similarly, when extracting hot water in the daytime, the cold water added to the tank from the bottom of the tank is usually different from the water temperature of the same layer in the tank, thus, forming a negative buoyancy jet buoyant jet. As time goes on, the water temperature entering the water tank gradually increases, resulting in temperature stratification along the height direction in the water tank [1]. The stratification of temperature can reduce the water temperature discharged from the bottom of the water tank into the collector, improve the heat transfer efficiency and the overall thermal performance of the solar water heating system. However, the buoyant jet entering the water tank (or the cold water entering the water tank) will convolute and mix with the water layer in the water tank due to its buoyant jet characteristics, which will destroy the thermal stratification in the water tank, resulting in the decrease of the temperature of the available water and the weakening of the overall thermal performance of the system. Obviously, in order to reduce the destructive effect of buoyant jet on temperature stratification after entering the heat storage tank, it is necessary to have a deep understanding of its convolution and mixing characteristics, so as to reasonably optimize the design of the characteristics parameters of the heat storage tank (such as the inlet height of hot or cold water, inlet size, inlet speed, etc.), so as to improve the thermal performance of the solar water heating system. According to the report of Goppert et al. [2], if better thermal stratification is generated and maintained during the dynamic operation cycle of the solar thermal storage tank, the output of the solar system may be higher.

There are also some representative studies on the dynamic characteristics of thermocline. Dragsted et al. [3] immersed 15 temperature sensors in a vertical tank to collect the necessary data on temperature, and obtained the experimental results of thermocline evolution when the heat storage and water discharge cycles were applied. Rosen [4] studied various layered temperature distribution models, including linear, stepped, continuous linear, general linear, general three-zone and basic three-zone. For each model, Rosen also derived the calculation formulas of temperature distribution, mixing temperature and equivalent temperature. Hess and Miller [5] studied the influence of water storage tank wall on thermocline dynamics by using laser Doppler velocimetry (LDV) experiment and found that eddy current formation was observed near the top of water storage tank wall, which was caused by buoyancy and pressure gradient caused by reaction convection.

Han and Wu [6] established a model based on differential equation, which controlled the heat and mass flow in the viscous entrainment drive inlet of water storage tank and solved it by implicit finite difference method. Through TRNSYS solar simulation, it is found that when the inlet temperature is lower than the temperature of the top water storage tank, the viscous entrainment process causes the temperature at the bottom of the water storage tank to rise rapidly, which is unacceptable. Therefore, when using other models, aspect ratio and tank capacity will not have enough impact on the overall performance, so the tank performance is overestimated. Chan et al. [7], Cabelli [8] and Guo and Wu [9] have carried out numerical simulation of two-dimensional flow field and temperature field involving convolution and obtained some important conclusions.

In this paper, computational fluid dynamics (CFD) is used to simulate the thermal stratification in solar thermal storage tank under charging and discharging mode, and the thermocline analysis and influence of mixing in the tank were analyzed.

2 Computational Model

The structure of the mantle solar water exchange tank is shown in Fig. 1. The parameters used in the simulation are all actual design parameters, and the operating parameters are set according to actual working conditions and research content.

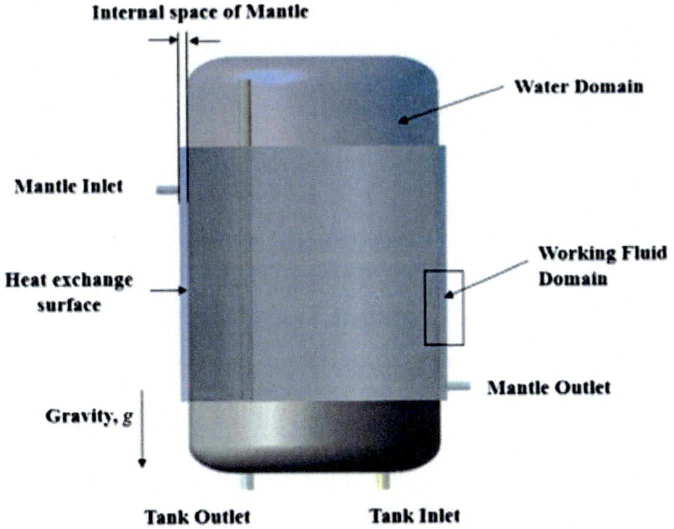

Fig. 1 Schematic of the boundary conditions used in the direct numerical simulations

3 Thermocline Dynamics

3.1 Thermocline Change in Charging Mode

Thermocline is part of the bulk fluid stored in a tank, which separates hot water from cold water due to density differences. Good sharp stratification cannot only suppress the mixing and entropy generation in various operation cycles, but also reduce the invalid volume or dead zone of the water storage tank which may account for 25% of the total volume [10].

As the heating time increases, the thickness of the thermocline gradually decreases with time during the heat storage stage, but with the gradual increase in temperature gradient, that is, the temperature stratification gradually increases. As the reduced area of the thermocline is replaced by more and more hot water, the high-temperature area becomes thicker and thicker. After 2 h in the charging stage, the temperature corresponding relationship between each stratification zone and the water tank can also be shown in Fig. 2. The abscissa is the dimensionless temperature, and the ordinate is the dimensionless height.

3.2 Thermocline Change in Discharging Mode

The temperature transition is not only the temperature distribution, but also a part of the dynamic change layer, which changes its shape and size every time the cycle reverses. In other words, contrary to the usual research scheme, in hot water research, the hot water and cold water are separated by steep (nearly zero thickness) thermocline, but in practice, the thermocline is often thicker, and its shape and

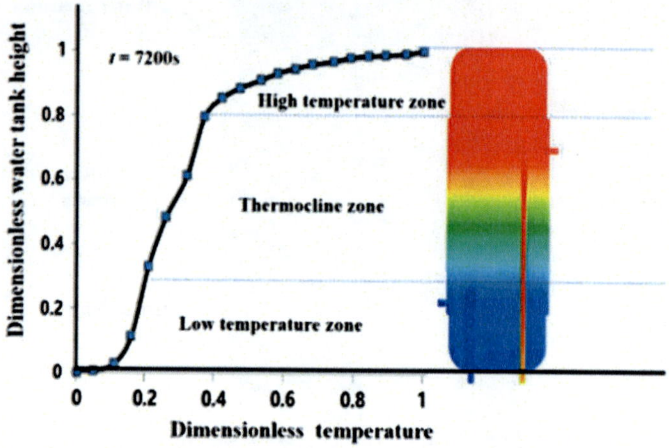

Fig. 2 Temperature distribution after 2 h of operation in charging mode

depth will change in both static and dynamic conditions. Therefore, it is necessary to strictly avoid the large variation of emission temperature difference due to the thermocline reaching near the outlet. In the process of energy removal, the earlier the temperature transition reaches the exit, it can be assisted by cycle reversal. Although it is a better control strategy than full heat storage and full water release, it still has some disadvantages on the bottom. First of all, due to the limited height of the hot line in a certain temperature range, it is forbidden to use anywhere near the outlet, which will lead to the inefficient utilization of the water tank capacity. Secondly, whenever water is added to the charging and discharging water cycle, the inlet water will take away the hotbed.

In the early stage of drainage, the mixing is mainly concentrated in the lower part of the water tank, while the thermocline is very thin. As more cold water replenishes the hot water discharged, the temperature stratification in the water tank becomes more obvious, the thermocline will move upward, and its thickness gradually becomes thicker and tends to be stable. Comparing the temperature field and the velocity field at mass flow 0.022 kg/s and t = 1800 s, we can find that at the height where the thermocline exists, the streamlines show signs of meandering movement, as shown in Fig. 3. In the low-temperature zone, due to the fall of the negatively buoyant jet, the descending fluid and the fluid at the bottom of the tank have entrained, and the mixing is more intense, which is reflected in the generation of larger vortices on the streamline diagram, as indicated by the red circle in the right picture of Fig. 3.

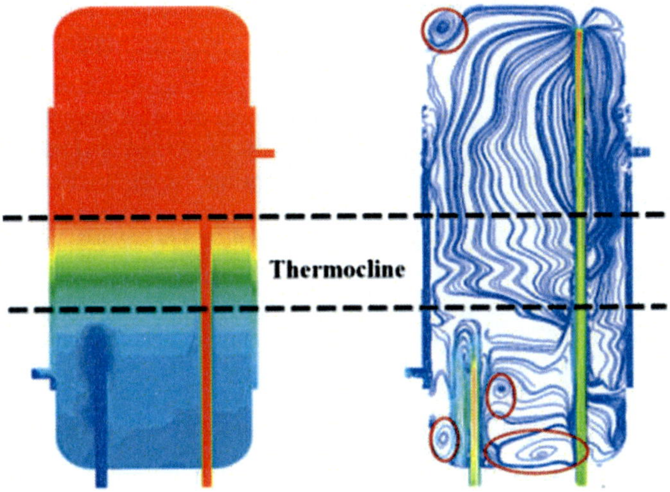

Fig. 3 Temperature contours (left) and streamlines (right) in the tank in discharging model

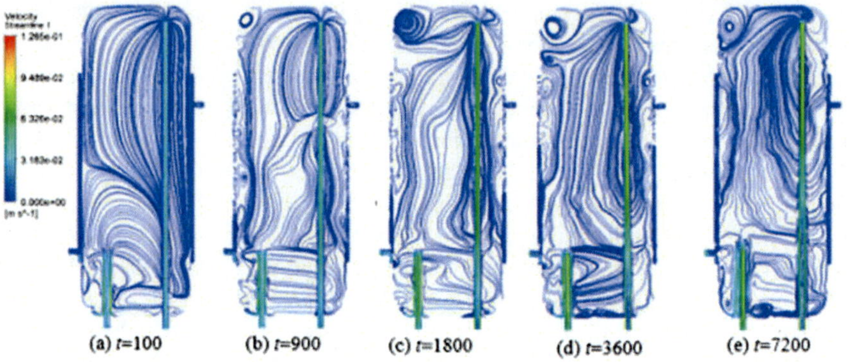

Fig. 4 Streamline contours in the tank in the discharging mode when the mass flow is 0.008 kg/s

4 Transient Dynamics of Mixing in Water Tank

During the off peak period of solar energy supply, the cooler collector return water is driven downward due to the higher density. When the incident cold water enters the bottom of the tank, it begins to entrain along the surrounding hot water layer. The whole process is called plume entrainment, as shown in the right picture of Fig. 3. In other words, plume entrainment is the process of heating the incident water, in which the plume rearranges its position in the water tank according to the matching available density of the surrounding fluid. This entrainment process increases the thickness of thermocline through Kelvin–Helmholtz instability or engulfment mechanism.

Due to the difference in water velocity and temperature, the inflow water is engulfed by the surrounding water. The previous K-H eddy and its viscous shear on the thermocline will also aggravate this situation, so the eddy advection will enter the thermocline region due to the viscous shear of the formed K-H eddy. After that the horizontal span and lateral vertical movement rate of the thermocline increase during the dynamic cycle.

When using a lower water flow state (mass flow 0.008 kg/s), compared with Fig. 3 (mass flow 0.022 kg/s), in the area above the maximum penetration height of the jet, the streamlines are very regular and the temperature is stratified relatively uniform. The characteristics of these negatively buoyant jets are more clearly seen in the velocity streamline contours of the cold water inlet area in Fig. 4.

5 Conclusion

In this paper, the direct numerical simulation method is used to simulate and analyze the instantaneous flow characteristics in the solar water heating system under two different operation modes (charging mode and discharging mode). The results show that:(1) The heat storage mode is absorbed, and the heat is transferred to the water tank through the jacket heat exchange device, so that the water in the water tank is gradually heated, and mainly through heat conduction, the water layer gradually forms three layered areas along the height direction: low-temperature area, thermocline zone and high-temperature zone. (2) In the drainage mode, the cold water injected into the water tank is entrained with the adjacent hot water area, forming a circulation area around it, and forcing the colder fluid to gradually move downward to form temperature stratification. This paper analyzes the main problems of improving energy efficiency in intelligent control solar water heating system, especially provides a detailed theoretical basis and basis for the application of energy Internet.

References

1. Gupta SK, Jaluria Y (1981) Transient thermal effects in an enclosed water body due to heated water discharge for heat rejection and solar energy storage. Energy Convers Manag 21:3–8
2. Goppert S, Lohse R, Urbaneck T et al (2009) New computation method for stratification pipes of solar storage tanks. Sol Energy 83:1578–1587
3. Dragsted J, Furbo S, Dannemand M et al (2017) Thermal stratification built up in hot water tank with different inlet stratifiers. Sol Energy 147:414–425
4. Rosen MA (2001) The exergy of stratified thermal energy storages. Sol Energy 71(3):173–185
5. Hess CF, Miller CW (1982) An experimental and numerical study on the effect of wall in a thermocline type cylinder enclosure—I experiments. Sol Energy 28:145–152
6. Han SM, Wu ST (1978) Computer simulation of a solar energy system with a viscous-entrainment liquid storage tank model. In: The proceedings of the third southeastern conference on application of solar energy, pp 165–182
7. Chan AMC, Smereka PS, Giusti D (1983) A numerical study of transient mixed convection flows in a thermal storage tank. ASME J Solar Energy Eng 105:246–253
8. Cabelli A (1977) Storage tanks—a numerical experiment. Sol Energy 19:45–54
9. Guo KL, Wu ST (1985) Numerical study of flow and temperature stratifications in a liquid thermal storage tank. ASME J Solar Energy Eng 107:15–20
10. Hoogendoorn CJ, Bakker RG, Herweijer EJJ et al (1985) Thermal stratification in water vessels for energy storage. In: INTERSOL 85: the proceedings of the 9th biennial congress international solar energy society. Montreal, Canada, pp 829–833

Research on Data Acquisition Algorithm Based on ZYNQ Biometric Signal

Shaoquan Jiang, Xuebing Cao, Tao Jiang, Yonghang Tai, and Chao Zhang

Abstract Under the irradiation of laser, biological particles emit fluorescence and scattered light. The spectral signals composed of fluorescence and scattered light can be used to classify and recognize biological particles. The acquisition of these two kinds of optical signals needs to be converted into electrical signals for acquisition. For this electrical signal, A/D converter with extremely fast speed is needed. In the data acquisition process, ZYNQ is used to select and classify the collected signals, and the classified data is displayed on the LCD screen.

Keywords Biometric spectroscopy · High-speed A/D conversion · ZYNQ · Data collection

1 Introduction

With the development of scientific research in recent years, the real-time monitoring technology of atmospheric bioaerosols has become a hot spot in the research field. The detection technology based on biological spectra has the advantages of fast and accurate, etc. The spectra of biological particles are mainly fluorescence spectra and scattered light spectra, fluorescence spectra can visually describe the type of particles, and scattered light spectra can describe the size of particles [1].

In high-speed data acquisition systems, since microcontrollers usually do not run as fast as we need, digital signal processors or FPGAs are usually used as microcontrollers to control the work of peripheral devices such as digital-to-analog converters. However, using a digital signal processor to acquire and analyze high-speed signals has major drawbacks, digital signal processors must be operated by software to implement various functions. So if the program is not stable enough, it will be very easy to run away and reset the program, although FPGA has a very high clock rate, low internal latency, and fast operation. All control logic is done by hardware, but FPGA is not flexible in data processing, especially in floating point

S. Jiang · X. Cao · T. Jiang · Y. Tai (✉) · C. Zhang
School of Physics and Electronic Information, Yunnan Normal University, Kunming, China
e-mail: taiyonghang@126.com; zhchynnu@foxmail.com

Z. Yu et al. (eds.), *Advancements in Mechatronics and Intelligent Robotics*,
Advances in Intelligent Systems and Computing 1220,
https://doi.org/10.1007/978-981-16-1843-7_28

operations. The new generation of FPGA launched by Xilinx solves these problems very well, and the chip is integrated with dual-core ARM cores, which can easily complete the data processing, and the chip has the parallel processing capability of FPGA, so this chip has a very good performance in the FPGA side of ZYNQ is referred to as the PL side and the ARM side of ZYNQ is referred to as the PS side.

The processed data is displayed on an LCD screen and the processed data is visualized in a number of ways, allowing us to view the data more intuitively.

2 Hardware Design

The hardware circuit mainly contains A/D conversion circuit, signal conditioning circuit, and op-amp power supply circuit. The fluorescence and scattered light of the organism are very weak light and the corresponding converted electrical signal is also very weak. Therefore, the components selected for conditioning this signal and the power supply to this circuit should have a very low noise level [2–4].

2.1 Selection of High-Speed A/D Converters

The high-speed A/D converter is the core device inside this system and plays a vital role in the classification and selection of the signal later. When making the selection of the high A/D converter, it is necessary to select the chip suitable for the system according to the signal and frequency to be acquired. An oscilloscope is connected to the output port of the biospectral signal to view the electrical signal waveforms converted from fluorescence and scattered light [5], as shown in Figs. 1 and 2.

Fig. 1 Fluorescence signal

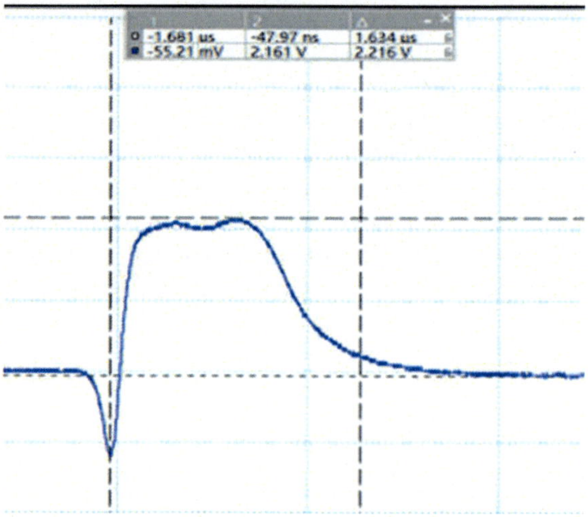

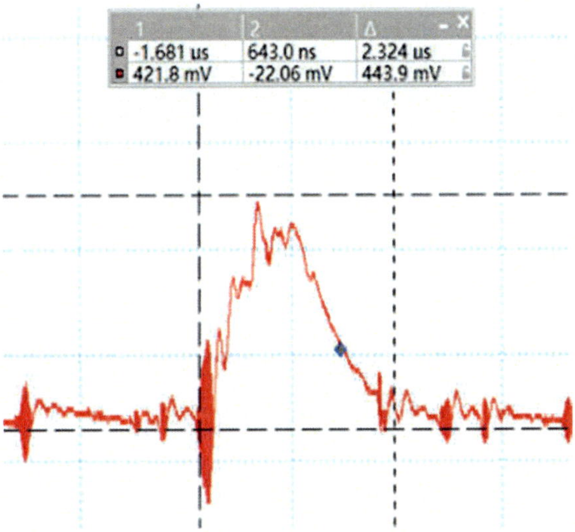

Fig. 2 Scattered light signal

The frequencies of these two signals are mostly around 1 MHz, and some can reach 2 MHz, so 2 MHz is selected as the highest frequency of the signal. According to the Nyquist sampling theorem, the sampling rate must be greater than or equal to twice the highest frequency, so we only need to select a device with a sampling rate greater than or equal to 4 MHz. Some of the signal amplitudes are very strong, reaching about 2 V, while others are only a few hundred millivolts. When selecting the device range and device resolution, both aspects need to be considered. The resolution of the A/D converter determines the accuracy of the signal sampling, and the range determines the range of the signal input. Both of these aspects play a key role in the subsequent signal classification and selection. In the case of weighing these factors, the AD9224 high-speed AD conversion chip of ADI was selected.

2.2 *High-Speed A/D Conversion Circuit*

The sampling resolution of AD9224 is 12 bits, and the sampling frequency is up to 40 MHz. The power supply adopts analog and digital power supply separately, which realizes the mutual non-interference of digital and analog, and guarantees the collection of small signals and the stable output of digital signals.

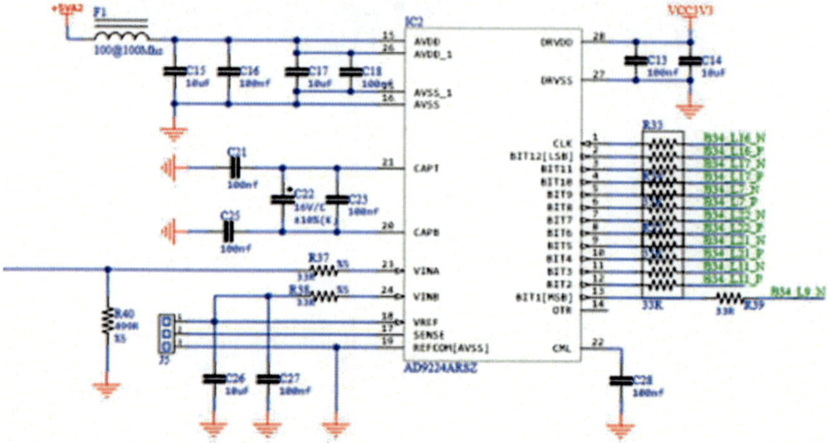

Fig. 3 Acquisition circuit AD9224

We also need to consider the range of signal input when acquiring signals, so as not to exceed the range of the chip when retesting, and also need to consider the stability of digital signal transmission. Therefore, a 33R matching resistor is connected in series with the digital signal output port of AD9224 to enhance the stability during signal transmission. The circuit design is shown in Fig. 3.

2.3 Signal Conditioning Circuit

Since the biological spectrum signal is converted into an electrical signal, the amplitude is very small and accompanied by a lot of noise, and it is necessary to simply remove noise and amplify the small signal. When designing the filter, it was found that some high-frequency signals with higher frequencies were the main cause of serious signal interference. When designing the filter, a low-pass filter was designed to filter out high-frequency signals. When amplifying the signal, in order to ensure that the signal is not distorted and the amplified signal is not infected by its own noise, and it is necessary to select a device with low noise and sufficient slew rate and gain bandwidth. TI's OPA659 is widely used in many small signal amplifier circuits with its excellent low noise, high bandwidth, and high slew rate. This system also uses this excellent amplifier chip. The circuit is shown in Fig. 4.

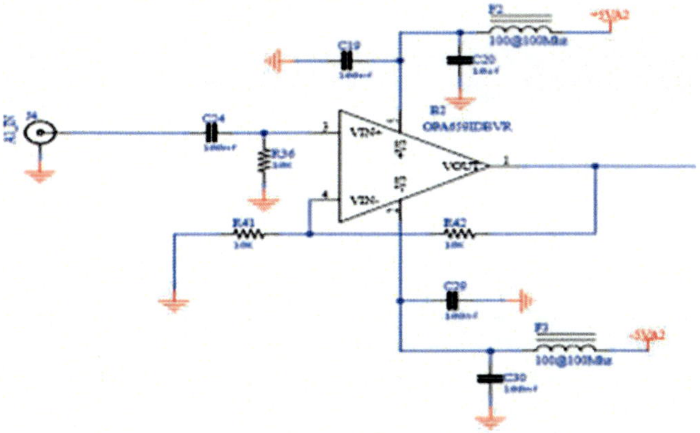

Fig. 4 Signal conditioning

Table 1 Data of 0.5 μm particles

≥0.3 μm	≥0.5 μm	≥1.0 μm	≥3.0 μm	≥5.0 μm	≥10.0 μm
3687	3652	694	110	1	0

Table 2 Data of 3.0 μm particles

≥0.3 μm	≥0.5 μm	≥1.0 μm	≥3.0 μm	≥5.0 μm	≥10.0 μm
3390	3344	3110	2957	110	0

3 Overall System Testing and Analysis

When testing the system, first detect the scattered light affected by the particle diameter. In the particle generator, 0.5 μm particles and 3.0 μm particles are generated, and the size distribution of scattered light caused by these two particles is observed, as shown in Tables 1 and 2.

Observing the data in Tables 1 and 2, we can find that most of the particle data of 0.5 μm are concentrated on 0.5 μm, and the particles of 3.0 μm are also concentrated on 3.0 μm. The spectral signal formed by the scattered light can accurately reflect the size of the particles, which verifies the feasibility and accuracy of the scheme.

In the particle generator, biological particles are generated. The LCD screen displays the corresponding waveform of the electrical signal converted by the fluorescence and scattered light generated by a single biological particle, as shown

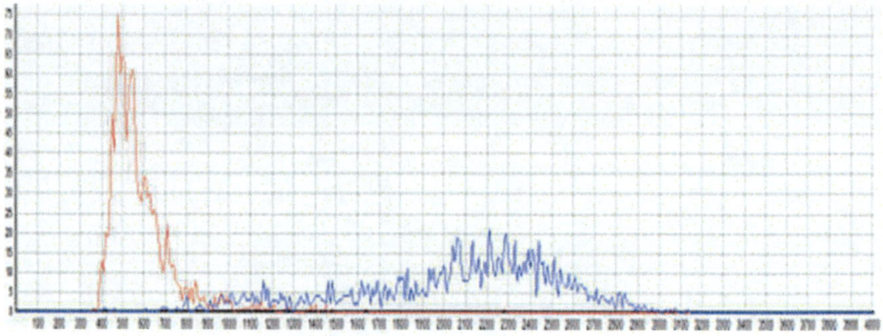

Fig. 5 Fluorescence and scattered light spectral distribution

in Fig. 5. The red is the scattered light signal of the biological particle. The yellow ones are the fluorescent signals of biological particles.

It can be seen from the waveform that the scattered light of biological particles is stronger than fluorescence. According to this feature of the biological particle spectrum, the collected data is selected and classified, and finally, the fluorescence spectrum distribution and scattered light spectrum distribution of biological particles are drawn, as shown in Fig. 5. Red is the fluorescence distribution, and blue is the scattered light distribution.

Observing the distribution diagram of the data, it can be seen that the system can collect the fluorescence spectrum signal and the scattered light spectrum signal of the biological particle and can perform some processing on the biological spectrum signal.

4 Conclusion

This paper is based on ZYNQ, high-speed ADC, and LCD display biological spectrum signal data acquisition system. Compared with the traditional FPGA + MCU system, the difficulty of this system design is mainly in the high-speed data transmission design between the high-speed ADC and ZYNQ. Data buffering, ZYNQ parallel processing technology, and LCD display in ZYNQ. Through the actual test verification after inputting high-speed signals, the feasibility and robustness of the system scheme under high-speed sampling meet the design requirements. This system has broad application prospects in the fields of biological recognition and biological spectrum signal analysis.

References

1. Rao Z, Hua D, He T et al (2016) The performance of bioaerosol measurement lidar based on intrinsic fluorescence. Acta Phys Sinica 65(20)
2. Cai S (2012) Research on bioaerosol detection technology based on intrinsic fluorescence measurement. University of Chinese Academy of Sciences; Graduate University of Chinese Academy of Sciences
3. Cai S, Zhang P, Zhu L et al (2012) Research on bioaerosol detection technology based on intrinsic fluorescence measurement of tryptophan. Acta Optica Sinica (05):127–132
4. Chen C (2018) Research on characteristics and application of photomultiplier tube. Digital Commun World (07):136–137
5. Yang J, Xu P, Chi C, Zhao W (2019) Development and application of gated microchannel plate photomultiplier tube. Photoelectronic Technol 39(01):34–38

Research Progress on the Flow Behavior of Negative Buoyancy Jet

Qiong Li, Xiaoqiao Huang, Yonghang Tai, and Wenfeng Gao

Abstract Negative buoyancy jet is the main internal physical process in many practical engineering applications. The study of its instantaneous transport mechanism is of scientific significance for intelligent control in industrial applications. In this paper, the flow behavior of circular and planar/linear negative buoyant jets in homogeneous environment and stratified environment, especially the maximum penetration height of negative buoyant jets, is introduced and summarized in detail.

Keywords Intelligent control · Negative buoyancy jet · Flow behavior

1 Introduction

In the process of building ventilation/refrigeration/heating and air conditioning system application, when the door is opened in winter, the cold air injects into the room and interacts with the surrounding hot air to form negative buoyancy jet phenomenon. Negative buoyant jet (fountain) is a phenomenon formed when the incident fluid with higher density shoots upward into the environment fluid with lower density, or when the fluid with lower density injects downward into the environment fluid with higher density. Negative buoyancy jet phenomenon exists widely in nature and industrial environment. In terms of environment and air pollution, industrial waste gas and pollution gas are commonly emitted from industrial chimneys, cooling towers of power plants, chemical plants, and salty wastewater from seawater desalination in the offshore area [1]. In the field of building environment and energy utilization, this phenomenon widely exists in the

Q. Li · W. Gao
Solar Energy Research Institute, Yunnan Normal University, Kunming 650500, Yunnan, China

X. Huang · Y. Tai (✉)
School of Physics and Electronic Information, Yunnan Normal University, Kunming 650500, Yunnan, China
e-mail: taiyonghang@126.com; zhchynnu@foxmail.com

Z. Yu et al. (eds.), *Advancements in Mechatronics and Intelligent Robotics*,
Advances in Intelligent Systems and Computing 1220,
https://doi.org/10.1007/978-981-16-1843-7_29

ventilation/refrigeration/heating and air conditioning system of buildings [2] and in the solar hot water supply system. Negative buoyancy jet is widely used in engineering and technical applications including the above application examples due to its unique flow behavior. Therefore, in-depth study and full understanding of negative buoyancy jet flow behavior can provide theoretical basis for intelligent control of ventilation and air conditioning system and dilution and diffusion of pollutants, which has important scientific significance.

2 Research Progress of Negative Buoyancy Jet in Homogeneous Environment

There are many kinds of negative buoyant jets interacting with environmental fluids in nature and artificial environment. There are many factors that affect the shape and flow behavior of negative buoyancy jet, including the mixing state and strength (laminar or turbulent, strong or weak); the geometry of the incident fluid (circular or planar/linear); the direction of the incident fluid (perpendicular or inclined to the ambient fluid); and the physical characteristics of the incident fluid and the ambient fluid properties (including relative density, viscosity, quiescence/motion, whether mixing or reaction is allowed), etc. Here, only the common negative buoyancy jets are classified from these aspects.

2.1 Circular Negative Buoyancy Jet

Early studies on negative buoyant jets focused on the maximum rise (penetration) height of turbulent negative buoyant jets in ambient fluids. Morton [3] was the first to study the flow behavior of negative buoyancy jet. Using the similarity theory, he made a semiempirical theoretical analysis of the maximum rising height of the negative buoyant jet and the convolution between the jet and the ambient fluid in the process of flow and obtained the correlation between the two characteristic quantities and Reynolds number (Re) and Froude number (Fr). Turner [4] modified Morton's empirical formula based on the similarity theory, dimensional analysis and experimental results in homogeneous fluid, and the following relation formula were obtained.

$$z_m = C\,\mathrm{Fr} \tag{1}$$

Here, z_m is the dimensionless maximum rising height of the steady-state jet and C is the proportional constant. Turner [4] also used a series of experimental results of negative buoyant jets with Fr numbers in the range of 2–30 in a uniform ambient fluid to give $C = 2.46$. A large number of follow-up studies found that C also concentrated in 2.46. In the past 20 years, research work has focused on the flow behavior of weak and negative buoyant jets in the transition region from laminar

Table 1 Summary of the value of *C* from various studies

Author	Fr	Re	Source fluid/ambient fluid	*C*
Turner [4]	2–30	/	Saline water/fresh water	2.46
Seban et al. [5]	6.6–53.5	894–1923	Hot air/ambient air	2.52
Mizushina et al. [6]	3.0–257.7	870–2710	Fresh water/heated fresh water	2.34
James et al. [7]	24–110	1550–11000	Saline water/fresh water	2.46
Baines et al. [8]	5–200	/	Saline water/fresh water	2.46
Baines et al. [8]	31.6–102.7	/	Fresh water/saline water	2.46
Cresswell [9]	3.2	2500	Hot water/cold water	2.46
Zhang [10]	≥7	850–6000	Saline water/fresh water	3.06
Pantzlaff [11]	15.8–78.0	1250–10500	KCl solution/water	2.12
Papanicolaou [12]	1.4–83.2	770–5840	Fresh water/saline water	2.46
Wang et al. [13]	≥6.0	1000–1500	/	2.46
Burridge and Hunt [14]	≥2.8	969–4022	Saline water/fresh water	2.46
Burridge et al. [15]	≥4	≥750	Saline water/fresh water	2.46

flow to turbulent flow. The *C* values obtained from these studies are summarized and listed in Table 1.

2.2 Plane/Linear Negative Buoyancy Jet

Compared with the research of circular negative buoyancy jet, the research results of planar/linear negative buoyancy jet are much less reported. There are two main reasons, one is that the practical application of plane/linear negative buoyancy jet is less, the other is that it is more difficult to carry out research from the experimental aspect. This is because in order to eliminate the influence of the boundary, the length of the jet inlet should be long enough, which is usually a more specific and challenging work. Of course, it is easier to study the flow behavior of plane/linear negative buoyant jet theoretically (mainly dimensional analysis and scaling analysis) and numerically. Although it is still a great challenge for numerical simulation to greatly weaken and eliminate the influence of boundary.

Similar to circular negative buoyancy jets, Hunt and Coffey [16] classified planar/linear negative buoyancy jets as follows:

Very weak plane fountains: when $\mathrm{Fr} \leq 2.3$;
Weak plane fountains: when $2.3 \leq \mathrm{Fr} \leq 5.7$;
Forced plane fountains: when $\mathrm{Fr} \geq 5.7$.

However, this classification does not consider the influence of Re number. For the case of larger Re number and higher Fr number, the study shows that the influence of Re number can be ignored. But when Re number and Fr number are not large, it is found that the influence of Re number is significant (Table 2).

Table 2 Summary of correlation for z_m from various studies on plane/line fountains

Author	Method	Fr	Re	Scaling relation
Campbell (1989)	Experimental	5.6–51	–	$z_m \propto Fr^{4/3}$
Baines [8]	Analytical/experimental	5–1000	325–2700	$z_m \propto Fr^{4/3}$
Zhang [10]	Analytical/experimental	0.62–6.5 10–113	894–1923	$z_m \propto Fr^2$ $z_m \propto Fr^{4/3}$
Lin (2000)	Numerical/analytical	0.2–1.0	200–800	$z_m \propto Fr$
Lin (2000)	Numerical/analytical	0.0025–0.2	5–800	$z_m \propto Fr^{2/3}/Re^{2/3}$
Lin (2003)	Numerical/analytical	0.2–1.0	5–200 200–800	$z_m \propto Fr/Re^{1/2}$ $z_m \propto Fr^{4/3}$
Hunt (2009)	Analytical	≤ 2.3 2.3–5.7 ≥ 5.7	–	$z_m \propto Fr^{2/3}$ $z_m \propto Fr^2$ $z_m \propto Fr^{4/3}$
Srinarayana (2010)	Experimental	0.47–11.57	2.1–121	$z_m \propto Fr/Re^{1/2}$
Srinarayana (2013)	Numerical	0.25–2.0 2.25–3.0 3.0–10.0	100	$z_m \propto Fr$ $z_m \propto Fr^{1.15}$ $z_m \propto Fr^{4/3}$

3 Research Progress of Negative Buoyancy Jet in Stratified Ambient Fluid

Although there are numerous applications of negative buoyancy jet into stratified environment fluid, the research work in this field is much less than that in homogeneous environment fluid. As mentioned above, due to the stratification of the ambient fluid, the flow behavior of the negative buoyancy jet is controlled not only by the Re number and Fr number but also by the stratification degree of the ambient fluid. The stratification of environmental fluid can be single component, such as temperature stratification flow in reservoir or fresh water lake, salinity stratification flow caused by saltwater intrusion in ocean and estuary, and concentration stratification flow caused by sewage discharge into water body. It can also be two-component or multi-component coexisting water body stratification flow, such as temperature stratification flow in reservoir and fresh water lake, which often exist at the same time. There are concentration stratified flows caused by microorganisms (such as cyanobacteria), temperature stratified flows caused by sunlight in addition to salinity stratified flows in the ocean, temperature and salinity double stratified flows in solar salt ponds, and stratified flows of other components in addition to temperature and salinity stratified flows in saltwater lakes and oceans, such as microorganisms, oxygen, and dioxins carbon oxide, and so on.

3.1 Circular Negative Buoyancy Jet

There are two ways to form the negative buoyancy jet in stratified environment fluid: one is that the buoyancy flux at the entrance of the jet is zero, that is, the density of the jet at the entrance is the same as that of the environment fluid at the bottom; the other is that the buoyancy flux at the entrance of the jet is not zero. No matter how the negative buoyancy jet is formed, the jet will present three different maximum rising heights, namely the initial maximum rising height, the final rising height and the expansion height. For the case of zero buoyancy flux at the jet entrance, Fischer et al. [17] obtained the scaling formula of the maximum rising height of the jet, that is,

$$Z_m = \text{CM01}/4\text{N1}/2 \tag{2}$$

where M0 is the momentum flux at the jet entrance, $N = -[(g/\rho)(d\rho/dz)]1/2$ is the buoyancy frequency, and C is the proportional constant. Bloomfield and Kerr [18] used the experimental results to get the values of C corresponding to the initial maximum rise height, the final rise height and the expansion height are 3.25, 3.00 and 1.53, respectively. For the case that the buoyancy flux B0 at the jet entrance is not zero, Bloomfield and Kerr [18] proposed a new scaling relation, that is,

$$Z_m = f(\sigma*)\text{M03}/4/\text{B01}/2 \tag{3}$$

where σ^* = M02N2/B02 is a dimensionless parameter introduced by Fischer et al. [17]. If the dimensionless temperature stratification parameter is s, then σ^* = Fr2s. Therefore, the scaling relation of dimensionless maximum rise height of negative buoyant jet in linear stratified environment fluid is

$$z_m = f(\text{Fr2s})\text{Fr} \tag{4}$$

where is the function of Fr2s. In addition, there are also some studies on the flow behavior of negative buoyancy jet penetrating the interface between two layers of fluid.

3.2 Plane/Linear Negative Buoyancy Jet

Compared with circular negative buoyancy jet, the research on the flow behavior of planar/linear negative buoyancy jet in stratified environment is very limited. Bloomfield and Kerr [18] found that the flow behavior of planar/linear negative buoyancy jet in stratified environment is similar to that of circular negative buoyancy jet in stratified environment. However, the initial maximum penetration height of the jet in stratified ambient fluid is not significantly reduced due to the interaction between ascending flow and descending flow. Bloomfield and Kerr [18] obtained the scaling relation of the maximum rising height of the jet as:

$$Z_m = \begin{cases} CM_0^{1/3}N^{2/3}, & B_0 = 0 \\ f(\sigma)M_0B_0^{2/3}, & B_0 \neq 0 \end{cases} \quad (5)$$

where C is the proportional constant, B_0 is the buoyancy flux at the jet entrance. They use the experimental results to get the values of C corresponding to the initial maximum rise height, the final rise height, and the expansion height are 2.46, 2.27, and 1.07, respectively.

where σ = M02N2/B02 is a dimensionless parameter introduced by Fischer et al. [17]. If the dimensionless temperature stratification parameter is s, similarly, σ = Fr2s. Therefore, the scaling relation of the dimensionless maximum rising height of the plane/linear negative buoyant jet in the linear stratified environment fluid is z_m = *f*(Fr2s)Fr4/3, where is the function of Fr2s.

Lin and Armfield [19] carried out scaling analysis and numerical simulation on the flow behavior of weak plane/linear negative buoyancy jet in linear stratified environment fluid and obtained the following scaling relationship

$$z_{m,s} \sim \frac{Fr^{2/3}}{Re^{1/3}s^{1/3}} \quad (6)$$

4 Conclusion

In the aspect of the research progress of negative buoyancy jet flow behavior, firstly, a variety of negative buoyancy jets are systematically classified from the aspects of incident fluid shape, incident direction, Fr/Re number and environmental fluid. Secondly, the flow behavior of circular and planar/linear negative buoyant jets in homogeneous **environment**, especially the maximum penetration height of negative buoyant jets, is introduced and summarized in detail. Finally, the research work on the flow behavior of circular and planar/linear negative buoyant jets in stratified environment is also introduced and summarized.

This paper summarizes the main research progress of negative buoyancy jet flow behavior, which provides a detailed basis for the understanding of the difficulties and development trend in these aspects, especially for the use and intelligent control of negative buoyancy phenomenon in industry and ventilation of building.

References

1. Oliver CJ, Davidson MJ, Nokes RI (2013) Predicting the near-field mixing of desalination discharges in a stationary environment. Desalination 309:148–155
2. Hunt GR, Linden PF (2005) Displacement and mixing ventilation driven by opposing wind and buoyancy. J Fluid Mech 527:27–55
3. Morton BR (1959) Forced plumes. J Fluid Mech 5:151–163
4. Turner JS (1966) Jets and plumes with negative or reversing buoyancy. J Fluid Mech 26:779–792
5. Seban RA, Behnia MM, Abreau KE (1978) Temperatures in a heated jet discharged downward. Int J Heat Mass Transf 21:1453–1458
6. Mizushina T, Takeuchi FOH, Ikawa H (1982) An experimental study of vertical turbulent jet with negative buoyancy [J]. Wärme-und Stoffübertragung 16:15–21
7. James WP, Kim IVK (1983) Dilution of a dense vertical jet [J]. J Environ Eng 109:1273–1283
8. Baines WD, Corriveau AF, Reedman TJ (1993) Turbulent fountains in a closed chamber [J]. J Fluid Mech 255:621–646
9. Cresswell RW, Szczepura RT (1993) Experimental investigation into a turbulent jet with negative buoyancy [J]. Physics of Fluids A: Fluid Dynamics 5(11):2865–2878
10. Zhang H, Baddour RE (1998) Maximum penetration of vertical round dense jets at small and large Froude numbers [J]. J Hydraul Eng 124 550–553
11. Pantzlaff L, Lueptow RM (1999) Transient positively and negatively buoyant turbulent round jets [J]. Exp Fluids 27(2):117–125
12. Papanicolaou PN, Kokkalis TJ (2008) Vertical buoyancy preserving and non-preserving fountains, in a homogeneous calm ambient [J]. Int J Heat Mass Transfer 51(15–16):4109–4120
13. Wang RQ, Law AWK, Adams EE et al (2011) Large-eddy simulation of starting buoyant jets [J]. Environ Fluid Mech 11(6):591–609
14. Burridge HC, Hunt GR (2012) The rise heights of low- and high-Froude-number turbulent axisymmetric fountains [J]. J Fluid Mech 691:392–416
15. Burridge HC, Mistry A, Hunt GR (2015) The effect of source Reynolds number on the rise height of a fountain [J]. Phys Fluids 27(4):297–317
16. Hunt GR, Coffey CJ (2009) Characterising line fountains. J Fluid Mech 623:317–327
17. Fischer HB, List EJ, Koh C et al (1979) Mixing in inland and coastal waters. Academic Press, New York
18. Bloomfield LJ, Kerr RC (1998) Turbulent fountains in a stratified fluid. J Fluid Mech 358:335–356
19. Lin W, Armfield SW (2002) Weak fountains in a stratified fluid. Phys Rev E 66(6):066308

Machine Learning

Numerical Simulation of Discrete-Time SIR Epidemic Models with Dispersion

Fang Zheng

Abstract In this paper, the discrete-time S-I-R model was used to obtain two stable equilibrium points considering the spread of disease enhancement and disease inhibition between two patches. In this paper, we discuss the important influence of disease transmission on the change of population behavior. Enhancing disease spread plays a key role in the production and support of multiple attractors, which increases the likelihood of disease persistence. In short, spread can enhance the persistence of the disease.

Keywords Discrete · S-I-R model · Dispersion · Bistability

1 Introduction

In a broad sense, the different modes of transmission of infectious diseases are important to the intensity of the epidemic, the pathogen and the host of the infection [1]. In the past 20 years, research on disease dynamics has focused on models of human hosts [2, 3]. Regardless of host type, it is rare to incorporate models of diffusion into populations and diseases. The local (single plaque) disease dynamics model modifies the translation results by excluding disease-induced deaths from the S-I-S epidemiological process. It is impossible, animal diseases have an impact on host mortality. However, the main content we choose is the simple discrete time infectious disease model bistable. Will diffusion cause change or even chaos in the biological world? The discrete S-I-S infectious disease model of a single patch can produce complex (chaotic) dynamics, which cannot be obtained by traditional continuous time infectious disease models [4]. Nonconstant infectivity can make the model produce multiple stable equilibria. An example of an epidemic model with this behavior is first the continuous time epidemic model developed. More recently, some of the authors proposed a simpler continuous time model [4, 5].

F. Zheng (✉)
Department of Basic, Air Force Engineering University, Xi'an, China
e-mail: zhengfangkgd@163.com

Z. Yu et al. (eds.), *Advancements in Mechatronics and Intelligent Robotics*,
Advances in Intelligent Systems and Computing 1220,
https://doi.org/10.1007/978-981-16-1843-7_30

The findings of Castillo-Chavez [5, 6]. It has a significant impact on the formulation of effective infectious disease and public health prevention policies.

2 Two-Patch S-I-R Epidemic Models with Dispersion

The S-I-R epidemic model with dispersion in double patches is the main mechanism of species diversity production and support [2, 7]. Population movements increase the likelihood of disease transmission [7]. The important role of diffusion in human disease dynamics has been studied in influenza [8, 9] and tuberculosis [10]. n host-parasite interactions, parasites have a strong influence on the host's behavior, potentially making them more vulnerable to predators. For example, ascaris parasitize in the intestinal tract, take a large number of nutrients, and affect the absorption function of the intestinal tract, causing host malnutrition. Based on our work [11], we established and analyzed an S-I-R epidemic model of population, spread between two patches, patches 1 and 2. In patch $i \neq j \in \{1,2\} S_i(n)$ denotes the population of susceptibles; $I_i(n)$ denotes the population of the infected, assumed infectious; $R_i(n)$ denotes the population of recoved; $N_i(n) = S_i(n) + I_i(n) + R_i(n)$ denotes the total population size at generation n; $N_{i\infty} = \lim_{n\to\infty} N_i(n)$ represents the total number of population when the environment, resources, and other aspects are stable. We assuming the patch 1 and 2 with a simple exchange of a fixed fraction of the population per generation and assume there is no dispersal in recovered population. For each patch $i \neq j \in \{1,2\}$, let d_{is}, d_{iI}. They are the proportion of susceptible and infected people who spread from plaques patch i to j separately. We obtained the following two patches infectious system:

$$\begin{cases} S_1(n+1) = (1-d_{1S})\tilde{S}_1(n) + d_{2S}\tilde{S}_2(n) \\ S_2(n+1) = d_{2S}\tilde{S}_1(n) + (1-d_{2S})\tilde{S}_2(n) \\ I_1(n+1) = (1-d_{1I})\tilde{I}_1(n) + d_{2I}\tilde{I}_2(n) \\ I_2(n+1) = d_{1I}\tilde{I}_1(n) + (1-d_{2I})\tilde{I}_2(n) \\ R_1(n+1) = rR_1(n) + r(1-\delta)I_1(n) \\ R_2(n+1) = rR_2(n) + r(1-\delta)I_2(n) \end{cases} \tag{1}$$

where

$$\tilde{S}_i(n) = f_i(N(n)) + r_i S_i(n) g_i(y),$$
$$\tilde{I}_i(n) = r(1-g_i(y))_i S_i(n) + r_i \delta_i I_i(n),$$

If patch 2 is empty and there is no dispersion between the two patches

$$d_{1S} = d_{2S} = d_{1I} = d_{2I} = \tilde{S}_2 = \tilde{I}_2 = 0,$$

then system (1) reduces to the single patch model system:

$$\begin{cases} x(n+1) = \frac{\mu}{\mu+r} + \frac{r}{\mu+r}x(n)(1-y^2(n)) \\ y(n+1) = \frac{r}{\mu+r}y^2(n)x(n) + \frac{r\delta}{\mu+r}y(n) \end{cases} \tag{2}$$

Without diffusion, the system (1) simulates two independent patches. Next, we illustrate the potential role of propagation over patches with the same local dynamics by a simple example.

3 Identical Local Patch Dynamics

As we all know, the diffusion between patches can change the behavior of local areas [12, 13]. A disease that is doomed to extinction in patch i, local basic reproductive number $R_{0i} < 1$, can persist in the infectious disease model as long as the spread continues to occur. We come to the following conclusion.

Theorem 1 In system (1), the set $\{(I_1, I_2, S_1, S_2) | I_1 = I_2, S_1 = S_2\}$ is invariant if

$$r_1 = r_2, \delta_1 = \delta_2, f_1 = f_2, G_1 = G_2, r_1 = r_2,$$
$$\delta_1 = \delta_2, f_1 = f_2, d_{1I} = d_{2I}, d_{1S} = d_{2S}, G_1 = G_2.$$

Proof Note that $S_1(n) = S_2(n), I_1(n) = I_2(n)$ implies that

$$\tilde{S}_1(n) = \tilde{S}_2(n),$$

And $\tilde{I}_1(n) = \tilde{I}_2(n)$
whenever $r_1 = r_2, \delta_1 = \delta_2, f_1 = f_2, G_1 = G_2,$ Consequently, $(1 - d_{1S})\tilde{S}_1(n) + d_{2S}\tilde{S}_2(n) = S_1(n+1) = S_2(n+1),$
And $\{(I_1, I_2, S_1, S_2) | I_1 = I_2, S_1 = S_2\}$ is an invariant set.

4 Expansion of Disease Enhancement and Inhibition

Whether changes in behavior caused by the disease increase the likelihood that the disease will persist? Let us consider an extreme case, the susceptible people are confined to a patch do not spread, while the infectives are allowed to disperse between patches in modeled using system (1) with $d_{iI} > 0, d_{iS} = 0,$

Disease-controlled dispersal occurs in system (1) whenever $d_{iI} > d_{iS}$, while disease-enhanced dispersal occurs when $d_{iS} < d_{iI}$. If $d_{iI} = 0$ while $d_{iS} > 0$, the susceptible people are allowed to move freely between the two patches, while infected people are confined to their respective patch and are not allowed to spread. This controlled spread of the disease from patches 1 to 2 leads to the following system:

$$\begin{cases} x_1(n+1) = (1-d_{1S})\tilde{x}_1(n) \\ x_2(n+1) = d_{1S}\tilde{x}_1(n) + \tilde{x}_2(n) \\ y_1(n+1) = \tilde{y}_1(n) \\ y_2(n+1) = \tilde{y}_2(n) \end{cases} \tag{3}$$

where $0 < d_{1S} < 1$,

$$\tilde{x}_i(n) = \frac{\mu_i}{\mu_i + r_i} + \frac{r_i}{\mu_i + r_i} x_i(n)(1 - y_i^2(n)),$$

and

$$\tilde{y}_i(n) = \frac{r_i}{\mu_i + r_i} y_i^2(n) x_i(n) + \frac{r_i \delta_i}{\mu_i + r_i} y_i(n).$$

Model (3) assumes a geometric increase in the population. Whenever $d_{1S} = 0$, system (3) reduces to system (2). System (3) has a locally asymptotically stable disease-free equilibrium E_0 at

$$(1, 1, 0, 0)$$

The Jacobian matrix at E_0 is

$$J(E_4) = \begin{pmatrix} \frac{r_1}{r_1+\mu_1} & 0 & 0 & 0 \\ 0 & \frac{r_2}{r_2+\mu_2} & 0 & 0 \\ 0 & 0 & \frac{r_1\delta_1}{r_1+\mu_1} & 0 \\ 0 & 0 & 0 & \frac{r_2\delta_2}{r_2+\mu_2} \end{pmatrix} \tag{4}$$

The matrix (4) has four characteristics:

$$\lambda_1 = \frac{r_1}{r_1 + \mu_1} < 1,$$
$$\lambda_2 = \frac{r_2}{r_2 + \mu_2} < 1,$$
$$\lambda_3 = \frac{r_1\delta_1}{r_1 + \mu_1} < 1,$$
$$\text{and} \quad \lambda_4 = \frac{r_2\delta_2}{r_2 + \mu_2} < 1.$$

So, the disease-free equilibrium E_0 is locally asymptotically stable.

$$\text{Let} \quad p_i = \frac{\mu_i}{\mu_i + r_i}, q_i = \frac{r_i}{\mu_i + r_i}.$$
$$\text{If} \quad d_{1S} = 1 - \frac{4(1 - q_1\delta_1)^2}{q_1p_1^2 + 4q_1(1 - q_1\delta_1)^2} = \frac{2\sqrt{p_2q_2}(\frac{1}{q_2} - \delta_2)^2 - p_2}{p_1 + q_1x_1(1 - y_1^2)}$$

then system (3) has a unique unstable endemic equilibrium population size at the point E_1

$$\left(\frac{\mu_1(1 - d_{1S})}{2(\mu_1 + r_1 - r_1(1 - d_{1S}))}, \frac{d_{1S}(p_1 + q_1x_1(1 - y_1^2) + p_2)}{2p_2}, y_1, y_2\right)$$
$$\text{where} \quad x_iy_i = \frac{1}{q_i} - \delta_i.$$

$$\text{If} \quad d_{1S} < 1 - \frac{4(1 - q_1\delta_1)^2}{q_1p_1^2 + 4q_1(1 - q_1\delta_1)^2}$$
$$\text{and} \quad d_{1S} > \frac{2\sqrt{p_2q_2}(\frac{1}{q_2} - \delta_2)^2 - p_2}{p_1 + q_1x_1(1 - y_1^2)}$$

system (3) has two endemic equilibria E_2 and E_3.

So whether a disease persists or not really depends on its initial conditions. Therefore, system (3) can support the coexistence of two disease equilibria with locally asymptotically stable disease-free equilibria, with or without dispersion between patches.

We assume that the dispersion of the system is produced by disease, reduces (3) to the following system of equations:

$$\begin{cases} x_1(n+1) = \tilde{x}_1(n) \\ x_2(n+1) = \tilde{x}_1(n) \\ y_1(n+1) = (1 - d_{1I})\tilde{y}_1(n) + d_{2I}\tilde{y}_1(n) \\ y_2(n+1) = d_{1I}\tilde{y}_1(n) + (1 - d_{2I})\tilde{y}_2(n) \end{cases} \tag{5}$$

System (2.5.25) reduces to system (2.4.14) whenever $d_{1I} = d_{2I} = 0$. System (3), like the single-patch model system (3), has a locally asymptotically stable disease-free equilibrium at (1,1,0,0).

Regardless of the parameter values, some positive initial populations lead to death in patch 1 and patch 2. In system (5), like (3), with or without diffusion, if the population grows at a geometric rate, the population is bistable.

Even in the absence of spread, it is possible for the disease to persist in a multi-plaque system with spread from plaque to plaque, even if the disease is near extinction.

If $B^2 > C$ system has two points with infectives in patch 2, and there no infectives in patch 1 at the following levels:

$$\begin{pmatrix} \frac{\mu_1(1-d_{1S})}{2(\mu_1+r_1-r_1(1-d_{1S}))} \\ \frac{q_2(p_2+p_1+\frac{p_1 q_1(1-d_{1S})}{1-q_1(1-d_{1S})})\pm\sqrt{B^2+C}}{2p_2 q_2} \\ 0 \\ \frac{2p_2(1-q_2\delta_2)}{q_2(p_2+p_1+\frac{p_1 q_1(1-d_{1S})}{1-q_1(1-d_{1S})})\pm\sqrt{B^2+C}} \end{pmatrix}^{\mathrm{T}}$$

where

$$B = q_2(p_1 + p_2 + q_1 x_1),$$
$$C = 4p_2 q_2(1 - q_2\delta_2)^2$$

5 Numerical Simulation

In this section, the theoretical analysis of the stability motioned above will be simulated by MATLAB.

$$\text{Let} \quad r_1 = r_2 = 0.3, \mu_1 = \mu_2 = 0.9, \delta_1 = \delta_2 = 0.9,$$

We obtain the local basic reproduction number in each patch $i \in \{1, 2\}$ is:

$$R_0 = \frac{r_i\delta_i}{\mu_i + r_i} = 0.225 < 1 - \frac{\sqrt{r_i\mu_i}}{2(\mu_i + r_i)} = 0.783.$$
$$R_0 = \frac{r_i\delta_i}{\mu_i + r_i} = 0.225 < 1 - \frac{\sqrt{r_i\mu_i}}{2(\mu_i + r_i)} = 0.783.$$

If there is no spread, the disease will die out in the respective patch. In a single-patch model without diffusion, adding diffusion to the model produces bistability, Example with $d_{1S} = 0.6$ supports two stable equilibria at (0.2,1.8,0,0) and (0.2,0.4,0,1.09) at the same time.

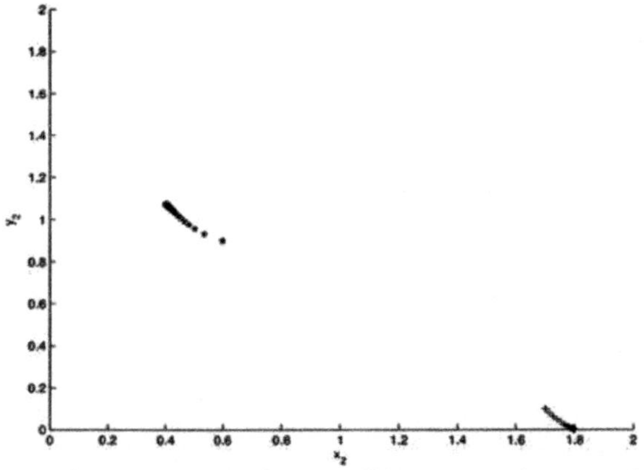

Allowing the spread of susceptible persons from patch 1 to patch 2 in the model can change the local pattern of plaques. For plaques where no disease is allowed to spread, adding spread can cause the disease to persist.

6 Conclusion

In this paper, we consider the important role that diffusion plays in the spread of disease and discuss two special cases: one in which spread occurs only among the susceptible and one in which spread occurs only among the sick. In the single-spot discrete S-I-R model, when we assume that infection is modeled as a possion process, bistable bifurcation will occur. In the two-plaque discrete-time S-I-R model, the diffusion of disease enhancement and disease inhibition may produce two stable equilibrium points. The results of numerical simulation were found to be consistent with the theory, which indicated that diffusion played an important role in the spread of disease and provided an effective tool for the control of infectious disease or epidemic situation. In other words, dispersion can enhance persistence.

References

1. Barrera JH, Cintron-Arias A, Davidenko N, Denogean LR, Franco-Gonzalez SR (1999) Dynamics of a two-dimensional discrete-time SIS model, Biometric Department, MTBI Cornell University Technical Report
2. Anderson RM, Anderson RM (1992) May, infectious diseases of humans: dynamics and control. Oxford University, Oxford
3. Bailey NTB (1975) The mathematical theory of disease, 2nd edn. Griffin, London

4. Dushoff J, Huang W, Castillo-Chavez C (1998) Backwards bifurcations and catastrophe in simple models of fatal diseases. J Math Biol 36:227
5. Huang W, Cooke KL, Castillo-Chavez C (1992) Stability and bifurcation for a multiple-group model for the dynamics of HIV/AIDS transmission. SIAM J Appl Math 52:835
6. Feng Z, Castillo-Chavez C, Capurro A (2000) A model for tuberculosis with exogenous reinfection. Theoret Popul Biol 57:235
7. Levin SA (2000) Encyclopedia of biodiversity. In: Levin SA (ed) vol. L. Academic Press, New York
8. Baroyan VO, Rvachev LA (1967) Deterministic epidemic models for a territory with a transport network. Kibernetika 3:67
9. Castillo-Chavez C, Capurro A, Zellner M, Velasco-Hernandez JX (1998) El transporte publico y la dinamica de la tuberculosis a nivel poblacional. A portaciones mathematicas, Ser. Commun. 22:209
10. Castillo-Chavez C, Yakubu A-A (in press) Discrete-time S-I-S models with complex dynamics. In: Proceedings WCNA 2000, nonlinear Anal. TMA
11. Castillo-Chavez Carlos, Yakubu Abdul-Aziz (2001) Dispersal, disease and life-history evolutioln. Math Biosci 173:35–53
12. Arreola R, Crossa A, Velasco MC (2000) Discrete-time S-E-I-S models with disoersal between two patches. Biometric Department, MTBI Cornell University Technical Report
13. Best J, Pasour V, Tisch N, Castillo-Chavez C (2001) Delayed density dependence and the dynamic consequences of dispersal between patches. Preprint

Research on the Impact of Fifth-Wheel Damping Coefficient on the Lateral Stability of Tractor-Semitrailer

Chuan Jin Ou and Bing Tao Li

Abstract In order to analyze the lateral stability of tractor-semitrailer, a vehicle dynamics model including fifth-wheel damping characteristics was established. Single sinewave steering input tests were carried out. Yaw velocity responses of the tractor and the semitrailer were revealed through time-domain analysis, and system stability was analyzed through changing fifth-wheel damping coefficient. Moreover, the simulation analysis was carried out using TruckSim software. The results show that the increase of fifth-wheel damping coefficient can improve the stability of the system, which provides a reference for the research on the lateral stability of tractor-semitrailer.

Keywords Fifth-wheel · Damping coefficient · Lateral stability · Tractor-semitrailer

1 Introduction

With the development of China's economy, combination vehicles have become the main force of trunk road transportation. The tractor-semitrailer is composed of tractor and semitrailer through the coupling effect of the mechanical connection of the fifth-wheel and hitch pin and the electrical connection device. The electric and mechanical devices between the tractor and the semitrailer are used to control the driving, steering, and braking of the semitrailer. Compared with single vehicle, the coupling effect between tractor and semitrailer makes the system more complex. Vehicle structure, operation and match characteristics between tractor and semitrailer lead to many new characteristics of tractor-semitrailer, such as motor coor-

C. J. Ou (✉)
Research Institute of Highway, Ministry of Transport, Beijing 100088, China
e-mail: cj.ou@rioh.cn

B. T. Li
College of Mechanical Engineering, Jiamusi University, Jiamusi 154007, China

Z. Yu et al. (eds.), *Advancements in Mechatronics and Intelligent Robotics*, Advances in Intelligent Systems and Computing 1220,
https://doi.org/10.1007/978-981-16-1843-7_31

dination, connection reliability, lateral vibration at high speed, response to lateral disturbances during driving and folding phenomenon during braking.

While tractor-semitrailers promote social and economic development, they have also triggered a series of traffic accidents, resulting in huge economic losses and serious casualties. The lateral stability is particularly critical among the safety performances of the tractor-semitrailer. Lateral instability caused by unreasonable structural parameter matching is the research focus and difficulty of tractor-semitrailer system dynamics. Many researchers have also conducted related studies, but have not analyzed the impact of fifth-wheel characteristics on the system [1–4]. Therefore, a dynamic model including the damping characteristics of the fifth-wheel is established, and simulation analysis is performed by using simulation software to change the damping coefficient of fifth-wheel.

2 Dynamic Model of Tractor-Semitrailer

The tractor and the trailer are independent of each other, and there is a mutual motion relationship through the coupling of the hitch pin in the fifth-wheel, which constitutes a very complex multi-degree-of-freedom space motion system. During the analysis on the dynamics of tractor-semitrailer, the mathematical model is simplified as a monorail dual-centroid model. The model includes longitudinal motion, lateral motion, and yaw motion, adding a steering system model. The model is shown in Fig. 1 and Table 1.

For the tractor, its differential equations of motion are:

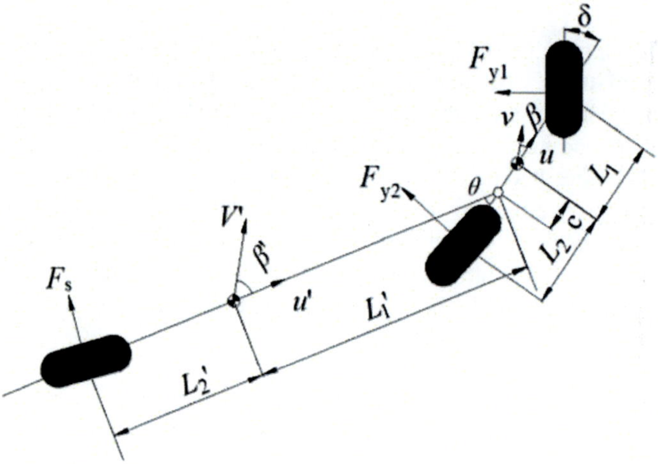

Fig. 1 Monorail double center of mass model of tractor-semitrailer

Table 1 Model parameters of handling stability of tractor-semitrailer

Symbols	Physical meanings	Reference value	Symbols	Physical meanings	Reference value
M	Mass of tractor	8800 kg	a_s	Sideslip angle of semitrailer rear wheel	–
m	Mass of semitrailer	30,000 kg	ω	Yaw velocity of tractor	–
I_z	Moment of inertia around tractor centroid	46,120 kg m^2	ω'	Yaw velocity of semitrailer	–
I'_z	Moment of inertia around semitrailer centroid	21,2500 kg m^2	δ_f	Front-wheel angle	–
L_1	Distance between tractor centroid and tractor front axle	2.062 m	θ	Articulated angle	–
L_2	Distance between tractor centroid and tractor rear axle	2.728 m	β	Sideslip angle of tractor	–
L'_1	Distance between semitrailer centroid and hitch point	7.675 m	β'	Sideslip angle of semitrailer	–
L'_2	Distance between semitrailer centroid and semitrailer rear axle	3.985 m	α_1	Sideslip angle of tractor front wheel	–
c	Distance between tractor centroid and hitch point	2.215 m	α_2	Sideslip angle of tractor rear wheel	–
c_1	Cornering stiffness of tractor front wheel	382 kN/rad	F_{y1}	Cornering force of tractor front wheel	–
c_2	Cornering stiffness of tractor rear wheel	733 kN/rad	F_{y2}	Cornering force of tractor rear wheel	–
c_s	Cornering stiffness of semitrailer rear wheel	884 kN/rad	F_s	Cornering force of semitrailer rear wheel	–
u	Longitudinal velocity of tractor	–	F_p	Component of hitch point lateral force in tractor coordinate system	–

(continued)

Table 1 (continued)

Symbols	Physical meanings	Reference value	Symbols	Physical meanings	Reference value
u'	Longitudinal velocity of semitrailer	–	F_p'	Component of hitch point lateral force in semitrailer coordinate system	–
v	Lateral velocity of tractor	–	C_p	Damping coefficient of fifth wheel	–
v'	Lateral velocity of semitrailer	–	M_p	Damping moment of fifth wheel	–

$$\begin{cases} Mu(\dot{\theta}+\dot{\beta}) = F_{y1}+F_{y2}+F_p \\ I_z\ddot{\theta} = -F_{y1}L_1 - F_{y2}L_2 - F_p \end{cases} \tag{1}$$

For semitrailer:

$$\begin{cases} mu'(\dot{\theta}+\dot{\beta}) = F_s - F_p' \\ I_z'\ddot{\theta} = -F_sL_2' - F_p'L_1 \end{cases} \tag{2}$$

The velocity relationship of the hinge points is shown in Fig. 2.

$$\begin{cases} v' \cos\beta' = u\cos\beta\cos\theta - (v\sin\beta - c\omega)\sin\theta \\ v' \sin\beta' + L_2\omega' = v\cos\beta\sin\theta + (v\sin\beta - c\omega)\cos\theta \end{cases} \tag{3}$$

The angle between the longitudinal axis of the tractor and the semitrailer is

$$\theta = \theta_o + \int_0^t (\omega - \omega')\mathrm{d}t \tag{4}$$

Obtain:

$$\dot{\theta} = \omega - \omega' \tag{5}$$

The linear tire model is adopted, and the wheel cornering force can be expressed as

$$F_{y1} = -c_1\alpha_1 \tag{6}$$

$$F_{y2} = -c_2\alpha_2 \tag{7}$$

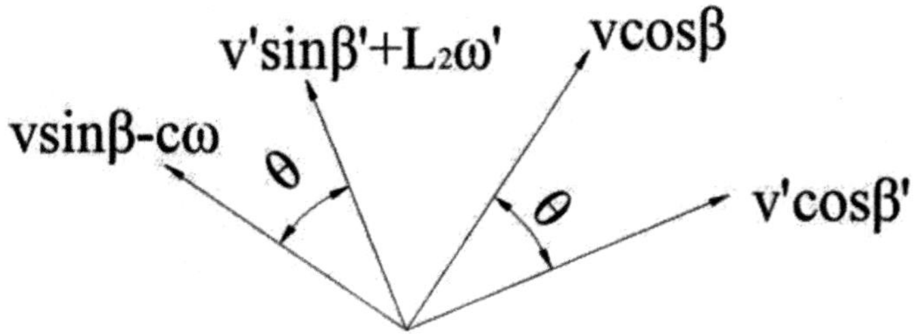

Fig. 2 The velocity relationship of the hinge points

$$F_s = -c_s \alpha_s \tag{8}$$

The slip angle of each wheel is

$$\alpha_1 = \frac{v + L_1 \omega}{u} - \delta_f \tag{9}$$

$$\alpha_2 = \frac{v - L_2 \omega}{u} \tag{10}$$

$$\alpha_S = \frac{v - L_2' \omega'}{u'} \tag{11}$$

Then

$$F_{y1} = c_1 \left(\delta_f - \beta - L_1 \omega / u \right) \tag{12}$$

$$F_{y2} = c_2 (-\beta + L_2 \omega / u) \tag{13}$$

$$F_s = c_s \left(L_2' \omega' / u - \beta' \right) \tag{14}$$

The cornering force of the rear wheel of a semitrailer is expressed as

$$F_s = c_s \left[-\theta + \frac{c\omega - u\beta + \omega' \left(L_1' + L_2' \right)}{u} \right] \tag{15}$$

Since $u' \approx u$, $\sin \beta' \approx \beta'$, and comparing formula 12 and 13 to obtain:

$$\beta' = \beta + \theta - \omega c / u - \omega' L_1' / u \tag{16}$$

The relative rotational resistance moment M_p is

$$M_p = C_p\dot{\theta} \tag{17}$$

In order to analyze the transient characteristics of tractor-semitrailer, the following spatial state matrix is constructed:

$$A\frac{\mathrm{d}X}{\mathrm{d}t} + BX = C\delta_f \tag{18}$$

In the formula:

$$X = [\beta, \omega, \omega', \theta]^{\mathrm{T}}$$

$$A = \begin{bmatrix} (M+m)u & -mc & -mL_1' & mu \\ muc & I_z & 0 & C_p \\ mL_1'u & 0 & I_z & -C_p \\ 0 & 0 & 0 & 1 \end{bmatrix}$$

$$B = \begin{bmatrix} c_1 + c_2 + c_s & Mu + (c_1L_1 - c_2L_2 - c_sc)/u & mu - c_s(L_1' + L_2)/u & c_s \\ c_1(L_1 + c) - c_2(L_2 - c) & Muc + \left[(L_1^2 + L_1c)c_1 + (L_2^2 - L_2c)c_2\right]/u & 0 & 0 \\ (c_1 + c_2)L_1' - c_sL_2' & MuL_1' + \left[c_1L_1L_1' - c_2L_1'L_2 + L_2'cc_s\right]/u & c_s(L_1'L_2' + L_2'^2) & -c_sL_2' \\ 0 & -1 & 1 & 0 \end{bmatrix}$$

$$C = [c_1 \quad c_1(L_1 + c) \quad c_1L_1' \quad 0]^{\mathrm{T}}$$

3 Motion Characteristics Transient Analysis and Fifth-Wheel Characteristics Analysis

3.1 Time Domain Characteristics Analysis

According to the standard ISO 14791, a single lane-change test is carried out. The steering wheel angle input of the tractor is a sine function with a period of 4 s and

Fig. 3 Front-wheel angle

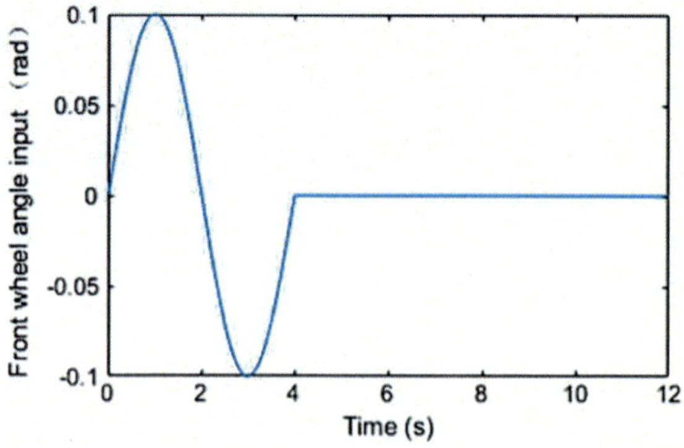

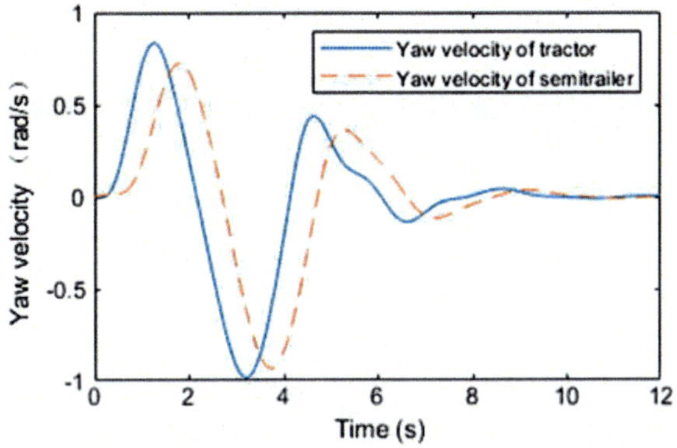

Fig. 4 Yaw velocity

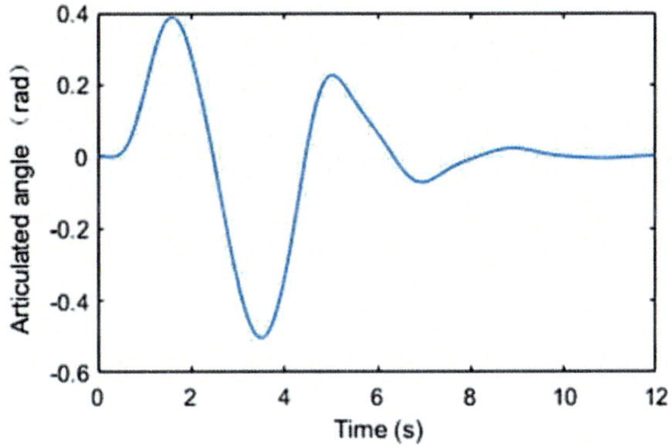

Fig. 5 Articulated angle

an amplitude of 0.1 rad, as shown in Fig. 3. The vehicle longitudinal speed u is 20 m/s. Other parameters are shown in Table 1. The transient response of the system is obtained through simulation. Figures 4 and 5 are the response curves of the yaw velocity and articulated angle of the tractor and semitrailer, respectively.

It can be seen from Fig. 4 that there is a phase difference between the yaw velocity response curve of the tractor and the yaw velocity response curve of the semitrailer, but the motion characteristics of the two are roughly the same. The yaw velocity of the tractor reached a maximum of 0.985 rad/s at about 3.2 s, and the yaw velocity of the semitrailer reached a maximum of 0.932 rad/s at about 3.8 s. The motion response of the semitrailer lags the tractor by about 0.5 s, and the yaw amplitude of the semitrailer is 5.4% lower than that of the tractor. The system stabilized in about 10 s.

The other responses of the system are similarly expressed in other literature [5, 6], and this article will not discuss them further.

3.2 Study on Fifth-Wheel Parameter Matching

The vehicle longitudinal speed *u* is 20 m/s, and 4 kinds of rotational resistance moments M_p are applied. The first time is 1 N m/rad, and the magnification is 100 times each time, that is, 1 N m/rad, 100 N m/rad, 10,000 N m/rad, 1,000,000 N m/rad, through calculation and analysis, the articulation angle change curve (Fig. 6) and the maximum value of the articulation angle θ_{max} (Table 2) are obtained. It can be seen from Fig. 6 that with the increase of M_p, the amplitude of θ_{max} becomes smaller, which indicates that the increase of fifth-wheel damping can slow down the yaw motion of tractor-semitrailer. But the value does not change much.

The M_p value increased from 1 N m/rad to 100 N m/rad, a 99 times increase, but θ_{max} only decreased by 0.04%, and the two curves basically coincided. The M_p value increased from 100 N m/rad to 10,000 N m/rad, and θ_{max} decreased by only 3.41%. The M_p value increased from 10,000 N m/rad to 1,000,000 N m/rad, and θ_{max} decreased significantly, a decrease of 81.93%.

In the TruckSim simulation software, set the damping characteristic of the fifth-wheel to "Linear coefficient", as shown in Fig. 7, and set the M_p values as above, and perform simulation calculations to obtain the hinge angle change curve

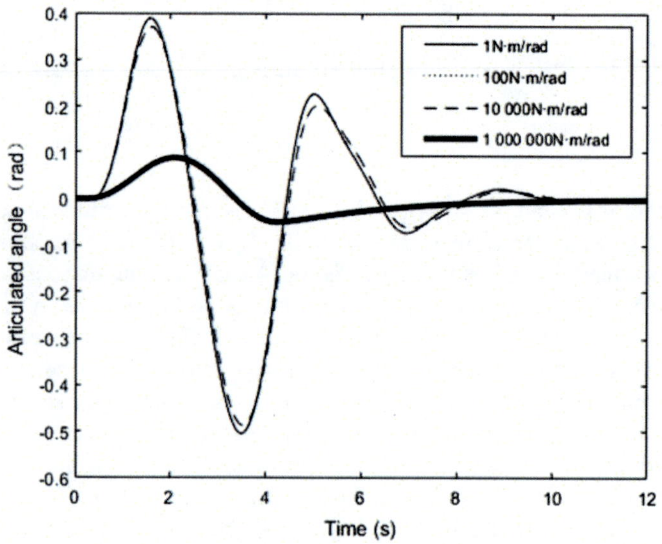

Fig. 6 Calculated change curve of articulation angle

Table 2 Calculated the maximum value of the articulation angle

M_p (N m/rad)	1	100	10,000	1,000,000
θ_{max} (rad/s)	0.5044	0.5042	0.4870	0.0880

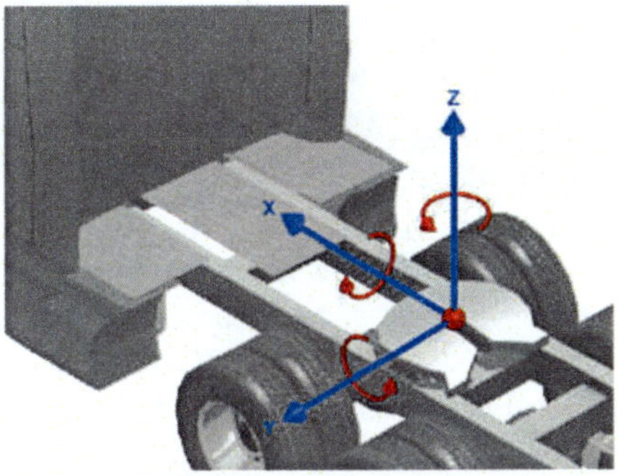

a) Tractor fifth-wheel coordinate settings

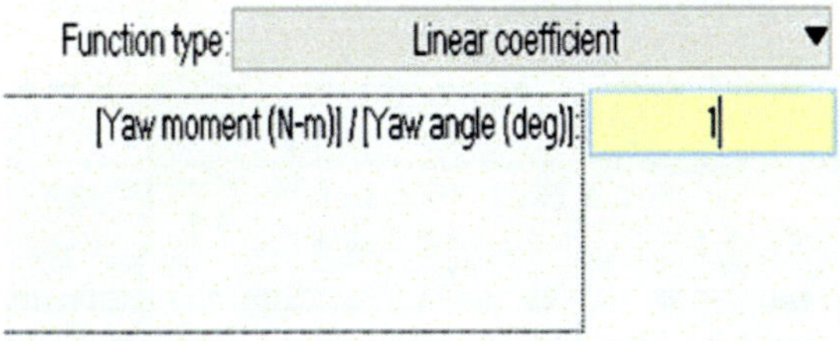

b) Damping characteristic of the fifth-wheel

Fig. 7 Fifth-wheel parameter setting

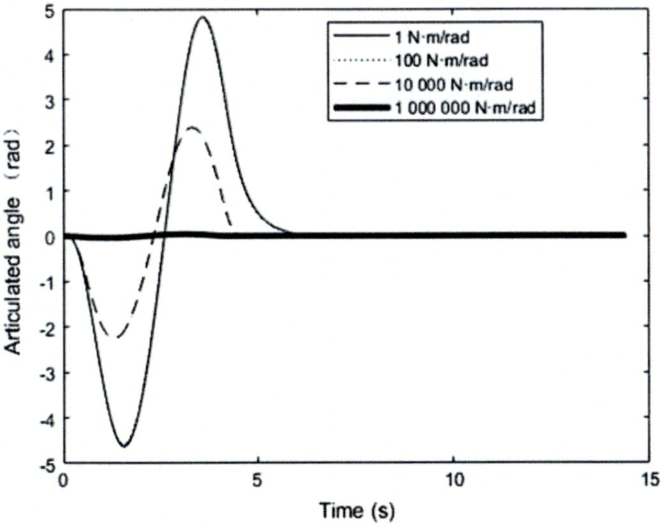

Fig. 8 Change curve of articulation angle in TruckSim

Table 3 The maximum value of the articulation angle in TruckSim

M_p (N m/rad)	1	100	10,000	1,000,000
θ_{max} (rad/s)	4.8187	4.7789	2.3887	0.0420

(Fig. 8) and the maximum value of the hinge angle θ_{max} (Table 3). The curve change trend is similar to Fig. 6, that is, when the M_p value is small, the increase of the M_p value has little effect on θ_{max}. When the M_p value is large, the change of the M_p value causes the θ_{max} value to decrease significantly. The test results show that the application of proper damping moment at the fifth-wheel can reduce the relative steering movement of the tractor and semitrailer. However, during the test, it was also found that while the yaw motion was attenuated, the steering flexibility of the tractor-semitrailer would deteriorate.

4 Conclusion

A mathematical simulation model and a TruckSim simulation model are established, and the influence of the fifth-wheel damping parameters on the yaw motion of tractor-semitrailer is analyzed. Results show that the motion of the semitrailer has a lag with respect to the tractor. The increase of the fifth-wheel damping coefficient can improve the lateral stability of the system. When the fifth-wheel damping coefficient is small, the increase of its value has little effect on system stability. Only when the damping coefficient is large, the increase of the damping coefficient can obviously reduce the articulation angle, but a deeper parameter sensitivity analysis is needed. At the same time as the yaw motion is attenuated, the steering flexibility of the tractor-semitrailer will be deteriorated.

Acknowledgements This work is sponsored by the National Key Research and Development Project "Research on Logistics Basic Modulus and Application Standards" (No. 2017YFF0208701).

References

1. Winkler CB, Fancher PS, Macadam CC (1983) Parametric analysis of heavy duty truck dynamic stability. Trans. Res. Inst., University of Mich
2. Bareket Z, Fancher P (1991) Truck or bus dynamic modeling for a driving simulator. Trans. Res. Inst., University of Mich
3. Van Zanten AT, Rainer E, Georg P (1995) VDC, the vehicle dynamics control system of Bosch. SAE paper 950759, pp 1419–1436

4. Liu H-f (2001) The experiment study on the simulation of transverse sway of automobile-trailer in a high speed. Jilin University
5. Hong-guo X, Hong-fei L, Zeng-liang. Y (2006) Overview of tractor-trailer lateral stability study. J Highw Transp Res Dev 23(2):141–150
6. Chen C, Tomizuka M (2000) Lateral control of commercial heavy vehicles. Veh Syst Dyn 33(2):391–420

A Method of Multi-Information Perception Collection for Power Equipment Disaster Damage

Min Xu, Jianhua Yang, Jian Yang, Lin Peng, Shiyang Tang, Nige Li, Zhansheng Hou, Zhimin He, Gang Wang, He Wang, Xingchuan Bao, Hai Yu, Liang Zhu, Zehao Zhang, Jing Li, Tianxiong Gu, Yang Yang, and Dailiang Ye

Abstract In this paper, aiming at the problem that the on-site rescue personnel can not quickly prepare to perceive the operation damage of the equipment in the disaster damage environment, the invention integrates the machine vision intelligent recognition and wireless sensor acquisition technology and innovatively designs the on-site environment and the equipment perception acquisition method. Firstly, the multi-band point cloud environment map is created by using machine vision method, and the damage state of power equipment is identified from the perspective of emergency disposal to realize the accurate perception and information association of the site environment; secondly, the power field equipment identification based on mobile terminal is studied to realize the identification of the site power grid equipment and related components; finally, the wireless communication modules such as ZigBee are used To realize the fast networking and data acquisition of power equipment sensors. By using augmented reality technology, the real-time operation monitoring data collected by the equipment and the physical equipment are integrated and displayed, and emergency workers can quickly obtain the operation status of the corresponding parts and assist the on-site workers to more accurately and comprehensively view and judge the operation of the whole equipment and parts.

Keywords ZigBee · Cloud environment map · Machine vision

M. Xu (✉) · L. Peng · S. Tang · Z. Hou · Z. He · G. Wang · H. Wang · X. Bao · H. Yu · L. Zhu · Z. Zhang
Global Energy Interconnection Research Institute, Nanjing 210003, Jiangsu, China

M. Xu · N. Li
State Grid Key Laboratory of Information & Network Security, Nanjing 210003, China

J. Yang · J. Yang · J. Li · T. Gu · Y. Yang · D. Ye
State Grid ZheJiang Electric Power Co., Ltd., Hangzhou 310000, Zhejiang, China

Z. Yu et al. (eds.), *Advancements in Mechatronics and Intelligent Robotics*,
Advances in Intelligent Systems and Computing 1220,
https://doi.org/10.1007/978-981-16-1843-7_32

1 Introduction

Recent years, a variety of natural disasters occur frequently and have a growing trend. Disasters such as typhoon, heavy rainfall, geological earthquake, rain, and snow freezing have brought huge losses to the power grid, seriously affecting the safe and stable operation of the power grid and the normal production and operation of the company.

Since 2008, State Grid Corporation of China has started to build the emergency command information system, which has played an active role in the emergency response of the company in recent years. The system focuses on the management of emergency plan, emergency team and other emergency information, and provides insufficient support in the field of information collection, information exchange, information fusion, and other emergency command and auxiliary decision-making. The efficiency of emergency field information collection is not high. The traditional method of disaster damage information collection mainly relies on manual field investigation and reporting. When major emergencies come, there are problems such as difficult data acquisition, slow acquisition, large limitations, and few data sources, which affect the scientific decision-making of emergency response. It is unable to collect the damaged data of the site equipment quickly and comprehensively. The efficiency of data collection needs to be improved.

2 Multi-Information Perception Acquisition Technology of Power Equipment Disaster Damage

In this paper, a set of effective methods is proposed to deal with the above problems, integrating machine vision intelligent recognition and wireless sensor acquisition technology, and innovatively designing the field environment and equipment perception acquisition methods. Firstly, the multi-band point cloud environment map is created by using machine vision method, and the damage state of power equipment is identified from the perspective of emergency disposal to realize the accurate perception and information association of the site environment; secondly, the power field equipment identification based on mobile terminal is studied to realize the identification of the site power grid equipment and related components; finally, the wireless communication modules such as ZigBee are used to realize the fast networking and data acquisition of power equipment sensors. By using augmented reality technology, the real-time operation monitoring data collected by the equipment and the physical equipment are integrated and displayed, so that the on-site emergency workers can more intuitively understand the damage of the equipment, quickly obtain the operation status of the corresponding parts, and assist the on-site workers to more accurately and comprehensively view and judge the operation of the whole equipment and parts. Improve the efficiency of on-site emergency repair of large power equipment and the recovery speed of on-site equipment failure. It includes the following steps:

Step 1: The flow is as shown in Fig. 1. First, rgb-d sensor and multi-band sensor (infrared detector and ultraviolet detector) are used to obtain the visual point cloud information of indoor environment in real time, and the binary form of orb feature operator is used to extract the features of each subspace of RGB image; then the image features are described as binary form of vision according to the characteristics of orb operator. Words are stored in a tree structure model to build a binary visual dictionary of spatial information fusion. The closed-loop detection of the two constraints of time continuity and geometric consistency is carried out to determine whether the closed-loop conditions are met [1]. If the closed-loop conditions are not met, the room is obtained by combining depth information and 3D spatial mapping. For the three-dimensional point cloud of the internal environment, the RANSAC algorithm and ICP algorithm are used to splice the point cloud to the three-dimensional map; if the closed-loop conditions are met, the built three-dimensional map is optimized; finally, the intelligent point cloud identification of the on-site power equipment is realized, and the background operation data of the equipment are quickly retrieved after obtaining the equipment number [2]. Field operators need to use the intelligent identification function to identify power grid equipment and related components, so as to lay the foundation for data acquisition.

Step 2: Some physical quantities or states in power equipment can be obtained by visual method, and the tested object can be analyzed by image processing technology, so that the physical quantities or states of power equipment can be measured or recognized, and abnormal phenomena and potential faults can be

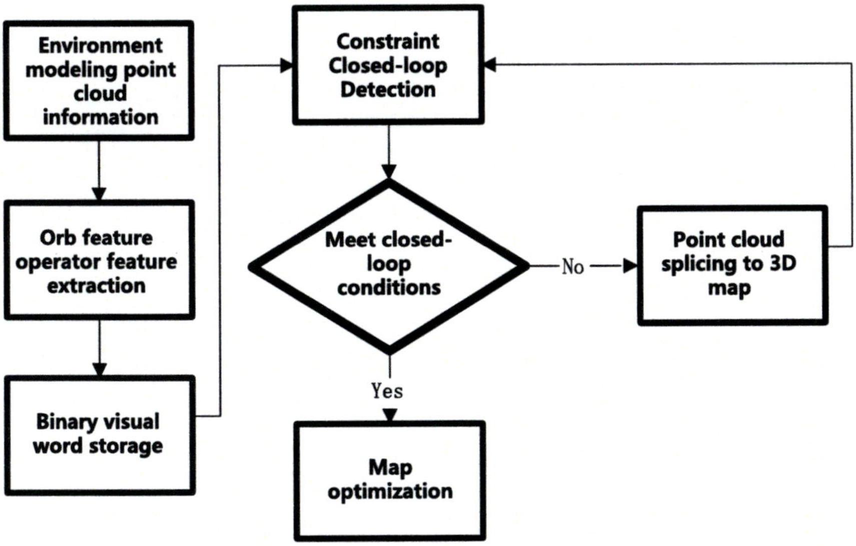

Fig. 1 Environment scanning modeling process

found in time. At the same time, the changes of subtle images that are difficult to be distinguished by human eyes can be found by multi-band image technology to realize equipment fault diagnosis and state early warning [3].

As shown in Fig. 2, the recognition process of electrical equipment with multi-band image information is as follows: (1) image preprocessing: the image acquisition process and transmission process cause noise and other factors that are not conducive to image analysis, so the first step of image analysis is preprocessing to improve image quality. It mainly uses low-pass filter to remove image noise and improve image quality. (2) Image registration: image registration is not only required for electrical equipment from infrared image to visible image recognition. In the comparison between the patrol image and the historical database image, due to the difference of shooting angle and local area, in order to facilitate the later feature extraction, it is also necessary to register these deviated images. Based on the method of feature matching, SIFT algorithm can extract stable feature points, and deal with the matching problem in the case of translation, rotation, affine transformation and perspective transformation between two images. It has good robustness to the change of light and can match with a high probability. (3) Image feature extraction: image feature refers to the original characteristics or attributes of image field. Some of them are natural features directly felt by the image [4], and some of them are artificial features that can be obtained by transformation or measurement. In patrol inspection, the color, shape, and texture of the image can be regarded as natural features, while the gray, histogram, and infrared temperature difference can be regarded as human features.

Step 3: After identifying the specific equipment, the field terminal translates the identified equipment into the grid equipment code and uses the satellite transmission module to send the data to the background server. The background server obtains the real-time operation data and historical operation data of the equipment

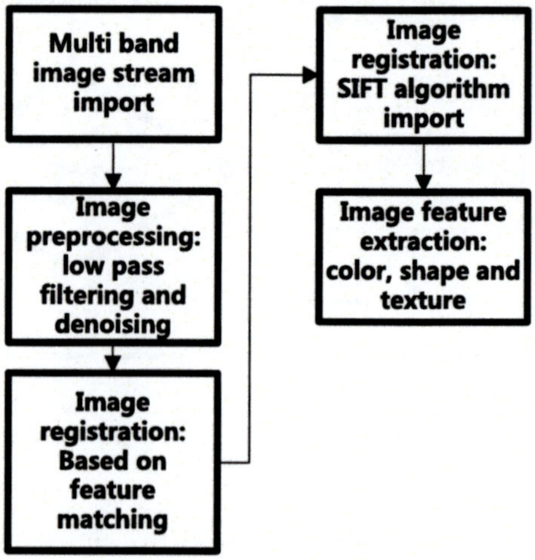

Fig. 2 Identification process of electrical equipment

through the integrated grid online monitoring system and sends them to the field terminal equipment.

Step 4: Use ZigBee and other wireless communication modules to realize the rapid networking and data acquisition of power equipment sensors. The field emergency equipment is equipped with FPGA based acquisition terminal, ZigBee transmission network and upper computer of data center, which can communicate with the field power equipment wireless sensor network. ZigBee has three standards: ZigBee coordinator, ZigBee router and ZigBee terminal equipment. The coordinator is responsible for initialization, maintenance and control of the network; the router is responsible for data collection, relay messages and provide routing information; the terminal node is responsible for data collection. Each network must have only one coordinator, the coordinator and router must be FFD, and the terminal node can be FFD or RFD. ZigBee standard supports star, tree and mesh network topologies. ZigBee Tree Network is the most commonly used topology type. In this topology, the coordinator initializes the network, the router forms the network branch and relays messages and the terminal node as the leaf node does not participate in message routing. The acquisition terminal is responsible for collecting and preprocessing the power information of the substation switch cabinet, and sending the data to the upper computer of the data center by using ZigBee's multi-hop technology. The data center completes the analysis and processing of the data, displays the final processing results in the form of graphs and reports, and stores the data according to the needs, so as to realize the rapid networking and data collection of field equipment sensors [5].

Step 5: The real-time and historical operation information of the identified equipment is displayed and integrated with the physical scene of the equipment by using the augmented reality technology, so that the field operators can obtain the detailed real-time operation data and damage situation of the physical equipment in the way of augmented reality and assist the on-site emergency personnel to quickly understand the disaster situation of the power equipment and carry out rescue and repair. First of all, the collection object, real-time operation data, historical operation, and maintenance information of the power equipment operation and maintenance monitoring system are extracted. Based on the individual characteristics of the extracted monitoring information, the field monitoring data are divided into operation data and equipment status data. Secondly, the acquisition data are accurately registered, and the monitoring data of the field power equipment and the logical connection between the equipment components are established. Combined with the historical data, the bottom equipment monitoring data are de-noising to avoid the interference of random disturbance to the sensing data as much as possible. Finally, the real-time operation data after the fusion processing are classified and fused with the coordinates of the on-site physical equipment components, which are superimposed on the relevant parts of the on-site power equipment, so as to realize the combination of the panoramic equipment components and their corresponding monitoring data, so that the on-site operators can more intuitively understand the internal and external structure of the equipment and the overall operation status of the equipment and components.

3 Conclusions

In view of the low efficiency of information collection in the emergency scene, the traditional method of disaster damage information collection mainly relies on the report after the artificial field investigation, and there are many problems such as difficult data acquisition, slow acquisition, large limitations, and few data sources when the major emergency occurs, and this paper puts forward an effective method for processing, integrating the machine vision intelligent identification and wireless sensor acquisition technology and innovatively designs. Taking into account the field environment and equipment perception acquisition method, firstly, the multi-band point cloud environment map is created by using machine vision method, and the damage state of power equipment is identified from the perspective of emergency disposal to realize the accurate perception and information association of the site environment; secondly, the power field equipment identification based on mobile terminal is studied to realize the identification of the site power grid equipment and related components; finally, the wireless communication modules such as ZigBee are used to realize the fast networking and data acquisition of power equipment sensors. By using augmented reality technology, the real-time operation monitoring data collected by the equipment and the physical equipment are integrated and displayed, so that the on-site emergency workers can more intuitively understand the damage of the equipment, quickly obtain the operation status of the corresponding parts, and assist the on-site workers to more accurately and comprehensively view and judge the operation of the whole equipment and parts. Improve the efficiency of on-site emergency repair of large power equipment and the recovery speed of on-site equipment failure.

Acknowledgements This work was financially supported by the science and technology project to State Grid Corporation "Research on the key technologies of power grid disaster intelligent perception and emergency command (5700-202019185A-0-0-00)."

References

1. Galvez-López D, Tardos JD (2012) Bags of binary words for fast place recognition in image sequences. IEEE Trans Rob 28(5):1188–1197
2. Angeli A, Filliat D, Doncieux S et al (2008) A fast and incremental method for loop-closure detection using bags of visual words. IEEE Trans Rob 24(5):1027–1037
3. Cummins M, Newman P (2007) Probabilistic appearance based navigation and loop closing. In: IEEE international conference on robotics and automation. IEEE, pp 2042–2048
4. Cummins M, Newman P (2009) Highly scalable appearance-only SLAM-FAB-MAP 2.0. Rob Sci Syst. https://doi.org/10.15607/rss.2009.v.039
5. Mouragnon E, Lhuillier M, Dhome M et al (2006) Real time localization and 3D reconstruction. In: IEEE computer society conference on computer vision and pattern recognition. IEEE computer society, pp 363–370

An Algorithm for Calculating the Contribution of Acoustic Features in Speaker Recognition

Yu Quan Qu, Hua Long, and Ying Duan

Abstract In order to solve the problem that the model performance will be different if different acoustic features are selected in the speaker recognition field, an algorithm for calculating the contribution degree of acoustic features in speaker recognition is proposed. First of all, the use of increase or decrease in weight method to calculate contribution of each dimension, and then, with a fisher ratio than the acoustic features of each dimension fisher ratio calculation, to undertake the corresponding percentage of fusion after can get the characteristics in the contribution of speaker recognition model. Taking the acoustic feature GFCC and the speaker model *i*-vector as an example, this paper calculates the contribution rate of each component of GFCC under the speaker recognition by using the proposed new algorithm, analyzes the differences of the three methods, and provides a new method for subsequent feature processing.

Keywords Contribution · Speaker recognition · Increase or decrease component method

1 Speaker Recognition

Speaker recognition (SR), also known as voiceprint recognition, is a biometric technique. This technology is mainly used to automatically identify the speaker's identity by the feature information contained in the human voice [1]. It is an important thing to select appropriate features in the speaker recognition field. The quality of feature representation ability is related to the performance of the whole model. Of course, it is necessary to choose an appropriate dimension. In the

Y. Q. Qu · H. Long (✉) · Y. Duan
Kunming University of Science and Technology, Kunming 650000, Yunnan Province, China

H. Long
Yunnan State Key Laboratory of Computer Science, Kunming 650000, Yunnan Province, China

Z. Yu et al. (eds.), *Advancements in Mechatronics and Intelligent Robotics*, Advances in Intelligent Systems and Computing 1220,
https://doi.org/10.1007/978-981-16-1843-7_33

dimensions of acoustic features, some dimensions are positive for speaker recognition and some are indeed side effects, all of which are important to implement an algorithm to calculate the contribution of acoustic features.

2 Evaluation of the Relative Importance of SR

Contribution [2] is a measure of the relative importance of feature dimensions in the system. There are two methods for evaluating the contribution of acoustic features to speaker recognition.

2.1 *Increase or Decrease Component Method*

The idea of the increase or decrease component method (IDCM) [3] is to combine several adjacent components of acoustic features and then carry out speaker recognition experiments to obtain the relative importance of each component.

$$Q(i) = \frac{1}{n}\left((\sum_{j>i} (P(i,j) - P(i+1,j)) + \sum_{j<i} (P(j,i) - P(j,i-1)) \right) \quad (1)$$

$P(*)$ function stands for any performance indicator that measures the speaker model, and here we use equal error rate.

2.2 *Fisher Ratio*

Fisher ratio (FR) is to transform high-dimensional eigenvectors into low-dimensional eigenvectors, that is, to ensure that the inter-class distance between the speaker feature parameters is as small as possible and the inter-class distance is as large as possible.

$$D^i_{\text{fisher}} = \frac{\sigma^i_{\text{between}}}{\sigma^i_{\text{within}}} \quad (2)$$

D_{fisher} is the Fisher ratio of each component of the characteristic parameter, σ_{between} is the inter-class dispersion, the mean of variances of the components of different characteristic parameters, and σ_{within} is the dispersion within the class, the mean of the variance of the components of the same characteristic parameter.

$$\sigma_{\text{between}} = \sum_{i=1}^{M} (\mu_k^{(i)} - \mu_k^{(i)})^2 \tag{3}$$

$$\sigma_{\text{within}} = \sum_{i=1}^{M} \left[\frac{1}{n} \sum_{j=1}^{n_i} (x_k^{(j)} - \mu_k^{(i)})^2 \right] \tag{4}$$

Assume that M represents the total number of speech samples, n_i represents the number of speech samples of one segment of speech signal s, and $\mu_k^{(i)}$ is the mean of the component of the characteristic parameter of the kth dimension of speaker i. The average value of the kth dimension eigenparameter component of all speakers is expressed as μ_k. The characteristic parameter component of the kth dimension of speaker j is represented as $x_k^{(j)}$.

2.3 *Merge IDCM and FR (MIF)*

The relative importance of each feature component is obtained by IDCM through the identification experiment of the combination of several adjacent components of the feature, but most of the components of the acoustic feature are interdependent with the components of the accessory. Fisher ratio, which is commonly used in statistics, calculates the ratio of intra-class distance and inter-class distance of each feature component to illustrate the distinguishing ability of each feature component to the target. This paper proposes a new method to calculate the acoustic characteristic component by merge IDCM and Fisher ratio (MIF). The final importance of imp is calculated:

$$\text{imp}_i = \lambda \times Q(i) + (1 - \lambda) \times D_{\text{fisher}}^i \tag{5}$$

λ is constant. Set the threshold value Ω (there $N = 25$) which is used to select the optimal characteristic component:

$$\Omega = \frac{1}{N} \sum_{i=1}^{N} \text{imp}_i \tag{6}$$

3 Experiment

Based on the MATLAB 2019a experimental environment, the corpus was LibriSpeech, with a total of 317 people. The corpus is divided into a register set, a test set, and a training set. The registration set was 100 speakers, and the registration time was about 2 min. The test set corresponds to the registered set speaker, and the test duration is about 10 s. The remaining speakers are training sets. Feature preprocessing is set as: frame length 25 ms, frame shift 10 ms, and preweighting coefficient 0.97. The *i*-vector [4] speaker model is set to: The total variability space matrix dimension is 400, and the LDA dimension is 200.

So, let us do the first step: Calculate the IDCM of each component. The EER obtained from all sequence combinations of GFCC 1st to 25th was calculated, respectively, as shown in Table 1. According to the calculation method given in Formula (1), the $Q(i)$ of each component is calculated, as shown in Fig. 2.

We proceed to the second operation to calculate the Fisher ratio value (it should be noted here that too small FR value is not conducive to the judgment of the algorithm, so we will scale up here) according to Formulae (2–3).The result is shown in Fig. 3. Finally, we calculate the results of Figs. 2 and 3 according to Formula (5), and then the contribution degree of the component is reached. If this is just a calculation for the next step, then calculate the threshold according to Formula (6) to extract the components you need, and the optimal dimension combination is selected. For acoustic characteristics such as GFCC transformed by discrete cosine transform, we suggest $\lambda = 0.5$.

From the results in Fig. 1, we set the threshold, and the optimal GFCC component is concentrated in the range between 2nd and 18th. From the results of Fig. 2, we also set thresholds artificially based on the FR value, and the optimal GFCC component is concentrated in the range from 2nd to 22nd. What is worse than IDCM, FR is that some components in the optimal range have very small FR values. If we adopt the method of feature bending and discontinuously extract the optimal features, this will cause the loss of continuous inherent properties of speech. It interferes with the content originally expressed by the voice and further interferes with the recognition of the speaker system. It can be seen from Fig. 3 that the optimal interval of GFCC exists between the 2nd and 20th, and the MIF distinguishes the difference between the advantages and disadvantages to the greatest extent, which makes it easier to find the interval you want to choose (Fig. 4).

4 Conclusion

In this paper, a new method for calculating the importance of acoustic characteristic components is proposed by combining the method of increasing and decreasing components with the Fisher ratio. Experiments show that the new method is more accurate in calculating the importance of acoustic features. This lays a foundation for the subsequent work in the speaker recognition field, such as feature parameter selection and feature fusion.

Table 1 GFCC used IDCM to calculate each component EER

EER	1	2	3	4	5	6	7	8	9	10	11	12	13	14	15	16	17	18	19	20	21	22	23	24	25
1	39	12	8.5	5.3	5.0	4.1	3.7	2.9	2.3	2.0	1.6	1.4	1.1	1.6	1.7	1.9	1.4	1.8	2.4	2.6	2.8	3.2	4.4	4.5	5.9
2		33	12	6.4	5.5	3.8	3.5	4.0	2.1	2.1	2.0	1.7	1.2	1.0	1.2	1.0	1.0	1.2	1.1	1.2	1.1	1.9	2.5	2.7	3.0
3			28	14	13	6.3	4.3	3.8	4.6	2.7	1.5	1.1	0.8	1.3	0.9	0.7	1.4	1.4	1.6	1.5	1.5	1.9	2.5	2.8	3.4
4				27	14	8.6	6.5	5.9	5.3	3.2	2.2	1.2	1.1	1.6	1.6	1.2	2.2	2.1	1.2	2.2	1.4	2.4	2.9	2.5	3.9
5					37	15	9.7	6.2	7.2	4.1	3.2	2.7	1.9	1.6	1.2	1.2	1.0	1.2	1.6	2.6	2.9	3.0	2.9	2.9	4.4
6						37	14	8.8	9.0	5.7	3.7	3.4	3.1	2.0	2.2	1.2	1.3	1.5	1.3	1.4	1.8	2.0	2.1	2.5	2.4
7							28	16	13	10	6.9	4.9	4.4	3.1	4.3	2.0	1.3	2.1	2.5	2.3	2.2	2.9	3.7	3.4	3.5
8								27	16	9.9	7.4	6.0	4.2	3.9	4.1	3.1	1.5	2.2	2.1	1.7	2.5	2.9	3.0	3.1	3.5
9									28	17	10	7.5	5.9	5.4	5.4	4.0	3.2	3.1	2.3	2.7	2.9	3.2	2.7	2.7	3.0
10										30	14	9.9	6.5	5.8	6.9	3.5	4.0	3.9	3.2	1.7	1.6	3.1	3.5	3.7	3.9
11											26	14	9.2	7.2	7.0	5.1	4.4	3.1	3.6	2.7	2.5	2.9	2.7	2.5	3.4
12												35	12	8.1	8.0	5.9	6.4	5.1	5.9	4.8	3.4	3.0	2.4	2.5	2.5
13													37	13	11	9.6	6.2	5.7	6.3	3.8	4.1	3.3	4.1	3.1	3.4
14														37	14	11	7.2	5.8	5.9	4.1	4.3	5.0	4.4	3.1	3.0
15															33	16	10	6.9	7.1	5.2	5.1	5.2	5.2	4.1	4.1
16																39	16	12	9.2	7.6	10	8.1	7.1	6.7	5.2
17																	36	14	11	8.7	8.1	6.6	7.1	6.1	7.1
18																		39	12	13	8.6	8.2	9.5	8.0	7.8
19																			37	16	12	10	10	7.5	7.7
20																				33	17	13	12	8.0	8.4
21																					39	17	15	10	7.9
22																						38	19	14	13
23																							35	18	15
24																								39	17
25																									40

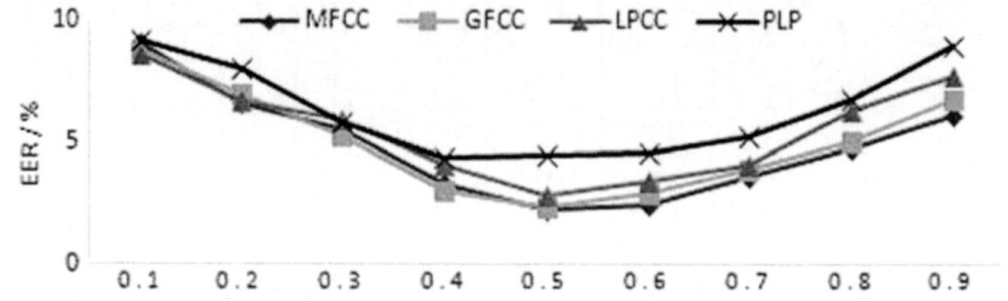

Fig. 1 Contribution of each dimension of GFCC calculated by IDCM

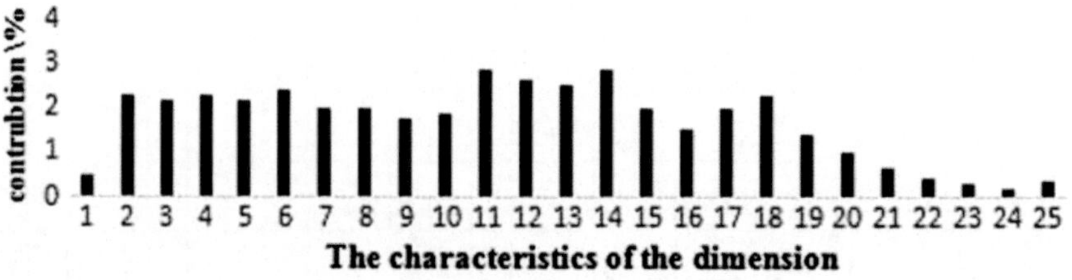

Fig. 2 Contribution of each dimension of GFCC calculated by IDCM

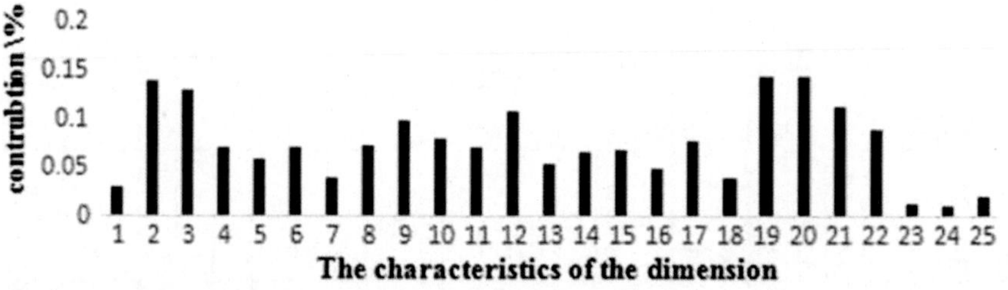

Fig. 3 Contribution of each dimension of GFCC calculated by Fisher ratio

Fig. 4 Contribution of each dimension of GFCC calculated by MIF

References

1. Liu J, Philip Chen CL, Li T, Zuo Y, He P (2019) An overview of speaker recognition. Trends Comput Sci Inf Technol 4(1):001–012
2. Hinton G, Deng L, Yu D et al (2012) Deep neural networks for acoustic modeling in speech recognition: the shared views of four research groups. IEEE Signal Process Mag 29(6):82–97
3. Zhen b, Wu X, Liu Z et al (2001) The relative importance of cepstrum components in speech recognition and speaker recognition. J Peking Univ Nat Sci Ed 37(3):371–378
4. Dehak N, Kenny PJ, Dehak R et al (2010) Front-end factor analysis for speaker verification. IEEE Trans Audio Speech Lang Process 19(4):788–798

Research on Multi-perception Data Analysis Model for Power Grid Emergency Services

Liang Zhu, Jianhua Yang, Jian Yang, and He Wang

Abstract The power grid disaster response system is an important part of the power grid, and the multi-perception data is an important basis for data perception and in-depth decision-making of the power grid disaster situation. Through the establishment of the multi-perception data analysis model, it can meet the needs of the grid emergency business to deal with various emergencies quickly, efficiently, accurately and flexibly.

Keywords Power emergency · Multiple perception · Data analysis

1 Introduction

In recent years, a variety of natural disasters occurs frequently and has a growing trend. Disasters such as typhoon, heavy rainfall, geological earthquake, rain and snow freezing have brought huge losses to the power grid, seriously affecting the safe and stable operation of the power grid and the normal production and operation of the company. Since 2008, State Grid Corporation of China has started to build the emergency command information system, which has played an active role in the emergency response of the company in recent years. The system focuses on the management of emergency plan, emergency team and other emergency information, and provides insufficient support in the field of information collection, information exchange, information fusion, situation analysis and other emergency command and auxiliary decision-making.

L. Zhu (✉) · H. Wang
Global Energy Interconnection Research Institute Co. Ltd, Nanjing, Jiangsu, China
e-mail: zhuliang19881003@126.com

L. Zhu · H. Wang
State Grid Key Laboratory of Information & Network Security, Nanjing, Jiangsu, China

J. Yang · J. Yang
State Grid Zhejiang Electric Power Co., Ltd., Hangzhou, Zhejiang, China

Z. Yu et al. (eds.), *Advancements in Mechatronics and Intelligent Robotics*,
Advances in Intelligent Systems and Computing 1220,
https://doi.org/10.1007/978-981-16-1843-7_34

The rapid collection of multi-element information of power equipment disaster emergency scene needs to consider the factors such as state, environment, task, safety regulation and so on to complete the task cooperatively, which involves the research of collaborative technology. Cooperative awareness mainly studies the social characteristics and behaviors of each agent, including task decomposition, task allocation, communication and interaction among agents, as well as negotiation and resolution of conflicts. It mainly studies the physical or logical distributed intelligent entities to achieve the purpose of problem solving through mutual cooperation and parallel operation [1].

In order to improve the performance of data collection and analysis in case of emergency, this paper studies the multi-element data collection technology for power grid emergency services, which provides strong support for the stable operation of smart grid.

2 Power Emergency Multi-information Collection

The technology of power emergency environment and equipment information acquisition, including machine vision intelligent identification acquisition and wireless sensor acquisition technology (Fig. 1). Firstly, using the method of machine vision to create indoor environment map, and from the perspective of emergency disposal to identify the damage state of power equipment, to achieve the accurate perception of the scene environment and information correlation. Secondly, the power situation research based on mobile terminal Field equipment identification, realize the identification of field power grid equipment and related components, and finally use ZigBee and other wireless communication modules to realize the rapid networking and data collection of power equipment sensors [2].

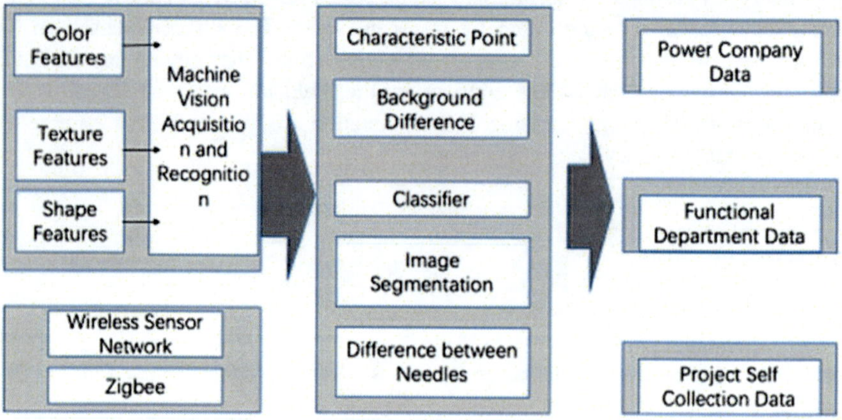

Fig. 1 Structure chart of power emergency multi-information collection

2.1 Machine Vision Intelligent Recognition Acquisition

Through the machine vision intelligent identification and acquisition technology, it mainly completes the identification of the overall damage of the equipment in the power facilities [3]. The research of identification algorithm will mainly be carried out from the perspective of emergency disposal demand; that is, the results of algorithm identification will be used to estimate the amount of resources needed to complete power equipment maintenance and follow-up research work, which will provide direct reference information for emergency decision-making.

2.2 Wireless Sensor Network Acquisition

In this project, Zigbee standard is used to realize wireless data acquisition and sharing between wireless sensors of power equipment. Zigbee standard is a wireless sensor network standard for short distance, low speed, low power consumption and low cost. Zigbee protocol adopts IEEE 802.15.4 standard in physical layer and link layer, and adds network layer, security module and application support sublayer module, so as to realize large area group [4]. According to its communication ability, the devices in IEEE802.15.4 network can be divided into two types: full device (FFD) and reduced device (RFD). FFDs have all the functions defined in IEEE802.15.4 and can play any role in the network, while RFD can only communicate with FFD because of its limited functions.

The field emergency equipment is equipped with FPGA-based acquisition terminal, Zigbee transmission network and upper computer of data center, which can communicate with the field power equipment wireless sensor network. Zigbee has three standards: Zigbee coordinator, Zigbee router and Zigbee terminal equipment. The coordinator is responsible for initialization, maintenance and control of the network; the router is responsible for data collection, relays messages and provides routing information; the terminal node is responsible for data collection. Each network must have only one coordinator, the coordinator and router must be FFD, and the terminal node can be FFD or RFD. Zigbee standard supports star, tree and mesh network topologies. Zigbee tree network is the most commonly used topology type. In this topology, the coordinator initializes the network, the router forms the network branch and relays messages, and the terminal node as the leaf node does not participate in message routing [5]. The acquisition terminal is responsible for collecting and preprocessing the power information of the substation switch cabinet and sending the data to the upper computer of the data center by using Zigbee's multi-hop technology. The data center completes the analysis and processing of the data, displays the final processing results in the form of graphs and reports, and stores the data as needed.

2.3 UAV Line Disaster Investigation and Identification

The UAV automatic detection system is used to identify the power disaster scene and realize the automatic identification and exploration of the disaster scene. UAV line exploration mainly involves UAV automatic detection system and UAV image recognition technology. It can quickly obtain the damage of large power facilities [6].

The UAV automatic detection is mainly used to detect the damage of power facilities and equipment from a long distance quickly, and identify the serious scenes such as UHV collapse and the overall submergence of substation quickly.

3 Multivariate Data Analysis Model

3.1 Multiple Emergency Data Fusion

The terminal-level integration of emergency data can be realized by using the integrated multi-sensor mobile terminal, which integrates 3G/4G module, GPS, Wi-Fi, Bluetooth, Zigbee, RFID, data processing module (including CPU and RAM), USB interface, UART interface, SD card slot and other modules [7].

1. The designed data fusion mobile terminal leaves a part of clear space for wireless communication of sensor module. Four modules, 3G/4G module, GPS module, Zigbee module and RFID module, are mobile modules, which can be removed from the mobile terminal when not needed to reduce signal interference.
2. The wired communication interface and wireless sensor are divided into upper and lower parts in the mobile terminal, especially the modules that are easy to produce interference, and a certain distance is reserved between them to reduce the interference between them.

The data collected by the terminal shall be time-stamped according to the time of data collection. The data processing system shall further classify and filter the data, and store the data in the terminal according to different categories for emergency analysis.

In addition, by summarizing and researching the standardized interface of electric power emergency data system, using the experience of standardization work related to emergency business data, emergency plan data, emergency data collection of energy industry, designing the data access interface of emergency disaster data system and the interface standard of pushing data to emergency data system, standardizing the emergency data of electric power industry from the public interface and data standard of information collection.

3.2 *Integrated Model of Intelligent Collection and Sharing of Public Emergency Information in Disaster-Damage Areas*

By studying the technology of intelligent collection and sharing of social public emergency information related to disaster-damage area, the integration of disaster-damage information is supported, and the integration with government emergency management big data platform is realized [8]. Through information exchange and resource sharing with emergency platforms of all regions and departments, it has the ability to connect the overall emergency plan with special plan, department plan, local plan and grassroots plan, vertically to the bottom, involving the country, province, city, county and grassroots level, horizontally to the edge, involving natural disasters, accident disasters, public health and social insurance, achieve the coordinated response between multiple departments and various localities.

The information acquisition model of disaster-damage area is different from data acquisition in general sense. It only pays attention to signal access and ignores automatic processing from signals to obtain processed information [9]. In addition, it needs to have preliminary information analysis and processing functions. In image monitoring, image monitoring system is used to automatically monitor the fire conditions of large space, outdoor large places and urban areas. When a fire occurs, the system will automatically give an alarm signal and automatically switch the scene image of the fire to the control center. Urban traffic image monitoring system is used to automatically analyze the license plate number at the key points in and out of the urban area. In addition, more and more data sensing networks and image monitoring networks are integrated organically. After abnormal pressure, temperature, concentration and other signals appear, the related images on site are immediately sent to the control center automatically for processing and confirmation by the personnel on duty; using satellite remote sensing image, we can realize precipitation analysis, typhoon analysis, fog monitoring, sandstorm information, water condition monitoring, sea ice monitoring, vegetation change, dry early monitoring, forest fire information, snow information, urban heat island, estuary sediment, land shortage analysis, etc., for the above analysis. Some can be done automatically, and some need more in-depth data processing and analysis [10]. In this project, the data distributed in the relevant departments of electric power emergency work will be exchanged through the data center, and then data fusion will be completed through the data exchange and sharing system; the warning and early warning data of government professional departments (such as relevant data of meteorological, land, forest and other government functional departments) will be accessed through the data access platform, and then through the data access platform data exchange and sharing system complete data exchange, audit and other work, and provide effective technical solutions for data integration of the system. System data is obtained from ECS system in a unified way to realize information interaction and linkage with ECS system.

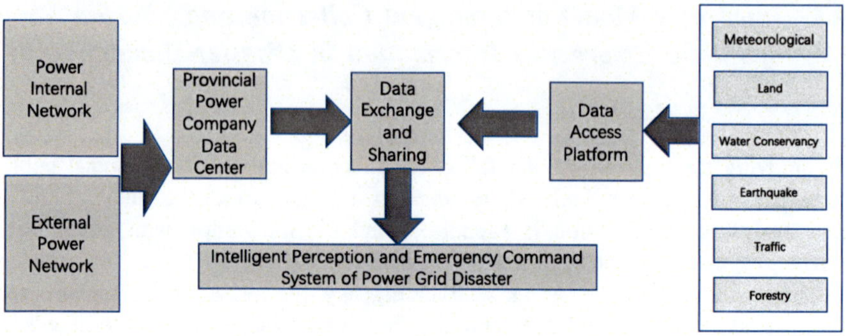

Fig. 2 Integrated analysis model of intelligent collection of public emergency information

4 Summary

Through the application of cooperative sensing and multi-heterogeneous information fusion, this paper can solve the problems of power grid emergency business, such as the operation status monitoring, distribution equipment detection and identification, and has the following performance advantages: (1) Increase the system's viability. When some sensors cannot be used or interfered, or a certain target is not in the coverage area, there are always some sensors that can provide information, so that the system can operate continuously without interference, weaken the fault and increase the detection probability. (2) Expand space coverage. Through multiple overlapped sensor areas, the space coverage is expanded. Some sensors can detect the places that other sensors cannot detect, thus increasing the space monitoring range and detection probability of the system.

In this paper, the emergency services in the power grid, in view of the various conditions of the scene, provide a reference for the in-depth study of the application of multivariate data processing technology at home and abroad.

Acknowledgements First of all, I would like to thank my colleague Min Xu, who has provided me with various helps in the creation of the paper, especially in the verification of data collection, so that I can fully mine various applications of multi-source sensor data. In addition, I also want to thank my company for providing me with the basic conditions for continuous research.

This paper was supported by the science and technology project in State Grid Corporation, which name is 'Research on the Key Technologies of Power Grid Disaster Intelligent Perception and Emergency Command (5700-202019185A-0-0-00).'

References

1. Narasimhan SG, Nayar SK (2002) Vision and the atmosphere. Comput Vis 233–254
2. Narasimhan SG, Nayar SK (2003) Contrast restoration of weather degraded images. IEEE Trans Pattern Anal Mach Intell (06):713–724. https://doi.org/10.1109/tpami.2003.1201821

3. He K, Sun J, Tang X (2011) Single image haze removal using dark channel prior. IEEE Trans Pattern Anal Mach Intell (12):2341–2353
4. Shwartz S, Namer E, Schechner YY (2006) Blind haze separation. Proc IEEE CVPR (01):1984–1991
5. Omer I, Werman M (2004) Color lines: image specific color representation. In: Proceeding of IEEE computer vision and pattern recognition, vol 06, pp 946–953
6. Wang F (2006) Quantitative methods and applications in GIS. CRC Press, Boca Raton, Florida, pp 77–96
7. Yan Z, Parent C, Spaccapietra S et al (2010) A hybrid model and computing platform for spatio-semantic trajectories. In: Proceedings of the 7th extended semantic web conference (ESWC), Heraklion, Greece, Cited:1
8. Bajwa S, Chung E, Kuwahara M (2005) Performance evaluation of an adaptive travel time prediction model. In: Proceeding of the 8th international IEEE conference on intelligent transportation systems, Vienna, Austria
9. Aamodt A, Plaza E (1994) Case-based reasoning: foundational issues, methodological variations and system approaches. AI Commun 7(1):39–59
10. Kwon J, Barkley T, Hranac R et al (2010) Decomposition of travel time reliability into various sources: incidents, weather, work zones, special events, and base capacity. In: Proceedings of 90th transportation research board (TRB) annual meeting, Washington DC

Noise Classification Algorithm Based on Short-Term Energy and Zero-Crossing Rate

Si-Yang Luo and Hua Long

Abstract Due to the problems of dimensionality disaster caused by excessive extraction of feature parameters and time-consuming during noise classification, this paper proposes a noise classification algorithm based on short-term energy and zero-crossing rate. Firstly, the short-term energy and zero-crossing rate of noise audio are extracted. Then, the maximum number of frames that conforms to the main distribution range of the short-time energy and zero-crossing rate of the six kinds of noise is counted. Finally, the classification of the six kinds of noise is realized by calculating the probability that the maximum number of frames accounts for the total frame number of the audio segment. The experimental results show that the classification accuracy of 468 audio samples of six kinds of noise is 98.08%, and the running time including feature extraction and classification is only 3.82 s.

Keywords Noise classification · Short-term energy · Zero-crossing rate

1 Introduction

As the most common way of human communication, speech is often disturbed by noise. Therefore, speech enhancement has received much attention as a necessary part of speech signal processing in noisy environments [1]. However, each kind of noise has its own characteristics. If the noise can be classified and different solutions can be adopted for different noises, the speech enhancement algorithm can have better performance in different noise environments [2].

The difficulty of audio classification mainly lies in the selection of characteristic parameters and classifiers. Literature [2] extracted a total of five types of feature

S.-Y. Luo (✉) · H. Long
Faculty of Information Engineering and Automation, Kunming University of Science and Technology, Kunming 650000, Yunnan Province, China

S.-Y. Luo · H. Long
Yunnan Key Laboratory of Computer Technology, Kunming 650000, Yunnan Province, China

Z. Yu et al. (eds.), *Advancements in Mechatronics and Intelligent Robotics*, Advances in Intelligent Systems and Computing 1220,
https://doi.org/10.1007/978-981-16-1843-7_35

parameters, such as linear prediction coding coefficients and local residual energy, and combined GMM to classify noise. Literature [3] extracted the 22-dimensional Bark domain energy of the noise and combined SVM to classify the noise. Literature [4] extracted the 64-dimensional MFCCs of noise combined with DNN to classify the noise. Literature [5] extracted 648-dimensional features of noise, such as MFCCs and zero-crossing rate, and classified the noise with CNN. The above classification methods use more dimensional feature parameters and classifiers. When there are too many feature parameters in the classification, the time required for classification will increase exponentially, resulting in the so-called dimensional disaster [6]. Generally, the use of classifiers will lead to greater complexity and time-consuming classification process [7].

Inspired by literature [7], this paper applies the extraction method of short-term energy and zero-crossing rate of speech signals to noise signals, aiming at the problems of extracting more feature parameters and complexity of the classification process in the noise classification process. Firstly, the short-term energy and zero-crossing rate of noise audio are extracted. Then, the maximum number of frames that conforms to the main distribution range of the short-time energy and zero-crossing rate of the six kinds of noise is counted. Finally, the classification of the six kinds of noise is realized by calculating the probability that the maximum number of frames accounts for the total frame number of the audio segment.

2 Audio Feature Extraction

2.1 *Short-Term Energy*

Feature extraction is a key step in signal recognition [8]. After framing the signal, the short-term energy of a frame of signal is defined as follows:

$$E = \sum_{n=0}^{L-1} y^2(n) \tag{1}$$

$y(n)$ represents the amplitude of the audio signal at the sampling point, and L represents the frame length.

According to the analysis, the short-term energy of white is less than 0.065, and the short-term energy of most frames of hf channel is greater than 2. The short-term energy of factory1 is mainly located in the range of (0.065,1.5). The short-term energy of babble and volvo is mainly in the range of (0.065,2). The short-term energy of m109 is mainly located in the range of (0.065,4).

2.2 Short-Term Zero-Crossing Rate

After framing the signal, its short-term zero-crossing rate calculation method is defined as follows:

$$Z = \sum_{n=0}^{L-1} |\mathrm{sgn}[y(n)] - \mathrm{sgn}[y(n-1)]| \quad (2)$$

According to the analysis, the zero-crossing rate of most frames of volvo is less than 10. The zero-crossing rate of hf channel and white is more than 100. The zero-crossing rate of m109 is mainly located in the range (10,30). The zero-crossing rate of babble is mainly located in the range (30,46). The zero-crossing rate of factory1 is mainly located in the interval (40,60). The analysis results show that the short-term zero-crossing rates of the six types of noise are quite different, and the difference between the short-term energy and short-term zero-crossing rates provides conditions for noise classification.

3 Algorithm Steps

The flowchart of noise classification algorithm based on short-term energy and zero-crossing rate is shown in Fig. 1.

Step 1. The audio signals of six kinds of noise are segmented into audio segments whose integer length is N sampling points, and then the channel number, sampling rate and accuracy of the audio segment are normalized.

Step 2. The processed audio segment was read in batches, and then extract the two features of short-term energy E and zero-crossing rate Z for each frame of signal.

Step 3. Count the number of frames in noisy audio that meet the following conditions: (1) Calculate the number of frames a_1 in the entire audio segment that

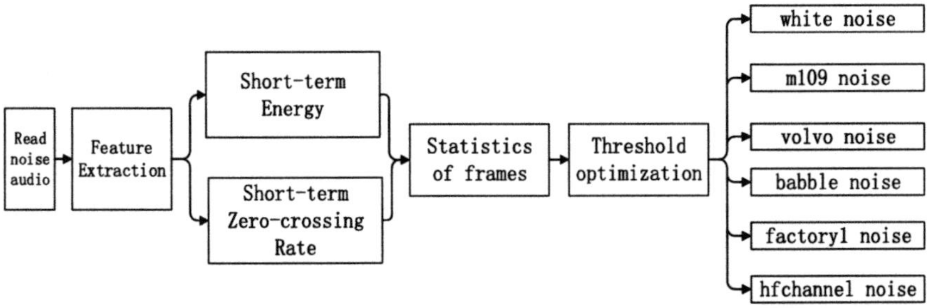

Fig. 1 Flowchart of noise classification algorithm

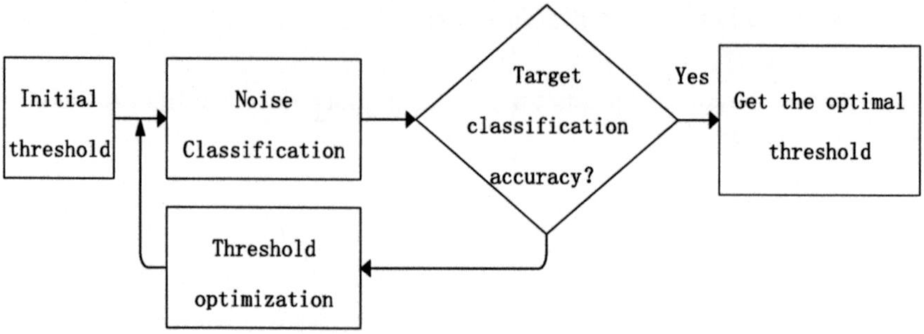

Fig. 2 Flowchart of threshold optimization

satisfy the conditions $0 < E < 0.065$. (2) Calculate the number of frames a_2 in the entire audio segment that satisfy the conditions $0.065 < E < 2$ and $Z < 10$. (3) Calculate the number of frames a_3 in the entire audio segment that satisfy the conditions $0.065 < E < 4$ and $10 < Z < 30$. (4) Calculate the number of frames a_4 in the entire audio segment that satisfy the conditions $0.065 < E < 2$ and $30 < Z < 46$. (5) Calculate the number of frames a_5 in the entire audio segment that satisfy the conditions $0.065 < E < 1.5$ and $40 < Z < 60$. (6) Calculate the number of frames a_6 in the entire audio segment that satisfy the conditions $2 < E$ and $100 < Z$.

Step 4. Set six initial thresholds $n_1 \sim n_6$ according to the number of frames $a_1 \sim a_6$ in Step 3, and distinguish six types of noise based on the current initial threshold and the number of frames, respectively. If the initial thresholds are optimized until the target classification accuracy rate is reached, the optimal threshold for classification is finally obtained. Figure 2 shows a flowchart of threshold setting.

Step 5. (1) When $a_1 > n_1 * fn$ is satisfied, the audio segment is considered to be white noise. (2) When $a_2 > n_2 * fn$ is satisfied, the audio segment is considered to be volvo noise. (3) When $a_3 > n_3 * fn$ is satisfied, the audio segment is considered to be m109 noise. (4) When $a_4 > n_4 * fn$ and $a_5 < n_5 * fn$ are satisfied, the audio segment is considered to be babble noise. (5) When $a_5 > n_5 * fn$ is satisfied, the audio segment is considered to be factory1 noise. (6) When $a_6 > n_6 * fn$ is satisfied, the audio segment is considered to be hf channel noise.

4 Experimental Result

The algorithm in this paper is implemented by MATLAB 2018a, and the noise data uses six noise audios of babble, factory1, hf channel, m109, volvo and white in the NOISEX-92 dataset. Using the audio processing tool sox, each noise signal is divided into 3 s segments. There are 78 segments of each type of noise and a total of 468 samples of six types of noise signals. Then, use sox to normalize each noise signal into a single-channel signal, the sampling frequency is normalized to

8000 Hz, and the precision is normalized to 16 bits. In this experiment, the frame length is $L = 256$, the frame shift length is inc = 128, and the number of frames is $fn = 186$.

After extracting the two types of features of the audio segment, the number of frames that meet the condition can be calculated. The calculation method follows Step 3 in the algorithm step and then sets the initial threshold $n_1 \sim n_6$, according to Fig. 2. The setting process shown in Fig. 2 adjusts the threshold optimally, and finally the threshold is adjusted to obtain the optimal threshold, where n_1=0.7, n_2=0.35, n_3=0.5, n_4=0.35, n_5=0.5, n_6=0.3.

The classification of the noise segment can be completed if the following conditions are met: (1) When $a_1 > 0.7 * fn$ is satisfied, the audio segment is considered to be white noise. (2) When $a_2 > 0.35 * fn$ is satisfied, the audio segment is considered to be volvo noise. (3) When $a_3 > 0.5 * fn$ is satisfied, the audio segment is considered to be m109 noise. (4) When $a_4 > 0.35 * fn$ and $a_5 < 0.5 * fn$ are satisfied, the audio segment is considered to be babble noise. (5) When $a_5 > 0.5 * fn$ is satisfied, the audio segment is considered to be factory1 noise. (6) When $a_6 > 0.3 * fn$ is satisfied, the audio segment is considered to be hf channel noise.

The algorithm in this paper is used to classify the audio files with 78 segments of each of the six kinds of noise, and the classification accuracy is shown in Table 1. In this paper, the classification accuracy of the total 468 noise audio samples is 98.08%.

In order to prove the accuracy and operation efficiency of the algorithm in this paper, the pre-training model based on CNN network and SVM classifier is used to compare with the algorithm in this paper. The training data of the two comparison algorithms select 40 audio segments of each noise, and the test data are the remaining 38 audio segments of each noise. The results are shown in Table 2.

The experimental results show that the algorithm of this paper not only achieves 98.08% classification accuracy for the six types of noise, but also the running time including feature extraction and classification is only 3.82 s, which reflects the advantages of this algorithm.

Table 1 Noise classification accuracy

Classification accuracy/%						
	white	volvo	m109	babble	factory1	hf channel
white	100	0	0	0	0	0
volvo	0	100	0	0	0	0
m109	0	0	98.72	1.28	0	0
babble	0	0	0	97.44	2.56	0
factory1	0	0	0	7.69	92.31	0
hf channel	0	0	0	0	0	100

Table 2 Experiment result

Algorithm	Accuracy (%)	Time (s)
CNN pre-training model	93.42	33.70
SVM classifier	94.74	22.56
Algorithm in this paper	98.08	3.82

5 Conclusion

In this paper, the features of short-term energy and zero-crossing rate of noisy audio are used to classify six types of noise, and a noise classification algorithm based on short-term energy and zero-crossing rate is proposed. The experimental results show that the algorithm in this paper can correctly classify 78 audio samples of white, volvo and hf channel noise, with a classification accuracy of 100%. The classification accuracy of 468 audio samples of six kinds of noise is 98.08%, and the running time including feature extraction and classification is only 3.82 s. At the same time, it also shows that the algorithm of this paper not only has a high classification accuracy rate but also has a small amount of calculation when using two types of features and no classifier, which improves the efficiency of the noise classification algorithm. In the future work, we should consider using other features or classification methods to further improve the accuracy of noise classification and combine noise classification algorithms with speech enhancement algorithms to improve the performance of speech enhancement algorithms.

References

1. Loizou PC (2013) Speech enhancement: theory and practice. CRC press
2. Choi JH, Chang JH (2012) On using acoustic environment classification for statistical model-based speech enhancement. Speech Commun 54(3):477–490
3. Yuan W, Lin J, Chen N, Wang Y (2013) A noise classification method based on noise energy distribution in bark domain. J East China Univ Sci Technol Nat Sci 39(4):472–476 (in Chinese)
4. Meng X (2016) Speech enhancement algorithm based on noise characteristics. Sci Technol Eng 16(33):244–248 (in Chinese)
5. Ma J-F (2018) Research on speech enhancement algorithm based on convolutional neural network. South China University of Technology
6. Baohua R (2012) Research on audio classification method based on minimum distance. Electroacoust Technol 36(11):46–51
7. Zhang X-Y (2019) Speech music classification algorithm based on zero-crossing rate and spectrum. J Yunnan Univ Nat Sci Ed (5)
8. Jia CH, Guo M (2011) Feature extraction and classification of fruit fly sounds based on HHT method. J Yunnan Univ Nat Sci Ed 33(2):152–157

Interactive Multi-Model Tracking Based on Time Difference Information

Si Miao Liu, Jing Min Tang, and Jin Wen Zheng

Abstract Target location and tracking technology are widely used in military and civilian fields. In recent years, development of related technology illustrates higher level of requirements for the positioning and tracking. Due to the constraints of the environment and equipment, the target state model presents a highly nonlinear state in practical applications that make the solution more difficult. In this paper, a tracking algorithm based on *time difference of arrival* (*TDOA*) is studied in detail. It is proposed to use an interactive multiple model algorithm improved by unscented Kalman filter to track it and to improve the shortcomings of *Kalman filter (KF)* in the traditional *interacting multiple model* (*IMM*) algorithm.

Keywords TDOA location · Maneuvering target tracking · IMM algorithm

1 Introduction

With recent years development of China's navigation system, the target location and tracking technology have also become a hotspot of research which widely used in wireless sensor networks, radar communication and other directions. In the military field, modern society becomes more inclined to electronic warfare [1], and the positioning and tracking of the signal source have become the main means of electronic warfare detection; in the civilian field, the current positioning navigation, positioning carding, wild search and rescue, location acquisition and other aspects have more extensive use [2]. In the traditional algorithm, the use of time difference of arrival (TDOA) method is highly fault-tolerant, high precision, and the measurement environment has no more restrictions, which has been an important research direction of positioning. The factors affecting positioning accuracy are generally the following: The number of receiving stations, the way of deploying

S. M. Liu (✉) · J. M. Tang · J. W. Zheng
School of Information Engineering and Automation, Kunming University of Science and Technology, Kunming 650500, China
e-mail: lsmsher@163.com

Z. Yu et al. (eds.), *Advancements in Mechatronics and Intelligent Robotics*,
Advances in Intelligent Systems and Computing 1220,
https://doi.org/10.1007/978-981-16-1843-7_36

receiving stations, the time difference measurement error and the error of deploying stations [3], etc.

Most of the observation targets we need in reality are in motion, so in addition to positioning, we also need to track the targets in motion. The essence of tracking is the process of estimating the target's position, trajectory and motion using the data obtained from our own sensor measurements. Numerous experiences show a good and suitable target model can help and promote the extraction of relevant information to a great extent. Good and appropriate target models can help and facilitate the extraction of relevant information to a large extent. Observational data in target tracking are very limited, and the model is even more important. The target model is a mathematical expression of the target state at each moment state, describing the functional relationship between any two adjacent state vectors when searching for a maneuver target. Only by choosing the appropriate motion model can the tracking performance be improved. For a maneuvering target, it may have multiple states during a movement [4]. And starting from the maneuvering target, we propose to track it using interactive multi-model algorithm improved by traceless Kalman filtering to improve the shortcomings of Kalman filter (KF) filtering in the traditional interacting multiple model (IMM) algorithm.

2 Interactive Multi-Model Tracking Based on Time Difference Information

2.1 IMM Algorithm

The interactive multi-model algorithm is equivalent to an intermediary of multiple motion models and is a framework for information fusion interactions that can match different models according to the variable motion states of the desired target [5]. A Markov matrix is used to describe the probability of switching motion models. By first information fusion and then filtering, the problem of estimating the state of the maneuvering target is made easier to solve [6].

The IMM algorithm is illustrated in the following steps.

Step 1 Input interaction.
Step 2 Nonlinear filtering operations.
Step 3 Updating the probabilities of the interactive model.
Step 4 Fusion of the output results. The flow of the IMM algorithm is shown in Fig. 1.

Interactive models, which the number is r, with model set $r = \{m_1, m_2, m_3 \ldots m_r\}$. Thus, the interactive model expression can be written as Eq. (1) $X_i(k)$ is the equation of motion and $Z_i(k)$ is the measurement equation.

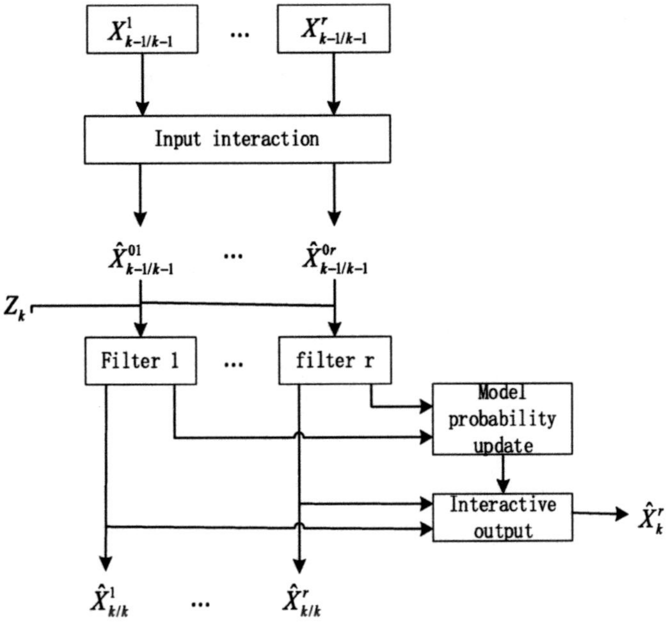

Fig. 1 IMM algorithm flowchart

$$\begin{cases} X_i(k) = \Phi_i X_i(k-1) + \Gamma_i w_i(k-1) & i = 1, 2, \ldots, r \\ Z_i(k) = h(X_i(k)) + v_i(k) \end{cases} \tag{1}$$

where k represents some specific moment, $w_i(k)$ is the white Gaussian noise vector with zero-mean covariance matrix Q_i, Φ_i is the state transfer matrix, Γ_i is the noise gain matrix, $h(*)$ is the measurement equation, and the covariance matrix for R_i is $v_i(k)$. Probability transfer matrix meets the needs below:

$$\begin{cases} \pi_{ij} = pr\big(m_j(k)/m_i(k-1)\big) \\ \sum_{j=1}^{r} \pi_{ij} = 1 & j = 1, 2, \ldots, r \end{cases} \tag{2}$$

Let be motion matching probability in the *i*th model as well as the *k*th moment is $\mu_i(k)$, then the model's predictive probability illustrates as Eq. (3).

$$\mu_{i,j}(k-1) = \frac{\pi_{ij}\mu_i(k-1)}{\sum_{j=0}^{r} \pi_{ij}\mu_j(k-1)} \tag{3}$$

The specific steps of the IMM algorithm can be summarized as three steps below.

Step 1 Enter the parameters of the interaction model and the estimated value of the state fusion as the initial state of the loop. The target motion state vector and the mixed covariance matrix can be expressed as Eqs. (4) and (5).

$$X^{0j}(k-1/k-1)=\sum_{i=1}^{r}\hat{X}_i(k-1)\mu_{ij}(k-1) \tag{4}$$

$$\begin{aligned}P^{0j}(k-1/k-1)=&\sum_{i=1}^{r}\mu_{ij}(k-1/k-1)\{P_i(k-1/k-1)\\&+\left[\hat{X}^i(k-1/k-1)-\hat{X}^{0j}(k-1/k-1)\right]\\&\left[\hat{X}^i(k-1/k-1)-\hat{X}^{0j}(k-1/k-1)\right]^T\}\end{aligned} \tag{5}$$

Step 2 Filtering. Let the combination of $\hat{X}^{0j}(k-1/k-1)$,$P^{0j}(k-1/k-1)$ and $Z(k)$ be the initial value of the kth moment. Calculating the $\hat{X}_j(k/k)$ and $P_j(k/k)$ in kth moment.

Step 3 The interactive model is updated with probabilities based on the likelihood function (see Eq. 6).

$$f_j(k)=\frac{1}{(2\pi)^{\frac{r}{2}}\left|S_j(k)\right|^{\frac{1}{2}}}\exp\left[-\frac{1}{2}v_j^T(k)S_j^{-1}v_j(k)\right] \tag{6}$$

where $v_j(k)$ is a residual information vector, and 2 is a covariance matrices vector (see Eq. 7).

$$\begin{aligned}v_j(k)&=Z(k)-h\left(\hat{X}^j(k/k-1)\right)\\S_j(k)&=HP^j(k/k-1)H^T+R\end{aligned} \tag{7}$$

Then, the system conditional model probability is:

$$\mu_j(k)=\frac{f_j(k)\sum_{j=0}^{r}\pi_{ij}\mu_j(k-1)}{\sum_{i=1}^{r}f_i(k)\left(\sum_{j=0}^{r}\pi_{ij}\mu_j(k-1)\right)} \tag{8}$$

Step 4 Output result fusion. Compute model probability according to Eq. (9) and output interactively.

$$\hat{X}(k) = \sum_{i=1}^{r} \hat{X}_i(k/k)\mu_j(k) \tag{9}$$

$$P(k/k) = \sum_{i=1}^{r} \mu_i(k)\left\{ P_i(k/k) + \left[\hat{X}_i(k/k) - \hat{X}(k/k)\right]\left[\hat{X}_i(k/k) - \hat{X}(k/k)\right]\right\} \tag{10}$$

where $\hat{\mathrm{X}}(\mathrm{k})$ is the state vector, and $P(k/k)$ is the covariance matrix: In the interactive multi-model algorithm, the filtered initial value in the model is the covariance of the model state estimate at the current moment by the state estimate of the previous moment and its covariance matrix. The algorithm is highly adaptable and plastic and is a target tracking model algorithm with stable performance and good results.

2.2 IMM-UKF Tracking Algorithm Based on Time Difference Information

In this section of the study, unscented Kalman filter (UKF) is used instead of KF in the traditional interactive multi-model. The computation of UKF is similar to the extended Kalman filter (EKF) algorithm, but the performance of UKF is better than EKF. The use of deterministic sampling strategy in the UKF algorithm can reduce the computational complexity and avoid the divergence at high order when EKF does not reasonably deal with it. Target tracking via IMM-UKF in two-dimensional plane state conditions using time difference information.

The system equation in the IMM-UKF algorithm based on the time difference information can be shown in the same Eq. (1).

$$Z_k = \begin{pmatrix} r_2 - r_1 \\ \vdots \\ r_N - r_1 \end{pmatrix} + v_k = \begin{pmatrix} c * S_{1,2} \\ \vdots \\ c * S_{1,N} \end{pmatrix} + v_k \tag{11}$$

The entire IMM-UKF refers to the IMM algorithm steps, where UKF is used for state filter estimation when performing the second filtering step of the algorithm (see Fig. 2).

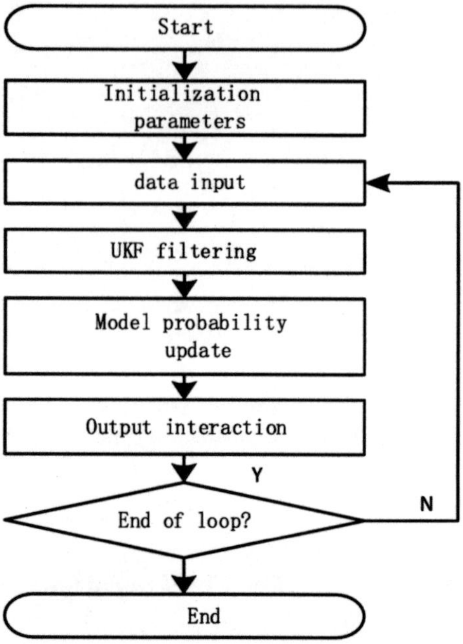

Fig. 2 IMM-UKF algorithm flowchart

3 Simulation Analysis

The effectiveness of IMM-UKF algorithm is verified by simulation experiments, and the performance of the algorithm is analyzed using tracking trajectories, model use probability and the mean square error of the algorithm as the evaluation index. A single model (constant velocity, CV) tracing is also used to compare the mean square error with IMM-UKF.

Experimental conditions: In the two-dimensional scene, the initial position of the target is (2, 1) km, the speed in the x-direction is 1 km/s, the speed in the y-direction is 2 km/s, the number of receiving stations is 4, and the coordinates are as follows: l1: (0, 0) km, l2: (0,10) km, l3: (10,0) km, l4: (10,10) km, where l1 is the main station. The CV model and the CT1 (right-turn model) and CT2 (left-turn model) are selected in the interactive multi-model algorithm, and the probability transfer matrices for the three models are as follows: $\Pi = \begin{bmatrix} 0.8 & 0.1 & 0.1 \\ 0.1 & 0.8 & 0.1 \\ 0.1 & 0.1 & 0.8 \end{bmatrix}$. 50 Monte Carlo simulations were performed for each model with an initial probability of $u = [0.9 \quad 0.05 \quad 0.05]^T$, a movement time of 400 s and a target trajectory as shown in Table 1.

There are five stages in the motion of the target, involving three motion models, the CV model and the CT1 and CT2 models. The CT1 has an angular velocity of 1°/s, and the CT2 has an angular velocity of −1°/s. The utilization probabilities of the three models are shown in Fig. 3. The joint analysis of Fig. 3 shows that the

Table 1 Target trajectory

Phase	Time	State
1	$1s \sim 80s$	Uniform linear motion
2	$81s \sim 160s$	Turn right at a constant speed, the turning angular velocity is $1°/s$
3	$161s \sim 240s$	Uniform linear motion
4	$241s \sim 320s$	Turn left at a constant speed, the turning angular velocity is $-1°/s$
5	$321s \sim 400s$	Uniform linear motion

target makes uniform linear motion at stages 1, 3 and 5, and therefore, the CV model is mainly utilized, while the tracking model with corresponding angular velocity is utilized at stages 2 and 4.

In Fig. 4, the mean square error is compared between the IMM-UKF algorithm and the CV model alone for tracking the target position. The mean square error of both algorithms in Fig. 5 increases significantly around the 80th, 160th, 240th and 320th seconds. Combined with the target motion trajectory, we can see that the model in IMM-UKF switches when the target motion state changes, and the tracking error is larger in this process. The use of a single motion model (CV) will not match the model and the motion state during the turning phase, and thus, the false tracking effect is poor, and the mean square error is too large. In terms of the whole motion process, the tracking using the three models included in the IMM-UKF algorithm is overall better than the tracking using a single model.

In Fig. 5, the mean square error is compared between the IMM-UKF algorithm and the CV model alone for tracking the target position. The mean square error of

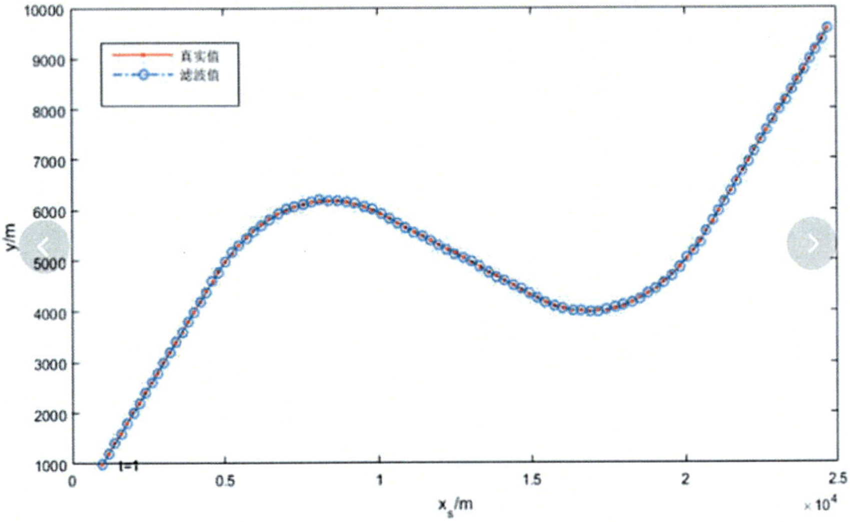

Fig. 3 Experimental target trajectory and tracking results

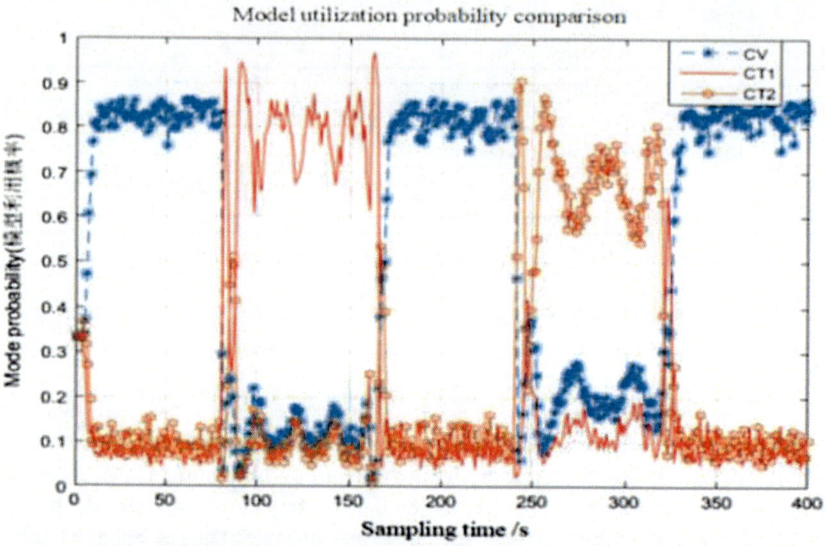

Fig. 4 Probabilities at different moments of the three models

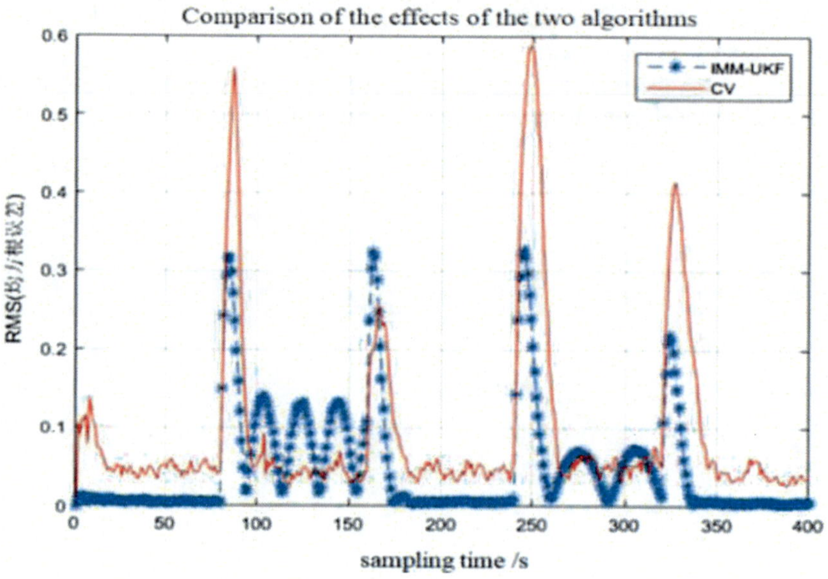

Fig. 5 Comparison of tracking error of the two algorithms

both algorithms in Fig. 5 increases significantly around the 80th, 160th, 240th and 320th seconds. Combined with the target motion trajectory, we can see that the model in IMM-UKF switches when the target motion state changes, and the tracking error is larger in this process. The use of a single motion model (CV) will not match the model and the motion state during the turning phase, and thus, the false tracking effect is poor, and the mean square error is too large. In terms of the whole motion process, the tracking using the three models included in the IMM-UKF algorithm is overall better than the tracking using a single model.

4 Conclusion

It is proposed to improve the shortcomings of KF filtering in the traditional IMM algorithm by tracking it using an interactive multi-model algorithm improved by traceless Kalman filtering in this paper. A detailed derivation of the IMM-UKF implementation flow based on time difference information is presented. The tracking of the maneuvering target by IMM-UKF is performed experimentally, and the probability of using each motion model when using the interactive multi-model algorithm is simulated and analyzed. The simulation analysis shows that the interactive multi-model algorithm has lower errors than the single-model algorithm, higher efficiency ratio and more reliable tracking performance than the single-model algorithm.

References

1. Qu X, Luo Y et al (2008) Electronic warfare target location method. Publishing House of Electronics Industry, Beijing
2. Wan P (2018) Research and application of key technologies for TDOA passive positioning. Xidian University
3. Zhu Y (2017) Research on multi-station passive location technology. Xidian University
4. Chen J (2014) Passive location and tracking technology based on time difference method. Xidian University
5. Du M, Bi D, Wang S (2018) Adaptive IMM maneuvering target tracking algorithm under clutter background. Modern Radar 40(07):47–53
6. Ghazal M, Doustmohammadi A (2017) A novel robust interacting multiple model algorithm for maneuvering target tracking. Adv Electr Comput Eng 17(3)

Study on the Relationship Between Seat Back Angle and Human Body Torso Angle

Yunpeng Cong, Changjiang Du, Lipeng Qin, and Zeyang Tian

Abstract The purpose of this study is to preliminarily investigate the numerical relationship between the car seat backrest angle and the body angle under the condition of driving a car. In the early stages, the measurement of the car seat was carried out in several cities across the country for comfort evaluation purposes, including the measurement of the three-dimensional scan size of the human body and the measurement of the car's attitude. Typical subjects were randomly selected in different regions and genders, after studying the correlation between the torso angle of the driving posture and the seat backrest angle. Different car types are also tested. It was found that value of the seat backrest angle could be estimated by the method of "torso angle + correction value".

Keywords Seat back angle · Torso angle · Correlation

1 Introduction

Relevant studies have shown that drivers are prone to fatigue during driving, and the comfort of car seats is one of the most important factors. At present, the dimensions of the dummy used to test the seat comfort in the automotive industry are derived from the Standard SAE J833, which is quite different from the body characteristics of the Chinese drivers. China Automotive Technology Research Center Co., Ltd. had carried out relevant research in recent years, committed to creating a basic database of China's car seat comfort, forming technical specifications, and providing guidance for industry seat comfort evaluation. This paper focuses on solving the relationship between the backrest angle of the car and the torso angle of the human body, looking for the calculation method of the seat back angle.

Y. Cong · C. Du (✉) · L. Qin · Z. Tian
China Automotive Technology and Research Center Co. Ltd., Tianjin 300300, China
e-mail: 15802285470@126.com

Z. Yu et al. (eds.), *Advancements in Mechatronics and Intelligent Robotics*,
Advances in Intelligent Systems and Computing 1220,
https://doi.org/10.1007/978-981-16-1843-7_37

2 Research Purpose

This study aims to explore the numerical relationship between the seat back angle and the limb angle in the state of driving the car. In the early stage, the project team conducted the measurement of the car seat comfort in 6 cities across the country and the test models included compact SUVs, medium and large SUVs, and medium-sized cars. That is to study the correlation between the torso angle and seat back angle of typical subjects in different regions, models and genders, so as to calculate the estimation of the seat back angle by means of "torso angle + correction amount" [1].

3 Research Method

3.1 *Obtain Point Cloud Data of Driver and Seat Posture*

The point cloud data of the subjects in simulated driving state was obtained by combining the handheld scanner and binocular scanner, and the coordinates of each bone marker point of each subject when driving 9 cars were obtained. The original file was imported into the measurement software for calculation.

3.1.1 Paste Mark Points

14 reflective markers were pasted on the subjects, and the specific positions of the mark points are shown in Table 1.

There are corresponding marking points on the door frames and seats of 9 test vehicles. The position of the marking points is different for different models. All the marking points (reference points) on the vehicle were marked by professionals from China Automotive Technology and Research Center Co., LTD. If these marker points were found to be dropped before scanning, the surveyors should fill in the original marking positions with red markers in time. Red marking points or reflective marking points are easier to be identified by the binocular scanner. The marked points when the driver simulates the driving posture are shown in Fig. 1.

3.1.2 Posture Scanning

After the scanning device was installed, it needed to be debugged and calibrated. According to the size of the vehicle, surveyors adjusted the distance between the camera and projector and the test vehicle, generally the distance is 2–3 m. According to the scene environment, surveyors adjusted the brightness of the projector and adjust the aperture and focal length of the camera until the best

Table 1 Definition of human body markers

Number	Mark point name	Definition
1	Lateral point of left eye	The outer corner of the left eye, the point where the upper and lower eyelids meet
2	Lateral point of left neck	At the outer neck triangle, the intersection of the front edge of the trapezius muscle and the curve connecting the neck fossa and the cervical spine point on the lateral part of the neck
3	Left acromion point	The outer end of the left shoulder, a protrusion that extends forward and outside
4	Left radius point	The highest point of the upper edge of the capitulum radii
5	Left ulnar styloid point	A downward tapered protrusion on the posterior medial side of the left ulna
6	Point of left fourth metacarpal bone	The most protruding point of the proximal end of the first phalanx of the fourth finger on the back of the left hand
7	The greater trochanter of the left femur	Ask the subject to abduct the right thigh and bend the thigh with the body outward. At this time, a dimple is formed in the skin of the greater trochanter, from which the measuring point can be detected
8	Lateral epicondyle point of left femur	The lateral process of the distal end of the left femur
9	The left external ankle point	The protruding point on the lower end of the left fibula
10	The left heel point	The most protruding point of the left heel
11	Medial epicondyle point of right point	The medial process of the distal end of the right femur
12	The right medial ankle point	The protruding point on the lower end of the right tibia
13	The right heel point	The most protruding point of the right heel
14	Medial point of right heel	Right heel point forward and inside point

imaging results were obtained. According to the physical characteristics of the subjects, surveyors adjusted the camera angle to scan all the marked points.

The next step was to perform scan calibration. The calibration board carried out standardized sampling at different positions on the left side of the driver's seat. At least 5 samples were collected. The overall standard deviation is 0.4 mm, and the allowable error is ±0.2 mm. Surveyors checked whether the scanner's camera was focused clearly and the scan range in real-time status; then clicked the light button to check whether the projector was normally lit and whether the lighting range was consistent with the scanning range (if the position is not illuminated, it will directly cause scanning defects). Besides, surveyors guided the subjects to adjust the seat to the most comfortable sitting posture for driving and then asked them to hold the steering wheel, step on the accelerator pedal with their right foot, keep the posture,

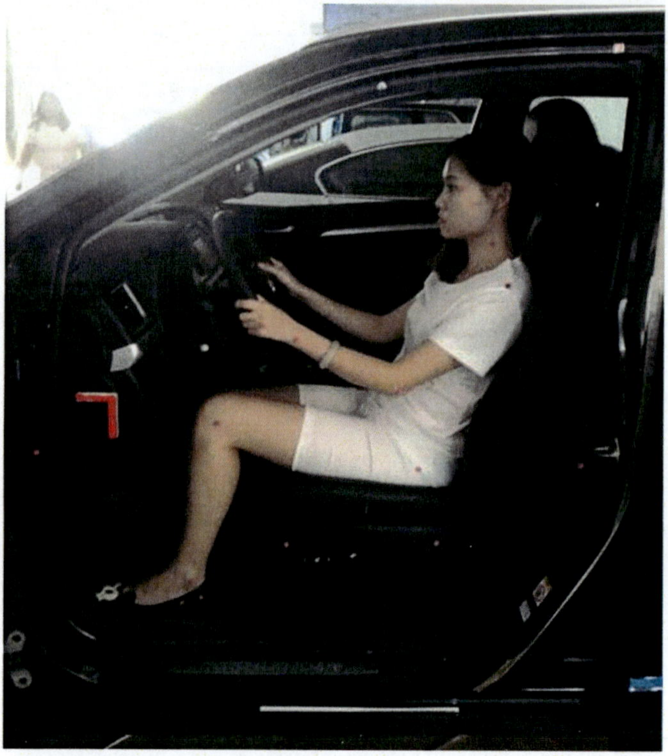

Fig. 1 A schematic diagram of marking points for sitting scanning

and started scanning. The scanning time was about 15 s. This scanning process was a static scan, and the subject must keep the posture still, otherwise scanned again (Fig. 2).

Due to the differences in physical characteristics and driving preferences of the subjects, if the subjects adjusted the driving seat too backward to collect data on the seat backrest and the upper body, after the first scanning, the subjects were invited to get off and the surveyors should adjust the driving seat to the front without changing the seat height and backrest angle, then scanned the empty seat [2].

3.1.3 Save the Data

Based on the subject number and vehicle number, the scanned data was named and saved. The data format obtained by the binocular scanner is different from the data format obtained by the handheld scanner. The coordinate information of all reference points and human body marker points was derived (It could be exported according to the actual measured coordinate system). The data collector defined the

Fig. 2 Sitting posture scanning scene pictures

code and number for each measurement point and outputted the coordinate information (x, y, z). The resulting point cloud image is shown in Fig. 3.

3.2 Calculate the Angle of Each Joint

The angle of backrest (SA1), the angle between torso and thighs (A2), and the angle between thighs and calves (A9) in the point cloud data measured by automobile measurement software are shown in Fig. 4.

3.3 Select Samples for Difference Calculation

A sample of 30 men and women are randomly selected from different cities and different test vehicles. The angle between the seat back and the underside is SA1 measured by the desktop protractor, as shown in Fig. 5, which is 112.86°. The difference value SA2 between the limb angle and the backrest angle was calculated,

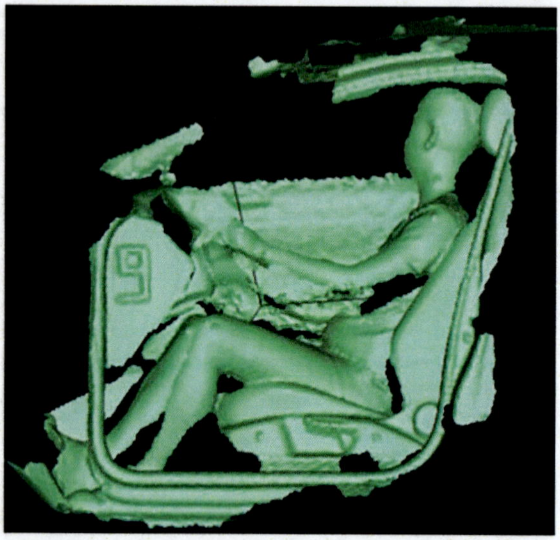

Fig. 3 Point cloud data obtained with a handheld scanner

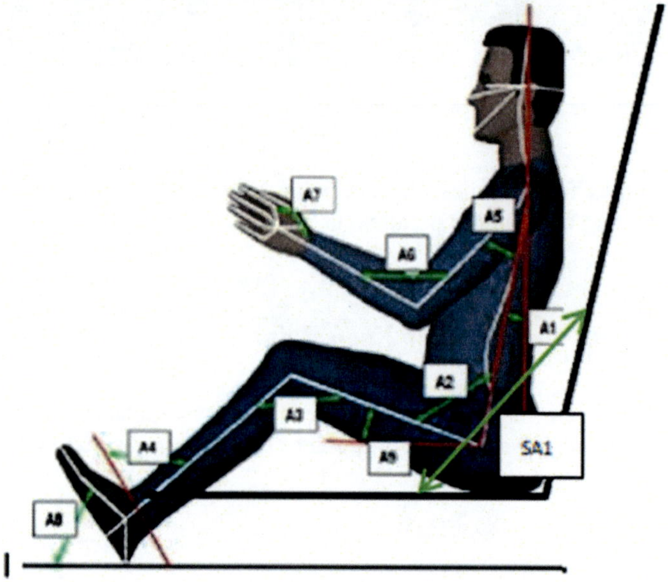

Fig. 4 A schematic diagram of each joint angle

as shown in Fig. 6, SA2 = A2 + A9 − SA1. And the median of SA2 in the sample was calculated, and use this median as the correction quantity of subdivision sample to estimate the backrest angle.

Fig. 5 Use the protractor to measure the backrest angle in the picture

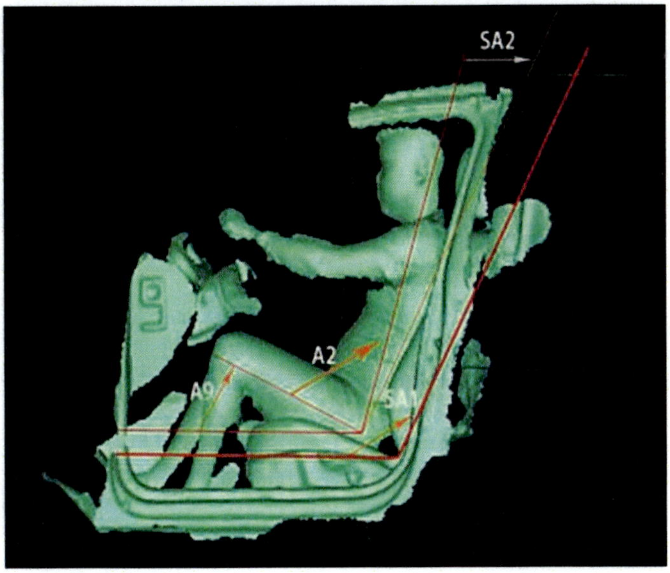

Fig. 6 A calculation diagram of SA2

4 Research Result

4.1 The Median of SA2

The median of SA2 for 30 men and women in 6 major cities and 9 models was calculated. The median value is shown in Table 2.

Table 2 The median value of SA2 in 6 cities

	City A		City B		City C		City D		City E		City F	
No	Male	Female	Male	Female	Male	Female	Male	Female	Male	Female	Male	Female
1	−7.715	−2.373	−7.715	−2.373	−7.715	−2.373	−3.52	−0.52	−7.715	−2.373	−0.161	0.407
2	−1.898	1.957	−1.898	1.957	−1.898	1.957	−7.046	0.24	−1.898	1.957	−1.077	−0.069
3	0.522	0.511	0.522	−8.77	0.522	0.511	−0.71	2.21	0.522	0.511	−3.92	−5.837
4	−1.898	1.957	−1.898	1.957	−1.898	1.957	−7.046	0.24	−1.898	1.957	−1.077	−0.069
5	−0.464	−3.703	−0.464	−3.703	−0.464	−3.703	−2.313	−2.079	−0.464	−3.703	−2.313	−2.079
6	−4.027	−3.422	−13.245	−5.96	−4.027	−3.422	−0.443	−3.236	−4.027	−3.422	−0.443	−3.236
7	0.434	−0.723	0.434	−0.723	−0.723	0.434	−1.596	−0.383	0.434	−0.723	−1.596	−0.383
8	−1.898	1.957	−1.898	1.957	−1.898	1.957	−3.332	−3.072	−1.898	1.957	−3.332	−3.072
9	−11.97	−12.94	−11.97	−12.94	−11.97	−12.94	−7.87	−5.606	−11.97	−12.94	−7.87	−5.606

4.2 Calculate the Backrest Angle

Backrest angle SA1 = A2 + A9 − SA2. For example, the median SA2 for males in No. 1 car in city F is −0.161, and the median SA2 for females is 0.407, so the backrest angle for males in No. 1 car in city F is A2 + A9 − (−0.161), the backrest angle for females in No. 1 car in city F is A2 + A9 − 0.407, and the backrest angles of subjects from 6 cities in 9 cars are calculated in turn.

4.3 The Outliers Test

For multiple measurement values of the same measurement item, the measurement item conforming to the normal distribution is tested by The Grubbs test. The obtained backrest angle is tested by normal test, and the data conforming to the normal distribution is tested by Grubbs test. If there are marked red dots, the values marked by the red dots will be regarded as an outlier, and then the original file will be reprocessed to ensure the correct coordinate system after conversion. The virtual bottom surface of the vehicle was moved to make it completely coincide with the real bottom surface. If it does not completely overlap, the reprocessing was continued until the bottom surfaces of the two vehicles completely overlap, and take the reprocessed result. The significance level is 0.05, and the data whose results do not conform to the normal distribution use box plots. The backrest angles of the subjects in car No. 1 in city F were tested for abnormal values, and the results are shown in Figs. 7 and 8 were obtained.

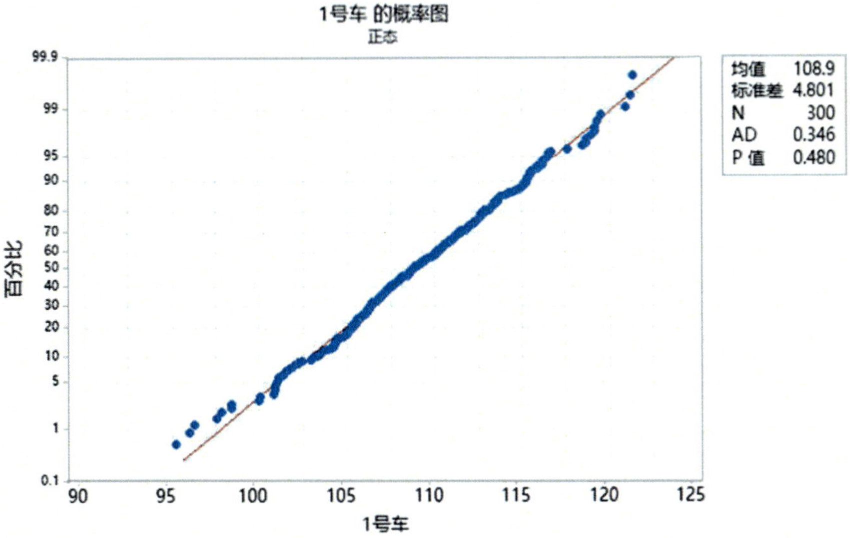

Fig. 7 Probability graph of normal test for the backrest angle of No. 1 car in city F

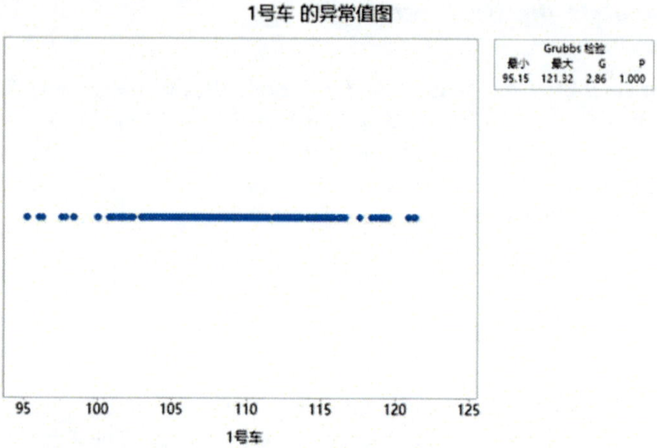

Fig. 8 Outlier graph tested by Grubbs for backrest angle of car No. 1 in city F

In Figs. 7 and 8, the normal test was performed on the SA1 of the subject of No. 1 car in city F, and it was found that $P > 0.05$. Therefore, SA1 conformed to the normal distribution. Using Grubbs test, there were no outliers in SA1.

Then, the SA1 of the subjects of No. 1 car in city F was tested for outliers by gender, and the results are shown below.

In Figs. 9 and 10, the normal test was performed on the SA1 of the male subjects of No. 1 car in city F, and it was found that $P > 0.05$. Therefore, SA1 conformed to the normal distribution. Using Grubbs test, there were no outliers in SA1.

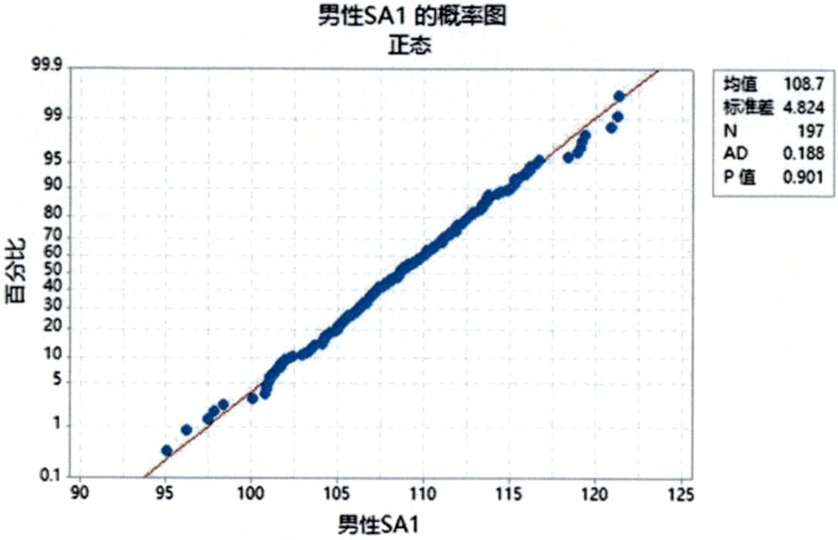

Fig. 9 The normal test probability graph of the male backrest angle of No. 1 car in city F

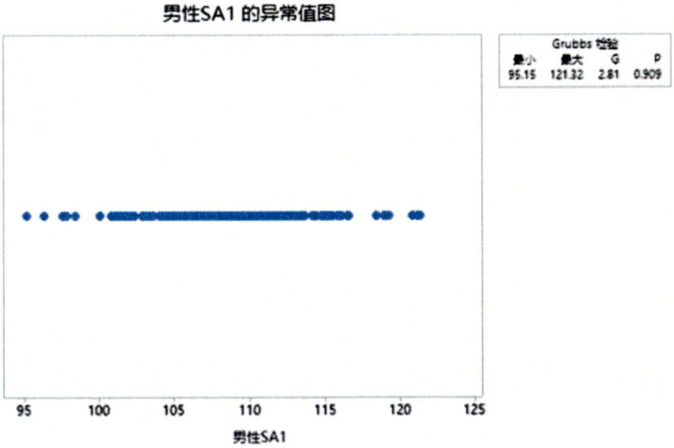

Fig. 10 Outlier graph tested by Grubbs for male backrest angle of No. 1 car in city F

In Figs. 11 and 12, the normal test was performed on the SA1 of the female subjects of No. 1 car in city F, and it was found that P > 0.05. Therefore, SA1 conformed to the normal distribution. Using Grubbs test, there were no outliers in SA1.

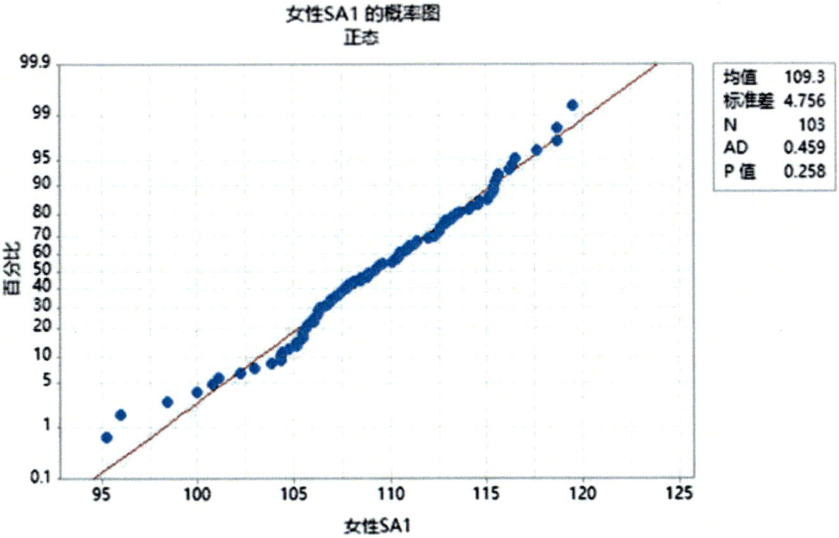

Fig. 11 The normal test probability graph of the female backrest angle of No. 1 car in city F

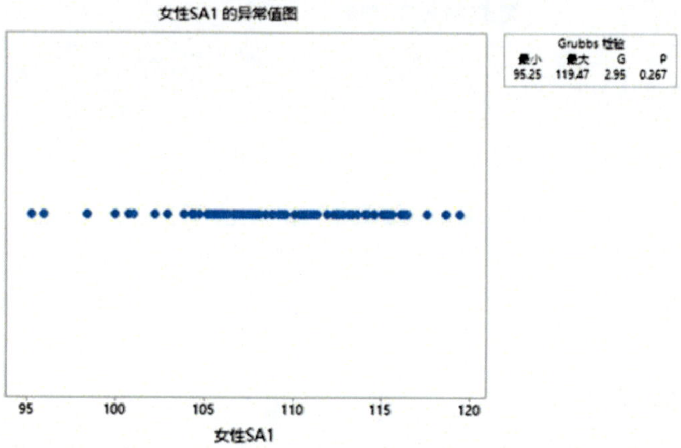

Fig. 12 Outlier graph tested by Grubbs for female backrest angle of No. 1 car in city F

5 Analysis of Research Results

5.1 Influencing Factors of Backrest Angle

As a direct marketing product, the target consumer groups of different models have a clear and clear positioning. However, the seat, as a component that the driver directly contacts the car, generally does not have a clear target consumer positioning for its comfort performance. Therefore, that will cause the seat comfort cannot match the user's expectations of vehicle positioning. The subjective comfort of car seats includes preference, visual comfort, operating comfort, static ride comfort, tactile comfort, and functional comfort. In the automobile industry, in order to meet the most basic comfort requirements of the driver, the seat back is usually designed to have an adjustment function, and its adjustment range can reach 15°. To meet comfort requirements, drivers and passengers will actively adjust the angle of the seat back to a comfortable position. At present, the comfort configuration of car seats is relatively abundant, which has obvious effects on improving the driving environment of cars. The value of the seat back angle is affected by the length of the lower limbs, the seat design and personal driving habits.

5.2 Calculation Method of Backrest Angle

Backrest angle is usually calculated by dummy model and lever protractor. From the collected data of simulated human driving status point cloud, it is found that most people in the state of driving simulation usually keep their back attaching to the seatback, so the backrest angle is close to the included angle of the torso and thigh. Then the value of the backrest angle can be roughly estimated by using the method of correcting the amount value.

6 Conclusion

The seat back angle is a major factor affecting drivers' driving comfort, and there is no unified calculation method. There is a certain quantitative relationship between the trunk angle and the seat back angle adjusted by a person in the driving state, and this correction amount can be used to estimate the backrest angle. The sample size of this study is small, and the individual differences are large, which cannot be used as a rigorous scientific conclusion. The next step is to increase the collection of samples, clarify the screening criteria of the subjects when screening the subjects, refine the car type, and explore the influence of the length of the lower limbs on the backrest angle. More research on the relationship between subjective perception and objective measurement data, determine design indicators for comfort development, verify design indicators, further improve and enrich the two databases, further modify design indicators for comfort development, and collect dynamic comfort indicators Experiments and data to establish a database of dynamic comfort experiments.

References

1. Manary MA, Reed MP (1998) ATD positioning based on driver posture and position [C]. SAE Paper 983163
2. Cao L, Dai H, Zhang R et al (2014) An analysis on the influence of seat backrest angle on occupant crash safety [J]. Automot Eng 12(36):1461–1465
3. Li J (2017) Research on seat static comfort based on body pressure distribution [D]. Hunan University
4. Zhang Y (2018) The design and study of the comfort degree of the seat surface of the car driving seat [D]. Shaanxi University of science and technology

Multi-Population Genetic Algorithm Based on Adaptive Learning Mechanism

Jiawen Pan, Qian Qian, Yong Feng, and Yunfa Fu

Abstract Traditional genetic algorithm has some disadvantages, such as slow convergence, unstable, and easy to fall into local extreme. In order to overcome these disadvantages, an improved genetic algorithm is proposed in the present study. First, based on the analysis of advantages and disadvantages of learning mechanisms in literature, new improvements of learning mechanisms under the multi-population parallel GA are made. In previous studies, gene patterns from which other individuals can learn will be extracted from the excellent individuals of the population, this study improved the learning mechanism by adaptively changing the related control parameters, and dynamically controlling the process of the learning mechanism. Simulation results show that the new algorithm has a great improvement in many aspects of the global optimization, such as convergence speed, the accuracy of the solution, and stability.

Keywords Genetic algorithm · Adaptive · Multi-population parallel mechanism · Learning mechanism

1 Introduction

Genetic algorithm (GA) [1] is a global random search algorithm proposed in the early 1970s, which draws on the natural selection ideas and natural genetic mechanism of the biological world. GA has been widely used in Machine Learning, Control, Optimization, and other fields [2, 3].

J. Pan · Q. Qian (✉) · Y. Feng · Y. Fu
Faculty of Information Engineering and Automation, Kunming University of Science and Technology, Kunming 650500, Yunnan, China
e-mail: qianqian_yn@126.com

J. Pan · Q. Qian · Y. Feng
Yunnan Key Laboratory of Computer Technology Applications, Kunming University of Science and Technology, Kunming 650500, Yunnan, China

Z. Yu et al. (eds.), *Advancements in Mechatronics and Intelligent Robotics*,
Advances in Intelligent Systems and Computing 1220,
https://doi.org/10.1007/978-981-16-1843-7_38

GA has great parallel optimization ability due to it can search multiple points by using fitness function as the search criterion. However, the algorithm also has defects such as "premature" convergence and slow convergence speed. Many methods have been proposed to solve these problems. It is an effective method to introduce learning mechanisms into GA. Hinton et al. [4] are the first researchers to introduce learning mechanisms to guide the evolution of GA algorithms. Candidates of the learning mechanisms are the optimization algorithms with strong local search capabilities, such as simplex method [5], hill-climbing method [6], etc. Literature [7] pointed out that most learning methods can be summarized as Lamarckian learning mechanism, Baldwinina learning mechanism, and the combination of these two to form a new learning mechanism. Literature [8] proposed a model learning mechanism under the guidance of the Lamarckian learning mechanism and the pattern theorem [1], but this mechanism uses a linear method to control related parameters, which may lead to the invalidity of the learning mechanism during operation. Based on previous studies, the present paper proposes Multi-population Genetic Algorithm based on Adaptive Learning Mechanism (ALMGA), which can dynamically control the calculation process of the learning mechanism under a multi-population GA according to the different conditions of the individual fitness during the evolution. Finally, simulation results show that ALMGA has great improvements in global optimization, local search, and convergence speed.

2 Almga

2.1 *Multi-Population Parallel Mechanism*

Multi-population parallel genetic algorithm is a good method to improve the performance of genetic algorithm [9], this mechanism can be divided into three basic types: master–slave model, fine-grained model, and coarse-grained model.

In this paper, the parallel mechanism based on coarse-grained model is introduced to perform independent evolution of multiple populations. Each population is configured with different parameters and initial values, and data migration is performed regularly. Specifically, at the beginning of each evolutionary generation, the population exchanges the optimal individual and gene pattern with the public area (see Sect. 1.2 for details).

2.2 *Basic Concepts of Learning Mechanism*

Definition 1 Let $H = \{H_1, H_2, ..., H_m\}$ be the collection of all populations, M is the number of populations, N is the number of chromosomes in each population.

Definition 2 Age and overage. For any population H_m and $\forall x_i \in H_m$, let $x_{ig.\text{age}}$ represents the age of the chromosome x_i in g generation. For chromosome x_i, its initial age is 0, the age update formula (1) is as follows:

$$x_{ig+1.\text{age}} = x_{ig.\text{age}} + \alpha\left(\frac{H_{mg.\text{ave}}}{x_{i.\text{fit}}}\right) \tag{1}$$

Among them, $i = 1 \sim N$, α represents a control parameter. $H_{mg.\text{ave}}$ represents the average fitness of the population H_m in g generation, $x_{i.\text{fit}}$ represents the fitness of the chromosome x_i. If the fitness of the chromosome is lower than the average fitness of the population for a long time, its age will be high, which means the "evolution potential" of the chromosome is worse. When the age exceeds a predefined age value, the chromosome becomes an overage chromosome and will be eliminated in subsequent evolution. $H_{m.\text{old}}$ represents the number of overage individuals in population H_m.

Definition 3 Excellent rate. For any population H_m, H_e represents the excellent rate of H_m (see 1.4 for the calculation formula), the number of excellent individuals in H_m is $r = H_e * N$, the r individuals with highest fitness in H_m are called excellent individuals.

Definition 4 Gene pattern and pattern extraction. For any population H_m, let any two chromosomes of the population as $x = \{x_1, x_2, \ldots, x_j\}$ and $y = \{y_1, y_2, \ldots, y_j\}$, $x_j, y_j \in \{0, 1\}$. The chromosome generated by the above two chromosomes through the formula $z = x \times y = \begin{Bmatrix} x_j, x_j = y_j \\ *, x_j \neq y_j \end{Bmatrix}$ are called gene pattern, $z_j \in \{0, 1, *\}$, "*" represents undetermined genes. In the present algorithm, the gene pattern is obtained by extracting all excellent individuals of the population in the current generation, $H_{mg.\text{scheme}}$ represents the gene pattern of the population H_m in the g generation.

Definition 5 pattern learning. For any population H_m, let any chromosome of the population as $x = \{x_1, x_2, \ldots, x_j\}$, the chromosome and gene pattern $H_{mg.scheme} = \{z_1, z_2, \ldots, z_j\}$ perform pattern learning behavior through formula: $x \odot H_{mg.scheme} = \begin{Bmatrix} x_j, x_j = z_j \text{或} z_j = * \\ z_j, x_j \neq z_j \text{且} z_j \neq * \end{Bmatrix}$ get a new chromosome $x_{.\text{new}}$. For example, when $x = 1010110$ and the gene pattern $H_{mg.\text{scheme}} = 1{*}0100{*}$, then after learning $x_{.new} = 1001000$.

Definition 6 The weight of gene pattern. For any population H_m, $H_{mg.\text{weight}}$ represents the weight of $H_{mg.\text{scheme}}$, the higher the weight, the more "excellent" $H_{mg.\text{scheme}}$ is, the lower the weight, the more "inferior" $H_{mg.\text{scheme}}$ is (see Sect. 1.7 for the calculation formula).

Definition 7 Best gene pattern. For any population H_m, $H_{m.bscheme}$ represents the most "excellent" (with highest weight) gene pattern of the population until current generation.

Definition 8 Global best gene pattern. During the evolution of the algorithm, the best gene pattern (with highest weight) among all populations is called the global best gene pattern, which is represented by $H_{.bscheme}$.

Definition 9 Common areas. Multi-population parallel GA requires regular data migration and exchange. In the present algorithm, a common area is used to receive the gene patterns and optimize the chromosomes shared by all populations, the common area will compare the weights of different gene patterns to select $H_{.bscheme}$ and compare the fitness values to select the global optimal individual. Common area constantly monitors all populations and transfers $H_{.bscheme}$ to the population that requests learning between populations.

2.3 *Operation Process of Learning Mechanism*

Learning mechanism includes two learning method: learning within populations (LWP) and learning between populations (LBP), which are controlled by possibilities P_i and P_o, respectively.

The process of LWP is as follows:

(a) Calculate H_e of the population;
(b) Select r chromosomes as excellent individuals of the population according to the fitness values;
(c) $H_{mg.scheme}$ is obtained by pattern extraction from the excellent individuals, compare $H_{mg.scheme}$ with $H_{m.bscheme}$ of the population and update $H_{m.bscheme}$;
(d) P_i controls chromosomes of the population to conduct pattern learning based on $H_{m.bscheme}$, and update fitness values and age of chromosomes.
(e) If it reaches the required generation of information migration, transfer $H_{m.bscheme}$ and the best individual in the common area.
(f) The common area compares the $H_{m.bscheme}$ and optimal chromosomes passed from all populations with the existing $H_{.bscheme}$ and global optimal individuals, and updates the $H_{.bscheme}$ and global optimal individuals.

The process of LBP is as follows:

(a) P_o controls whether the population to conduct LBP and with a positive result, LBP sends a learning request to the common area;
(b) Common area sends $H_{.bscheme}$ to the above population;
(c) The population first deletes overage individuals, then generates new chromosomes to learn from $H_{.bscheme}$ and to fill the vacancy of eliminated overage individuals, finally update fitness values and age of chromosomes.

In LWP, individuals can improve their own traits in less generations by learning the common genes of excellent individuals, by this way, the evolution directions of the individuals can be guided. In contrast, LBP guides the evolution of the population, a new chromosome is produced to learn from $H_{.bscheme}$ and replace overage individuals, thus improving the "evolution potential" of population by improving the diversity of the population. Together, the LWP and the LBP make full use of their own advantages and improve the performance of the algorithm.

2.4 Adaptively Change the Value of the Excellent Rate

Excellent rate H_e selects the excellent individuals of the population at each iteration, and guides the evolution direction of the population through the gene pattern obtained from these excellent individuals. Paper [8] uses a linear formula to calculate H_e, the value of H_e will increase linearly when the average fitness of the population increases with the number of iterations. A high value means that there are more excellent individuals. Therefore, the gene model obtained by pattern extraction will be full of non-deterministic gene, making the gene pattern invalid to guide the evolution. In order to avoid the condition that the gene patterns are full of undetermined genes, this paper proposes a calculation formula (2) to adaptively control the excellent rate values based on evolutionary generations:

$$H_e = \begin{cases} H_{e1} + \frac{H_{e2}-H_{e1}}{2}\sin\left(\frac{g}{G}\pi\right) & g \geq \frac{G}{2} \\ H_{e1} + \frac{H_{e2}-H_{e1}}{2}\left(1 - \sin\left(\frac{g}{G}\pi\right)\right) & g < \frac{G}{2} \end{cases} \tag{2}$$

In the formula, (2): H_{e1} is the minimum excellent rate; H_{e2} is the maximum excellent rate; g is the current evolution generation number; G is the maximum evolution generation number of the algorithm.

In the Fig. 1, the dotted line represents the variation of H_e in the paper [8], which follows a linear transformation method, the solid line indicates the change curve of the H_e following current formula (2). As we can see, H_e has a greater value than linear transformation at the beginning of the algorithm. This is because in the early stage of evolution, the average fitness of the population is relatively low, and excellent individuals contain more inferior genes, so formula (2) adaptively increases the value of H_e to increase the number of excellent individuals. As a result, the pattern extraction based on a large number of excellent individuals will reduce the probability of the inferior genes entering the gene pattern. During the late stage of the algorithm, the average fitness of the population and the similarity between the genes of the excellent individuals are relatively high, the value of H_e is adaptively reduced to ensure that only the truly excellent genes can be extracted into the gene pattern. In addition, adaptively change the value of the excellent rate can also avoid the phenomenon that the gene pattern in the middle and late stages of the algorithm is full of undetermined genes so that the gene pattern can better guide the evolution direction of the population.

2.5 Adaptively Change the Value of P_i

Learning within populations (LWP) relies on P_i to control the chromosomes of the population to learn $H_{m.bscheme}$. Paper [8] uses a formula that follows a linear transformation to calculate P_i. Such formula does not fully consider the evolution situation of the population. Since the learning frequency gradually increases along with the increase of the average fitness of the population, the population may fall into "premature" convergence. In the present algorithm, both the generations and the evolution situations (i.e., fitness values) of the population are considered. In the early stage of the algorithm, the quality of individuals in the population is generally poor, so the guidance to the population should be strengthened by increasing the value of P_i. As the evolution continues, P_i should be gradually reduced to maintain the diversity of the population, so as to increase the converging probability to the best solution. In addition, the population with low fitness values should take a large value of P_i to strengthen the guidance, and the population with high fitness values should reduce P_i, so as to ensure the diversity of the population, avoiding the phenomenon of "premature" convergence. Based on the above reasons, this study designs a formula for calculating P_i as follows:

$$P_i = P_{i1} + (P_{i2} - P_{i1})\frac{1}{1 + \exp\left(\frac{g}{G} - 1\right)}\left(1 - \left(\frac{H_{m.\mathrm{ave}}}{H_{m.\mathrm{best}}}\right)\right) \tag{3}$$

In formula (3): P_{i1} is the minimum probability of LWP; P_{i2} is the maximum probability of LWP; g is the current evolution generation; G is the maximum

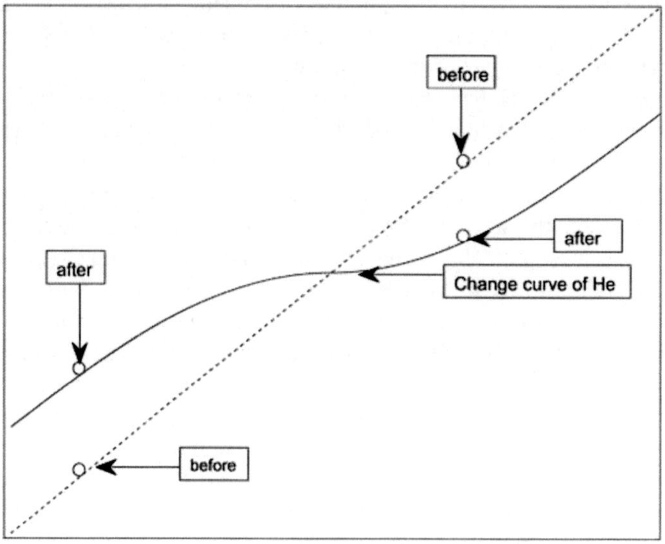

Fig. 1 Adaptive curve of excellent rate

evolution generations; $H_{m.\text{ave}}$ is the average fitness of the population; $H_{m.\text{best}}$ is the maximum fitness of the population.

2.6 Adaptively Change the Value of P_o

Learning between populations (LBP) controlled by P_o. Paper [8] does not fully consider the influence of overage individuals, so that there may be a situation in which too many overage individuals are eliminated, resulting in instability of the algorithm, or there may be no overage individuals at all, thus making learning invalid. In all, when there are many overage individuals or when the population has a low fitness value, P_o should be increased. Based on the above reasons, formula (4) is used to adaptively adjust the probability P_o of LBP according to the number of overage individuals in the population and the evolution situations of the population:

$$P_o = 4\sqrt[3]{\frac{H_{m.\text{old}}}{N}}(P_{o1} + (P_{o2} - P_{o1}))\cos\left(1 - \left(\frac{\max(H_{m.\text{ave}}) - H_{m.\text{ave}}}{\max(H_{m.\text{ave}}) - \min(H_{m.\text{ave}})}\right)\right) \tag{4}$$

In the formula (4): P_{o1} is the minimum probability of LBP; P_{o2} is the maximum probability of LBP; g is the current evolution generation; G is the maximum evolution generation; $H_{m.\text{old}}$ is the number of overage individuals in the population; N represents the total number of chromosomes in the population; $\max(H_{m.\text{ave}})$ is the maximum average fitness value of all populations; $H_{m.\text{ave}}$ is the average fitness of the population; $\min(H_{m.\text{ave}})$ is the minimum average fitness value of all populations.

According to formula (4), the number of overage individuals in the population determines the value of P_o. As the evolution continues, the number of overage individuals in the population continues to increase and the P_o value also gradually increases. Therefore, every time LBP is performed, the number of overage individuals in the population will be always maintained in a reasonable number, thus avoiding the situations that there are no overage individuals or too many overage individuals in the population. Finally, a cosine function is introduced to nonlinearly control the learning probability of populations, which can make the learning smooth and maintain the stability of the algorithm. Specifically, for populations whose average fitness value at the range of $[\min(H_{m.\text{ave}}), (\max(H_{m.\text{ave}}) - \min(H_{m.\text{ave}}))/2]$, the population should catch up with other populations, so it has a larger P_o value; for populations whose average fitness value at the range of $[(\max(H_{m.\text{ave}}) - \min(H_{m.\text{ave}}))/2, (\max(H_{m.\text{ave}}) + \min(H_{m.\text{ave}}))/2]$, the quality of the genes is good, so the P_o value of the population is slightly increased to get better learning results; in other conditions, a small P_o value is assigned, so as to emphasize the learning within populations. By nonlinearly changing the learning probability of different populations, LBP can better play the role of guiding the evolution direction and strengthening the local search ability of the population.

2.7 The Weight of the Gene Pattern

The learning mechanism can guide the evolution direction of the population, but it is not ensured that the gene pattern of current generation is better than the gene pattern of previous generation. Therefore, an evaluation function is designed in the present algorithm to calculate the weight of the gene pattern, so that the algorithm can use the weight to compare the quality of different gene patterns. The calculation formula is as follows:

$$H_{mg.\text{weight}} = \bar{f}\left(1 - \frac{3G - 2\sqrt{g}}{5G}\right) + H_{m.\text{best}} \frac{3G - 2\sqrt{g}}{5G} \quad (5)$$

In the formula (5): $H_{m.\text{best}}$ is the fitness value of the best individual in the population; $\bar{f}$ is the average fitness value of the remaining excellent individuals within the population except $H_{m.\text{best}}$; g is the current evolution generation; G is the maximum evolution generations.

In the early stage of evolution, the average fitness of the population is relatively low, most chromosomes in the population contain "inferior" genes. Under such conditions, the ability of the best individuals in the population to guide the evolution direction should be strengthened, and the impact of other excellent individuals should be weakened. As the algorithm continues, the average fitness of the population gradually increases, and the similarity between the best individuals and other excellent individuals also increases, in order to avoid the misguided evolution direction by the local optimal solution, the impact of $H_{m.\text{best}}$ should be reduced and the impact of $\bar{f}$ should be increased.

2.8 The Flow of ALMGA

The algorithm execution process is as follows:

Set the control parameters of the algorithm, including population number, chromosome length, age, etc., initialize the population and public area.

Calculate the fitness of individuals in the population.

All populations independently evolution according to their own evolution parameters, perform genetic operations and update individual fitness and age.

All populations calculate the number of excellent individuals according to formula (2) and perform pattern extraction, update $H_{m.b\text{scheme}}$, and pass the optimal individual and $H_{m.bscheme}$ to the common area. The common area receives data from various populations and updates the global optimal individual and $H_{.bscheme}$.

Individuals in the population determine whether to perform LWP by Eq. (3).

Each population determines whether to perform LBP by Eq. (4).

Check the termination condition, and go to (8) if it meets, otherwise go to (2).

The optimal solution is outputted.

3 Simulation Test

In order to verify the performance of the ALMGA algorithm proposed in this article, this section conducts simulation tests based on four standard test functions given in Table 1. The HGA proposed in paper [6], the PGABL proposed in paper [8] and the ALMGA algorithm proposed in this paper are compared.

Both HGA and ALMGA use the same parameters, and the parameters of PGABL are consistent with that in the paper8. The above algorithms all use binary coding, fitness proportion selection, single-point crossover, and multi-point mutation. The control parameters of the algorithm are as follows: the population number is 4, the chromosome size of each population is 50, the chromosome length is $D * 40$, and the number of algorithm generations is 400. In HGA, $p_{m1} = 0.2$, $p_{c1} = 0.8$, $p_{m2} = 0.05$, $p_{c2} = 0.5$, $p_{m3} = 0.1$, $p_{c3} = 0.6$. In PGABL $a = 1$, life = 15, $P_c = 0.8$, $P_m = 0.06$. In ALMGA, α=1, life = 15, $P_{m1} = 0.2$, $P_{c1} = 0.8$, $P_{m2} = 0.05$, $P_{c2} = 0.5$, $P_{m3} = 0.1$, $P_{c3} = 0.6$, $P_{e1} = 0.1$, $P_{e2} = 0.2$, $P_{i1} = 0.3$, $P_{i2} = 0.5$, $P_{o1} = 0.1$, $P_{o2} = 0.15$.

Table 2 shows the test results after each of the three algorithms are executed 30 times. It can be seen from the test results that compared with HGA and PGABL, the optimal value of the function solved by ALMGA is closer to the theoretical optimal value of the function, which indicates the algorithm has high accuracy. The number of convergences of ALMGA is close to the total number of experiments, among them, the number of convergences for solving functions f_1 and f_2 is consistent with the total number of experiments, which indicating the algorithm has good stability. It can be seen from the result of PGABL to solve function f_1 that although the number of convergence is low, but the average convergence iterations is also low, which indicating the learning mechanism has given the algorithm better optimization capabilities. The PGABL to solve function f_2 has less convergence generation to meet convergence accuracy, this is because the global optimal value of f_2 is surrounded by the local optimal value. PGABL is easy to fall into the local optimal value during the algorithm optimization process, it shows that the algorithm has poor ability to jump out of the local optimal value, and also shows that ALMGA has a better ability to jump out of the local optimal value. There are two extreme cases when solving $f_3 - f_4$, PGABL solves $f_3 - f_4$ without convergence, while HGA has fewer times of convergence, and ALMGA has more times of convergence, which indicating ALMGA has better global optimization and convergence capabilities.

Table 1 Test functions

	Function name	Function expression	Domain	Optimal value
f_1	*Camle*	$\left(4 - 2.1X_i^2 + \frac{X_i^4}{3}\right)X_i^2 + X_iX_{i+1} + (-4 + 4X_{i+1}^2)X_{i+1}^2$	$[-10, 10]$	−1.031628
f_2	*haystack*	$\left[\frac{3}{0.05 + (X_i^2 + X_{i+1}^2)}\right]^2 + (X_i^2 + X_i^2)^2$	$[-5.12, 5.12]$	3600
f_3	*Rastigrin*	$\sum_{i=1}^{D}(X_i^2 - 10\cos 2\pi X_i + 10)$	$[-10, 10]$	0
f_4	*Ackley*	$-20\exp\left(-0.2\sqrt{\frac{1}{D}\sum_{i=1}^{D}X_i^2}\right) - \exp\left[\frac{1}{D}\sum_{i=1}^{D}\cos 2\pi X_i\right] + 20 + e$	$[-2.048, 2.048]$	0

Table 2 Comparison results of function test

	Accuracy	Algorithm	Optimal value	Best convergence iterations	Average convergence iterations	Convergence times	Convergence rate
f_1	10^{-5}	HGA	−1.03162470255	131	181	20	0.667
		PGABL	−1.03162615926	114	163	13	0.433
		ALMGA	−1.03162845151	24	61	30	1.000
f_2	10^{-3}	HGA	3599.999004362	166	258	14	0.467
		PGABL	3599.999216859	131	155	2	0.067
		ALMGA	3599.999993133	41	72	30	1.000
f_3 (D = 10)	10^{1}	HGA	6.057019493037	274	340	7	0.233
		PGABL	25.86890531471			0	0
		ALMGA	2.484773167789	149	268	26	0.867
f_4 (D = 25)	10^{0}	HGA	0.753118330563	292	331	3	0.133
		PGABL	2.018254436453			0	0
		ALMGA	0.401447729784	137	192	28	0.933

4 Conclusion

In order to overcome the shortcomings of traditional genetic algorithms, the Multi-population Genetic Algorithm Based on Adaptive Learning Mechanism is proposed in the present study. The algorithm introduces multi-population parallel mechanisms to strengthen the global optimization capabilities of genetic algorithms, and introduces learning mechanisms to guide the evolution, and strengthens local search capabilities of the algorithm. At the same time, it makes adaptive improvements to the learning mechanism, so that the algorithm can dynamically control the process of the learning mechanism. Finally, through simulation experiments on four standard test functions, the results show that the ALMGA proposed in this paper significantly improved the searching ability and convergence ability of the algorithm.

References

1. Holland HJ (1975) Adaptation in natural and artificial systems. Ann Arbor Michigan Univ Press 6(2):126–137
2. Zhou Y, Zhou L, Wang Y et al (2017) Application of multiple-population genetic algorithm in optimizing the train-set circulation plan problem. Complexity
3. Li L, Tang Y, Liu J et al (2013) Application of the multiple population genetic algorithm in optimum design of air-core permanent magnet linear synchronous motors. Proc CSEE 33 (15):69–77
4. Hinton EG, Nowlan JS (1987) How learning can guide evolution. Complex Syst 1(43):495–502
5. Tang W, Chen Y, Zhang M (2019) An energy balanced routing algorithm with simplex method. Comput Technol Dev 29(3):55–59
6. Zheng M, Zhuo M, Zhang S et al (2017) Reconstruction for gene regulatory network based on hybrid parallel genetic algorithm and threshold value method. J Jilin Univ Eng Technol Ed 47 (2):624–631
7. Zhibo L, Qitao H, Hongzhou J (2009) Mixed application of two learning mechanisms in genetic algorithm. Syst Eng Electron 31(8):1985–1989
8. Zhang G, Wu Z, Liu X (2005) Parallel genetic algorithm based on learning mechanism. Comput Appl 25(2):374–376
9. He W, Wang J, Hu L (2009) The improvement and application of real-coded multiple-population genetic algorithm. Chin J Geophys 52(10):2644–2651

Influence of Channel Synergy and Channel Conflict on Channel Performance in Omni-Channel Retailing

Ling Ke, Weiting Yang, Zongjing Wang, Yulin Li, and Xuefang Zhang

Abstract Various retail enterprises have started strategic layout in omni-channel this year. The number of omni-channel customers has been exploded. Retail enterprises begin their attempt of improving the comprehensive customer experience in all channels in order to improve their channel performance. This study uses empirical methods to reveal that channel synergy has a positive influence on channel performance, while channel conflict has no significant impact on channel performance; meanwhile, it verifies that comprehensive customer experience plays an intermediary role in the influence of channel synergy and channel conflict on channel performance.

Keywords Channel synergy · Channel conflict · Customer comprehensive experience · Channel performance

1 Introduction

With O2O omni-channel retailing crept into Chinese market, a clear commercial path was formed gradually. Omni-channel retailing must consider the influence of the synergy and conflict between different channels on customers as well as the enterprises [1–3]. Ning Yangyang et al. [4] believed that the channel synergy system with high integration level would contribute to consolidate customer belief and reduce their risk perception in shopping. That is to say, the higher the channel synergy degree customer perceived in omni-channel, the more they ignored the channel conflict and obtained a pleasant shopping experience as well as increased the purchase frequency. As a result, the channel performance would be improved. A channel is not only the path of product circulation, but also with the function of information transmission. Therefore, if the channel functions and roles changed, the

L. Ke · W. Yang · Z. Wang · Y. Li · X. Zhang (✉)
School of Economics and Management, Hubei Polytechnic University, Huangshi 435003, Hubei, China

Z. Yu et al. (eds.), *Advancements in Mechatronics and Intelligent Robotics*,
Advances in Intelligent Systems and Computing 1220,
https://doi.org/10.1007/978-981-16-1843-7_39

research on channel synergy and conflict would no longer be limited to the study of channel members' behaviors.

Therefore, this study focuses on solving these two problems: (1) whether the channel synergy and channel conflict would make a difference on the channel performance of retail enterprises from the perspective of customers; (2) whether the comprehensive customer experience plays an intermediary role in the influence of channel synergy and conflict on channel performance.

2 Theoretical Basis and Research Hypothesis

2.1 Channel Synergy

Channel synergy is to analyze customer basic information and customer service track information, and realize "one trigger, full response" through information sharing among channels [5]. When a customer chose to contact with retail enterprises from a certain channel while it did not possess the ability to meet customer needs, then the channel should be rebuilt or reorganized; if it already be equipped with the ability to meet certain demand, while that channel lied in the secondary choice of the retailer, then the key point would be how to effectively intervene the channel selection behavior of customers. Lanlan et al. [6] argued that the main purpose of channel synergy should be cultivating customers' usage habits of various channels and the service efficiency of channel resources. Therefore, channel synergy should focus on solving the concerns from the perspective of omni-channel customers.

2.2 Channel Conflict

Channel conflict happened when various channel members fought for their own interests. Lin Bingkun et al. [7] elaborated that information asymmetry among members was likely to result in channel conflict, while information integration of various channels could help to solve this problem. While Cuellar and Brunamonti [8] analyzed that the primary cause of the conflict was the inconsistency between the rational purchase behavior of customers and the target products they got interested in during shopping.

2.3 Comprehensive Customer Experience

Comprehensive customer experience in omni-channel retailing. Omni-channel includes not only the traditional channel, but also the information channel, logistics channel and customer mobile channels. Compared with single-channel and multi-channel, omni-channel is not just a simple superposition of quantities, it is rather an integration of all channels and a joint force to meet the comprehensive needs of consumers. Exploring the follow-up effect of various channel integration can help enterprises further explore the impact of channel integration on consumer behaviors. The comprehensive customer experience is a multi-dimensional and personalized psychological state reflected in the process of purchasing decision-making, which belongs to the experience dimension of subjective perception as well.

Based on the theory of customer experience dimension, this paper studies the role of comprehensive customer experience in omni-channel retailing from the perspective of customers.

According to the above analysis, this study proposed the following hypotheses:

H1: Channel synergy has a significant positive influence on comprehensive customer experience in omni-channel retailing.

H2: Channel conflict has a significant negative impact on comprehensive customer experience in omni-channel retailing.

2.4 Channel Performance

Channel performance is the performance of channel members in operating channel flow. It can be measured from the perspective of single-channel member, or from the perspective of all channel members, including the market performance and financial performance of the channel. Channel performance does not solve the problem during the transactions through authority, standards or laws, but more depends on mutual trust and synergies.

At present, scholars mainly measured channel performance from the perspective of channel members. While in this study, we measured channel performance through customer perception on the degree of synergy among channels.

The Influence of Channel Synergy and Channel Conflict on Channel Performance. In omni-channel retailing, the level of the dependence among channels can directly affect the sales performance of retail enterprises. In addition, there is a complementary substitution effect between online and offline channels [9]. Through data mining, Xue Hong et al. [10] found that some retail enterprises have created a new contact point for online retail during the process of customer attraction in offline store, which was a good supplement to the existing channels, and provided a new basis for the positive influence of channel synergies on channel performance.

The impact of channel conflict is mainly reflected in the promotion or reduction of channel performance. Resolving the channel conflict could be benefit on improving the channel vitality. However, once the channel conflict was out of control, the channel performance would be reduced. According to the above analysis, this study proposed the following hypotheses:

H3: Channel synergy has a significant positive influence on channel performance in omni-channel retailing.

H4: Channel conflict has a significant negative impact on channel performance in omni-channel retailing.

The Influence of Comprehensive Customer Experience on Channel Performance. The omni-channel integration of retail enterprises means that it can reduce the risk of channel switching for customers, which will help to increase the frequency of purchase and improve the channel performance. Channel performance could be measured by financial indicators or non-financial indicators. This study selected non-financial indicators and proposed the following hypothesis:

H5: Comprehensive customer experience has a significant positive influence on channel performance in omni-channel retailing.

Based on the above reviews and hypotheses, the research model was established (see Fig. 1).

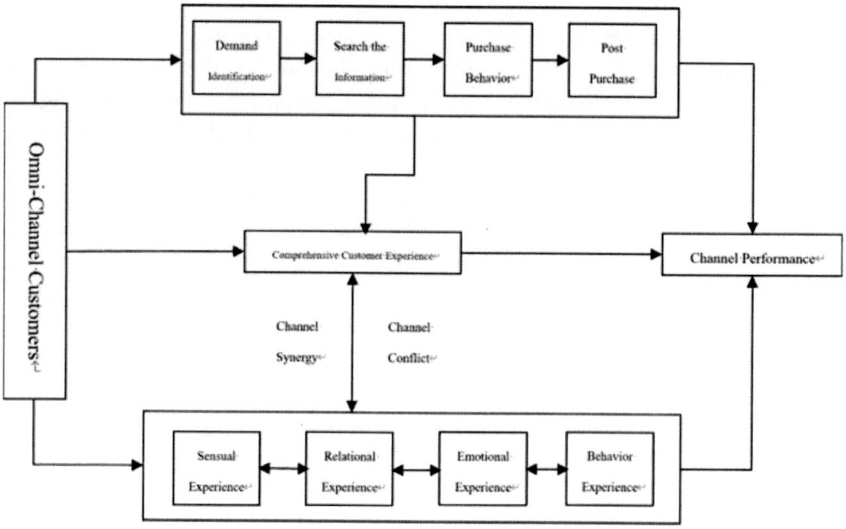

Fig. 1 Research model

3 Research Design

Online survey was used in this study and a questionnaire was designed to collect data. The survey lasted from September to December of 2019. Through the virtual community channel, 800 questionnaires were released in the form of points reward on www.wjx.cn. 719 questionnaires were collected and invalid questionnaires were eliminated. 686 valid questionnaires were obtained with an effective rate of 85.8% in the end.

The focus group in this study are omni-channel customers. During the progress of the questionnaire design, I conducted in-depth interviews with 12 senior omni-channel customers at first. I found that the omni-channel customers repeatedly switched between different channels in the process of demand identification, search the information, purchase behavior and post-purchase behavior, so as to complete their comprehensive experience and other key information. I integrated them into the domestic and foreign mature measurement scales and complemented the questionnaire. The questionnaire sets up screening options to identify omni-channel customers.

4 Empirical Analysis

4.1 Descriptive Statistical Outcomes

The descriptive statistics of the questionnaire showed that male and female accounted for 48.6% and 51.4%, respectively. Among them, 22.4% were senior high school or below, 28.5% were junior college students, 41.8% were undergraduate students and 7.3% were postgraduate students or above degree; in addition, 41.4% were customers with 3–5 years online shopping history, and 23.6% of them had more than 5 years online shopping experience; furthermore, 63.5% said their monthly online shopping expenses accounted for more than 10% of their income per month. All the subjects of valid samples had omni-channel purchase behaviors, and the sample statistics were quite close to the statistics on the development of the Internet in China.

4.2 Correlation Test Outcomes

The correlation of variables was tested, and the correlation coefficient was shown in Table 1. The correlation coefficient of each variable was almost between 0.2 and 0.5. After that, variance inflation factor (VIF) was used for verification. All VIFs were less than 5, and the correlation range of each variable did not exceed the standard. Therefore, the variables could be selected.

Table 1 Variable Pearson correlation coefficient matrix

	Channel synergy	Channel conflict	Sensual experience	Related experience	Emotional experience	Behavior experience	Channel relationship experience
Channel synergy	1						
Channel conflict	0.475[b]	1					
Sensual experience	0.328[b]	0.305[a]	1				
Relational experience	0.395[b]	0.412[b]	0.481[b]	1			
Emotional experience	0.299[b]	0.345[b]	0.349[b]	0.385[b]	1		
Behavior experience	0.451[b]	0.425[b]	0.458[b]	0.363[b]	0.449[b]	1	
Channel relationship performance	0.439[b]	0.402[b]	0.429[b]	0.434[b]	0.443[b]	0.513[b]	1

Note [a, b], respectively, indicate that the regression coefficient was significant at the statistical level of 0.1 and 0.05

4.3 Reliability and Validity Test Outcomes

SPSS software was used for exploratory factorial analysis on the population samples and orthogonal rotation was used on the loading matrix of the outcomes. The measurement scheme was modified through the analysis of combination reliability (CR) of each variable and latent variables Cronbach's α value with the internal consistent reliability method and Corrected Item Total Correlation method (CITC). After calculation, the CITC values of all variables in this study were greater than 0.6 ($P < 0.01$), and the CR values and Cronbach's α values of latent variables were greater than 0.7, indicating that there was certain reliability of the data. Then the validity was analyzed by AVE index. Compared AVE value with the square of correlation coefficient of each variable and all recommended AVE values were greater than 0.5, indicating that there was certain discrimination validity of the data (Table 2).

4.4 Study Results of Model and Hypothesis

Structure Model. AM0S21.0 was used to calculate the fitting degree between the sample data and the model as shown in Table 3. All the indexes have met the requirements of the standard model, therefore the degree of fitting could be considered as reasonable. And then use Amost21.0 to calculate the corresponding path coefficient as shown in Table 4.

It can be seen from Table 4 that the regression standardized path coefficient of channel synergy on comprehensive customer experience was 0.753, and its significance probability P value passed the significance test at the statistical level of 1%, and H1 has been verified. The path coefficient of channel conflict on comprehensive customer experience was −0.204, which was opposite to the hypothesis, while P value was greater than 5% and H2 has not been verified. The path coefficient of channel synergy on channel performance was 0.274, and the P value passed the significance test at the statistical level of 5%, so H3 has been verified. However, due to its small path coefficient, it indicated that channel synergy had relatively small influence on channel relationship performance. And the path coefficient of channel conflict on channel performance was −0.205, and P value was greater than 5%, so H4 has not been verified. While the path coefficient of comprehensive customer experience on channel relationship performance was 0.686, P value passed significance test at statistical level of 1%, so H5 has been verified.

Table 2 Reliability and validity test outcomes

Concept (Latent variable)	Measuring item (Explicit variable)	Factor loading value	α	AVE	CR
Channel synergy	Customers can freely switch between various channels of retail enterprises without feeling barriers	0.758	0.702	0.516	0.815
	Online orders, customers go to offline stores to pick up, return and exchange goods	0.755			
	Online orders, offline store door-to-door measurement, delivery, installation and after-sales service	0.767			
Channel conflict	There are differences in product quality between various channels	0.859	0.811	0.719	0.885
	There are differences in product quality between various channels	0.872			
	There are differences in product quality between various channels	0.812			
Sensual experience	The appearance design of various channels products are good-looking	0.762	0.750	0.609	0.756
	The product description of various channels products are well illustrated and clearly organized	0.798			
Relational experience	The products of various channels can meet the needs of target customers	0.779	0.773	0.598	0.816
	The products of various channels can induce the self-thinking of the customers	0.762			
	Customers can be interested in new products for their previous satisfying purchasing experience	0.798			
Emotional experience	Spending in various channels makes customers feel valued	0.818	0.792	0.641	0.842
	Customers consider certain retail enterprise as a trustworthy one	0.784			
	There is a strong sense of belonging of certain retail enterprise	0.798			
Behavior experience	Customers will buy new products	0.816	0.727	0.684	0.812
	Customers will form good public praise publicity	0.838			
Channel relationship performance	The omni-channel strategies of retail enterprises can increase the lifetime value of customers	0.704	0.858	0.586	0.895
	Various channels can actively explore and respond to customer needs	0.804			
	Various channels can actively respond to environmental changes	0.805			

Table 3 Test results of fitting degree

Fitting indexes	χ^2/df	GFI	CFI	NFI	AGFI	RSEA
Model indexes	2.375	0.937	0.929	0.951	0.918	0.074
Criteria indexes	<3	>0.9	>0.9	>0.9	>0.9	<0.08
Passed or not	Passed	Passed	Passed	Passed	Passed	Passed

Table 4 Results of hypotheses testing

Hypotheses	Path	Standard value	C.R	S.E	P	Verified or Not
H1	Channel synergy → Comprehensive customer experience	0.753	5.365	0.252	[b]	Verified
H2	Channel conflict → Comprehensive customer experience	−0.204	−1.014	0.176	0.064	Not verified
H3	Channel synergy → Channel relationship performance	0.274	2.451	0.089	[a]	Verified
H4	Channel conflict → Channel relationship performance	−0.205	−1.015	0.175	0.063	Not verified
H5	Comprehensive customer experience → Channel relationship performance	0.686	6.255	0.072	[b]	Verified

Note [a, b], respectively, indicate that the regression coefficient was significant at the statistical level of 1% and 5%

5 Conclusion, Inspiration and Limitation

5.1 Conclusion

This study focused on the influence of channel synergy and channel conflict on channel relationship performance from the perspective of omni-channel customer and discussed whether the comprehensive customer experience played an intermediary role in such influence on channel relationship performance.

The Influence of Channel Synergy on Comprehensive Customer Experience and Channel Performance. Channel synergy has a significant positive influence on comprehensive customer experience and channel performance. And the comprehensive customer experience acts a partial mediating role in the relationship between channel synergy and channel performance. The stronger the omni-channel synergy ability of the retailers, the more customers can feel appropriate for the promotion, price and service in various channels, and therefore obtain a benign comprehensive customer experience. In consequence, the omni-channel design of the channel structure can realize a brilliant synergy among channels, produce good comprehensive customer experience and improve channel performance.

The impact of channel conflict on comprehensive customer experience and channel performance. The hypothesis that channel conflict has significant negative impact on comprehensive customer experience and channel performance was not tenable. Customers usually pay no attention to the channel conflict, they tended not being confused by the inconsistency of product promotion information among different channels, and instead, they craved to choose their own purchase path among different channels for the best discounts. In addition, the object of this study was mainly mature retail enterprises, whose disadvantages in channel conflict were not that obvious. Customers were often attracted more by the advantages of products and prices, which in a way weakened the perception of channel conflict. However, retail enterprises could not be allowed the channel conflict being increasingly intense. Since once it was out of control, it would affect the comprehensive experience of customers, thus affecting the channel performance.

5.2 *Inspiration*

A high degree of synergy among channels should be achieved. Retail enterprises often rely on the advantage of lower price to attract a large number of customers to transfer their purchasing behavior from offline to online. However, some retail enterprises adopt multiple standards for the product quality of various channels, which was under the suspicion of deceiving customers. This kind of practice will affect the reputation of enterprises. Retail enterprises must change their management concept and break down channel barriers at the same time, attract customers to conduct cross-purchase in various channels. For example, support online shop ordering, offline store door-to-door measurement, installation and after-sales services.

Comprehensive customer experience strategy should be well-designed. Retail enterprises should integrate sensory, relational, emotional and behavioral elements, and attach importance to mutual benefit with customers. For example, the key point to attract customers to choose channels is to satisfy customers' shopping psychology of novelty, cheapness, convenience and quickness. They can use VR (Virtual Reality) technology to promote product information online. In addition, they can actively build a communication platform with customers to weaken the phenomenon of information asymmetry.

Public Praise Marketing Should Be Strengthened. Current customers have got a strong sense of safeguarding their own rights and emphasize the maximization of their own interests. Therefore, enterprises should obtain and deal with customers' opinions and suggestions in time, while the authenticity and openness of customer evaluation should be kept at the same time. In addition, enough attention should be paid to after-sales services. Normally, enterprises should support customers to return and exchange goods free of charge within a certain number of days even if there lies no quality problem after purchasing under the premise of not affecting the secondary sale, so as to eliminate the last scruples of customers.

5.3 *Limitation*

There were certain limitations in this study. The sample selection of this study is mainly mature retail enterprises which have well-organized countermeasures in channel conflict, which leads to the insignificant sample difference. In the future, other types of retail enterprises can be added, or the industry can be segmented to make the research results more valuable for reference.

Acknowledgements This project is supported by the Scientific research program of Hubei Provincial Department of education in 2020 (B2020202) and Innovation Training Program of Hubei Polytechnic University for College Students in 2020 (S202010920075).

References

1. Li Y, Li G, Kumar TG, Cheng TCE (2019) Omni-channel retailing: do offline retailers benefit from online reviews. Int J Prod Econ 218:43–61
2. Verhoef PC, Kannan PK, Inman JJ (2015) From multi-channel retailing to omni-channel retailing: introduction to the special issue on multi-channel retailing. J Retail 91(2):174–181
3. Hubner A, Wollenburg J, Kuhn H (2016) Last mile fulfilment and distribution in Omni-channel grocery retailing: a strategic planning framework. Int J Retail and Distrib Manag 44(3):228–247
4. Ning Y, Liu P, Yi Z (2019) Research on the relationship between channel synergy performance and consumer interaction behavior based on the context of consumer channel migration. Bus Econ Res 01(9):54–57
5. Kozlenkova IV, Hult TG, Lund DJ, Mena JA, Kekec P (2015) The role of marketing channels in supply chain management. J Retail 91(4):586–609
6. Cao L, Li L et al (2015) The impact of cross-channel integration on retailers' sales growth. J Retail 91(2):198–216
7. Lin B, Lu Q, Yang M (2016) Research on online and offline marketing strategies synergy of multi-channel retailers. Soft Sci 12:135–139
8. Cuellar SS, Brunamonti M (2014) Retail channel price discrimination. J Retail Consum Serv 21(3):339–346
9. Carlson J, O'Cass A, Ahrholdt D (2015) Assessing customers' perceived value of the online channel of multichannel retailers: a two country examination. J Retail Consum Serv 27(11): 90–102
10. Xue H, Zhang P, Li W et al (2017) Research on decision-making synergy of omni-channel consumer behaviors. Comput Eng Sci 39(8):1570–1575

Prospects for the Application of Artificial Intelligence (AI) Technology in the Power Grid

Shijun Zhang, Yaowen Ye, and Juan Yang

Abstract Artificial intelligence is a technical science that studies and develops theories, methods, technologies, and applications used to simulate, extend, and expand human intelligence. The main research areas include expert systems, machine learning, evolutionary computing, fuzzy logic, computer vision, natural language processing, recommendation systems, etc. At present, artificial intelligence has become an important content in the seven major sections of the "new infrastructure," which has a significant role in promoting 5G base station construction, UHV, new energy vehicle charging piles, and industrial Internet technology, especially to lay the foundation for digital and intelligent transformation in many fields. In the power field, it has been applied to the operation, control and management of the power system, and has achieved phased results. The application of artificial intelligence technology in the power system has greatly improved its automation and intelligence level and promoted the intelligent upgrade of the power industry. The following will start from three aspects: the application scenarios, key premises, and application prospects of artificial intelligence technology in the power system.

Keywords Artificial intelligence · New infrastructure · Power field

1 Introduction

Artificial intelligence technology has always attracted great attention in the electric power field since its inception. Traditional artificial intelligence methods such as expert systems, artificial neural networks, fuzzy set theory, and heuristic search

S. Zhang (✉)
State Grid Info-Telecom Great Power Science and Technology CO., LTD, Xi'an 710048, China
e-mail: zhangshijun@sgitg.sgcc.com.cn

Y. Ye · J. Yang
Wuhan Maritime Communication Research Institute, Wuhan, China

Z. Yu et al. (eds.), *Advancements in Mechatronics and Intelligent Robotics*,
Advances in Intelligent Systems and Computing 1220,
https://doi.org/10.1007/978-981-16-1843-7_40

have been widely used in electric power systems. With the high proportion of new energy terminals with multiple characteristics of energy production, storage, and consumption such as distributed power, electric vehicles, and distributed energy storage, the modern power system presents the characteristics of nonlinearity, uncertainty, and temporal and spatial differences, leading to many challenges in the prediction, dispatch, and transaction forms of traditional artificial intelligence methods in the power system.

The new generation of artificial intelligence technology, mainly represented by advanced machine learning theory, big data, and cloud computing, has strong optimization processing capabilities and strong learning capabilities to deal with high-dimensional, time-varying, and nonlinear problems. In order to break through the above technical bottlenecks and provide solutions, the following figure shows the application framework of artificial intelligence technology in key areas of smart grid.

2 Application Scenarios in Power Systems

2.1 Power System Security and Control Field

Power system stability assessment. Traditional machine learning methods often adopt the method of "extracting features first, then classifying and evaluating" when evaluating power system stability [1]. Feature extraction mainly relies on manual experience. When solving complex classification problems, generalization ability is restricted. Deep learning can use the powerful learning ability of the deep model to automatically extract data features and complete classification evaluation to achieve the organic unity of feature extraction and classification analysis [2, 3]. The technical framework of artificial intelligence technology in the power field is shown in Fig. 1.

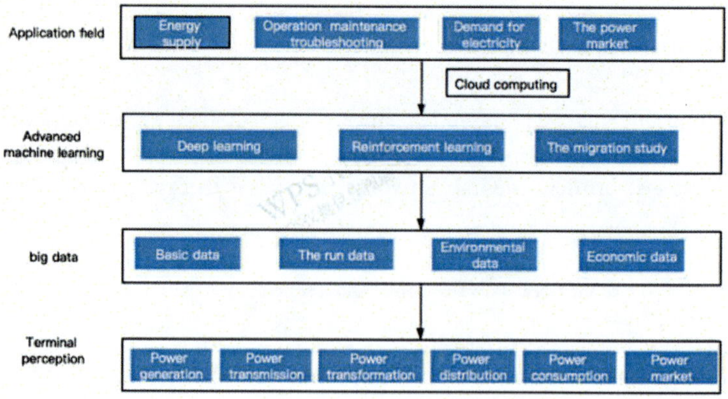

Fig. 1 Application framework of artificial intelligence in key areas of power grid

Power system dispatch and energy dispatch. Power system dispatching and energy dispatching are coordinated optimization decision-making problems with multiple time scales throughout the power generation, transmission, distribution, and power utilization links. Artificial intelligence technology plays an important role in the realization of intelligent dispatch. In order to adapt to the development trend of smart grid decentralized autonomy, studies have proposed a consistent migration learning method, which effectively utilizes historical optimization information and greatly reduces the dispatch risk of large-scale complex power grids [4, 5].

2.2 *Operation and Maintenance and Fault Diagnosis Field*

Power system fault diagnosis. In the field of power system fault diagnosis, traditional methods are based on manual feature extraction and empirical judgment, which increases the complexity and uncertainty of the method. When dealing with complex fault information and large-scale network fault diagnosis, its scope of application is limited. The deep learning method can learn the inherent structural features of the data in depth and integrate the learned feature information into the model construction, thereby reducing the inadequacy of artificial design features and the complexity of traditional feature extraction [6, 7].

Safety risk monitoring on the job site. Video analysis of job site safety risk monitoring covers personnel safety and equipment safety in the power industry. When an abnormal situation occurs, prompt alarms to effectively assist safety personnel in dealing with risks, minimize false alarms and omissions, liberate repetitive labor, continuously improve the level of production safety protection technology, and reduce safety supervision and equipment inspection cycles [8].

Intelligent substation inspection. Automatically and intelligently analyze and judge the inspection data obtained from robot terminals and video surveillance terminals. It can discover hidden dangers of safety accidents in time, including analysis of the status, defects, and personnel behaviors of the equipment in the station, and improve the efficiency and reliability of substation operation and maintenance management, which realizes the coverage of 100% routine inspections and 80% comprehensive inspections of substations and UHV stations, which can support the customization of inspection mission strategies, greatly increase the replacement rate of on-site operations, and reduce the burden on front-line teams [9–11].

2.3 *Electricity Demand Field*

Load Forecasting. Due to the increase in space load, the development of electric vehicles and the promotion of electricity markets, data sources, and data volumes have increased significantly, and the difficulty of load forecasting has increased.

Accurate load forecasting is essential for the safe and economic operation of power systems, and it has always been one of the most widely used scenarios of artificial intelligence technology. With the help of deep learning, the impact of complex factors on the load can be captured, and the prediction accuracy and generalization ability of the model can be improved. On the problem of comprehensive energy demand forecasting, some studies [12, 13] have proposed a short-term electricity, heat, and gas load joint forecasting method based on deep structure multi-task learning.

Demand response. Because both economic dispatch and demand response participate in the energy market and always meet the equilibrium constraints, it is necessary to comprehensively consider the mutual influence of the two to establish a supply–demand interaction model. However, traditional centralized methods are difficult to solve such problems. Some studies [14] have proposed a new deep transfer reinforcement learning algorithm, which uses deep learning to achieve high-precision nonlinear transfer learning by storing effective information in historical optimization tasks, and with the advantages of distributed computing, high-quality optimal solutions can be quickly obtained.

Smart charging. The free charging behavior of large-scale electric vehicles will have a negative impact on the safety and stable operation of the power system. The use of advanced machine learning theory can effectively deal with the uncertainty of electric vehicle charging and speed up the solution of control strategies. Some studies [15] combine dynamic programming with deep learning methods to control the charging of electric vehicles according to real-time electricity prices to minimize the overall cost. The results show that methods based on deep neural networks can approach the optimal strategy.

2.4 *Electricity Market*

Market transactions and bidding. After the reform of the electricity market, it is necessary to consider the game between the operation of power generation, transmission, distribution, and the energy consumption of users. Research shows that reinforcement learning algorithms have certain advantages in solving game problems with uncertainty, and the complexity of the model has a small effect on the efficiency of the algorithm. Some studies [16, 17] have built a multi-agent-based game framework for market players based on multi-agent market game issues and used reinforcement learning methods to optimize the purchase and sale prices of power generators and e-commerce vendors in bilateral contract markets and centralized trading markets. Some studies [18, 19] have also proposed a power generation bidding strategy based on reinforcement learning, and the results show that the benefits of power generation companies that use reinforcement learning bidding strategies have always been better than those that use simple non-learning bidding strategies.

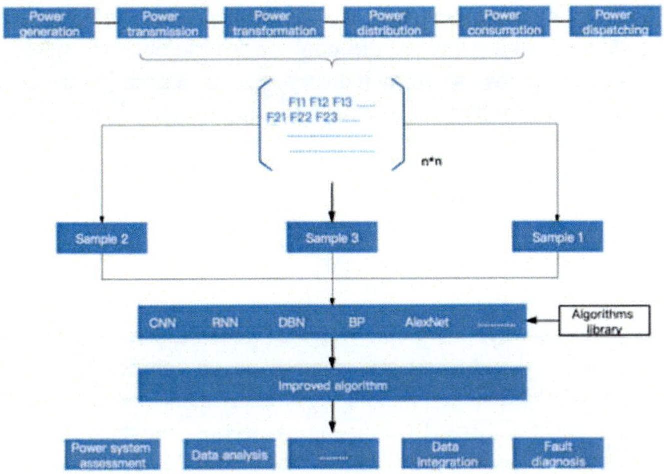

Fig. 2 Application scenarios of artificial intelligence technology in the power field

This part mainly introduces eight application scenarios of artificial intelligence technology in the electric power field. The specific logical relationship is shown in Fig. 2. It can be clearly seen from the figure that the first step in the application of artificial intelligence technology is to extract the characteristics of each link of the power grid. It can be clearly seen from the figure that the first step in the application of artificial intelligence technology is to extract the characteristics of each link of the power grid. Finally, the trained model can be used for power system stability assessment, data disturbance analysis, fault diagnosis, etc.

3 Key Pre-requisites for Application in Smart Grid

Algorithms, data, and computing power are the three elements of the rapid development of artificial intelligence and are also the key prerequisites for the application of artificial intelligence technology in smart grids.

3.1 Algorithm

Algorithm is the premise. In recent years, breakthroughs in algorithms such as deep learning and reinforcement learning have provided significant opportunities for the development of smart grids. However, fields such as fault diagnosis, intelligent dispatching, and power trading have strong industry characteristics, and the specific application scenarios of the power industry are complex. Taking substation operation and maintenance personnel as an example, it is difficult to track and identify the research object due to the unobvious difference in appearance characteristics. Therefore, how to form industry-specific algorithms is an important scientific problem that needs to be solved in the application of artificial intelligence technology in smart grids [20].

3.2 Data

Data is the core. Massive data can provide valuable information for artificial intelligence technology and smart grid, thereby driving artificial intelligence technology to achieve better recognition and universality. Data comes from information perception and collection. Smart sensors are collection terminals for electrical and state quantities of the power system [21]. IoT technology can solve data transmission, and blockchain technology is used to ensure data credibility. How to build a power information physical fusion system through the integration of artificial intelligence, smart sensors, the Internet of Things, blockchain, and other technologies and use all structured data as training and test sets to improve the generalization ability of artificial intelligence which is an urgent problem to solve.

3.3 Computing Power

Computing power is the foundation. The computing power composed of a large number of high-performance hardware can meet the advanced application requirements of artificial intelligence technology in the complex smart grid. It is urgent to study special computing components for artificial intelligence technology for smart grids to improve computing power. The current mainstream methods in the industry: One is to deploy a large number of computing resources such as GPUs and TPUs; the other is to use front-end edge computing modules to process them on the spot as much as possible to improve response speed.

4 Outlook in Future Power Grid

At present, artificial intelligence has made certain progress in safety hazard investigation, labor cost saving, and production efficiency improvement, but it is still in the exploratory stage in solving the integration problem of multiple professional fields. Based on the current development and application status of artificial intelligence technology, it is suggested that the application in future power grid can be divided into the following three stages.

4.1 The First Stage

This stage is aimed at short-term development, that is, the deepening application stage of perceptual intelligence. Continue to deepen the intelligent substitution of simple businesses in the power system, cultivate a team of talents who are both familiar with business and understand artificial intelligence technology, and build artificial intelligence basic experiments and public service platforms. The goal of this stage is to replace repetitive, single, mechanical working methods with intelligent working methods, reduce labor costs, and improve work efficiency and reliability. In addition, the work in this phase will also provide big data resources, infrastructure support, and talent reserves for the next phase.

4.2 The Second Stage

This stage is aimed at the mid-term development route, the cognitive intelligence development stage. Through artificial intelligence technology, the intelligent analysis and decision-making of complex problems in the professional field are gradually realized, and the algorithm model library that can handle the business in the professional field is gradually built. The goal of this stage is to use artificial intelligence technology to effectively integrate various systems, play the synergy between systems, explore the potential value of the system, and achieve management optimization. At the same time, with the help of artificial intelligence technology, it analyzes various complex problems of the smart grid in the context of big data to obtain prediction and analysis capabilities that surpass the intelligence of a single human brain.

4.3 The Third Stage

This stage is aimed at the long-term development vision, that is, the stage of creating intelligence, and gradually building a smart energy system that can realize independent discovery of knowledge, laws, and autonomous control. Throughout the historical process, the current development of artificial intelligence is still in the stage of transition from perceptual intelligence to cognitive intelligence, and there is still a long way to go from "strong artificial intelligence" with the ability to think, create, and affect. The goal of this stage is to build an energy Internet smart system, so that the entire energy system has autonomous cognition capabilities and can interact autonomously with other external smart systems. The independent coordinated control of each link in the system can be realized based on the incentives of external environmental factors, so as to improve energy utilization efficiency and save energy production costs.

5 Conclusion

In short, the article analyzes the four main scenarios in the power system in detail and elaborates on how to maximize the role of artificial intelligence. In the future, after the three basic capabilities of algorithms, data, and computing power are greatly improved, artificial intelligence will exert greater effects in many areas of the power system.

Acknowledgements This work was supported in part by Natural Science Foundation of China (61672433), in part by the Science and Technology Projects of State Grid Gansu Electric Power Company (52272219100Q).

References

1. Cayir A, Yenidoan I, Dag H (2018) Feature extraction based on deep learning for some traditional machine learning methods. In: 3rd international conference on computer science and engineering UBMK'18, pp 494–497
2. Paterakis NG, Mocanu E, Gibescu M (2017) Deep learning versus traditional machine learning methods for aggregated energy demand prediction. In: IEEE Pes innovative smart grid technologies conference Europe. IEEE
3. Hu W, Zheng L, Min Y (2018) Research on power system transient stability assessment based on deep learning of big data technique. Power Syst Technol 41(10):3140–3146
4. Meng F, Akella R, Crow ML (2010) Distributed grid intelligence for future micro grid with renewable sources and storage. North American Power Symposium. IEEE, pp 1–6
5. Zhang XS, Li Q, Yu T (2017) Collaborative consensus transfer Q-learning based dynamic generation dispatch of automatic generation control with virtual generation tribe. Proc CSEE 37(5):1455–1467
6. Yang T, Huang JK, Xu K, Wu JR, Chen SJ (2018) Diagnosis method of power transform-er fault based on deep learning. Power Syst Big Data 21(6):23–30
7. Wang Y, Liu M, Bao Z (2016) Deep learning neural network for power system fault diagnosis. In: Control conference. IEEE, pp 6678–6683
8. Wei D, Gong QW, Lai WQ: Research on internal and external fault diagnosis and fault-selection of transmission line based on convolutional neural network. Proc CSEE 36(S1): 21–28
9. Sun YR, He W (2017) Online prevention and control of power grid operation security risk based on smart grid scheduling technology. Mod Ind Econ Inform 21:55–61
10. Chen LM, Gong WZ, Yi L (2009) Application of intelligent robot iterative inspection system in substation. Power Syst Clean Energy 25(9):30–32
11. Lu SY, Zhang Y, Li JX (2017) Application of mobile robot in high voltage substation. High Voltage Eng 43(1):276–284
12. Shi J, Tan T, Guo J, Liu Y, Zhang J (2018) Multi-task learning based on deep architecture for various types of load forecasting in regional energy system integration. Power Syst Technol 42(3):698–706
13. Kong XY, Zhang F, Zhi J (2018) Short-term load forecasting based on deep belief network. Autom Electric Power Syst 42(5):133–139
14. Rahimi F, Ipakchi A (2010) Demand response as a market resource under the smart grid paradigm. IEEE Trans Smart Grid 1(1):82–88
15. Lopez KL, Gagne C, Gardner MA (2019) Demand-side management using deep learning for smart charging of electric vehicles. Smart Grid IEEE Trans 0(3):2683–2691

16. Zeng J, Zhao X, Li J, Li G, Li K, Wei Z (2017) Game among multiple entities in electricity market with liberalization of power demand side market. Autom Electric Power Syst 41(24): 129–136
17. Wang L, Zhang Z, Chen J (2017) Short-term electricity price forecasting with stackedde-noising autoencoders. IEEE Trans Power Syst 32(4):2673–2681
18. Cheng L, Yu T (2018) Typical scenario analysis of equilibrium stability of multi-group asymmetric evolutionary games in the open and ever-growing electricity market. Proc CSEE 38(19):5687–5703
19. Bai QG (2013) Using reinforcement learning to study the features of the participants' behavior in wholesale power market. Hunan University, Changsha
20. Dai Y, Wang L, Li Y, Yan Y, Han J, Wen F (2018) A brief survey on applications of new generation artificial intelligence in smart grids. Electric Power Constr 39(10):1–11
21. Ju P, Zhou X, Chen W (2018) Yu Y (2018) "Smart Grid Plus" research overview. Electric Power Autom Equip 38(5):2–11

BPSO Algorithm with Opposition-Based Learning Method for Association Rule Mining

Qianyi Zhong, Qian Qian, Yong Feng, and Yunfa Fu

Abstract Traditional association rule mining algorithm does not operate efficiently when processing large and high-dimensional data. With an opposition-based learning method, a binary particle swarm optimization algorithm is proposed to improve the association rule mining problem. The proposed method uses a binary particle swarm algorithm to search for association rules and does not need to manually specify support and confidence thresholds. In addition, opposition-based learning is introduced, and the primary and the secondary opposition-based learning methods are used to reduce the probability of the algorithm falling into local extreme and improve the convergence accuracy of the algorithm. The experimental results show that the improved algorithm converges faster than existing algorithms and balances multiple indexes of reliability, correlation, and comprehensibility, thus mining more effective association rules.

Keywords Association rule mining · Binary particle swarm optimization · Opposition-based learning

1 Introduction

Association Rule Mining (ARM) was first proposed by Agrawal [1] to solve the shopping basket problem. At present, the related research problem is usually solved by precise algorithms or intelligent algorithms. For example, Apriori [1], FP-growth [2], and Eclat algorithms [3] are all typical precise algorithms. However, when dealing with a large number of item sets and complex data, precise algorithms will bring serious time and storage cost, and most of these algorithms need to set support and confidence thresholds manually, resulting in redundant rules or ignorance of important rules. Therefore, intelligent algorithms, such as particle swarm opti-

Q. Zhong · Q. Qian (✉) · Y. Feng · Y. Fu
Yunnan Key Laboratory of Computer Technology Applications, Kunming University of Science and Technology, Kunming 650500, China
e-mail: qianqian_yn@126.com

Z. Yu et al. (eds.), *Advancements in Mechatronics and Intelligent Robotics*, Advances in Intelligent Systems and Computing 1220,
https://doi.org/10.1007/978-981-16-1843-7_41

mization (PSO) algorithm, are utilized because of its simple implementation, fast convergence, and strong searchability. For example, Sarath and Ravi [4] used binary PSO algorithm to obtain association rules from data. Moslehi et al. [5] mixed genetic and PSO algorithms and proposed a multi-standard method to mine association rules from data. The above research shows that the PSO can get effective rules within a reasonable time range, but when the scale of data is large, the PSO has some defects such as premature and low searching accuracy. Therefore, this paper improves the existing PSO algorithm in association rule mining problem. Specifically, opposition-based learning method is introduced to reduce the probability of falling into local optimum and increase the global search ability of the algorithm. In addition, the improved algorithm takes the degrees of support, confidence, promotion, and understanding as the criteria to promote the effectiveness of the rules.

2 Basic Concept of Association Rule

In the problem of association rule mining, let distinct items $I = \{i_1, i_2, \ldots, i_k\}$ be a collection of k items, k is the length of data items, the transactional database $D = \{t_1, t_2, \ldots, t_n\}$ is a collection of n transactions, and $t_n \neq \emptyset$, every transaction t_n corresponds to a subset of I, and $t_n \subseteq I$. If $X \subseteq I$, $Y \subseteq I$, also $X \cap Y = \emptyset$, then $X \rightarrow Y$ is an association rule, where X is called the antecedent, and Y is called the consequent of the rule.

The support of association rule $X \rightarrow Y$ refers to the probability that transactions X and Y appear in dataset D at the same time.

$$\text{support}(X \rightarrow Y) = f_{\text{sup}}(X \rightarrow Y) = \text{support}(X \cup Y) = P(X \cup Y) \tag{1}$$

The confidence of association rule $X \rightarrow Y$ refers to the probability that transaction Y appears in transaction X.

$$\text{confidence}(X \rightarrow Y) = f_{\text{conf}}(X \rightarrow Y) = P(Y|X) = \frac{P(X \cup Y)}{P(X)} \tag{2}$$

3 OBL-BPSO Algorithm for Association Rule Mining

3.1 Binary Particle Swarm Optimization (BPSO) Algorithm

PSO algorithm [6] is a group-based random search algorithm, but the basic PSO is only suitable for solving continuous optimization problems. In order to solve discrete optimization problems, Kennedy and Eberhart [7] improved the basic PSO algorithm and proposed binary particle swarm optimization (BPSO) algorithm. It is assumed that $P(t)$ be the particle swarm of t-th iteration, N is the number of particles, K is the dimension of the search space. The position vector of i-th particle as $X_i^t = (x_{i1}, x_{i2}, \ldots, x_{iK})$, velocity vector as $V_i^t = (v_{i1}, v_{i2}, \ldots, v_{iK})$, local best solution ($Pbest_i$) as $P_i = (p_{i1}, p_{i2}, \ldots, p_{iK})$, global best solution ($Gbest$) as $P_g = (p_{g1}, p_{g2}, \ldots, p_{gK})$. The velocity update formula of the i-th particle in the next iteration is as follows:

$$V_i^{t+1} = \omega V_i^t + c_1 r_1 (P_i - X_i^t) + c_2 r_2 (P_g - X_i^t) \tag{3}$$

Different from the sigmoid function that is used by previous studies, the current formula for updating the position of binary particles is as follows:

$$\begin{gathered} T\left(V_{ij}^{t+1}\right) = \left|\frac{2}{\pi}\arctan\left(\frac{2}{\pi}V_{ij}^{t+1}\right)\right| \\ X_{ij}^{t+1} = \begin{cases} (X_{ij}^{t})^{-1}, & \text{rand} < T\left(V_{ij}^{t+1}\right) \\ X_{ij}^{t}, & \text{rand} \geq T\left(V_{ij}^{t+1}\right) \end{cases} \end{gathered} \tag{4}$$

where ω is the inertia weight, $\omega \in [0, 1]$; c_1 is the self-learning factor, c_2 is the social learning factor; $T(.)$ is the V-shaped function; r_1, r_2 and *rand* are random vectors, the elements of which are between 0 and 1. When the absolute value of the velocity is large, the probability of position change by a V-shaped function will be greater than a sigmoid function. And when the velocity is close to 0, the position will change less than sigmoid function, which makes it easier to approach the global optimal solution.

The value of parameter ω has important influence, so the fine-grained inertia weighting (FGIW) [8] is introduced to dynamically adjust the algorithm, it is a nonlinear adaptive strategy, and the calculation formula is as follows:

$$W_i^{t+1} = W_i^t - \left[(W_i^t - 0.4) \times e^{-\left(\left|Gbest^t - Pbest_i^t\right| \times \frac{\text{iter}}{\text{max\_iter}}\right)}\right] \tag{5}$$

where W_i is ω of the i-th particle, iter is the current iteration number, and max_iter is the max iteration number. W_i's value gradually decreases from 0.9 to 0.4.

3.2 Rule Encoding

In OBL-BPSO algorithm, each particle represents a separate rule contains two layers, and the value is 0 or 1. In the first layer, the value 1 means that the item X appears in the rule. In the second layer, the value 1 means that the item is the rule

Fig. 1 Rule encoding in OBL-BPSO

	i_4	i_1	i_2	i_5	i_3	i_6	i_7
V_1	0	0	0	1	0	1	0
V_2	1	0	0	1	1	0	1

antecedent; otherwise, the item is the rule consequence, and Fig. 1 shows an $i_5 \rightarrow i_6$association rule.

3.3 *Primary and Secondary Opposition-Based Learning*

Since the update of particle positions in the BPSO depends on the position of the particles, when the best solution is a local optimal, all over particles will converge to this local optimal solution, thus losing the diversity of the population and causing the algorithm to fall into the local optimum. To overcome these shortcomings, this paper introduces the opposition-based learning (OBL) method proposed by Tizhoosh [9], and the main idea of OBL is to evaluate the candidate solutions of the problem in opposite direction and select the better solution as the individuals of the next generation. Specifically, this paper proposes the primary and the secondary OBL methods to avoid the "precocity" of the algorithm. In the BPSO algorithm, the opposite number corresponding to a binary candidate solution is $(\bar{x}_1, \bar{x}_2, \ldots, \bar{x}_K)$.

In the early stage of the algorithm, when particles do not converge to the best value, the primary OBL is used to opposite *Gbest*, and the global worst solution (*Wbest*) is replaced by this opposite solution. This method can enhance the ability of particles to explore new solutions without changing the global optimal solution. In the latter stage of the algorithm, particles are becoming more and more similar, opposition of only one particle cannot help the particles to jump out of the local optimum. In this case, the secondary OBL is applied to all particles, to give particles the ability to jump out of the local optimum. The opposite probability (P_m) determines whether to opposite a certain dimension of a particle, the calculation formula of the secondary OBL is as follows:

$$P_m = \frac{1}{K} \quad x_{ij} = \begin{cases} \bar{x}_{ij}, \text{rand} < P_m \\ x_{ij}, \text{rand} \geq P_m \end{cases} \tag{6}$$

where K is the dimension of the particle, x_{ij} is the j dimension value of the *i-th* particle, and rand is the random number between 0 and 1.

3.4 Fitness Function

Some new standards for measuring the quality of rules have been proposed in recent studies, such as degrees of lift and comprehensibility. Lift standards indicate the correlation between transaction X and Y, which is the ratio of the probability that transaction Y appears in transaction X to the probability that transaction Y appears in D dataset:

$$\text{lift}(X \to Y) = f_{\text{lift}}(X \to Y) = \frac{P(Y|X)}{P(Y)}. \quad (7)$$

Comprehensibility [10] is used to measure whether a rule is concise, the shorter the rule, the easier it is comprehensible:

$$\text{Comprehensibility}(X \to Y) = f_{\text{comp}}(X \to Y) = \frac{\log(1 + |Y|)}{\log(1 + |X \cup Y|)} \quad (8)$$

where $|Y|$ and $|X \cup Y|$ are the number of the consequent contained in the post-rule and the total rule, respectively. In this paper, the fitness function is defined as follows:

$$F = \begin{cases} \alpha_1 f_{\text{sup}} + \alpha_2 f_{\text{conf}} + \alpha_3 f_{\text{comp}}, & f_{\text{lift}} > 1 \\ 0, & f_{\text{lift}} \leq 1 \end{cases} \quad (9)$$

where α_1, α_2, α_3 are the weights of support, confidence, and comprehensibility standards, and each weight can be selected according to actual needs.

3.5 OBL-BPSO Algorithm Steps

- Step 1 Perform dimension reduction preprocessing on transaction datasets. If the support value of the infrequent items is less than a minimal value, it will be filtered;
- Step 2 Initialize the particles randomly, the better particle among the original particle and the opposite particle is kept;
- Step 3 Scan the transaction database, calculate the fitness value of each particle by formula (9), and save *Gbest*, $Pbest_i$, and *Wbest*;
- Step 4 Calculate the opposite solution of *Gbest* and replace *Wbest*;
- Step 5 If the particles have the same *Gbest* solution five times during the iteration process, retaining the *Gbest* position, calculating the P_m by formula (6), when the random number generated is less than P_m, opposite a certain dimension of the particle;

Step 6 Use formula (3) to update particle velocity, formula (4) to update particle position, determine the next generation particle velocity and position;

Step 7 Judging whether the algorithm reaches the iteration termination condition, and if so, putting the *Gbest* into the association rule result database; Otherwise, go to step 2.

One best rule can be obtained in each running, to obtain multiple association rules, steps 2–7 need to be repeated *M* times. If the generated rule is the same as the previously obtained rule, then the rule is discarded.

4 Experiments and Discussion

Three datasets are used to test the effectiveness of the algorithm, including extended Bakery dataset [11], classic Groceries dataset [12], and Retail dataset [13]. Classical Apriori algorithm [1], FP-growth algorithm [2], and BPSO proposed in the literature [4] are compared with OBL-BPSO. Apriori and FP-growth algorithms have the same parameter settings, in Bakery dataset, minSup = 0.03, minConf = 0.7; in Retail dataset, minSup = 0.007, minConf = 0.5; in Groceries dataset, minSup = 0.001, minConf = 0.5. OBL-BPSO and BPSO use the same parameters in all datasets, and the number of repetitions is 20, PopSize = 50, max_iter = 30, $\omega = 0.7$, $c_1 = c_2 = 2$. Each group of experiments was run 10 times, and the average value was taken as the final result.

At first, the running time of OBL-BPSO was compared with that of Apriori and FP-growth algorithms, and the test data is Bakery dataset divided into 1000, 5000, 10,000, 15,000, and 20,000 datasets. It can be seen in Fig. 2 that with the expansion of data scale, the running time of Apriori and FP-growth algorithms increase greatly, while the time increment of the OBL-BPSO is small. When the data scale is large, there will be a certain degree of "combination explosion" when using precise

Fig. 2 Algorithm running time under different datasets

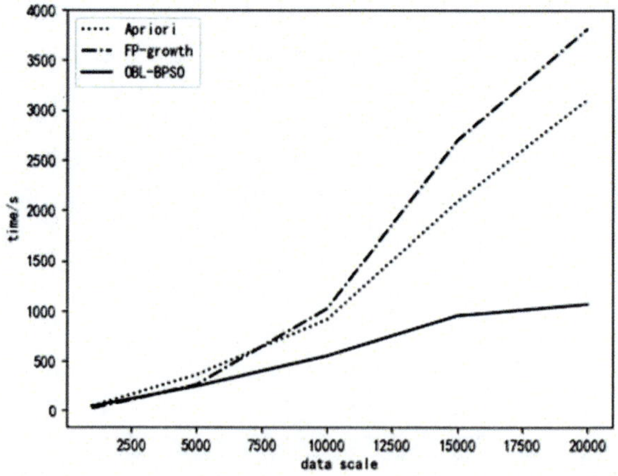

Table 1 Evaluation index results of different algorithms

Dataset	Algorithm	Num	AveSup	AveConf	AveLift	AveComp
Bakery5000	Apriori	29	0.034	0.886	9.887	0.516
	BPSO	13	0.041	0.639	7.501	0.617
	OBL-BPSO	17	0.028	0.939	20.683	0.578
Retail	Apriori	1	0.007	0.514	2.155	0.682
	BPSO	9	0.049	0.255	1.297	0.631
	OBL-BPSO	11	0.006	0.535	2.320	0.487
Groceries	Apriori	4	0.011	0.574	2.429	0.533
	BPSO	16	0.038	0.335	1.629	0.553
	OBL-BPSO	14	0.007	0.552	2.306	0.470

algorithm, because the database needs to be scanned many times, so the running time starts to increase exponentially, while OBL-BPSO can directly optimize rules without generating candidate sets, thus saving running time.

In order to compare the rule mining results of different algorithms, the following indicators are used: number of rules (Num), average support (AveSup), average confidence (AveConf), average lift (AveLift), and average comprehensibility (AveComp). Because Apriori and FP-growth get the same results on the same experimental dataset, only the evaluation index of Apriori is presented in Table 1.

According to Table 1, in strongly related dataset like Bakery 5000 dataset, compared with Apriori and BPSO algorithms, OBL-BPSO gets a reduced number of rules and slightly lower average support. However, the average confidence and the average promotion are high, indicating that OBL-BPSO can obtain rules with higher confidence and promotion, which has been neglected by other algorithms. In the Retail dataset and Groceries dataset, because they are weakly related datasets, the support and confidence of all transactions are low. Therefore, it is difficult to set appropriate thresholds when using Apriori algorithm, and the number of rules mined is limited. However, OBL-BPSO algorithm and BPSO algorithm do not need to set threshold, and more rules are mined in a reasonable time, and the rules mined by OBL-BPSO algorithm have higher confidence.

The above experiments show that the proposed OBL-BPSO algorithm takes less time, and the mined rules have higher reliability, correlation, and comprehensibility.

References

1. Agrawal R, Imieliński T, Swami A (1993) Mining association in large databases. ACM SIGMOD Rec 22(2):207–216
2. Han J, Pei J, Yin Y (2004) Mining frequent patterns without candidate generation: a frequent-pattern tree approach. Data Min Knowl Discov 8(1):53–87
3. Zaki MJ (2000) Scalable algorithms for association mining. IEEE Trans Knowl Data Eng 12(3):372–390

4. Sarath KNVD, Ravi V (2013) Association rule mining using binary article swarm optimization. Eng Appl Artif Intell 26(8):1832–1840
5. Moslehi F, Haeri A, Martínez-Álvarez F (2019) A novel hybrid GA–PSO framework for mining quantitative association rules. Soft Comput 24(6):4645–4666
6. Kennedy J, Eberhart R (1995) Particle swarm optimization. International conference on neural networks. IEEE press, Perth, Western Australia, pp 1942–1948
7. Kennedy J, Eberhart RC (1997) A discrete binary version of the particle swarm algorithm. IEEE international conference on systems. IEEE, Orlando, pp 4104–4108
8. Chauhan P, Deep K, Pant M (2013) Novel inertia weight strategies for particle swarm optimization. Memet Comput 5:229–251
9. Tizhoosh HR (2005) Opposition-based learning: a new scheme for machine intelligence. In: Proceedings of the IEEE international conference on computational intelligence for modelling, control and automation, 2005 and international conference on intelligent agents, web technologies and internet commerce. IEEE, Vienna, pp 695–701.
10. Ghosh A, Nath B (2004) Multi-objective rule mining using genetic algorithms. Inf Sci 163:123–133
11. Ganghishetti P, Vadlamani R (2014) Association rule mining via evolutionary multi-objective optimization. In: MIWAI 2014: proceedings of the 8th international workshop on multi-disciplinary trends in artificial intelligence. Springer, Cham, pp 35–46
12. Pandian A, Thaveethu J (2017) SOTARM: size of transaction-based association rule mining algorithm. Turk J Electr Eng Comput Sci 25:278–291
13. Hossain M, Sattar AHMS, Paul MK (2019) Market basket analysis using apriori and FP growth algorithm. 2019 22nd international conference on computer and information technology (ICCIT). IEEE, Dhaka, pp 1–6

Optimization of Privacy-Preserving in Agricultural Census by Value-Inserting and Reconstruction Algorithm

Yun Liu, Ziyu Wang, and Tian Xiao

Abstract As a major national survey, the National Agricultural Census provides an important basis for researching and formulating rural economic and social development policies and plans. As statistical data analysis methods become diverse, it is necessary to protect personal privacy when processing data. We proposed both value-inserting and reconstruction algorithm (VIR) to ensure the privacy of personal information attributes contained in data. Firstly, a class of composite attributes is defined based on the concepts of QID and sensitive attributes. Afterwards, in order to conduct anonymization, several attribute values are randomly inserted into each sensitive data for every record through an anonymization method based on t-closeness, the randomization parameters of the value-inserting procedure will be optimized simultaneously by solving the minimization problem of L2 distance. Finally, according to the Bayesian method, the distribution of the target combination of attributes is reconstructed. The simulation results show that the utility and run-time of algorithm VIR are optimal when compared with the algorithm UAD and algorithm BUREL.

Keywords Privacy-preserving · Composite Attribute · T-closeness

1 Introduction

The National Agricultural Census is a major national survey. The analysis and research conducted on the statistical data of census can provide significant information for agricultural producers and policymakers [1]. Furthermore, a higher degree of privacy protection for the census data should be ensured in case of the leakage of personal privacy information. Many existing traditional methods related to privacy protection separate sensitive attributes from Quasi-IDentifier (QID), they only protect the attribute values of sensitive attributes. In fact, the attribute values of

Y. Liu · Z. Wang (✉) · T. Xiao
Kunming University of Science and Technology, Kunming 650500, China
e-mail: infante_infanta@163.com

Z. Yu et al. (eds.), *Advancements in Mechatronics and Intelligent Robotics*,
Advances in Intelligent Systems and Computing 1220,
https://doi.org/10.1007/978-981-16-1843-7_42

some QIDs (such as age, census register, and present industry) may also be considered as private information when conducting data analysis [2, 3]. Therefore, both sensitive attributes and these composite attributes with QID features and sensitive characteristics should be considered when performing privacy protection processing on data.

Jin et al. [4] proposed a utility-aware decomposition algorithm (Utility-Aware Decomposition, UAD). The universal privacy rules defined by the UAD algorithm can handle attributes with both characteristics of QID and sensitive attributes. However, the correlation between attributes cannot be observed based on UAD when the value of the corresponding privacy parameter is less than 1. Since all records are random, it is a great challenge to analyze the attribute relevance of the data. Cao et al. [5] proposed a BUREL algorithm. The algorithm arranges the tuples into buckets, which are subsequently reallocated from the bucket to equivalence classes to achieve β-Likeness. Although this algorithm addresses the limitation of *t*-Closeness to a certain extent, its anonymized sensitive value distribution is still interfered by privacy parameter values more than the original distribution.

In order to protect the sensitive information of census data composite attributes while guaranteeing the availability, the Value-Inserting and Reconstruction Algorithm (VIR) is proposed in this paper. After defining the composite attributes, an anonymization method $(t_1, \ldots t_q)$-Closeness is proposed, based on which, to anonymize data with randomized interpolation and optimize randomization parameters, generate anonymous records in aggregated expressions, and finally reconstruct the distribution of attribute target combinations based on Bayesian methods. Utility metrics and runtime metrics are optimal compared to UAD and BUREL algorithms.

2 Definition

2.1 Definition of Identifier

A wide range of existing privacy protection methods divide the attribute within data into three types, including explicit identifiers, quasi-identifiers (QID), and sensitive attributes [6, 7]. With the aim of protecting the sensitive information of all attributes in the data, this paper defines the composite attribute based on the original classification. The attribute types of personal data are divided into "explicit identifier", "sensitive composite attribute", "non-composite sensitive attribute" and "non-sensitive attribute". "Explicit identifier" is an attribute that could precisely determine an individual's identity, and "sensitive composite attribute" is an attribute that can potentially be combined with other records to determine an individual's identity. "Sensitive composite attributes" and "sensitive non-composite attributes" are classified as attributes involving personal private information during data analysis, with attribute values referred to as "sensitive values".

2.2 Definition of Method

Let T and T^* denote the original database and the anonymous database respectively. We introduce a conventional privacy protection method based on the definition of explicit identifiers at the beginning. It could be assumed that T contains only explicit identifiers, QIDs, and sensitive attributes.

Definition 1 (*Equivalence Class*)
It represents a set of records containing all the same QID values as an equivalence class.

Definition 2 (*t-Closeness*)
Let d denotes the number of values for a sensitive attribute. Let the distribution of sensitive values in the entire database be $A = (a_1, \ldots, a_d)$, and the distribution of sensitive values in an equivalence class be $B = (b_1, \ldots, b_d)$.

The anonymous database T^* satisfies t-Closeness if and only if the following inequality holds for all equivalence classes:

$$\mathfrak{D}[A, B] \leq t \tag{1}$$

where $\mathfrak{D}[A, B]$ denotes the earth mover's distance of distribution A and B [8].

$\mathfrak{D}(A, B)$ can measure the cost of converting A distribution to B distribution by the distribution quality of moving A distribution and B distribution. The cost of converting a unit mass from element i of A to element j of B is defined as the ground distance d_{ij} between i and j.

Definition 3 (*Earth Mover's Distance (EMD)*)
Let $A = (a_1, \ldots, a_d)$ and $B = (b_1, \ldots, b_d)$ denote the probability distribution, and let d_{ij} be the ground distance between element i of A and element j of B, which represents the cost of converting a unit mass from element i of A to element j of B. Expect to find $F = [f_{ij}]$ to minimize the overall cost function:

$$\text{WORK}(A, B, F) = \sum_{i=1}^{q} \sum_{j=1}^{q} d_{ij} f_{ij} \tag{2}$$

where f_{ij} represents the flow from A_i to B_j.

Equation (2) is subject to below constraints:

$$f_{ij} \geq 0,\ 1 \leq i \leq q, 1 \leq j \leq q$$

$$a_i - \sum_{j=1}^{q} f_{ij} + \sum_{j=1}^{q} f_{ji} = b_i,\ 1 \leq i \leq q$$

$$\sum_{i=1}^{q} \sum_{j=1}^{q} f_{ij} = \sum_{i=1}^{q} a_i = \sum_{i=1}^{q} b_i = 1$$

When an optimal F is found by solving the optimization problem, EMD is defined as:

$$\mathfrak{D}[A, B] = \text{WORK}(A, B, F) \tag{3}$$

Based on the proposed identifier definition, the mathematical concepts mentioned in the following content are redefined. To facilitate discussion, the anonymization analysis assumes that the database T merely consists of explicit identifiers and sensitive composite attributes.

We suppose the number of records in T is N. Let S denotes the set of sensitive composite attributes in T, q denotes the size of S (i.e. the number of sensitive composite attributes, $q = |S|$), S_j denotes the jth sensitive composite attribute of set S, $D(S_j)$ denotes the range of possible values of S_j, and d_j denote the size of $D(S_j)$. r_i could be regarded as the ith record of T, and $E(r_i, S_j)$ is deemed to be the value of sensitive composite attribute S_j of record r_i.

Definition 4 (*Equivalent Class of S_j*)
Let a set of records with the same sensitive composite attribute value as S_j be an equivalent class of S_j.

Definition 5 ($[\![t,j]\!]$-Closeness)
Let the distribution of elements of $D(S_j)$ in the entire database be $A_j = (a_1, \ldots, a_{d_j})$, and the distribution of sensitive values of S_j in an equivalent class of S_j in the anonymous database T^* be $B_j = (b_1, \ldots, b_{d_j})$. If and only if Inequality (4) holds for all equivalence classes of S_j in T^*, T^* satisfies $[\![t,j]\!]$-Closeness

$$\mathfrak{D}\left[A_j, B_j\right] \leq t \tag{4}$$

where $\mathfrak{D}\left[A_j, B_j\right]$ represents the EMD of distribution A_j and distribution B_j.

Definition 6 ($(t_1, \ldots t_q)$*-Closeness*)
If and only if all $j = 1, \ldots, q$ satisfy $[\![t,j]\!]$-Closeness, the anonymous database T^* satisfies $(t_1, \ldots t_q)$-Closeness.

This paper still uses EMD as the distance function to calculate $(t_1, \ldots t_q)$-Closeness.

Based on the definition of the sensitive composite attributes in Sect. 2.1, all the mathematical methods used for anonymization have been defined.

3 Value-Inserting and Reconstruction (VIR) Algorithm

3.1 Anonymization of Attributes

After removing the explicit identifier from the original database T, the algorithm first creates an empty set $R_{i,j}$ for each sensitive composite attributes S_j of each record r_i in the database and tosses a coin with probability p_j. If the coin is positive, we add the original value $E(r_i, S_j)$ of S_j and $\eta_j - 1$ different elements randomly extracted from $D(S_j)\backslash\{E(r_i, S_j)\}$ to the set $R_{i,j}$, otherwise, add η_j different elements randomly extracted from $D(S_j)$ to the set $R_{i,j}$.

p_j and η_j are two preset randomization parameters. The parameter setting will directly affect the data distribution of the anonymized database. The optimal parameter groups $p_1, \ldots, p_q$ and $\eta_1, \ldots, \eta_q$ are found by solving the L2 distance minimization problem between two distributions. The following content will analyze the parameter optimization process in detail.

Let U_j be a random variable, representing the $E(r_i, S_j)$ of record r_i. Let V_j be another random variable, representing the $E(r_i^*, S_j)$ of record r_i^*. Let the distribution of the value of S_j in the entire database be $A_j = (A_j[1], \ldots, A_j[d_j])$, and the distribution of the sensitive value of S_j in an equivalent class of S_j in the anonymous database T^* be $B_j = (B_j[1], \ldots, B_j[d_j])$, where:

$$A_j[m] = P(U_j = D(S_j)[m]), j = 1, \ldots, q \tag{5}$$

Let κ_c denote the probability of $E(r_i^*, S_j)$ containing the original value and the specified $\eta_j - 1$ different values, and κ_d denote the probability of $E(r_i^*, S_j)$ not containing the original value but containing specified η_j different values. The values of κ_c and κ_d are:

$$\kappa_c = \frac{p_j + (1 - p_j)\eta_j / d_j}{{}_{d_j - 1}C_{\eta_j - 1}} \tag{6}$$

$$\kappa_d = \frac{1 - (p_j + (1 - p_j)\eta_j / d_j)}{{}_{d_j - 1}C_{\eta_j}} \tag{7}$$

where the value range of η_j is 1 to $\tilde{\eta}$. $\tilde{\eta}$ is a preset parameter, and its maximum value is set to d_j.

According to Bayes' theorem:

$$\begin{aligned} &P(U_j = D(S_j)[m] \big| V_j = E(r_i^*, S_j)) \\ &\quad = \frac{P(V_j = E(r_i^*, S_j) \big| U_j = D(S_j)[m]) P(U_j = D(S_j)[m])}{P(V_j = E(r_i^*, S_j))} \end{aligned} \tag{8}$$

here are:

$$P(U_j = D(S_j)[m]) = A_j[m]$$

$$P(U_j = D(S_j)[m] \big| V_j = E(r_i^*, S_j)) = B_j[m] \tag{9}$$

and

$$P(V_j = E(r_i^*, S_j) \big| U_j = D(S_j)[m]) \\ = \begin{cases} \kappa_c & (D(S_j)[m] \in E(r_i^*, S_j)) \\ \kappa_d & (\text{otherwise}) \end{cases} \tag{10}$$

According to the formula of total probability:

$$P(V_j = E(r_i^*, S_j)) \\ = \sum_{\beta=1}^{d_j} P(V_j = E(r_i^*, S_j) \big| U_j = D(S_j)[\beta]) P(U_j = D(S_j)[\beta]) \tag{11}$$

According to the formula (9), (10) and (11):

$$B_j[m] = \frac{\Upsilon(E(r_i^*, S_j), D(S_j)[m])}{\sum_{\beta=1}^{d_j} \Upsilon(E(r_i^*, S_j), D(S_j)[\beta])} \tag{12}$$

among it:

$$\Upsilon(E(r_i^*, S_j), D(S_j)[m]) = \begin{cases} \kappa_c A_j[m] & (D(S_j)[m] \in E(r_i^*, S_j)) \\ \kappa_c A_j[m] & (\text{otherwise}) \end{cases} \tag{13}$$

After calculating all B_j for $\eta = 1, \ldots, \tilde{\eta}$, the maximum value of p_j can be obtained by solving the $t_j = \mathfrak{D}[A_j, B_j]$ about p_j. Since the equation is difficult to solve, the algorithm finds the value of p_j through a binary search. The following Theorem 1 can be used to reduce the search space. After obtaining the parameters η_j and the corresponding p_j, the two parameters are optimized based on the L2 distance.

Because the L2 distance is the square root of d_j times MSE [9], the expected L2 distance for S_j is:

$$E_{L2} = \sqrt{\frac{(1 - d_j)(p_j^2 d_j + \eta_j - (p_j^2 + d_j)\eta_j)}{p_j^2 d_j N(d_j - \eta_j)}} \tag{14}$$

Using the parameters η_j and p_j to calculate the expected L2 distance from Eq. (14), insert the parameter and the expected L2 distance into the associative array ParaCombs and bind the parameter to the expected L2 distance. The algorithm selects the optimal parameter $\langle \widehat{p_j}, \widehat{\eta_j} \rangle$ that minimizes the expected L2 distance for each sensitive composite attributes s_j of record r_i. Repeat the above calculation process for each $j = 1, \ldots, q$ to obtain the optimized parameter set $(\langle \widehat{p_1}, \widehat{\eta_1} \rangle, \ldots, \langle \widehat{p_q}, \widehat{\eta_q} \rangle)$ for r_i. Based on the optimization parameter group, create a set $\{\widehat{R_{i,1}}, \ldots, \widehat{R_{i,q}}\}$ for r_i, and finally insert the aggregated expression as an anonymous record r_i^* into T^*, and perform the anonymization process on each record of the original database T to obtain the anonymous database T^*.

Since the combination $\langle \widehat{p}_j, \widehat{\eta}_j \rangle$ of p_j and η_j satisfies $\mathfrak{D}[A_j, B_j] \leq t_j$, it is known from Theorem 1 below that the maximum distance for any group of records in the entire distribution of the database cannot increase when two sets of records are combined. Therefore, after optimizing parameters for each record, the anonymous database still satisfies $(t_1, \ldots t_q)$-Closeness.

Theorem 1 *Let A, B and B′ denote distribution, and the following inequalities holds:*

$$\mathfrak{D}[A, (1-\lambda)B + \lambda B'] \leq (1-\lambda)\mathfrak{D}[A, B] + \lambda\mathfrak{D}[A, B'] \tag{15}$$

where $\lambda \in [0, 1]$.

Proof Let A, A', B and B' denote distributions.

Since EMD is absolutely homogeneous, namely:

For any $\lambda \geq 0$,

$$\mathfrak{D}[\lambda A, \lambda B] = \lambda\mathfrak{D}[A, B] \tag{16}$$

and EMD is sub-additive, namely:

$$\mathfrak{D}[A + A', B + B'] \leq \mathfrak{D}[A, B] + \mathfrak{D}[A', B'] \tag{17}$$

On the basis of (16) and (17), we can get:

$$\begin{aligned} &\mathfrak{D}[(1-\lambda)A + \lambda A', (1-\lambda)B + \lambda B'] \\ &\leq \mathfrak{D}[(1-\lambda)A, (1-\lambda)B] + \mathfrak{D}[\lambda A', \lambda B'] \\ &= (1-\lambda)\mathfrak{D}[A, B] + \lambda\mathfrak{D}[A', B'] \end{aligned} \tag{18}$$

where $\lambda \in [0, 1]$.

Replace the A' in Inequality (18) with A, and Inequality (15) is proved.

If the database T contains non-composite sensitive attribute and non-sensitive attribute in addition to explicit identifiers and sensitive composite attributes, for non-composite sensitive attributes, it can be regarded as sensitive composite attributes and be anonymized. The non-sensitive attributes will not be processed.

3.2 Reconstruction of the Number of Combination

After generating an anonymous database T^* with an aggregated expression, first determine the combination C of target attributes that need to be analyzed.

Definition 7 (*Target Combination*)
Let C denote all combinations of elements $D(S'_1), \ldots, D(S'_{q'})$, i.e.

$$C = D(S'_1) \times D(S'_2) \times \cdots \times D(S'_{q'}) \tag{19}$$

where S' is the set of attributes that need to be analyzed, q' is the number of attributes in S', and S'_j is the jth attribute of S'.

Let $c_m(m = 1, \ldots, |C|)$ denote the mth element of C, and $c_m[i]$ denote the ith attribute value of c_{m}. Let x_m denote the actual number of records classified as c_m based on the attribute value, and $\widehat{x_m}$ denote the estimated number of c_m which is reconstructed by the algorithm.

Assume $E(r_i^*, S'_j)$ is the value of the jth attribute of S' at the anonymous record r_i^* in the aggregated expression, and $\mathfrak{E}(r_i^*, S')$ denote the Cartesian product of $E(r_i^*, S'_j)\, j = 1, \ldots, q'$:

$$\mathfrak{E}(r_i^*, S') = E(r_i^*, S'_1) \times E(r_i^*, S'_2) \times \cdots \times E(r_i^*, S'_{q'}) \tag{20}$$

The combination of the values of the selected attributes in an original record r is classified as $c_a \in C$. The algorithm first calculates the probability of anonymizing r_i into r_i^*, where for each α and β, $\mathfrak{E}(r_i^*, S')$ of record r_i^* contains $c_\beta \in C$. For the jth element of c_a and c_β, the anonymous value of the jth attribute of S' contains the jth element of c_a with the probability $p_j + (1 - p_j)\eta_j/d_j$, and contains a specific element of $D(S'_j)$ with the probability $p_j \cdot (\eta_j - 1)/(d_j - 1) + (1 - p_j)\eta_j/d_j$ instead of the jth element of c_a. So, $\delta_{\alpha,\beta}$ can be expressed as $\delta_{\alpha,\beta} = \prod_{j=1}^{q'} F(\alpha, \beta, j)$, among which:

$$F(\alpha, \beta, j) = \begin{cases} p_j + (1 - p_j)\eta_j/d_j & (c_\alpha[j] = c_\beta[j]) \\ p_j(\eta_j - 1)/(d_j - 1) + (1 - p_j)\eta_j/d_j & (\text{otherwise}) \end{cases} \tag{21}$$

Equation (21) can calculate all combinations of $\alpha = 1, \ldots, |C|$ and $\beta = 1, \ldots, |C|$, the number of combinations with different results is $2^{q'}$. Only $2^{q'}$ results are calculated $\delta_{\alpha,\beta}$ here.

Let $Z_{\alpha,\beta}$ be a function to return a bit array $b_1, \ldots, b_{q'}$, where:

$$b_j = \begin{cases} 1 & (c_\alpha[j] = c_\beta[j]) \\ 0 & (\text{otherwise}) \end{cases} \tag{22}$$

The number of possible values of $Z_{\alpha,\beta}$ is $2^{q'}$. Therefore, Eq. (21) can be refined as:

$$\delta_{\alpha,\beta} = \prod_{j=1}^{q'} \{Z_{\alpha,\beta}[j](p_j + (1 - p_j)\eta_j/d_j) + (1 - Z_{\alpha,\beta}[j]) \cdot (p_j \cdot (\eta_j - 1)/(d_j - 1) + (1 - p_j) \cdot \eta_j/d_j)\} \tag{23}$$

where $Z_{\alpha,\beta}[j]$ represents the jth bit of $Z_{\alpha,\beta}$.

Assume the original state of the attribute value of a record in T be represented by the random variable U, and the state of each element of $\mathfrak{E}(r_i^*, S')$ be represented by the random variable V.

From the total probability formula, can get:

$$Pr(U = \alpha) = \frac{\sum_{\beta=1}^{|C|} Pr(V \ni \beta) Pr(U = \alpha | V \ni \beta)}{\sum_{\beta=1}^{|C|} Pr(V \ni \beta)} \tag{24}$$

According to Bayes' theorem, can get:

$$\begin{aligned} & Pr(U = \alpha | V \ni \beta) \\ & = \frac{Pr(V \ni \beta | U = \alpha) Pr(U = \alpha)}{\sum_{\gamma=1}^{|C|} Pr(V \ni \beta | U = \gamma) Pr(U = \gamma)} = \frac{\delta_{\alpha,\beta} \widehat{\chi_\alpha}}{\sum_{\gamma=1}^{|C|} \delta_{\gamma,\beta} \widehat{\chi_\gamma}} \end{aligned} \tag{25}$$

$Pr(U = \alpha)$ can be expressed as χ_α / N, which is an unknown value. By using the estimated value $\widehat{\chi_\alpha}$ instead of χ_α, the following formula can be elicited:

$$Pr(U = \alpha) = \widehat{\chi_\alpha} / N \tag{26}$$

The expression $Pr(V \ni \beta)$ in (25) represents the probability that $\mathfrak{E}(r_i^*, S')$ contains c_β. Because there are N records in the anonymous database and c_β occurs ω_β times, it can be derived that:

$$Pr(V \ni \beta) = \omega_\beta / N \tag{27}$$

Thus, it can be obtained according to Eqs. (24), (25), (26) and (27):

$$\widehat{\chi_\alpha}^{\#+1} \Leftarrow \sum_{\beta=1}^{|C|} \omega_\beta \frac{\delta_{\alpha,\beta} \widehat{\chi_\alpha}^{\#}}{\sum_{\gamma=1}^{|C|} \delta_{\gamma,\beta} \widehat{\chi_\gamma}^{\#}} \tag{28}$$

in which, one element of $\widehat{\chi_\alpha}^{\#} (\alpha = 1, \ldots, |C|)$ represents iteration at order #.

Set the initial values for $\widehat{\chi_\alpha}^0$ of all α as ω_α and repeat formula (28) until the difference between $\widehat{\chi_\alpha}^{\#}$ and $\widehat{\chi_\alpha}^{\#+1}$ of all α is small enough. Finally, represent $\sum_{\beta=1}^{|C|} Pr(V \ni \beta)$ as $\prod_{s' \in S'} \eta(s')$, it can be obtained that:

$$\widehat{\chi_\alpha} \Leftarrow \widehat{\chi_\alpha}^{\#+1} / \prod_{s' \in S'} \eta(s') \tag{29}$$

After reconstructing the quantity $\widehat{x_m}$ of each combination c_m $(m = 1, \ldots, |C|)$, the target attribute combination $\widehat{C}$ of quantity reconstruction is obtained.

3.3 Algorithm Pseudocode

Algorithm 1: Value-Inserting and Reconstruction (VIR)
Input: Privacy Parameter $t_1, \ldots t_q$,Original Database T
Output: Combination $\hat{C}$ of target attributes for quantitative reconstruction
Algorithm:
1.for r_i, $i = 1, \ldots, N$ do
2. for j=1,…,q do
3. $A_j \Leftarrow$ Frequency distribution of $D(S_j)$ in T
4. Create associative array ParaCombs
5. for $\eta = 1, \ldots, \tilde{\eta}$ do
6. $\mathfrak{S} \Leftarrow \{Z \mid Z \subseteq D(S_j) \wedge \|Z\| = \eta\}$
7. for each element in $\mathfrak{S}$ do
8. $B_j \Leftarrow$ (12)式
9. $p \Leftarrow t_j = \mathfrak{D}[A_j, B_j]$ Solution of p 10. $el2 \Leftarrow E_{L2}(N, d_j, p, \eta)$ 11. Insert $(\langle p, s \rangle, el2)$ into ParaCombs 12. end for 13. end for 14. $\langle \widehat{p_j}, \widehat{\eta_j} \rangle \Leftarrow$ make the L2 distance minimum in ParaCombs
15. end for 16. Create $\{\widehat{R_{i,1}}, \ldots, \widehat{R_{i,q}}\}$ for record r_i 17.end for 18.Anonymous database $T^* \Leftarrow r_i^*$ in aggregation expressions 19.Target combination $C = D(S_1') \times D(S_2') \times \ldots \times D(S_{q'}')$ 20.for c_m, $m = 1, \ldots, \|C\|$ do 21. $\widehat{\chi_\alpha}^0 = \omega_\alpha$ 22. $\widehat{x_m} \Leftarrow (29)$ 23.end for

The pseudocode of the VIR algorithm is shown in Algorithm 1. The algorithm first calculates the distribution A_j of attribute values from the database T (line 3). B_j is subsequently calculated for each η and the maximum value of p that satisfies

$\mathfrak{D}[A_j, B_j] \leq t_j$ is found by solving the equation $\mathfrak{D}[A_j, B_j] = t_j$ of p (lines 5–9). Calculate the expected L2 distance with the parameter p and η, insert the parameter and the calculated L2 distance into the associative array ParaCombs [10], and select the parameter group with the smallest expected L2 distance value (lines 10–14). Create a collection $\{\widehat{R_{i,1}}, \ldots, \widehat{R_{i,q}}\}$ for the record r_i based on the optimization parameter group, and an anonymous database T^* is created with anonymous record r_i^* represented in aggregated expression (lines 16–18). After determining the target attribute combination C, the quantity of cm per combination is calculated through an iterative process to obtain a quantitatively reconstructed target attribute combination $\widehat{C}$ (lines 19–23).

3.4 Utility Metric

L1 distance and L2 distance are very common in statistics [11, 12], in which L2 distance is widely used as an indicator of privacy protection [13, 14]. However, these two distances are calculated without regards to the relative size of the two values. Even if the two smaller values and the two larger values have the same absolute difference, the relative difference between the smaller values will be greater than the relative difference between the larger values.

Definition 8 (*L1 Distance*)

$$\text{L1 distance} = \sum_{m=1}^{|C|} |\chi_m - \widehat{\chi_m}| \tag{30}$$

Definition 9 (*L2 distance*)

$$\text{L2 distance} = \sqrt{\sum_{m=1}^{|C|} (\chi_m - \widehat{\chi_m})^2} \tag{31}$$

The Hellinger distance that retains the distance measurement attributes (non-negativity, coincidence, symmetry, and triangle inequality) [15] is often used to quantify the distance between two distributions [16]. The Hellinger distance can be regarded as the L2 distance of the square root of the distribution. Since the difference of the square roots of the two smaller values will be greater than the difference of the square roots of the two larger values (if their absolute differences are the same), the Hellinger distance, which can calculate the relative difference, can be used as another distance indicator to more accurately measure the difference between distributions.

Definition 10 (*Hellinger Distance*)

$$\text{Hellinger distance} = \frac{1}{\sqrt{2}}\sqrt{\sum_{m=1}^{|C|}\left(\sqrt{\chi_m} - \sqrt{\widehat{\chi_m}}\right)^2} \tag{32}$$

Use the L1 distance, L2 distance, and Hellinger distance to calculate the distance values between the two distributions, and use these three distance values as utility indicators to evaluate the data availability after privacy protection processing.

4 Simulation Analysis

4.1 Dataset

In this paper, the Third National Agricultural Census dataset about farmers, which is received from Survey Office of the National Bureau of Statistics in Yunnan [17], is used to evaluate the performance of the algorithm. The dataset contains 70 attributes, there are more than 20 million records in total left after deleting all the records with missing values. The size of the attribute domain varies from 2 to 39 following categorizing certain continuous attributes.

The simulation analysis includes the utility analysis and runtime analysis of three algorithms, VIR, BUREL, and UAD. Utility analysis compares the L1, L2, and Hellinger distances of the three algorithms, respectively.

For objective analysis, each privacy parameter $t_j(j = 1,\ldots,q)$ of $(t_1,\ldots t_q)$-Closeness is set to an identical parameter t. The value of t is taken from 0.1 to 0.5, which covers different privacy levels in existing studies [18]. When analyzing attribute combinations of anonymous databases T^*, the number of attributes q' is usually less than 4 [19]. To fully evaluate the performance of the VIR algorithm, q' is set to 4 according to the specific characteristics of dataset, and data attributes involved are census register, education, present industry, and actual cultivated area.

4.2 Utility Analysis

Figure 1 shows the average results of three algorithms, VIR, BUREL, and UAD, for L1, L2, and Hellinger distance based on National Agricultural Census dataset. It can be seen from Fig. 1a and b, although the L1 distance value and L2 distance value of the VIR algorithm are relatively large when t takes a small value (t = 0.1), it is still slightly smaller than the consequence of the BUREL algorithm and UAD algorithm. When t takes other values, the L1 and L2 distance values of the VIR algorithm are significantly lower than those of the BUREL algorithm and UAD algorithm. As can be seen from Fig. 1c, since the Hellinger distance contains the

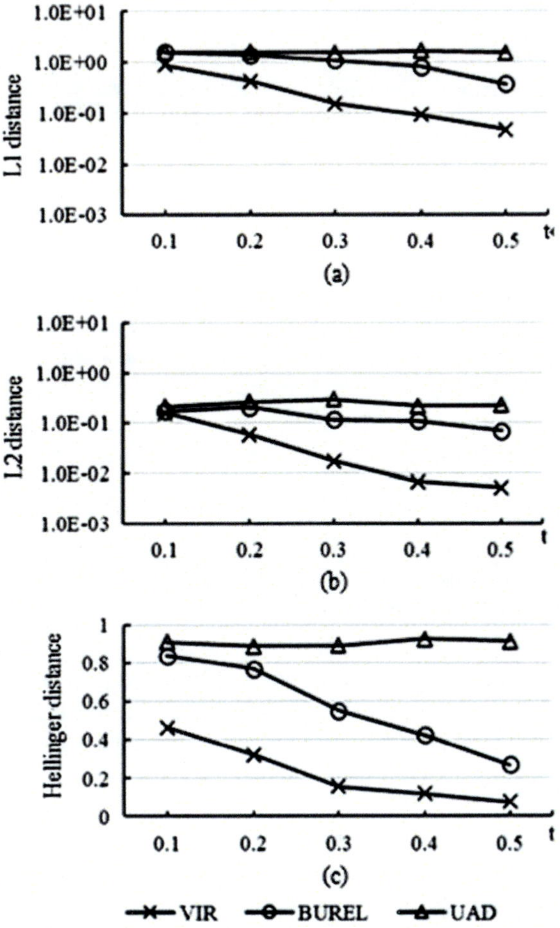

Fig. 1 Comparison of L1, L2, and Hellinger distance

relative difference of each value in the two distributions, for different values of *t*, the Hellinger distance value of the VIR algorithm should be smaller than those of the BUREL algorithm and the UAD algorithm.

It is known from Fig. 1 that while data anonymization is the parameter optimization based on the L2 distance, which may not minimize the L1 distance and the Hellinger distance, the algorithm can still greatly reduce the L1 distance and the Hellinger distance.

4.3 Runtime Analysis

Figure 2 shows the average results of the runtime of the three algorithms VIR, BUREL, and UAD on the National Agricultural Census dataset. It can be seen from

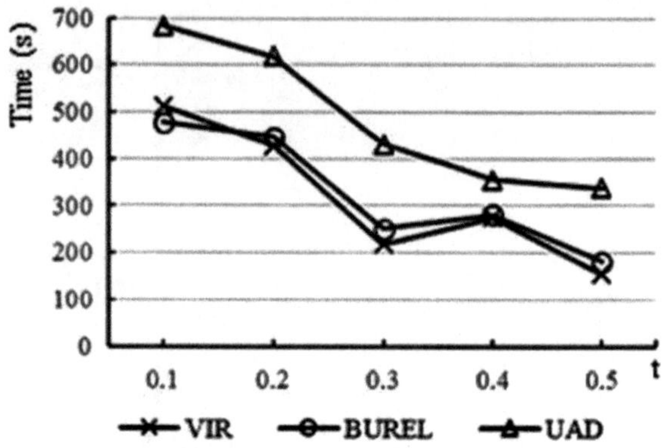

Fig. 2 Comparison of runtime

the diagram that for different values of *t*, the runtime of the VIR algorithm is almost similar to the BUREL algorithm, both of which have significantly less runtime than the UAD algorithm.

5 Conclusion

In order to protect data privacy information while guaranteeing data availability, this paper proposes value-inserting and reconstruction (VIR) based on composite attributes. A *t*-Closeness-based anonymization method is proposed originally after defining a set of composite attributes, followed by randomly adding attribute values to sensitive data to anonymize and optimize randomization parameters, and finally, refactoring the distribution of target attribute combinations according to Bayesian methods. The simulation result shows that the VIR algorithm is optimized in terms of utility metric and runtime. At present, there are various means of privacy protection, and the criteria for evaluating whether the value of utility metric is good may vary greatly due to different data analysis purposes. The direction of the further research is mainly to focus on the application of the proposed concepts in other privacy protection methods and to explore an approach to determining the criteria.

Acknowledgements The National Natural Science Foundation of China under Grant Nos. 61761025 Major Special Science and Technology Project of Yunnan Province under Grant Nos. 202002AD080002.

References

1. Zhang G, Xiao X, Biradar CM et al (2017) Spatiotemporal patterns of paddy rice croplands in China and India from 2000 to 2015. Sci Total Environ 579:82–92
2. Zhang H, Zhou Z, Ye L et al (2015) Towards privacy preserving publishing of set-valued data on hybrid cloud. IEEE Trans Cloud Comput 6(2):316–329

3. Sun G, Song L, Liao D et al (2019) Towards privacy preservation for "check-in" services in location-based social networks. Inf Sci 481:616–634
4. Jin X, Zhang M, Zhang N et al (2010) Versatile publishing for privacy preservation. The 16th ACM SIGKDD international conference on knowledge discovery and data mining. Association for Computing Machinery, Washington DC USA, pp 353–362
5. Cao J, Karras P (2012) Publishing microdata with a robust privacy guarantee. Proc Vldb Endowment 5(11):1388–1399
6. Wagner I, Eckhoff D (2018) Technical privacy metrics: a systematic survey. ACM Comput Surv (CSUR) 51(3):1–38
7. Cao MZ, Zhang LL, Bi XH et al (2018) Personalized (α, l)-diversity k-anonymity model for privacy preservation. Comput Sci 45(11):180–186
8. Park JJ, Florence P, Straub J et al (2019) Deepsdf: learning continuous signed distance functions for shape representation. The IEEE conference on computer vision and pattern recognition. IEEE Computer Society, Long Beach, pp 165–174
9. Xing K, Hu C, Yu J et al (2017) Mutual privacy preserving k-means clustering in social participatory sensing. IEEE Trans Ind Inf 13(4):2066–2076
10. Deng W, Zhao H, Zou L et al (2017) A novel collaborative optimization algorithm in solving complex optimization problems. Soft Comput 21(15):4387–4398
11. Zhou YP, Ye QL (2018) L1-norm distance based least squares twin support vector machine. Comput Sci 45(4):100–105, 130
12. Shen Q, Zhao Y (2018) Statistical feature hashing based on wavelet decomposition. J Appl Sci 36(02):247–254
13. Hassan MU, Rehmani MH, Chen J (2019) Differential privacy techniques for cyber physical systems: a survey. IEEE Commun Surv Tutor 22(1):746–789
14. Ren X, Yu CM, Yu W et al (2018) LoPub: high-dimensional crowdsourced data publication with local differential privacy. IEEE Trans Inf Forensics Secur 13(9):2151–2166
15. Bhatia R, Gaubert S, Jain T (2019) Matrix versions of the Hellinger distance. Lett Math Phys 109(8):1777–1804
16. Jin ZX, Fei SM (2018) Quantifying quantum coherence and nonclassical correlation based on Hellinger distance. Phys Rev a 97(6):062342
17. DCZDYN Homepage, https://www.dczd.yn.gov.cn/
18. Sfar AR, Natalizio E, Challal Y et al (2018) A roadmap for security challenges in the internet of things. Digit Commun Netw 4(2):118–137
19. Shen L, Zhang Z, Long Z (2017) Significant barriers to green procurement in real estate development. Resour Conserv Recycl 116:160–168

Adaptive Parallel Flower Pollination Algorithm

Xin Geng, Qian Qian, Yong Feng, and Yunfa Fu

Abstract Basic flower pollination algorithm (FPA) has some disadvantages, such as weak local search ability, slow convergence speed and low convergence accuracy. This paper proposes an improved adaptive parallel flower pollination algorithm. First, the parallel mechanism is introduced into the FPA to improve the shortcomings of insufficient diversity of a single population in the middle and late stages of the calculation. As a result, the parallel mechanism with different parameter values can further strengthen the algorithm's global search ability while keeping the same local search ability. Second, a nonlinear algorithm behavior conversion probability P is used, and a nonlinear Levi flight step scale factor is added. The former enables the algorithm to dynamically control global and local pollination behaviors according to the calculation period; the latter enables the algorithm to respond to the processing status and adaptively adjusts the jumping step length of pollen individuals in the solution space. In all, the two mechanisms together strengthen the search ability of the algorithm and the ability to jump out of the local optimum. The analysis of the optimization results of the test functions shows that compared with some other algorithms, the proposed algorithm greatly improved the accuracy of the optimal solution and the convergence speed of the optimization.

Keywords Flower pollination algorithm · Adaptive · Parallel mechanism · Coordinate

---

X. Geng · Q. Qian (✉) · Y. Feng · Y. Fu
Faculty of Information Engineering and Automation, Kunming University of Science and Technology, Kunming 650500, Yunnan, China
e-mail: qianqian_yn@126.com

X. Geng · Q. Qian · Y. Feng
Yunnan Key Laboratory of Computer Technology Applications, Kunming University of Science and Technology, Kunming 650500, Yunnan, China

Z. Yu et al. (eds.), *Advancements in Mechatronics and Intelligent Robotics*,
Advances in Intelligent Systems and Computing 1220,
https://doi.org/10.1007/978-981-16-1843-7_43

1 Introduction

Flower pollination algorithm (FPA) [1] is a computational intelligence meta-heuristic algorithm proposed by British scholar Yang in 2012. Its basic idea is to mimic the global and local pollination behaviors of flowers through a conversion probability P in searching of the optimal solutions. This algorithm has been widely used in control engineering, industrial production and other fields due to its advantages such as few parameters and easy adjustment. For instance, in the field of computer science, Pathak and Mahajan used a new pollination-based optimization method to optimize the genetic algorithm and verified the effectiveness of the method [2]; Jensi and Jiji mixed the basic FPA and K-means algorithm to form a new hybrid data clustering method, which solved the initialization difficulty problem of K-means algorithm and reduced the change of falling into local optimum [3]. In the field of imaging science, Rui Wang proposed a flower pollination algorithm with random position correction and applied it to medical image segmentation [4]. Despite its wide applications, the basic FPA has similar shortcomings as other intelligent algorithms, such as easy to fall into the local optimum, premature problem and slow later convergence speed. In order to overcome these shortcomings, many improvements are proposed. For example, Wang and Zhou introduced neighborhood search strategies, dimensional evaluation and change strategies and dynamic switching probability strategies to improve the optimization ability of the algorithm for high-dimensional problems [5]; El-Henawy and Ismail combined the particle swarm optimization algorithm with FPA and applied the hybrid algorithm to solve the integer programming problems and the global optimization problems with constraints [6].

2 Basic Principle of FPA

The FPA realizes the global pollination (global search) and local pollination (local search) behaviors of flowers through a behavior conversion probability $P = 0.8$ [1]. The basic steps are as follows: first initializing the pollen population and parameters, then calculating the fitness values and selecting the corresponding global search or local search methods to update the pollen position according to the conversion probability, repeating the selection until the maximum generation is reached.

Global pollination process: Pollinators pollinate flowers in a way that obeys Levy's flight, and the rule is as follows:

$$x_i^{t+1} = x_i^t + L(g_* - x_i^t) \tag{1}$$

where x_i^{t+1} and x_i^t are the positions of the i-th pollen at time $t+1$ and time t, g_* is the current global optimal pollen, and L is the Levy flight step length function. The calculation formula is as follows:

$$L \sim \frac{\lambda \Gamma(\lambda) \sin(\pi\lambda/2)}{\pi} \frac{1}{s^{1+\lambda}}, \quad (s \gg s_0 > 0) \tag{2}$$

where λ equals 1.5, and $\Gamma(\lambda)$ is the standard gamma function.

Local pollination process rule is as follows:

$$x_i^{t+1} = x_i^t + \varepsilon(x_j^t - x_k^t) \tag{3}$$

where x_j^t and x_k^t are any two pollen j and k in the group at time t, and ε is a random number in the range [0,1].

3 APFPA Algorithm

3.1 Multi-group Parallel Mechanism

Guo et al. [7] proved that the population diversity of the basic FPA continues to decrease during the execution process, which limits the search range of the algorithm in the solution space. High level of population diversity is the basic condition for searching for the global optimal solution. Single population evolution is difficult to overcome the contradiction between algorithm convergence speed and population diversity [8]. Sun and Wang [9] introduced the idea of parallelism into genetic algorithms. Zhou et al. [10] proposed parallel evolution method of multiple subpopulations to maintain the diversity of individuals. Pan et al. [11] proposed an improved multi-group parallel mechanism to control the diversity of the population by changing the parameters of genetic manipulation to achieve different evolutionary functions of the population. This paper proposes a multi-group parallel mechanism in FPA to change the global and local search behaviors by assigning different conversion probabilities, thus maintaining the diversity of the population and improving the performance of the algorithm.

First, initialize three populations N_1, N_2, N_3 and then use three strategies to operate on the populations, respectively, as follows:

Adaptive strategy: Population N_1 uses adaptive conversion probability p_1 (explained in 3.2);

Global search strategy: Population N_2 uses a larger conversion probability p_2 to increase the global search power and search for a better solution in the entire solution space;

Local search strategy: Population N_3 uses a smaller conversion probability p_3 to increase the local search power and tries to find a better solution near the current solution.

When the population falls into a local extremum, the populations N_1 (mainly during the early period of the searching) and N_2 together can make the algorithm jump out of the local extremum and explore new solution space. When the algorithm needs to find a solution with high accuracy, N_1 (mainly during the latter period of the searching) and N_3 together can increase the local search ability, in other words, strengthening the use of the existing solution of the population. The three populations evolve independently, but at each iteration, the global optimal solution is used to update their respective populations to achieve the purpose of communication.

Figure 1 shows the execution process of APFPA's multi-group parallel mechanism. The three populations evolve independently using their own conversion probabilities. The multi-group parallel mechanism not only enriches the diversity of the population, but also can jump out of the local extreme to find better individuals.

3.2 Adaptive Strategy

Adaptive conversion probability: In the basic FPA, a fixed conversion probability of 0.8 [1] is used. For FPA, large probability is required at the early stage to strengthen the global search, but a small probability is required at the latter stage to strengthen the local search. Therefore, the following problems will arise with a fixed conversion probability:

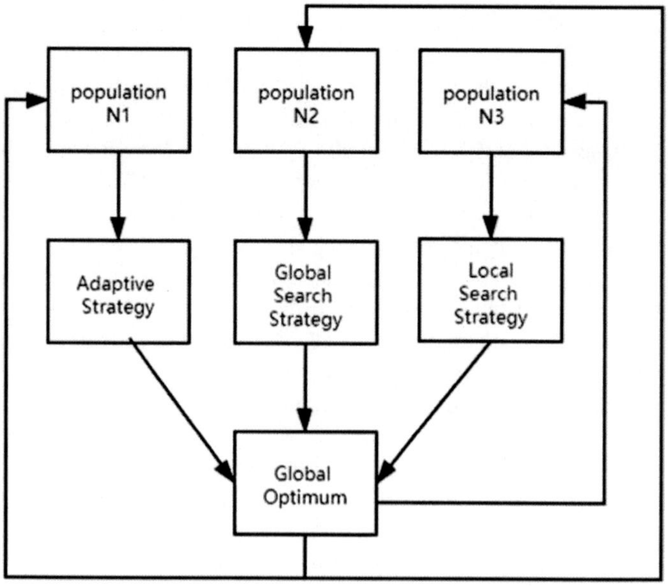

Fig. 1 Multi-group parallel mechanism

If the conversion probability is too large, the global search will be strengthened, which will lead to slow convergence in the later stage of the algorithm.

If the conversion probability is too small, the local search is strengthened, which will cause the algorithm to fall into local extremes during the early stage of the algorithm.

Therefore, in order to better control global search and local search behavior, an adaptive nonlinear conversion probability is adopted. With the increasing of processing iterations, the conversion probability is constantly adjusted to follow the optimization state. The formula is as follows:

$$p_1 = p_3 + (p_2 - p_3)\frac{1}{1 + \exp\left[\eta\left(2\frac{\text{it}}{N\_\text{iter}} - 1\right)\right]} \tag{4}$$

where p_2 and p_3 are the conversion probabilities used by the global search strategy and the local search strategy, respectively. η is the parameter that controls the curvature of the curve, it is the current iteration numbers, and N_iter is the maximum iterations.

Adaptive flight step scale factor: The global search of the basic FPA is relying on the Levy flight to update the individual positions. Although Levy flight can move randomly and generate a random jump step length, its certainty is too low. If more small steps are generated in the whole process of the algorithm, the convergence speed will be slowed down; if there are more large jump steps in the whole algorithm running process, the global optimal value may be missed. Therefore, an adaptive step scale factor is used in present paper: In the early stage of the algorithm, step sizes tend to be larger, so that the algorithm can converge to the optimal solution as soon as possible; when the algorithm executes to the later stage, step sizes tend to be smaller to strengthen the search near the optimal solution to improve the accuracy of the convergence. Therefore, the following formula is used to redefine formula (1):

$$x_i^{t+1} = x_i^t + \varphi L(g_* - x_i^t) \tag{5}$$

$$\varphi = b + (a - b) * \left(\arccos\left(\frac{\text{it}}{N\_\text{iter}}\right) - \frac{N\_\text{iter} - \text{it}}{N\_\text{iter}}\right) \tag{6}$$

where b is the minimum value of the flight step scale factor, a is the maximum value of the flight step scale factor, it is the current iteration numbers, and N_iter is the maximum iterations. The scale factor φ decreases nonlinearly with the increase of the number of iterations. The change trend is shown in Fig. 2.

3.3 Algorithm Execution Flow

Step 1: Initialize populations N_1, N_2, N_3, conversion probability p_2, p_3, scale factors a, b, parameter η and maximum iteration number N_iter;

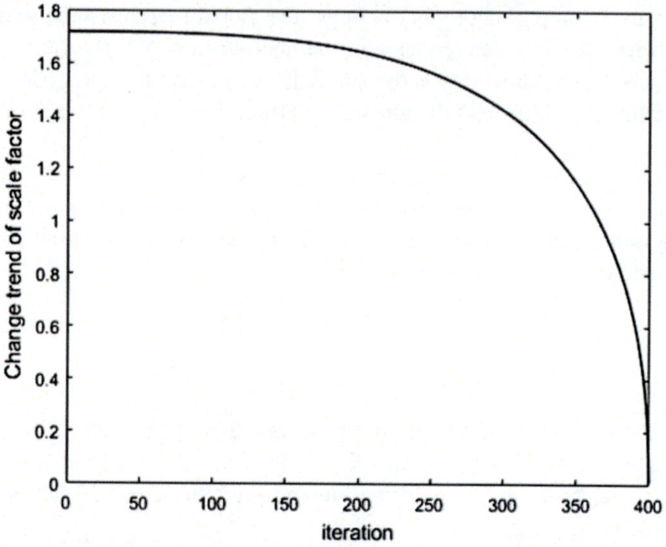

Fig. 2 Change trend of scale factor

Step 2: Calculate the fitness value of each individual of the three populations according to the fitness function and calculate the conversion probability p_1 according to formula (4);
Step 3: Generate a random number in the range [0, 1] and compare the conversion probabilities corresponding to the populations N_1, N_2 and N_3, respectively. If the number is less than the probability, perform the global search operation according to formula (5); otherwise, perform local search operation according to formula (3);
Step 4: Select a global optimum from the three current optimal solutions among the populations N_1, N_2 and N_3 to update the worst individual in the three populations and update the corresponding fitness value;
Step 5: Judge whether the iteration number is less than N_iter, if true, then repeat steps 2, 3 and 4; otherwise, output the optimal pollen position and fitness value.

4 Experimental Simulation and Analysis

In order to verify the effectiveness of the present algorithm, four representative standard test functions are selected for testing, among which f_1 and f_2 are unimodal functions, f_3 and f_4 are multimodal functions. Beetle Antennae search algorithm (BAS) [12] and FPA have common features such as few parameters and simple execution. This paper selects FPA and BAS as the comparison algorithm. The test function is shown in Table 1, where D is the dimension of the variable.

In APFPA, $p_2 = 0.8, p_3 = 0.4, a = 3, b = 0.01, \eta = 10, N\_iter = 400$. In FPA, $p = 0.8$, $N\_iter = 400$. In BAS, step = 1, $N\_iter = 400$.

Table 2 shows the test results of the three algorithms after 50 independent executions. It can be seen from the test results that compared with the basic FPA and BAS algorithms, the APFPA algorithm can get a better optimal solution. The standard deviation of APFPA is smaller than the other two algorithms, which also reflects the robustness of the algorithm. Especially in the optimization of the f_2 function, the other two algorithms are trapped in local extreme, while the APFPA algorithm can converge to 10^{-13}, which shows that the APFPA algorithm has a better ability to jump out of the local extremes.

Table 1 Test function

	Function name	Function expression	Domain	Optimal value
f_1	Sphere	$\sum_{i=1}^{D} x_i^2$	$[-100, 100]^D$	0
f_2	Step	$\sum_{i=1}^{D} (x_i + 0.5)^2$	$[-10, 10]^D$	0
f_3	Alpine	$\sum_{i=1}^{D} \lvert x_i \sin x_i + 0.1x_i \rvert$	$[-10, 10]^D$	0
f_4	Giunta	$\sum_{i=1}^{D} [\sin(\frac{16}{15}x_i - 1) + \sin(\frac{16}{15}x_i - 1)^2 + \frac{1}{50}\sin(4(\frac{16}{15}x_i - 1)) + 0.268]$	$[-10, 10]^D$	0

Table 2 Comparison results of test functions

	Algorithm	Optimal value	Average value	Worst value	Standard deviation
f_1 (D = 20)	APFPA	3.1520e−05	1.3632e−04	3.5405e−04	8.1743e−05
	FPA	1.8318e+01	8.1512e+01	2.1311e+02	4.6171e+01
	BAS	2.6911e+05	4.8261e+05	7.5535e+05	1.0058e+05
f_2 (D = 10)	APFPA	2.7104e−13	1.3355e−09	1.4145e−08	2.7250e−09
	FPA	8.2229e−07	3.1338e−03	9.1589e−02	1.3168e−02
	BAS	5.7792e+02	4.4602e+02	1.6797e+03	5.7792e+02
f_3 (D = 20)	APFPA	3.2753e−04	3.0050e−01	2.7638e+00	5.8364e−01
	FPA	1.4789e+00	4.5932e+00	9.5015e+00	2.0276e+00
	BAS	2.5791e+01	6.5684e+01	1.2120e+02	2.1846e+01
f_4 (D = 20)	APFPA	1.9918e−02	4.3138e−01	6.6285e−01	2.2260e−01
	FPA	5.5620e−01	1.0067e+00	1.4202e+00	2.0595e−01
	BAS	1.6040e+00	4.5405e+00	8.1105e+00	1.5294e+00

5 Conclusion

In order to overcome the shortcomings of basic FPA, this paper proposes a flower pollination algorithm based on parallel adaptive mechanism. The algorithm introduces a variety of group parallel mechanisms to enhance the global optimization capability and introduces an adaptive strategy to guide iterations, so that the algorithm can dynamically adjust the algorithm search capability. Finally, through simulation experiments on four standard test functions, the results show that the algorithm proposed in this paper significantly improves the search ability and convergence ability of the algorithm.

References

1. Yang X (2012) Flower pollination algorithm for global optimization. Springer, Berlin, pp 240–249
2. Pathak P, Mahajan K (2015) A pollination based optimization for load balancing task scheduling in cloud computing. Int J Adv Res Comput Sci 6(7):7–12
3. Jensi R, Jiji GW (2015) Hybrid data clustering approach using K-means and flower pollination algorithm. arXiv preprint arXiv:1505.03236
4. Wang R, Zhou Y, Zhao C, Wu H (2015) A hybrid flower pollination algorithm based modified randomized location for multi-threshold medical image segmentation. BioMed Mater Eng 26(s1):S1345–S1351
5. Wang R, Zhou Y (2014) Flower pollination algorithm with dimension by dimension improvement. Math Probl Eng 2014:
6. El-Henawy I, Ismail M (2014) An improved chaotic flower pollination algorithm for solving large integer programming problems. Int J Digital Content Technol Appl 8(3):72–79
7. Guo Q, Hui X, Zhang J, Li Z (2018) Improved flower pollination algorithm based on multimodal function optimization. J Beijing Univ Aeronaut Astronaut 44(04):828–840
8. He WW, Wang JL, Hu LS (2009) The improvement and application of real-coded multiple-population genetic algorithm. Chin J Geophys 52(10):2644–2651
9. Sun Y, Wang Z (1995) Parallel genetic algorithm. Syst Eng (02):14–16+52
10. Zhou Y, Lu Y, Shi C (1998) Adaptive parallel genetic algorithm based on overcoming premature convergence. J Tsinghua Univ (Sci Technol) (03):3–5
11. Pan J, Qian Q, Fu Y, Feng Y (2021) Multi-population genetic algorithm based on optimal weight dynamic control learning mechanism. J Front Comput Sci Technol 1–17
12. Jiang X, Li S (2017) BAS: beetle antennae search algorithm for optimization problems. Int J Robot Control 1(1)

Research on Optimization Technology of Multi-data Center for Multi-site Integration

Fei Xia, Hu Song, and Yuanhan Du

Abstract This paper designs a "end-cloud" resource coordination mechanism for efficient scheduling of edge computing and cloud computing, designs a "end-cloud" resource negotiation and allocation mechanism, mainly defines a "end-cloud" collaborative computing model, and designs terminal resource access to the cloud. The "end-cloud" collaborative computing model can be abstracted into three layers, namely the computing layer, the middle layer, and the user layer. This paper studies the efficient scheduling technology of edge computing and cloud computing and designs the "end-cloud" collaborative task allocation algorithm to achieve resource optimization scheduling. In this paper, we design a multi data center data consistency optimization and resource utilization improvement implementation scheme, and design a data model application method that satisfies various associated data linkage states when a normal model event occurs. Through the research of this paper, a complete set of multi-data center optimization technical solutions for multi-site integration has been formed, multi-site collaborative optimization effects can be exerted, hidden dangers can be discovered intelligently, optimization strategy development can be assisted, and the actual needs of front-line operation and maintenance personnel of power grid can be met.

Keywords Multi-station fusion · Multi-data center collaboration · Optimization technology

1 Introduction

The business form of power grid companies is in continuous development, and traditional operation and maintenance services can no longer meet the requirements of power grid construction [1]. It requires the grid infrastructure architecture and operation and maintenance capabilities to be able to adapt to the new ubiquitous

F. Xia (✉) · H. Song · Y. Du
Information and Telecommunication Branch, State Grid Jiangsu Electric Power Co., Ltd., Nanjing 211106, Jiangsu, China

Z. Yu et al. (eds.), *Advancements in Mechatronics and Intelligent Robotics*, Advances in Intelligent Systems and Computing 1220,
https://doi.org/10.1007/978-981-16-1843-7_44

form, with the coordination of "edge" and "cloud", and adapt to the multi-level management environment requirements of many multi-site converged data center stations. When the network and resources are complex and the number of jobs and resources is relatively large, intelligent scheduling algorithms are required to be able to adapt to a variety of big data batch processing and stream processing and to play the role of distributed data center and core data center computing. A distributed and highly concurrent data collaborative scheduling method is the key point of this project.

Aiming at the main problems faced by the collaborative operation and intelligent operation and maintenance of multiple data centers for multi-site integration, this paper investigates and analyzes the overall structure and characteristics of the distributed multi-site fusion data center and studies the multi-site fusion-oriented. The data center optimization technology realizes that the power grid data center operation and maintenance capabilities can adapt to the new ubiquitous form, have the coordination of "edge" and "cloud", adapt to the multi-level management environment requirements of many multi-site integrated data center stations, which is adapted to the operation and maintenance challenges of the Internet architecture, and has the architecture maintenance capabilities of multi-center, multi-function, distributed coordination, and integrated operation and maintenance.

At present, the research on resource management of multiple data centers in the world mainly has two directions: One is to divide large hardware resources into smaller virtual resources or containers for users to use [2, 3]; the other is to use cluster management software to form more scattered resources into a logic larger entity on the Internet [4], providing users with more powerful processing capabilities than stand-alone performance. Typical representatives are Amazon's elastic computing cloud and YARN and Mesos resource management systems [5]. The more famous domestic cluster management systems include Typhoon (typhoon) and Feitian. Typhoon is a cloud computing platform that integrates distributed storage and distributed computing developed by Tencent. It is composed of a highly reliable distributed file system XFS, a cluster scheduling system Torca, a distributed semi-structured storage system XCube, a distributed computing framework MapReduce, and an authentication component Taas [6].

2 "End-Cloud" Resource Coordination Mechanism for Efficient Scheduling of Edge Computing and Cloud Computing

The use of terminals, multiple data centers, and networks to unify the "end-cloud" resources can make full use of idle terminal computing resources and improve resource utilization. We design a "end-cloud" resource negotiation and allocation mechanism, mainly define the "end-cloud" collaborative computing model, and design the process of terminal resources accessing the cloud computing resource pool.

The "end-cloud" collaborative computing model can be abstracted into three layers, namely the computing layer, the middle layer, and the user layer, as shown in Fig. 1. The computing layer is composed of multiple heterogeneous resources and services, including cloud computing resources and terminal computing resources. The middle layer realizes the access of terminal heterogeneous resources, the coordinated allocation of "end-cloud" computing tasks, and the "end-cloud" collaborative computing fault tolerance through multi-data center collaborative schedulers and completes the collaborative computing, integration of multiple heterogeneous resources, and efficient resource allocation. The uppermost layer is the user layer. Users can submit task requests through the service interface provided by the cloud platform and interact with the middle layer. The negotiation and collaborative computing process between the terminal and cloud resources is completely transparent to the user.

The "end-cloud" collaborative computing model mainly solves how to reasonably add terminal resources to the cloud computing resource pool and realize the collaborative allocation of resources and collaborative computing. When a user submits a large number of computing tasks, certain computing tasks can be transferred to the terminal for execution, thereby reducing the computing pressure on the cloud.

By deploying related services in idle distributed data centers, it can complete the cloud platform access, access and calculation of related tasks. In the "end-cloud" collaborative computing environment, the terminal and the cloud jointly provide computing resources. Cloud resources can be directly managed by the core data center. The management of terminal resources requires the end-cloud collaborative scheduler to maintain a computing resource pool. The computing capacity (processing time of a unit task), node network bandwidth, and node MAC address are

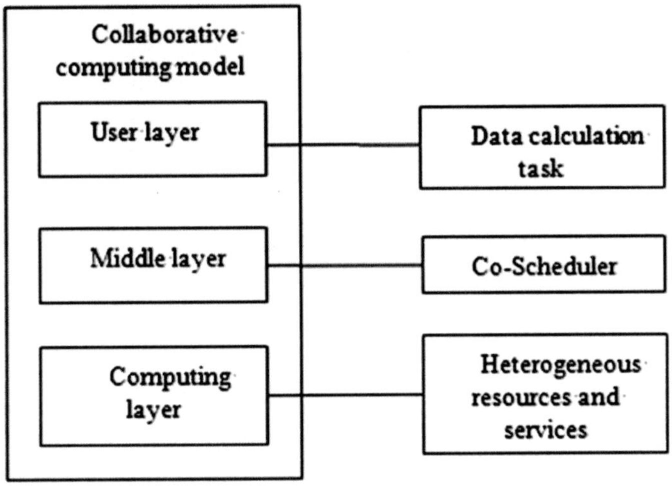

Fig. 1 Collaborative computing model diagram

used as the main storage information of each computing node in the resource pool record. The main goal is to provide a basis for resource allocation for "end-cloud" task scheduling.

3 Efficient Scheduling of Edge Computing and Cloud Computing

In the "end-cloud" collaborative computing environment, the diversity and instability of resources are more obvious than traditional distributed computing environments. Generally, when the network and resources are complex and the number of jobs and resources is relatively large, the scheduling algorithm needs to be able to adapt to the management structure of multiple heterogeneous resources, to ensure the optimal scheduling results and "end-cloud" collaboration. The core of task scheduling method design is to realize a reasonable "end-cloud" collaborative task allocation algorithm.

The cooperative scheduler uses a queue control strategy that combines first-come, first-served, and priority judgment from the task queue of all tasks submitted by the user, takes out the computing tasks that need to be executed most at present, and then allocates reasonable resources to the tasks. The resource scheduling process is shown in Fig. 2.

1. Resource collection acquisition

The scheduler obtains all available resource computing nodes in the system by querying real-time resource pool records. If there is no terminal computing node, it directly allocates computing resources in the cloud for task calculation, and if there are terminal resources, it allocates resources globally;

2. Scheduling algorithm execution

After the cooperative scheduler obtains the combination of all available computing resources in the system, it executes the corresponding scheduling algorithm to calculate the amount of tasks that each computing node should allocate;

3. Task distribution and calculation

According to the execution result of the above scheduling algorithm, the total computing task is divided into multiple subtasks and then distributed to each

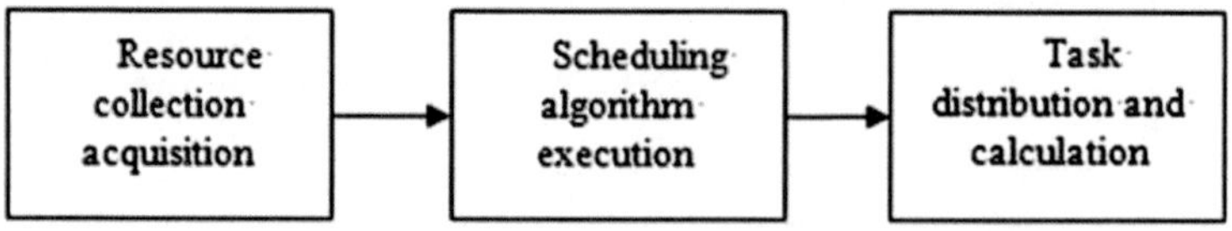

Fig. 2 Resource scheduling process diagram

computing node in a certain order. When the calculation of the subtasks on all computing nodes is completed, the final calculation result can be obtained.

Common scheduling algorithms are mainly divided into the following types:

1. Common scheduling algorithms: such as first-come-first-served strategy, rotation strategy, shortest deadline first, lottery scheduling algorithm (lottery scheduling), etc.
2. Graph-based scheduling algorithm: graph minimum cut algorithm, critical path-based scheduling algorithm, etc.

This project uses common scheduling algorithms for simple tasks and graph-based scheduling algorithms. The resource scheduling of "end-cloud" collaborative tasks can regard all tasks as a directed acyclic graph. In this paper, the minimum cut assignment algorithm based on graph theory is used to deal with the directed acyclic graph, and the optimal task set is obtained. For different task sets, each set is allocated to a physical node to run, so that the smaller the total cost paid, the less network resources the task consumes.

4 Multi-data Center Resource Optimization and Improvement Based on Graph Data Query

This paper designs multiple data center data consistency optimization and resource utilization improvement implementation schemes, which is shown in Fig. 3 and designs data model application methods that meet various related data linkage states when normal model events occur. We use probability calculation method to judge the potential of events, use mathematical calculation induction method to summarize historical rules, and use historical rules to analyze data consistency and resource utilization of multi data center, design rule conversion graph database query and analysis technology, and convert historical rule analysis functions into graph database query and retrieval functions, thereby improving consistency optimization and resources utilization rate improves efficiency.

For the research of graph database retrieval, currently it is mainly focused on structural exploration, such as graph query, shortest path query, confirmation and partial matching. The emergence of structured query language SQL has broadened the application fields of relational databases. In order to make the graph database

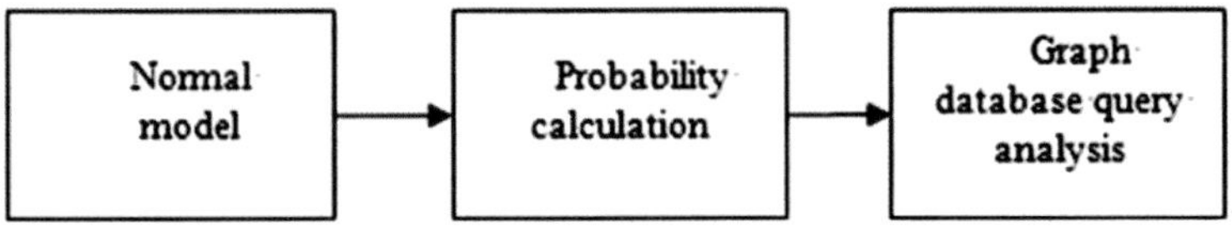

Fig. 3 Multi-data center resource optimization and improvement diagram

well support the storage of big data, the research of graph database index method is becoming a hot area of research by many experts and scholars. In addition, the research on graph database correlation mining (correlation mining) and RDF retrieval is also getting hotter.

This paper designs a data center data consistency optimization and resource utilization improvement plan based on graph database technology. The rapid query and retrieval of collaborative data in multiple data centers is mainly achieved through the original map projection of the graph database, subgraph merging technology, and efficient balanced load scheduling technology, and finally, the retrieval query results are returned to the user through the visualization platform. The reinvestment of the original image is to convert the data consistency optimization and resource utilization improvement plan into a graph analysis plan. The subgraph merging is to convert the graph analysis plan into the corresponding subgraph analysis and merge the subgraph analysis results into the complete plan analysis result. Efficient balanced load scheduling adjusts resource scheduling strategies based on the analysis results. Among them, the key point is the sub-picture merging technology.

According to the requirements of the above scheme, this paper designs a typical query mode based on the characteristics of the graph resource, which is the subgraph matching mode that uses MapReduce to reconstruct the subgraph. The map node merges adjacent nodes in the set into an incomplete subgraph. Reduce node combines incomplete subgraphs. Aiming at the large problem of reconstructed subgraphs caused by hot data in the graph, a data node load balancing algorithm for multi-data center analysis based on optimized MapReduce is proposed to realize subgraph balancing calculation based on rule base. Subgraph matching retrieval given a query graph Q and graph database $D = \{Gi\}$ finds the data graph Gi containing or approximately containing Q in the graph database and returns it to the user. In some cases, the specific location of the match needs to be returned.

5 Conclusion

This paper innovatively researches and analyzes the synchronization and coordination optimization mechanism between the distributed data center and the core data center and proposes the intelligent coordination and optimization technology scheme between the distributed data center and the core data center. It realizes the coordinated scheduling of cloud-side data center infrastructure resources, mainly including computing resources, storage resources, and network resources. It can perform on-demand adjustments and coordinated operation guarantees among multi-level centers to form a complete set of multi-site integration. We propose a complete set of multi-data center optimization technical solutions for multi-site integration. The solution takes full advantage of the multi-site collaborative optimization effect, intelligently discovers hidden dangers, assists in the formulation of optimization strategies, and can meet the actual needs of front-line operation and maintenance personnel of the power grid.

Acknowledgements This work is supported by Science and Technology Project of State Grid Corporation of China (Research and Application on Multi-Datacenters Cooperation and Intelligent Operation & Maintenance, No. 5700-202018194A-0-0-00).

References

1. Amokrane A, Faten M, Qi Z, Langar R, Boutaba R, Greenslater PG (2015) On satisfying green SLAs in distributed clouds. IEEE Trans Netw Service Manage 12(3):363–376
2. Harman M, Mansouri S, Zhang Y (2012) Search-based software engineering: trends, techniques and applications. ACM Comput Surv 45(1):11–15
3. Xia Q, Liang W, Xu Z (2017) The operational cost minimization in distributed clouds via community-aware user data placements of social networks. Comput Netw 112:263–278
4. Boettiger C (2015) An introduction to docker for reproducible research. ACM SIGOPS Operating Syst Rev 49(1):71–79
5. Jo M, Maksymyuk T, Strykhalyuk B et al (2015) Device-to-device-based heterogeneous radio access network architecture for mobile cloud computing. IEEE Wirel Commun 22(3):50–58
6. Cheng D, Zhou X, Wang Y et al (2018) Adaptive scheduling parallel jobs with dynamic batching in spark streaming. IEEE Trans Parallel Distrib Syst 29(12):2672–2685

A Review of Research on Incremental Reinforcement Learning of Dynamic Environment

Changli Hu, Jianfei Shao, Xiaowei Zhang, and Rongjian Wei

Abstract In this paper, incremental reinforcement learning is a new type of reinforcement learning. IRL is a emerging optimal policy to fuse the new state information into the existing knowledge, which makes the agent have a good exploration in the dynamic environment, and can adapt to and track the changing environment. This paper introduces the IRL framework model, which uses Q-learning to test the implementation of all algorithms. The application of IRL in continuous space: the use of policy relaxation and important weighting methods to deal with the changing environment, and the incremental sparse Bayesian method for online dialogue strategy learning research. In the future, we can try to explore the deceptive reward in the agent, apply it in the simulation environment, face the partial observable environment, and even solve the trajectory that is difficult to be robust by combining with the conditional target, which is used in the post disaster robot rescue activities.

Keywords Incremental reinforcement learning (RL) · Dynamic environments · Q-learning · Continuous spaces

1 Introduction

Recently, AlphaGo Zero, which is based on reinforcement learning, has defeated his predecessors, making reinforcement learning attract a large number of scholars' attention and research. Reinforcement learning is an important part of machine learning, and it has a wide application prospect in automatic control and operation research. It interacts with the environment through learning, constantly self-learning to seek the best policy and get the maximum reward.

C. Hu (✉) · J. Shao · X. Zhang · R. Wei
School of Information Engineering and Automation, Kunming University of Science and Technology, Kunming 650500, China
e-mail: 18296632792@163.com

Z. Yu et al. (eds.), *Advancements in Mechatronics and Intelligent Robotics*,
Advances in Intelligent Systems and Computing 1220,
https://doi.org/10.1007/978-981-16-1843-7_45

Since most previous researches on reinforcement learning cannot obtain good generalization performance in dynamic environment, a new learning method is needed to make agents explore effectively in dynamic environment. The new incremental reinforcement learning mechanism can fuse the new state information in the dynamic environment into the existing knowledge, so that the agent can continue to learn and explore. Wang et al. [1] first proposed an incremental learning algorithm for RL in a dynamic environment where reward functions may change over time. Incremental reinforcement learning is to detect the emerging environment and then use dynamic programming to update the value function for the detected environment. Finally, the agent uses the partially updated value function to start a new RL process to form a new optimal policy. Using IRL, we can continue to implement the process of specification learning and generate new optimal policy in dynamic environment, which can reduce the amount of calculation and save time to a certain extent. IRL can not only greatly improve the dynamic maze exploration in the simulation environment, but also use the real robot search and rescue task, intelligent warehouse, and so on. This paper is a summary of the applied direction of IRL and the direction that can be used in the future.

2 Brief Introduction of IRL Framework Principle

All IRQ algorithms are implemented in the test framework, and Env. drift refers to the change of environmental parameters over time. Floating environment generates new data and provides new information for agent. Drift environment is the collection of all state behavior pairs, and the return of agent in this environment is different from that of the original environment. Drifting environment is a space where the original environment and the new environment share part of the state behavior. In the drifting environment, supplementary learning is needed to integrate the new information into the existing knowledge system [2]. Detect drift is the detection of drifting environment [3], which detects part of the state behavior space where reward changes. At the moment, the behavior of the surrogate is not to update the state of the real environment, because the behavior of the surrogate is not to update the state of the environment. Since there is no fusion between the drift environment and the previous priority policy, the priority scanning process is used in the incremental learning process. Using the value function of the original environment and the feedback value function of the new environment, the dynamic programming method with the highest priority is used to update the state behavior space of the drift environment. The new information is fused into the previous optimal value function, that is, the value of the action space of adjacent states will also change. The whole framework is a trade-off between well-designed initialization and exploration. IRL seems to be a standard Q-learning, but in the same case, it is much better than the classical Q-learning exploration. Figure 1 shows the IRL idea design process.

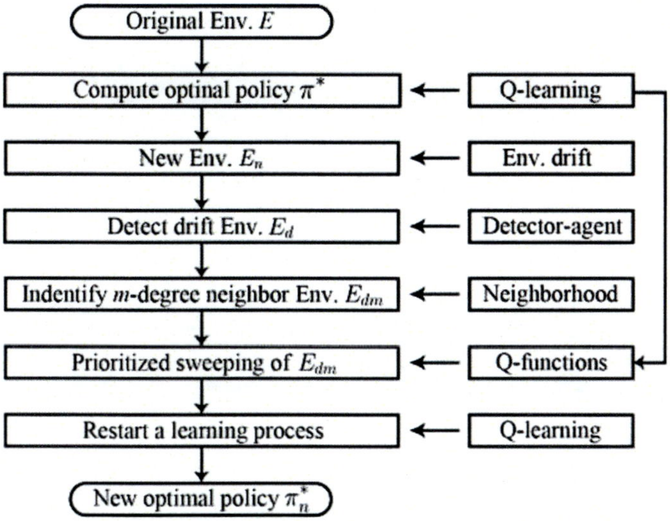

Fig. 1 Integrated IRL algorithm flowchart

3 The Application of IRL in Continuous Space Through Policy Relaxation and Importance Weight

Wang [4] and others studied IRL in continuous space and used nonlinear function approximation of policy search algorithm [5, 6] to deal with challenging reinforcement learning tasks in continuous state action space. It is represented by depth neural network, and parameters are updated by gradient descent method. When the agent updates the parameters from the original optimal value, the training data used by the algorithm may be limited to the data generated by the policy with good performance in the original environment, which makes the agent unable to effectively explore the new environment to find the more potential reward policy, which leads to the local optimal problem. Therefore, the author proposes the method of policy relaxation and importance weight allocation. The update idea is shown in Fig. 2. The dynamic environment of agent interaction $D = \{M_1 \text{ and } M_2, \ldots\}$. The optimal policy $\pi_{\theta_t^*}$ is obtained by training in new environment M_t, and η is the number of times to learn episode. In the learning episode of K, the agent is made to execute a relaxation policy. In order to make the relaxation policy feasible, the author adopts the importance sampling method of model Carlo method. In the process of parameter updating, the author will assign higher weight to the segments containing more information, so as to stimulate the former parameter optimization to integrate into the parameters of the new environment more quickly. This helps the algorithm avoid falling into the deceptive region adjacent to the last optimal parameter space. They solved the expectation of $r(\tau)$ under the distribution of $\tau \sim \pi\theta\ (\tau)$, and it ensures that the estimated policy has the same convergence as

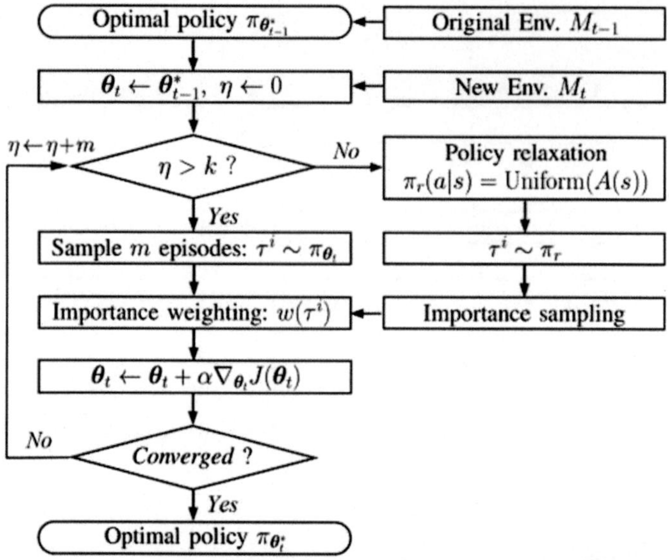

Fig. 2 Flow diagram of the integrated incremental RL algorithm in continuous spaces

the traditional policy gradient [7, 8]. They made a rough analysis of the complexity of the integrated algorithm. Since πr (τ) obeys uniform distribution and is calculated in the original policy gradient, no additional calculation is needed. According to the weight calculated by episodes return, the weight calculated in the calculation will effectively not generate additional cost. In general, the computational complexity is not much different from the classical gradient policy method.

Their proposed method was compared with three baselines (random baseline, pretrained baseline, policy reuse policy gradient baseline) in a two-dimensional navigation task in three dynamic environments (changing goal, puddle effects, changing both the goal and the puddles environments). The test results show that the agent exploration results of IRL in continuous space are better than those of three baselines. The method proposed by them adapts to the dynamic changes in the environment more quickly. By relaxing the policy and weighting the importance, the agent can better explore the new environment and form an effective optimal policy.

4 IRL for Linear Continuous Time Systems and Bayesian Methods

Bian [9] proposed incremental learning method to study online learning. The method is based on RDP and RLS, which obviously improves the efficiency and robustness of the IRL-based algorithm. In this paper, the generalization of robust optimal control problem is solved. Two examples of kinematic model control and power electronic system are presented to verify the effectiveness of the proposed method.

Lee et al. studied the incremental sparse Bayesian method for online dialogue policy learning. In this paper, we describe the probability linear model of MDP and value function approximation, which is extended to Bayesian framework with sparse coercion. Sparsity is an ideal characteristic in the matrix, and it is also a feasible method to control the complexity of the model. In order to avoid over fitting and make the budget calculation more effective and in order to make the coding more sparsity, the author uses Bayesian advantage to add an additional prior structure. The author studies the prior control sparsity, incremental hyperparametric learning, incremental basis function learning, incremental sparse Bayesian learning, and healing prevariance. For further information, please refer to the original [10]. Through the experiment of synthetic data, the paper explains how the incremental function works in all aspects and compares the method with the fixed parameters of Gauss and width, which shows the advantages of learning basis function and the importance of using the correct type of basis function. Gaussian kernel is used to represent the proximity between two belief states, and discrete kernel is used to represent the association of two system behaviors. The results of two kinds of training for simulated user dialogue and real user dialogue show that the paid tester cannot complete the task and accurately evaluate the result, the RL of uncertainty of return is not perfect, and there are also many meanings in the feedback of dialog results [11]. The method proposed in this paper can talk with real users quickly and learn other parameters based on incremental method, which has good effect. The learning speed of this method is faster than that of the baseline system. For re-active learning, the method using predictive uncertainty converges more rapidly than the method used for learning. The author points out that it is very interesting to explore other types of baseline besides Gaussian baseline and to use the policy of rapid adjustment in dialogue.

5 Conclusion

Through the brief introduction of the direction of IRL, IRL is one of the rising learning methods of reinforcement learning. It makes agent have a good exploration and utilization in dynamic environment. And through the method of policy relaxation and importance weight, the agent can form a good policy exploration in the new environment. IRL is also used in the continuous time system, which makes the agent have good exploration efficiency and algorithm robustness. In the sparse Bayesian framework, IRL makes the dialogue system have a good effect in human dialogue. In future, IRL can take advantage of its advantages in dynamic environment, try to explore deceptive reward in agent, apply it in simulation environment, face part of observable environment, even solve the track that is difficult to be robust by combining with conditional target, and apply it in post disaster robot rescue activities.

References

1. Wang Z, Chen C, Li H-X, Dong D, Tarn T-J (2019) Incremental reinforcement learning with prioritized sweeping for dynamic environments. IEEE/ASME Trans Mechatron 24(2): 621–632
2. Wang Z, Chen C, Li H-X, Dong D, Tarn T-J (2016) A novel incremental learning scheme for reinforcement learning in dynamic environments. In: Proceedings of 12th world congress on intelligent control and automation, pp 2426–2431
3. Moore AW, Atkeson CG (1993) Prioritized sweeping: reinforcement learning with less data and less time. Mach Learn 13(1):103–130
4. Wang Z, Li H-X, Chen C (2020) Incremental reinforcement learning in continuous spaces via policy relaxation and importance weighting. IEEE T Neural Netw Learn Syst 31(6):1870–1883
5. Hasselt HV (2012) Reinforcement learning in continuous state and action spaces. Reinforcement Learn 12:207–251
6. Yu Y, Chen S-Y, Da Q, Zhou Z-H (2018) Reusable reinforcement learning via shallow trails. IEEE Trans Neural Netw Learn Syst 29(6):2204–2215
7. Gu S, Lillicrap T, Turner RE, Ghahramani Z, Schölkopf B, Levine S (2017) Interpolated policy gradient: merging on-policy and off-policy gradient estimation for deep reinforcement learning. In: Proceedings of Advances in Neural Information Processing System, pp 3846–3855
8. Barreto A et al (2017) Successor features for transfer in reinforcement learning. In: Proceedings of Advances in Neural Information Processing System, pp 4055–4065
9. Bian T, Jiang Z-P (2019) Reinforcement learning for linear continuous-time systems: an incremental learning approach. IEEE/CAA JAutomatic 6(2):433–440
10. Lee S, Eskenazi M (2012) Incremental sparse Bayesian method for online dialog policy learning. IEEE J Sel Top Sign Process 6(8):903–916
11. Gasic M, Jurcıcek F, Thomson B, Yu K, Young S (2011) On-line policy optimization of spoken dialogue systems via interaction with human subjects. In: Proceedings of IEEE ASRU, pp 312–317

A Review of Image Classification Techniques Based on Meta-Learning

Jianjie Ji, Jianfei Shao, Peng Guo, Changli Hu, and Rong Pu

Abstract Image classification has a wide range of application scenarios, and it is difficult to collect enough data to train the model in many scenarios. Using meta-learning to classify images can solve the problem of small training data star. In this paper, the meta-learning image classification algorithms in recent years are reviewed in detail. In this paper, we summarize the datasets in the existing literature and calculate the existing data through the experimental results. Finally, the difficulties and future research trends of meta-learning image classification are discussed.

Keywords Meta-learning · Image classification · Small sample learning · Graph neural network

1 Introduction

Meta-learning, or learning to learning, is a new direction of deep learning and has become an important research branch after reinforcement learning. Artificial Intelligence → Machine Learning → Deep Learning → Deep Reinforcement Learning → Deep Meta-Learning. Meta-learning method can greatly improve the design of neural networks, such as learning. The relationship between deep learning, machine learning, and intensive learning is well explained in Fig. 1.

J. Ji · J. Shao (✉) · P. Guo · C. Hu · R. Pu
Faculty of Information Engineering and Automation, Kunming University of Science and Technology, Kunming 650000, Yunnan, China

Z. Yu et al. (eds.), *Advancements in Mechatronics and Intelligent Robotics*,
Advances in Intelligent Systems and Computing 1220,
https://doi.org/10.1007/978-981-16-1843-7_46

Fig. 1 The relationship between learning

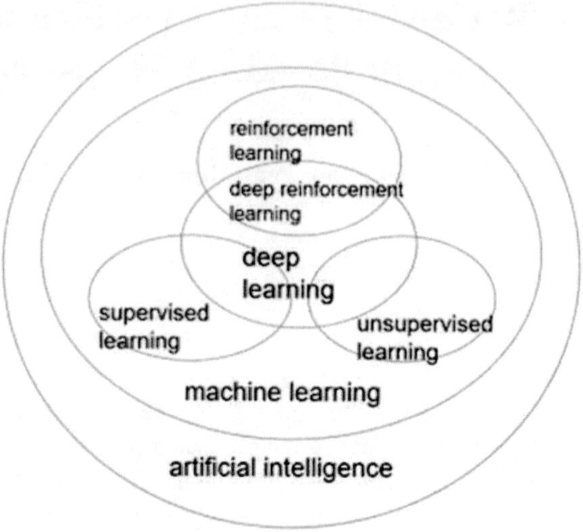

2 Meta-Learning Image Classification Method

First think about why we can learn quickly. Because we can use our previous experience to learn! It is really simple. Why can't deep learning learn fast now? Because we do not know much about making deep learning use of past experience. In most cases, we can only start training from scratch. Using fine-tune to learn a new task often does not work well. Therefore, to make in-depth learning fast, it is necessary to study how to make the neural network make good use of the previous knowledge, so that the neural network can adjust itself to new tasks. Meta-learning, one of the ways to achieve fast learning!

There are three common meta-learning methods:

2.1 Metric Learning

The main purpose of metric learning is to find the best distance measure to fit the geometric structure or similarity relationship of the current learning sample. Compared with fixed measures, such as Euclidean distance, the metric learned from the sample can more accurately reflect the proximity or similarity relationship between samples. Matching networks designed by others using cosine distance measures MN [1]. Inspired by a measurement-based learning based on depth characteristics and an external memory-enhanced neural network, new concepts can be quickly learned from small datasets without fine-tuning and is well suited for fine-gained image classification tasks. Matching networks use "episode." In this form, the tagged task set is sampled from the original data, and then the supporting set and target covers are obtained from the task set to form the meta-task. By training the supporting set, the error on the target set is minimized. Many meta-learning articles adopt "episode" for the dataset. By using the attention mechanism and two-way long and short-term memory network to more accurately extract image features. Calculate the mathematical expression for the test sample label, such as Formula (1),

$$\hat{y} = \sum_{1=1}^{k} a(\hat{x}, x_i) y_i \tag{1}$$

Where X_i, Y_i, is from the support set $s = (x_i, y_i)_{i=1}^{k}$ Samples and their corresponding labels, bifurcations represent test samples. It can be seen that the output of test sample labels is a linear combination that supports centralized labels, and a is a attention mechanism. By using Softmax on the cosine distance, it can be considered as a weighting factor for thousands of labels, which is used to measure.

Supports the correlation between centralized training samples and test samples. Bartunov and Vetrov [1] proposed a generative matching networks (GMN) in 2018, which considers that the generation of new samples follows a conditional probability distribution, which is used to generate new samples for data enhancement and to increase sample diversity.

This does not require that the training data itself be rich in diversity, and that a small amount of data can be used for small sample image classification tasks. Unlike previous matching networks, this method does not match samples directly. Instead, the samples are mapped to the semantic embedded space, in which the conditional likelihood function is used to match the semantic feature vectors of the samples, thus reducing the gap between the feature space and the semantic space. Cai et al. [2] proposed end-to-end memory matching networks. MMN is a meta-learning method that uses internal storage for memory encoding. It compresses the extracted image features into the memory gap with memory writing controller, and then uses context learner, two-way LSTM, to encode the memory gap, which not only improves the ability to represent image features, but also explores the relationship between categories.

System whose output is the embedded vector of the unlabeled sample. By reading in the embedded vector that supports the covert, the memory read-in controller multiplies the two points as the distance similarity most, which is simpler to compute than the cosine distance. The prototype network proposed by Snell et al. (Prototypical Networks, PN) [3] requires the calculation of class prototypes. By learning a measurement space, within which classifiers can classify samples based on the distance between samples and class prototypes. Each class prototype can be averaged by averaging the vectors of all samples in each class in the measurement space. The Euclidean distance is used to determine the category of the sample. Choi et al. used the non-fixed distance measure [4] to solve the problem of small sample image classification of sketch images and natural images, and a structured set matching networks (SCN) was proposed. From the point of view of task execution, this method belongs to thousand yuan learning. This paper calculates the local similarity of the local information corresponding to all the tags between images, combines the local features with the global features, and uses the multi-label data to enhance the interpretability of the image; but at the same time, it also increases the workload of labeling data. Human beings are used to comparing different things when they distinguish things, Sun et al. [5] proposed the end-to-end correlation network in 2018. The correlation network consists of two modules: the embedded module and the related module. The embedded module extracts the features of different samples, and the related module splices the features of different samples and then obtains the maximum score of correlation between different samples. Zhou et al. [6] proposed a visual analogy network based on the basis of thousand embedding regression to learn the low-dimensional embedding space and then learn the linear mapping function of classification parameters from the embedding space. When classifying a new class, the similarity between the new class sample and the embedded feature learned from the base class is measured.

2.2 *Using External or Internal Memory*

Learning from past experience, adding memory module to neural network Basic idea: since we want to learn from past experience, can we add memory to the neural network? Taking the article meta-learning with memory augmented neural networks as an example, let us take a look at his network structure (Fig. 2).

We can see that the input of the network takes the last y label as the input and adds external memory to store the last x input. This makes it possible for y label and X to establish a connection when the next input is back propagation, so that the subsequent x can obtain relevant images through external memory for comparison to achieve better prediction [8].

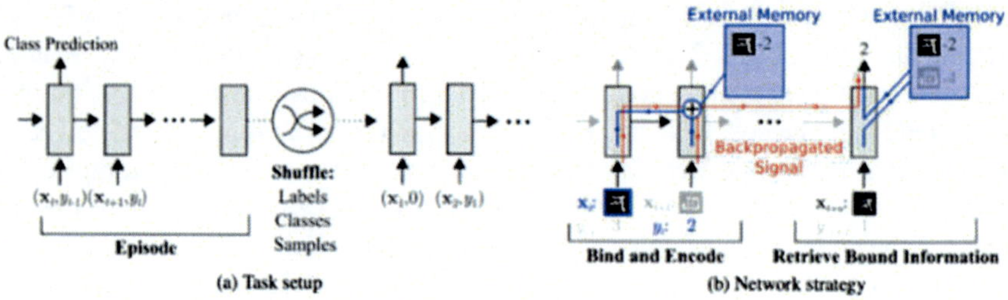

Fig. 2 Structure of memory neural network

2.3 Based on Finetune

After obtaining a certain amount of annotation data, fine-tuning is carried out based on a basic network. The method based on fine-tune has faster training speed and less training data, but the basic network needs to be obtained through datasets with a large number of tags.

Generally speaking, meta-learning is divided into two stages: slow learning of meta-level models on different tasks and fast learning of benchmark models executed in a single task. The purpose of meta-level model is to learn common knowledge in different tasks and then transfer it to the benchmark model to help learning on a single task. Munkhdalai and Yu [9] constructed a meta-networks using convolutional neural network in 2017, which used the convolutional neural network to construct a meta-network.

Learning, following the previous work, divides the meta-learning problem into two stages. It proposes a faster learning method, which uses one neural network to predict the parameters of another neural network. The generated parameters are called fast weights, which are used to accelerate the learning speed of the network (SG D). The training of the model is performed by the meta-level model and the benchmark model, including the acquisition of meta-information, the generation of fast weights and the updating of slow weights. The benchmark model uses the loss gradient information as the meta-information to obtain the meta-information in different meta-tasks and store it in the external devices. The fast weights are generated by prediction to accelerate the network learning. Zhou et al. [10] put forward the deep meta-learning model (DML) can learn in the concept space instead of in the traditional visual space. The model consists of three modules: concept generator, meta-learner, and concept decision maker. The concept generator uses deep residual network to capture advanced semantic concepts and then uses concept decision maker for information feedback and optimization. At the same time, the extracted features are used as concepts for further learning with thousand yuan learner. This method reduces the gap between visual space and semantic space by learning semantic concepts and uses deep residual network to build more complex models to adapt to complex data patterns. Santoro et al. [7] proposed to use RNN architecture and external storage memory. The use of external storage improves the accuracy of the model, but it also reduces the efficiency of the model and takes up a lot of storage space. MTL can effectively reduce the parameters of model updating by using transfer knowledge and build a deeper network to increase the complexity of the model. In 2018, a data enhancement method suitable for any meta-learning model was proposed. Aiming at the problem that the sample pattern captured by the generation model is insufficient, and the generation model is proposed to generate new samples. These samples are real samples and noise vectors passing through three layers of multi-layer perception with relu nonlinear activation function. The data enhancement method can be used in combination with any meta-learning algorithm. Since the generation network is used to generate new samples, the quality of the experimental results depends on the quality of the new samples

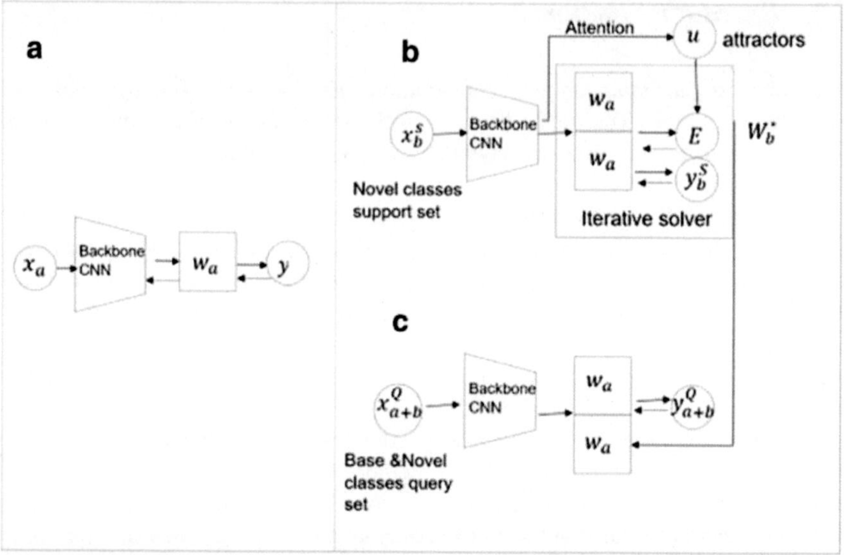

Fig. 3 Attention attractor networks structure

generated. Wang et al. [11] proposed a model regression networks. This method considers that there is a transformation between the classifiers learned from small sample data and those learned from large sample data, which can be learned through deep regression network. At the same time, as a kind of prior knowledge, the transformation can help the classification task on small sample data. In 2019, some researchers combined incremental learning with meta-learning and proposed the attention attractor networks (AAN) model, which not only shows good performance in the new class, but also can not forget the knowledge learned on the base class, A pre-training model is built on the base class to learn the classification parameter wa. Stage B combines attention mechanism and uses the classification parameter WB of a new task each time. Stage C takes Ba and w as the classification parameters WB* of the base class and the new class to test the meta-task. It represents the contribution of the task in the implementation of classification, which makes the classifier more flexible and applicable, and it is easier to classify a single new sample (Fig. 3).

3 Conclusion

This paper summarizes the current image classification algorithms based on meta-learning. Meta-learning takes into account the measurement method, model design, and initialization strategy. From the above analysis, we can see that meta-learning is in the ascendant, and various magical ideas emerge in endlessly, but the real killer algorithm has not yet appeared. We are looking forward to the future development!

References

1. Vinyals O, Blundell C, Lillicrap T, Koray K (2016) Matching networks for one shot learning. In: Proceedings of the 30th international conference on neural information processing systems. MIT Press, Barcelona, Spain, pp 3630–3638; Bartunov S, Vetrov D (2018) Few-shot generative modelling with generative matching networks. In: International conference on artificial intelligence and statistics. JMLR, Playa Blanca, Spain, pp 670–678
2. Cai Q, Pan Y-W, Yao T, Yan C-G, Mei T (2018) Memory matching networks for one-shot image recognition. In: Proceedings of the IEEE conference on computer vision and pattern recognition. IEEE, Salt Lake City, Utah, pp 4080–4088
3. Snell J, Swersky K, Zemel R (2017) Prototypical networks for few shot learning. In: Advances in neural information processing systems. MIT Press, Long Beach, pp 4077–4087
4. Choi J, Krishnamurthy J, Kembhavi A, Farhadi A (2018) Structured set matching networks for one-shot part labeling. In: Proceedings of the IEEE conference on computer vision and pattern recognition. IEEE, Salt Lake, USA, pp 3627–3636
5. Sun Q-R, Liu Y-Y, Chua T-S (2019) Meta-transfer learning for few-shot learning. In: Proceedings of the IEEE conference on computer vision and pattern recognition. IEEE, Long Beach, USA, pp 403—412
6. Zhou L-J, Cui P, Yang S-Q, Zhu W-W, Tian Q (2017) Learning to learn image classifiers with informative visual analogy. ArXiv preprint arXiv:1710.06177
7. Santoro A, Bartunov S, Botvinick M, Wierstra D, Lillicrap T (2016) Meta-learning with memory-augmented neural networks. In: Proceedings of the 33rd international conference on machine learning, pp 1842–1850
8. Mitchell TM, Thrun SB (1993) Explanation based neural network learning for robot control. In: Advances in neural information processing systems. MIT Press, Denver, USA, pp 287–294
9. Munkhdalai T, Yu H (2017) Meta networks. In: Proceedings of the 34th international conference on machine learning. ACM, Sydney, NSW, Australia, pp 2554–2563
10. Zhou F-W, Wu B, Li Z-G (2018) Deep meta-learning to learn in the concept space. Arxiv preprint arXiv:1802.03596
11. Wang Y-X, Hebert M (2016) Learning to learn: model regression networks for easy small sample learning. In: European conference on computer vision. Springer, Amsterdam, Netherlands, pp 616–634

Selective Heterogeneous Ensemble Method Based on Local Learning and Evolutionary Multi-objective Optimization for Wind Power Forecasting

Lixian Shi, Huaiping Jin, Biao Yang, and Huaikang Jin

Abstract To solve the problems of strong randomness and fluctuation in wind power forecasting, a novel wind power forecasting model using selective the heterogeneous ensemble (SHeE) method based on local learning and evolutionary multi-objective optimization (EMO) is proposed. First, Lasso regression is used to feature selection to remove irrelevant variables and redundancy. Then, a set of sample sets is constructed via a novel K-nearest neighbor (KNN)-based clustering method to enrich ensemble diversity from the perspective of data. To further inspire the diversity, three modeling methods, support vector regression (SVR), Gaussian process regression (GPR), and partial least squares (PLS) are applied to build the heterogeneous model library. Subsequently, ensemble pruning is performed by EMO. When a new sample arrives, the ensemble prediction result can be obtained by simple averaging. Finally, the effectiveness and superiority of the proposed method are demonstrated through a real wind farm dataset.

Keywords Wind power forecasting · Heterogeneous ensemble · Local learning · Multi-objective optimization

L. Shi · H. Jin · B. Yang
Faculty of Information Engineering and Automation, Kunming University of Science and Technology, Kunming 650500, China

L. Shi · H. Jin (✉) · B. Yang
Yunnan Key Laboratory of Artificial Intelligence, Kunming University of Science and Technology, Kunming 650500, China
e-mail: jinhuaiping@kust.edu.cn

H. Jin
Huaneng Renewables Co., Ltd. Yunnan Branch, Kunming 650000, China

Z. Yu et al. (eds.), *Advancements in Mechatronics and Intelligent Robotics*,
Advances in Intelligent Systems and Computing 1220,
https://doi.org/10.1007/978-981-16-1843-7_47

1 Introduction

Due to the randomness and intermittence of wind energy, the grid connection of wind farms will threaten the stability of the power system, and wind power forecasting is an effective way to solve this problem [1].

Despite the availability of prediction methods, it is still difficult to achieve accurate forecasting performance because of the fluctuation of wind energy. Hence, ensemble learning has been considered for wind power forecasting [2, 3]. To be specific, the basic idea of this method is to train and combine multiple diverse individual learners to achieve prediction tasks [4]. Unfortunately, the diversity of individual learners of traditional ensemble forecasting models is easily ignored, which leads to the poor performance of ensemble forecasting [5]. Furthermore, even if multiple individual learners have been generated, inevitably, there will be some redundant and low precision models. According to Zhou's finding [6], it may be better to integrate many instead of all individual learners, which means ensemble pruning is indispensable.

To address the above-mentioned issues, a novel wind power forecasting method using the selective heterogeneous ensemble (SHeE) method based on local learning and evolutionary multi-objective optimization is proposed. The main contributions of this paper can be summarized as follows:

(1) The proposed SHeE method stimulates the diversity of individual learners from both modeling method and sample perspectives.
(2) A novel ensemble pruning algorithm based on EMO is used.

The rest of the paper proceeds as follows. Section 2 presents the details of the proposed SHeE method. The case study is reported to verify the effectiveness and superiority of SHeE method in Sect. 3. Finally, the conclusion is drawn in Sect. 4.

2 Proposed SHeE for Wind Power Forecasting

2.1 Feature Selection Based on Lasso Regression

Wind power forecasting is a typical time series prediction problem in scientific research and relies heavily on historical wind farm data. Generally speaking, there are many factors affecting the output power of wind farms, e.g., wind speed, wind direction, temperature, atmospheric density, etc. However, many scholars' research indicates that wind direction, temperature, and atmospheric density have little impact on wind power forecasting. Therefore, in this paper, historical wind power and wind speed data are considered for modeling, and the structure of the wind power forecasting model [7] is described as follows:

$$\underbrace{\widehat{\mathrm{WP}}(t+h)}_{y} = f\,\underbrace{\mathrm{WS}(t), \mathrm{WS}(t-1), \ldots, \mathrm{WS}(t-l), \mathrm{WP}(t), \mathrm{WP}(t-1), \ldots, \mathrm{WP}(t-l)}_{X} \tag{1}$$

where h is the ahead step, $\widehat{\mathrm{WP}}$ is the predicted wind power, $f(\cdot)$ is the unknown function; WS and WP denote historical wind speed, and wind power data, respectively, l is the number of time lags. It is worth noting that some features may be redundant and uncorrelated, which will reduce the performance of the forecasting model. Hence, Lasso regression is used to select the suitable features of input samples.

The basic idea of Lasso regression is to introduce l_1 regularization into traditional linear regression [8]. Some regression coefficients equal to 0 and the features corresponding to the regression coefficient of 0 are deleted from the original feature space, which enables Lasso regression to realize feature selection. The loss function of Lasso regression is defined as:

$$\hat{\beta} = \arg\min_{\beta} y - X\beta + \lambda \sum_{i}^{d} |\beta_i| \tag{2}$$

where $\hat{\beta}$ is the optimal estimation of the regression coefficient, and λ is the penalty coefficient. In this paper, $\hat{\beta}$ is obtained by cross validation.

2.2 Construction of Local Domains

Clustering analysis has been widely applied for handling wind power data. Traditional hard partition clustering methods can divide the wind power data into several different types of fluctuation states, such as K-means [9] and Gaussian mixture model [10]. However, interestingly, there may be some samples that meet multiple states, which leads to insufficient fluctuation characteristics. Consequently, we propose a novel data partition method based on KNN to obtain several diverse and different states subsets for modeling.

Given a wind power training dataset $D = \{x_i, y_i\}_{i=1}^{n}$, K-means clustering is implemented to determine the state centers and a set of state centers $\{c_1, c_2, \ldots, c_m\}$ can be obtained. Then, based on the idea of KNN, any state center is regarded as a query point to obtain a local domain (LD) using a similarity measure. In this paper, the correlation coefficient is considered as similarity measure, which is calculated as

$$sim(x_i, c_j) = \exp\left(\frac{\sum_{r=1}^{d}(x_{i,r} - \bar{x}_i)(c_{j,r} - \bar{c}_j)}{\tau\sqrt{\sum_{r=1}^{d}(x_{i,r} - \bar{x}_i)^2}\sqrt{\sum_{r=1}^{d}(c_{j,r} - \bar{c}_j)^2}}\right) \tag{3}$$

where τ is the scaling factor. Subsequently, m local domains can be constructed. However, there is great redundancy information in those LDs. Consequently, a probabilistic analysis algorithm is used to remove redundant LDs. Suppose $\{\mathrm{LD}_m\}_{m=1}^{M-1}$ have been obtained, for a new state center c_{new}, a new local domain $\mathrm{LD}_{\mathrm{new}} = \{X_{\mathrm{new}}, y_{\mathrm{new}}\}$ is constructed. Subsequently, according to Bayesian inference, $p(\mathrm{LD_m}|c_{\mathrm{new}})$ can be calculated as follows

$$p(\mathrm{LD}_m|c_{\mathrm{new}}) = \frac{p(c_{\mathrm{new}}|\mathrm{LD}_m)p(\mathrm{LD}_m)}{p(c_{\mathrm{new}})} = \frac{p(c_{\mathrm{new}}|\mathrm{LD}_m)p(\mathrm{LD}_m)}{\sum_{i=1}^{M-1} p(c_{\mathrm{new}}|\mathrm{LD}_i)p(\mathrm{LD}_i)} \tag{4}$$

where $p(c_{\mathrm{new}}|\mathrm{LD}_m)$ is the conditional probability and $p(\mathrm{LD}_m) = \frac{1}{M-1}$. The redundancy index of $\mathrm{LD}_{\mathrm{new}}$ is defined as

$$J_{\mathrm{new}} = \max_{m=1,2,\ldots,M-1} p(\mathrm{LD}_m|c_{\mathrm{new}}) \tag{5}$$

Consequently, the condition for judging whether the new local domain is abandoned is shown as follows

$$\begin{cases} \gamma \times \mathrm{median}\{p(\mathrm{LD}_{\mathrm{new}}|x_i), x_i \in X_{\mathrm{new}}\} > J_{\mathrm{new}} \rightarrow \text{save } \mathrm{LD}_{\mathrm{new}} \\ \gamma \times \mathrm{median}\{p(\mathrm{LD}_{\mathrm{new}}|x_i), x_i \in X_{\mathrm{new}}\} < J_{\mathrm{new}} \rightarrow \text{abandon } \mathrm{LD}_{\mathrm{new}} \end{cases} \tag{6}$$

where γ is the scaling parameter and $\mathrm{median}\{\cdot\}$ denotes the median operator. Repeat the above steps until all the state centers have established LD and removed redundancy, and a set of diverse local domains $\{\mathrm{LD}_1, \mathrm{LD}_2, \ldots, \mathrm{LD}_M\}$ are obtained.

2.3 Construction of Heterogeneous Model Library

In Sect. 2.2, various local domains have been constructed. However, the diversity of samples is not enough. Therefore, this paper uses a variety of modeling methods to further improve the diversity of the ensemble.

For the ith local domain $\mathrm{LD}_i, i = 1, 2, \ldots, M$, SVR [11], GPR [12], and PLS [13] models are built. In this paper, the kernel function of SVR is RBF, and ε, c, and the number of principal components of PLS are determined by cross validation. Furthermore, the choice of covariance function is very important for GPR, and the selected covariance function in this paper is as follows

$$C(x_i, x_j) = \sigma_f^2 \exp\left(-\frac{1}{2l^2} x_i - x_j\right) + \frac{(x_i^T x_j + 1)}{l^2} + \sigma_n^2 \delta_{ij} \tag{7}$$

where σ_f^2, l and σ_n^2 are hyper-parameters. Subsequently, for M local domains $\{\mathrm{LD}_1, \mathrm{LD}_2, \ldots, \mathrm{LD}_M\}$, we can get the heterogeneous model library:

$$\mathrm{HML} = \{\{\mathrm{SVR}_1, \mathrm{GPR}_1, \mathrm{PLS}_1\}, \ldots, \{\mathrm{SVR}_M, \mathrm{GPR}_M, \mathrm{PLS}_M\}\} \tag{8}$$

where M is the number of local domains.

2.4 Selective Ensemble Based on the EMO Algorithm

In the constructed **HML**, it is inevitable that individual learners with poor prediction accuracy and high redundancy will appear, which will lead to the decline of ensemble performance. Hence, ensemble pruning is performed, and the optimization problem of ensemble pruning can be denoted as

$$\min[\mathrm{RMSE}_{\mathrm{sel}}, r_{\mathrm{sel}}] \tag{9}$$

where $\mathrm{RMSE}_{\mathrm{sel}}$ and r_{sel} are the objectives of forecasting performance and diversity, which are calculated in Eqs. (10–11)

$$\mathrm{RMSE}_{\mathrm{sel}} = \sqrt{\frac{1}{v}\sum_{i=1}^{v}\left(y_{\mathrm{val},i} - \hat{y}_i^{\mathrm{ens}}\right)^2} \tag{10}$$

$$r_{\mathrm{sel}} = \frac{\sum_{j=i+1}^{m_{\mathrm{sel}}}\sum_{i=1}^{m_{\mathrm{sel}}} r\left(e_i, e_j\right)}{\left(m_{\mathrm{sel}}^2 - m_{\mathrm{sel}}\right)/2} \tag{11}$$

where v is the number of validation samples, m_{sel} is the number of selected individual learners, and $\hat{y}_{\mathrm{i}}^{\mathrm{ens}}$, $r\left(e_i, e_j\right)$ can be calculated as follows

$$\hat{y}_i^{\mathrm{ens}} = \frac{\sum_{r=1}^{m_{\mathrm{sel}}} \hat{y}_i^r}{m_{\mathrm{sel}}} \tag{12}$$

$$r\left(e_i, e_j\right) = \frac{\mathrm{Cov}\left(e_i, e_j\right)}{\sqrt{\mathrm{Var}(e_i)\mathrm{Var}\left(e_j\right)}} \tag{13}$$

where $\hat{y}_i^r$ is the ith prediction value of the rth individual learner. To address the above optimization problems, an excellent EMO algorithm, non-dominated sorting genetic algorithm II (NSGA-II) [14] is performed. Of course, other EMO algorithms can also work well.

Through the above ensemble pruning operation, m_{sel} individual learners are selected for ensemble forecasting. When a new sample x_{new} comes, the final ensemble prediction result can be calculated by

$$\hat{y}_{\text{new}} = \frac{1}{m_{\text{sel}}} \sum_{i=1}^{m_{\text{sel}}} \hat{y}_{\text{new}}^{i} \tag{14}$$

2.5 Implementation of the SHeE Algorithm

In this paper, a novel wind power forecasting method named selective heterogeneous ensemble (SHeE) method is proposed. The whole process of the SHeE algorithm is shown in Fig. 1, and the implementation steps are described as follows:

(1) Collect the historical wind speed and power data as the modeling samples, which are divided into the training set D_{trn}, validation set D_{val}, and test set D_{test}.
(2) Carry out feature selection using Lasso regression on D_{trn} and save the indexes of selected features.
(3) Construct local domains $\{\text{LD}_1, \text{LD}_2, \ldots, \text{LD}_M\}$ and build the heterogeneous model library HML using SVR, GPR, and PLS.
(4) Select the most influential individual learners from HML by the NSGA-II algorithm.
(5) When a new sample x_{new} comes, the final prediction output is obtained by integrating the selected individual learners.

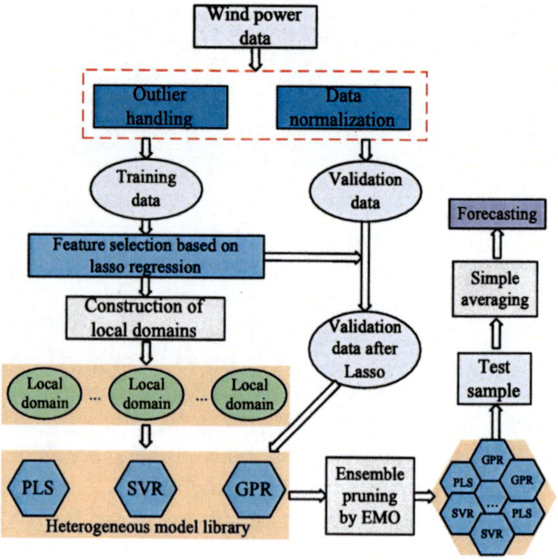

Fig. 1 The whole process of the SHeE algorithm

3 Case Study

In this section, the effectiveness and superiority of the proposed SHeE method for wind power forecasting are verified through a real wind farm dataset. The methods for comparison are as shown: (1) Persistence method: a classic method; (2) PLS: global PLS model; (3) GPR: global GPR model; (4) SVR: global SVR model; (5) HeE: heterogeneous ensemble without ensemble pruning; (6) SHeE: selective heterogeneous ensemble using the NSGA-II algorithm.

It is worth noting that the parameters of the proposed method have a great impact on the forecasting performance. The parameters determined by trial and error are shown: the number of samples of local domains is 500; the scaling factor τ of Eq. (3) is 1; the scaling parameter γ is assigned to 0.8; the population size and maximum generations of NSGA-II are set to 2000 and 200, respectively.

In this paper, root-mean-square error (RMSE) and coefficient of determination (R^2) are adopted to evaluate the prediction performance, which are shown in Eqs. (15–16):

$$\mathrm{RMSE} = \sqrt{\frac{1}{n_{\text{test}}}\sum_{i=1}^{n_{\text{test}}}(\hat{y}_i - y_i)^2} \tag{15}$$

$$R^2 = 1 - \frac{\sum_{i=1}^{n_{\text{test}}}(\hat{y}_i - y_i)^2}{\sum_{i=1}^{n_{\text{test}}}(y_i - \bar{y})^2} \tag{16}$$

where n_{test} is the number of the test set $\hat{y}_i$ and y_i denote the prediction and actual value, respectively, $\bar{y}$ represents the mean value of outputs.

3.1 *Dataset*

In this paper, original wind power and speed sequences are collected from a real wind farm in Yunnan Province, and the time interval is 15 min. According to Eq. (1), where l is set to 16 and h is set to 16, 5661 samples can be obtained and are further divided into the training set, validation set, and testing set with the sizes of 3500, 1000, and 1161, respectively.

3.2 *Comparison and Analysis of Experimental Results*

The forecasting performance of the proposed SHeE approach is compared to the other five methods, as shown in Table 1. It is clear that the proposed method SHeE achieves significantly better performance in both RMSE and R^2 than the other

Table 1 Forecasting performance of different methods

Method	RMSE	R^2
Persistence	15.2255	0.4785
PLS	14.4036	0.5333
SVR	14.3209	0.5386
GPR	15.9854	0.4251
HeE	14.0537	0.5557
SHeE	13.5406	0.5875

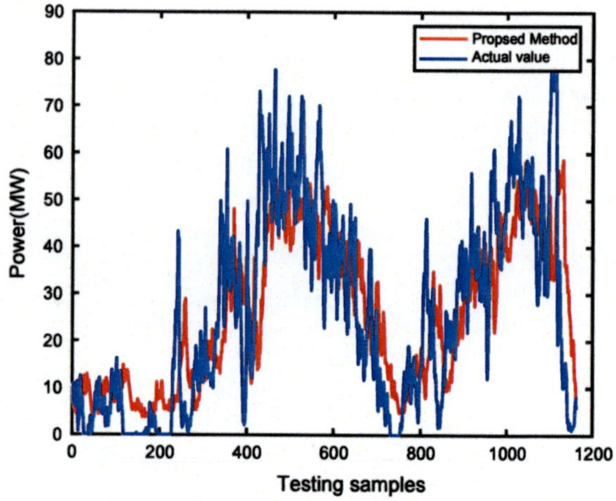

Fig. 2 The trend plot of the proposed method for wind power forecasting

methods. Compared with HeE, three global models PLS, SVR, and GPR cannot forecast well on test sample, which indicates that local learning and heterogeneous ensemble play a key role in improving prediction performance.

In addition, the comparison of HeE and SHeE suggests that ensemble pruning is effective to improve prediction accuracy. In terms of RMSE, the proposed method improves the prediction accuracy by 3.6%.

In Fig. 2, we can see that the forecasting trend of wind power using our proposed method on the test set. Unfortunately, the predicted curve and the real curve do not seem to fit very well. On the one hand, this is because the prediction interval is 4 h, which is relatively long. On the other hand, wind energy has strong randomness and volatility, which makes our proposed method can only generally catch the changing trend.

4 Conclusions

In this paper, a novel wind power forecasting method, i.e., SHeE, is proposed. This method first considers local learning to get diverse sample subsets and then constructs the heterogeneous model library. NSGA-II-based ensemble pruning is performed, and selected individual learners are integrated by simple averaging when a new sample comes. In Sect. 3, the experiment results show that SHeE is superior to other models. Furthermore, more machine learning methods should be considered in the future.

Acknowledgements This work was supported by the National Natural Science Foundation of China (grant numbers 61763020, 61863020), and Yunnan Fundamental Research Projects (grant number 2018FD040).

References

1. Georgilakis PS (2008) Technical challenges associated with the integration of wind power into power systems. Renew Sustain Energy Rev 12(3):852–863
2. Liu H, Chen C, Lv X, Wu X, Liu M (2019) Deterministic wind energy forecasting: a review of intelligent predictors and auxiliary methods. Energy Convers Manag 195:328–345
3. Tascikaraoglu A, Uzunoglu M (2014) A review of combined approaches for prediction of short-term wind speed and power. Renew Sustain Energy Rev 34:243–254
4. Xiao L, Dong Y, Dong Y (2018) An improved combination approach based on Adaboost algorithm for wind speed time series forecasting. Energy Convers Manag 160:273–288
5. Qu Z, Zhang K, Mao W, Wang J, Liu C, Zhang W (2017) Research and application of ensemble forecasting based on a novel multi-objective optimization algorithm for wind-speed forecasting. Energy Convers Manag 154:440–454
6. Zhou Z-H, Wu J, Tang W (2002) Ensembling neural networks: many could be better than all. Artif Intell 137(1–2):239–263
7. Wang J, Song Y, Liu F, Hou R (2016) Analysis and application of forecasting models in wind power integration: a review of multi-step-ahead wind speed forecasting models. Renew Sustain Energy Rev 60:960–981
8. Li F, Yang Y, Xing EP (2006) From lasso regression to feature vector machine. In: Advances in neural information processing systems, pp 779–786
9. Hartigan JA, Wong MA (1979) Algorithm AS 136: A k-means clustering algorithm. J Roy Stat Soc Ser C (Appl Stat) 28(1):100–108
10. Rasmussen C (1999) The infinite Gaussian mixture model. Adv Neural Inf Process Syst 12:554–560
11. Noble WS (2006) What is a support vector machine? Nat Biotechnol 24(12):1565–1567
12. Seeger M (2004) Gaussian processes for machine learning. Int J Neural Syst 14(02):69–106
13. Geladi P, Kowalski BR (1986) Partial least-squares regression: a tutorial. Analytica chimica acta 185:1–17
14. Deb K, Agrawal S, Pratap A (2000) A fast elitist non-dominated sorting genetic algorithm for multi-objective optimization: NSGA-II. In: International conference on parallel problem solving from nature. Springer, Heidelberg, pp 849–858

Research on Image Denoising Based on Wavelet Threshold

Jinyu Wang

Abstract Images are the main source of information for people. In the information age, digital image processing has become an important means for people to obtain, process, analyze and share information. It has penetrated into all aspects of life and achieved huge social and economic benefits. Image denoising is a large category of digital image processing technology, and it is a key step of the underlying processing. The quality of its processing results directly affects the performance of subsequent segmentation, recognition, and analysis. In traditional denoising methods, effective denoising and preservation of image detail information are very contradictory, and the denoising effect is not very satisfactory. Wavelet transform can successfully preserve the edge information of the image while denoising the image. Therefore, it is very necessary and important to study image denoising based on wavelet threshold and optimize and improve algorithm performance.

Keywords Wavelet · Threshold · Image denoising

1 Introduction

In reality, images are inevitably interfered by noise during the process of acquisition, transmission, and conversion. Noise deteriorates the image quality, blurs the image, and even drowns and changes the characteristics, which brings difficulties to image analysis and recognition. Therefore, it is necessary to denoise the image to make subsequent analysis and information extraction more accurate. In order to remove the noise, it will cause the blur of the image edge and the loss of some cultural details. On the contrary, the image edge is enhanced while the image noise is enhanced. In recent years, wavelet theory has been developed very rapidly, and image denoising technology based on wavelet analysis has also achieved better results with the continuous improvement of wavelet theory. Wavelet transform, as a

J. Wang (✉)
School of Information Engineering and Automation, Kunming University of Science and Technology, Kunming 650500, China

Z. Yu et al. (eds.), *Advancements in Mechatronics and Intelligent Robotics*,
Advances in Intelligent Systems and Computing 1220,
https://doi.org/10.1007/978-981-16-1843-7_48

brand-new mathematical theory analysis method, has achieved good application effects in many fields. Image processing is one of the earliest and most mature fields of its application. Wavelet transform has its own good time-frequency characteristics in image reduction. The field of noise has received more and more attention, opening up a precedent for using linear methods to reduce noise. Wavelet transform can better filter out Gaussian noise. Currently, image denoising technology based on wavelet analysis has become an important method for image denoising.

At present, image denoising algorithms mainly include denoising algorithms based on filtering, denoising algorithms based on multi-scale, and denoising algorithms based on partial differential equations. Filter-based denoising algorithms include median filtering, Kalman filtering, Wiener filtering, etc. These methods mainly achieve the purpose of denoising by filtering out the high-frequency part of the image, but while filtering out the noise, it will cause the image to be disturbed. The high-frequency part achieves the purpose of denoising. However, wavelet threshold denoising can better preserve the original image information while removing noise.

2 Noise

2.1 Input of Noise Image

Commonly used image input devices mainly include video cameras, digital cameras, scanners, etc. This article uses a video camera to store the captured images in digital format through A/D conversion. The images have the advantages of high definition, simple operation, and strong mobility. MATLAB image processing toolbox provides a wealth of image processing functions. Using these library functions can save time of writing complex underlying algorithm codes and focus on image processing methods. The digital image is a two-dimensional matrix, and MATLAB is good at matrix operations. In this article, MATLAB is used for image processing. Call the function imread in MATLAB to read in the image, the syntax of imread is imread ('filename'); the supported image file formats are TIFF, JPEG, GIF, BMP, PNG, XWD, etc., and the image generated by a general digital camera The format is usually JPEG, which can be read into the system through the imread statement.

2.2 The Significance of Image Denoising

The frequently used wavelet transform denoising algorithm has better results than other denoising algorithms. However, improper threshold selection can easily cause image distortion: improper threshold function settings can cause ringing, pseudo-Gibbs effect, or image blur due to defects. In this paper, through the research of image denoising based on wavelet threshold, the noise suppression effect is better than traditional methods, and the image information is more complete. It is not only suitable for ordinary images, but also for infrared images.

3 Matlab for Image Processing and Analysis

3.1 Wavelet Threshold Denoising Algorithm

By keeping the modulus maximum points under each scale in the wavelet change, setting other points to zero or reducing it to the greatest extent, and then performing wavelet inverse transformation on the processed wavelet coefficients, the purpose of noise suppression can be achieved. Generally speaking, image denoising based on wavelet threshold can usually be divided into the following three steps:

a. First, to obtain the final wavelet decomposition coefficients, it is necessary to first calculate the orthogonal wavelet basis containing the image noise and the number of wavelet decomposition layers N, and finally perform N-layer wavelet decomposition. Use the wavelet decomposition algorithm to perform N-layer wavelet decomposition on the noisy image. The wavelet decomposition algorithm processes the noisy image to get the coefficient result.
b. Use the high-frequency coefficients after wavelet decomposition to quantify the threshold. It is necessary to select an appropriate threshold and an appropriate threshold function to decompose the high-frequency wavelet coefficients obtained. For each layer from the first to the Nth layer, an estimated wavelet coefficient value is obtained.
c. Perform wavelet inverse transformation on the image signal after the a and b steps to obtain a denoised image. According to the low-frequency coefficients (scale coefficients) of the Nth layer after wavelet decomposition of the image and the high-frequency coefficients (wavelet coefficients) of each layer after threshold quantization, the wavelet reconstruction algorithm is used to reconstruct the image obtained [1].

3.2 Improved Threshold Wavelet Image Denoising

3.2.1 Wavelet Changes

The ability of wavelet to eliminate noise is mainly due to the low profile, multi-resolution characteristics of wavelet transform, decorrelation, and flexibility of basis functions.

3.2.2 Wavelet Denoising Process

See Fig. 1.

3.2.3 Threshold Calculation

In the denoising process, the wavelet threshold λ plays a decisive role: if a smaller threshold is selected, the wavelet coefficients can be retained as much as possible, which may retain more image information, but at the same time the noise is also retained. Conversely, if a larger threshold is set, more noise can be eliminated, and high-frequency information in the image will be lost, causing distortion.

Currently, the thresholds used include global thresholds and local adaptation thresholds. Most systems use the unified threshold $\delta = \sigma\sqrt{2\text{In}N}$ proposed by Donoho and Johnstone. Where σ is the standard deviation of the noise, and N is the size or scale of the signal. However, in the actual environment, the noise standard deviation of the image is impossible to know, so in the calculation method, an estimation method is used to determine the noise standard deviation. The most commonly used estimation methods are as follows:

$$\sigma = \frac{\text{median}(\text{d}j(k))}{0.6745}$$

In the formula, j is the wavelet decomposition scale, and median is the calculation command to find the median in Matlab [2].

3.2.4 Selection of Threshold Function

In threshold denoising, the threshold function embodies different strategies and different estimation methods for wavelet coefficient moduli that exceed and fall below the threshold.

There are two commonly used threshold functions: hard threshold function and soft threshold function. The hard threshold strategy retains wavelet coefficients larger than the threshold, and sets the wavelet coefficients smaller than the threshold to zero. The intuitive explanation is that it is either regarded as a useful signal (larger wavelet coefficients) or as noise (wavelet coefficients smaller than the

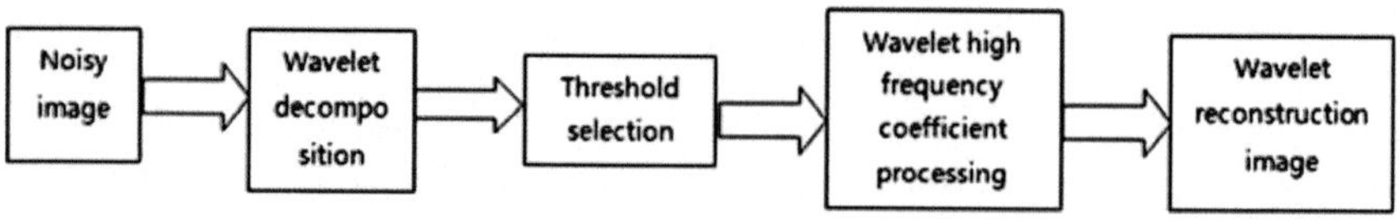

Fig. 1 The process of wavelet denoising

threshold). The soft threshold strategy sets the wavelet coefficients smaller than the threshold to zero, and subtracts the threshold from the absolute value of the wavelet coefficients larger than the threshold to remove the influence of noise.

Let ω be the original wavelet coefficients, $\eta(\omega)$ represents the thresholded wavelet coefficients, and λ is the threshold. The expressions of the two threshold functions are as follows:

(1) Hard threshold function:

$$\eta(\omega) = \begin{cases} \omega & |\omega| \geq \lambda \\ 0 & |\omega| < \lambda \end{cases} \quad (1)$$

The absolute value of the wavelet coefficient is not less than the set threshold, and keep it unchanged as the estimated wavelet coefficient, otherwise, let it be zero.

(2) Hard threshold function:

$$\eta(\omega) = \begin{cases} \operatorname{sign}(\omega)(|\omega| - \lambda) & |\omega| \geq \lambda \\ 0 & |\omega| < \lambda \end{cases} \quad (2)$$

The absolute value of the wavelet coefficient is not less than the set threshold, let it minus the set threshold as the estimated wavelet coefficient, otherwise, let it be zero [2].

4 Experimental Simulation and Result Discussion

4.1 Comparison and Simulation of Wavelet Denoising and Other Filtering Methods

```
det1 = detcoef2('compact',c,s,1);
sigma = median(abs(det1))/0.6745;
PSF = fspecial('average',3);
L = imfilter(J,PSF);
K = medfilt2(J,[3,3]);
M = wiener2(J,[3,3]);
[c,l] = wavedec2(J,2,'sym4');
a2 = wrcoef2('a',c,l,'sym4',2);
n = [1, 2];
p = [10.28,24.08];
nc = wthcoef2('t',c,l,n,p,'s');
mc = wthcoef2('t',nc,l,n,p,'s');
N = waverec2(mc,l,'db2');
```

The comparative analysis shows that the use of wavelet transform to denoise the image can retain more image details so that more useful information of the image can be retained, which is beneficial to the further processing of the image. The noise reduction effect of Wiener filtering is affected by the size of the filter bed. If the filter window size is selected appropriately, a more satisfactory result can be obtained, but if the selection is not appropriate, the noise reduction quality will be greatly reduced. It is not difficult to see from its denoising effect that median filtering is far inferior to wavelet denoising in terms of image clarity and detail preservation. Compared with other methods, the effect is not as good as without wavelet (Fig. 2).

It can be seen that, compared with other filtering methods, wavelet denoising is more in line with display denoising requirements.

4.2 *Processing of Wavelet Threshold Signal*

See Fig. 3.

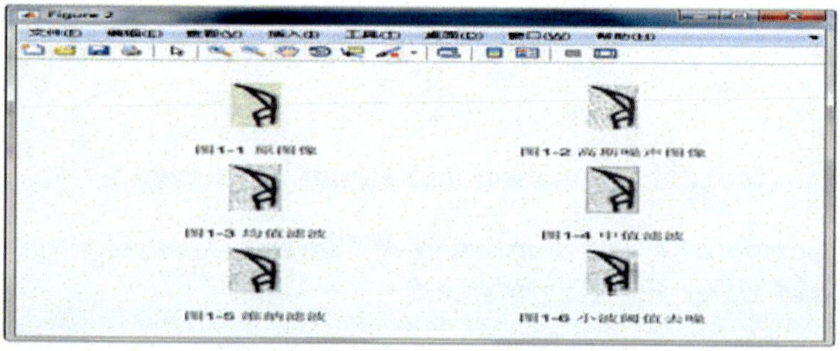

Fig. 2 Comparison and simulation of wavelet denoising and other filtering methods

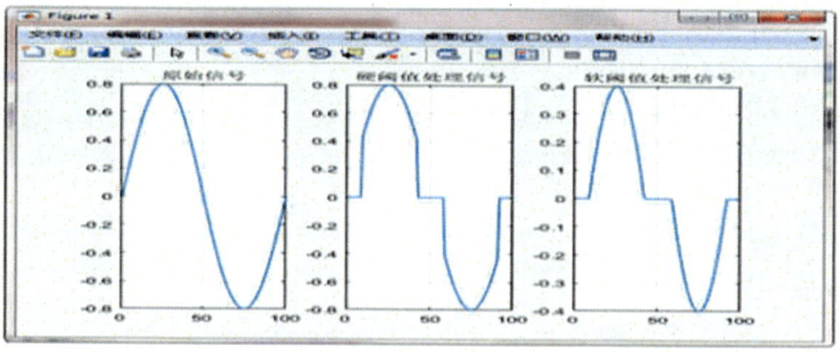

Fig. 3 The process of wavelet threshold signal

4.3 Oracle Image Processing Based on Wavelet Threshold

```
wname='sym4';lev = 2;
[c,l]=wavedec2(x,lev,wname);
sigma_s=0.054779;
alpha=2;
thr_s=wbmpen(c,1,sigma_s,alpha);
keepapp=1;
xds=wdencmp('gbl',x,wname,lev,thr_s,'s',keepapp);
sigma_h=0.62818;
thr_h=wbmpen(c,l,sigma_h,alpha);
xdh=wdencmp('gbl',x,wname,lev,thr_h,'h',keepapp);
```

4.4 Discussion of Results

The wavelet threshold image denoising method can remove most of the noise of the image and has good results. However, due to the problems of the threshold function and the threshold selection method, the set threshold cannot completely remove the image noise, and it will also be due to the threshold function. The problem makes the denoising image visual effect is not good, which requires continuous improvement of the threshold function and threshold selection method to obtain a wavelet threshold denoising method that can better remove image noise (Fig. 4).

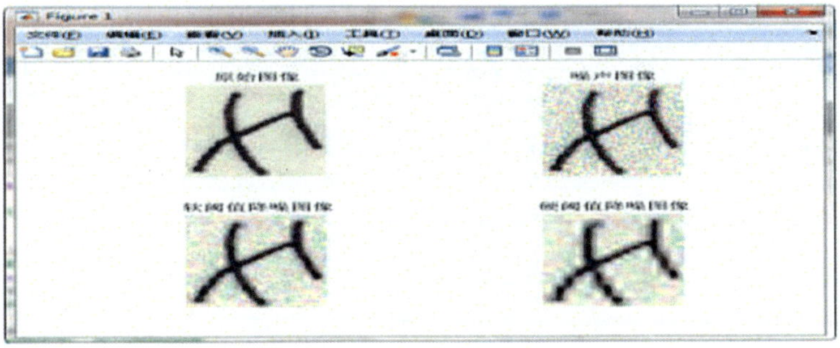

Fig. 4 The process based on wavelet threshold

5 Summary and Outlook

Image denoising is a very widely used technology in image preprocessing. Its function is to improve the signal-to-noise ratio and highlight the desired features of the image so that it can be processed at a higher level. Due to the excellent characteristics of wavelet transform, wavelet denoising plays an important role in the field of image denoising, especially wavelet denoising has attracted more and more attention.

The problem of image denoising has always been a difficult problem to solve. It is difficult to remove noise and retain the information of the original image. There are still many problems that need to be further explored and studied, such as how to combine the original image with wavelet transform. Features (such as the geometric trend of edges, cultural and scientific features, etc.) denoising, so that the original image information is not lost as much as possible while removing noise, and some existing denoising methods are combined with wavelet transform denoising. But what is certain is that the excellent characteristics of wavelet transform are unique. With the increasing development of wavelet theory, it will surely get more and more applications in the field of image noise reduction.

References

1. Dongsheng L (2013), Research on wavelet threshold image denoising method based on MATLAB [J]. Comput knowl technol 9(11):2662–2663 + 2677
2. Zhang C, Chen Q (2017) Image denoising based on wavelet transform [J]. Electron Sci technol 30(03):8–10

Research on Fault Diagnosis Method Based on Diffusion Map and Extreme Learning Machine

Yu Zhu Hu and Zhao Lin Zhang

Abstract There is a non-linear relationship between the vibration signal characteristics of the bearing equipment and the bearing health status, and the single feature of the bearing cannot fully reflect the bearing health status, and its multi-dimensional features have information redundancy, which affects the performance of the classifier. In response to this problem, Diffusion Maps (DM) and Extreme Learning Machine (ELM) are applied to bearing fault diagnosis, and the signal is denoised by Complementary Ensemble Empirical Mode Decomposition (CEEMD); then, the DM algorithm is used to perform high-dimensional features Dimensionality reduction processing, and establishment of an ELM fault diagnosis model to identify the health status of the bearing. Bearing experiments show that this method can effectively identify the health status of bearings and reduce the redundancy of high-dimensional information.

Keywords Bearing health status · Diffusion maps · Extreme learning machine

1 Introduction

With the development of science and technology, the number of highly integrated equipment is increasing. As one of the power components of the equipment, its damage will cause the stagnation of production and life and cause incalculable economic losses. Therefore, monitoring the health status of bearings is helpful to ensure the safety of industrial production and life, and it has also become one of the hotspots of scholars [1–3].

Y. Z. Hu
Aba Teachers University, Aba 623002, China

Z. L. Zhang (✉)
Kunming University of Science and Technology, Kunming 650500, China
e-mail: aeryszzl@163.com

Z. Yu et al. (eds.), *Advancements in Mechatronics and Intelligent Robotics*,
Advances in Intelligent Systems and Computing 1220,
https://doi.org/10.1007/978-981-16-1843-7_49

Due to the complexity of the industrial production environment, the denoising of signals has become one of the necessary processes to ensure accurate extraction of signals reflecting equipment status. Empirical Mode Decomposition (EMD) [4], as the current method of processing signals, has gone through processes such as Ensemble Empirical Mode Decomposition (EEMD) [5], Complementary Ensemble Empirical Mode Decomposition (CEEMD) [6], respectively, to solve the end effect, Decomposing instability and the difficulty of neutralizing white noise has become one of the mainstream methods of vibration signal processing.

Feature extraction is the prerequisite for accurate analysis of bearing health status. Complex equipment consists of multiple sub-parts, and there is a coupling relationship between each sub-part. It is difficult for a single feature to fully reflect the health status of the sub-parts; therefore, it is necessary to extract multi-dimensional features to extract signals, To ensure the effectiveness and completeness of feature extraction [7–9]. The performance of the classifier is limited, and it is difficult to solve the problem of information redundancy in high-dimensional information; Diffusion Maps (DM) [10, 11] can extract potential manifold information in non-linear high-dimensional space, with strong robustness The advantage of low algorithm complexity; the redundancy in the extracted high-dimensional features of the equipment can be reduced to prevent the impact of redundant information on the performance of the classifier. Extreme Learning Machine (ELM) [12] has certain advantages in generalization and learning speed compared to shallow learning systems such as SVM, which can ensure the timeliness of equipment health monitoring.

In summary, this article proposes a rolling bearing fault diagnosis method based on DM and ELM. The method first uses CEEMD to denoise the signal, filter out high-frequency information in the signal, and extract multi-domain high-dimensional features; then, DM is used to reduce the dimensionality of high-dimensional features to reduce redundancy; finally, the reduced features are input into the ELM model to diagnose equipment faults. The experimental results show that the method proposed in this paper can quickly and effectively identify the operating status of the equipment and ensure the safe operation of production.

2 Signal Denoising and Feature Extraction

2.1 CEEMD Decomposition

In order to reduce the error caused by white noise in EEMD, adding positive and negative white noise pairs to CEEMD can effectively reduce the number of noise collections and reduce the calculation time. The specific process is as follows:

Add the x group of positive and negative white noise pairs to the original bearing signal to obtain the P and T groups of signals.

$$\begin{bmatrix} P \\ T \end{bmatrix} = \begin{bmatrix} 1 & 1 \\ 1 & -1 \end{bmatrix} \begin{bmatrix} S \\ N \end{bmatrix} \tag{1}$$

Among them, S is the bearing signal, N is the white noise signal, and P and T are the two sets of signals obtained.

Perform EMD decomposition of P and T to obtain several IMF components, and solve the average value of each component as IMF, and the obtained score is the final IMF component.

$$\mathrm{IMF}_j = \frac{1}{2n} \sum_{i=1}^{2n} \mathrm{IMF}_{ij} \tag{2}$$

where IMF_{ij} is the j-th component of the i-th signal, and IMF_j is the j-th IMF component to be sought.

2.2 *Feature Extraction*

This paper selects 16 characteristic parameters such as mean, skewness, peak value in the time domain, and 13 characteristic parameters such as mean frequency and standard deviation in the frequency domain, totaling 29 dimensions. Some characteristic mathematical expressions are shown in Tables 1 and 2.

Table 1 Mathematical expression of time-domain characteristic parameters

Time-domain characteristics	Mathematical expression	Time-domain characteristics	Mathematical expression
Mean	$\bar{X} = \frac{1}{N}\sum_{i=1}^{N} x_i$	Peak-to-peak	$X_p = x_{\max} - x_{\min}$
Root mean square	$X_{\mathrm{rms}} = \sqrt{\frac{1}{N}\sum_{i=1}^{N} x_i^2}$	Volatility index	$S_f = \frac{X_{\mathrm{rms}}}{\lvert \bar{X} \rvert}$
Square root amplitude	$X_r = \lvert \frac{1}{N}\sum_{i=1}^{N} \sqrt{\lvert x_i \rvert} \rvert^2$	Peak index	$C_f = \frac{X_p}{X_{\mathrm{rms}}}$
Absolute mean	$X_{\mathrm{Mad}} = \frac{1}{N}\sum_{i=1}^{N} \lvert x_i \rvert$	Impulse indicator	$I = \frac{X_p}{\bar{X}}$
Skewness	$\alpha = \frac{1}{N}\sum_{i=1}^{N} x_i^3$	Margin Index	$L = \frac{X_p}{X_r}$
Kurtosis	$\beta = \frac{1}{N}\sum_{i=1}^{N} x_i^4$	Skewness index	$C_w = \frac{\alpha}{X_{\mathrm{rms}}^3}$
Variance	$\mathrm{Var} = \frac{1}{N-1}\sum_{i=1}^{N} (x_i - \bar{x})^2$	Kurtosis index	$K = \frac{\beta}{X_{\mathrm{rms}}^4}$

Table 2 Mathematical expression of frequency-domain characteristic parameters

No.	Mathematical expression	No.	Mathematical expression
1	$f_1 = \frac{1}{K}\sum_{k=1}^{K} s(k)$	2	$f_2 = \frac{1}{K-1}\sum_{k=1}^{K} (s(k) - f_1)^2$
3	$f_3 = \frac{\sum_{k=1}^{K} (s(k)-f_1)^3}{K\left(\sqrt{f_2}\right)^3}$	4	$f_4 = \frac{\sum_{k=1}^{K} (s(k)-f_1)^4}{K(f_2)^4}$
5	$f_5 = \frac{\sum_{k=1}^{K} s(k)f_k}{\sum_{k=1}^{K} s(k)}$	6	$f_6 = \sqrt{\frac{\sum_{k=1}^{K} s(k)\times(f_k-f_5)^2}{K}}$
7	$f_7 = \sqrt{\frac{\sum_{k=1}^{K} s(k)f_k^2}{\sum_{k=1}^{K} s(k)}}$	8	$f_8 = \sqrt{\frac{\sum_{k=1}^{K} s(k)f_k^4}{\sum_{k=1}^{K} s(k)f_k^2}}$
9	$f_9 = \frac{\sum_{k=1}^{K} s(k)f_k^2}{\sqrt{\sum_{k=1}^{K} s(k)\sum_{k=1}^{K} s(k)f_k^2}}$	10	$f_{10} = \frac{f_6}{f_5}$
11	$f_{11} = \frac{\sum_{k=1}^{K} (f_k-f_5)^3 s(k)}{K(f_6)^3}$	12	$f_{12} = \frac{\sum_{k=1}^{K} (f_k-f_5)^4 s(k)}{K(f_6)^4}$
13	$f_{13} = \frac{\sum_{k=1}^{K} (f_k-f_5)^{1/2} s(k)}{K(f_6)^{1/2}}$		

3 Feature Dimensionality Reduction

The Diffusion Maps algorithm seeks its geometric structure at different scales through the analysis of high-dimensional features. Its advantages include strong anti-noise ability and low algorithm complexity. The algorithm is as follows [13]:

Let the high-dimensional space be X_i, $i = 0, 1, \ldots, N - 1$. The kernel function is $k(x, y)$.

Step 1: Calculation $K_{i,j} = k(X_i, X_j)$;

Step 2: Calculate the diffusion matrix $P = D^{-1}K$;

where $D_t\left(X_i, X_j\right)^2 = \sum_{u\in X} |p_t(X_i, u) - p_t\left(X_j, u\right)|^2 = \sum_k \left|P_{ik}^t - P_{kj}^t\right|^2$, $P_{ij} = p\left(X_i, X_j\right)$.

Step 3: Calculate the eigenvalues and eigenvectors of the diffusion matrix:

$$Y_i := \begin{bmatrix} p_t(X_i, X_1) \\ p_t(X_i, X_2) \\ \vdots \\ p_t(X_i, X_N) \end{bmatrix} = P_{i*}^T \tag{3}$$

$$\|Y_i - Y_j\|_E^2 = \sum_{u\in X} |p_t(X_i, u) - p_t\left(X_j, u\right)|^2 = \sum_k \left|P_{ik}^t - P_{kj}^t\right|^2 = D_t\left(X_i, Y_j\right) \tag{4}$$

Step 4: The d dimension of feature mapping:

$$Y_i' = \begin{bmatrix} \lambda_1^t \psi_1(i) \\ \lambda_2^t \psi_2(i) \\ \vdots \\ \lambda_n^t \psi_n(i) \end{bmatrix} \tag{5}$$

Among them, $\psi_n(i)$ is the i-th element of matrix P.

Then the first d values of Y_i are the eigenvalues after dimensionality reduction.

4 ELM Algorithm

ELM is a typical single hidden layer neural network, and its expression can be expressed as:

$$\sum_{i=1}^{L} \beta_i \cdot (\omega_i \cdot x_j + b_i) = t_j \tag{6}$$

where $j = 1, 2, 3, \ldots, N$. ω_i, β_i and b_i are the weight and bias of the neuron, respectively.

If the activation function g(x) approximates N samples (X_i, t_i), and satisfies $\sum_{j=1}^{N} \|O_j - t_j\| = 0$. Then:

$$\sum_{i=1}^{L} \beta_i \cdot g(\omega_i \cdot x_j + b_i) = t_j, j = 1, 2, \ldots, N \tag{7}$$

The specific steps are:

Let the hidden layer output matrix H, weight matrix β, and expectation matrix T.

a. Given the number of neurons and initialize b_i and ω_i randomly;
b. Select the activation function and calculate H;
c. Calculate $\beta_i = H^+ T$, where H^+ is the Moore-Penrose generalized inverse of H.

5 Application of Diffusion Maps and ELM in Classification of Bearing Health Status

Because a single feature cannot fully reflect the health status of the equipment, and high-dimensional features have greater information redundancy, the performance of the classifier drops sharply, and the equipment health status cannot be accurately

Fig. 1 DM-ELM flowchart

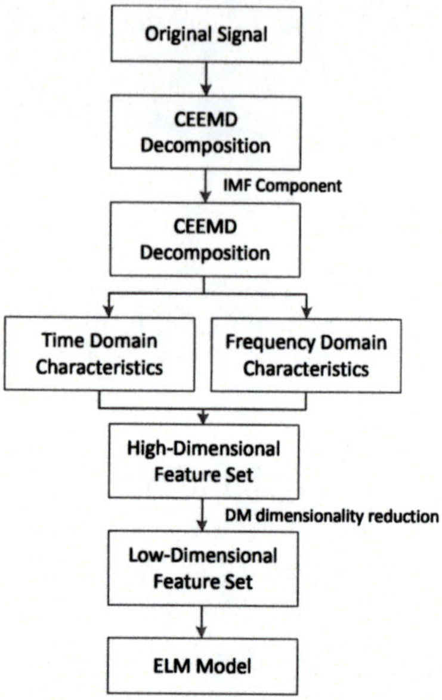

distinguished. This paper first uses CEEMD to denoise the vibration signal of the equipment, and extracts multi-domain and multi-dimensional features to construct a feature set; then, the DM algorithm is used to reduce the dimension of the feature set to reduce the information redundancy in the feature set; finally, use ELM establishes a fault diagnosis model to perform condition monitoring and health assessment of equipment. The flowchart is shown in Fig. 1.

6 Example Verification

This paper uses the bearing data set of the University of Paderborn in Germany [14], with a sampling frequency of 64,000 Hz and a rotation speed of 900 rpm. There are three fault states: normal, outer ring failure, and inner ring failure. There are 200 samples in each state, 150 of each type are used as the training set, 50 as the test set, and the number of sampling points for each sample is 1024. The three-state time-domain signal diagram is shown in Fig. 2. First, the signal is decomposed by CEEMD, the last component is removed, and the other components are added together to obtain the reconstructed signal. The time domain and frequency domain features of the reconstructed signal are extracted, totaling 29 dimensions; then, the eigen-dimension of the signal is estimated by the maximum likelihood value estimation method, and the eigen-dimension of the feature set is three; DM maps the 29-dimensional features, reduces the feature dimension to three dimensions, and uses ELM to establish a fault diagnosis model. The recognition results before and after dimensionality reduction are shown in Figs. 3 and 4. At the same time, when

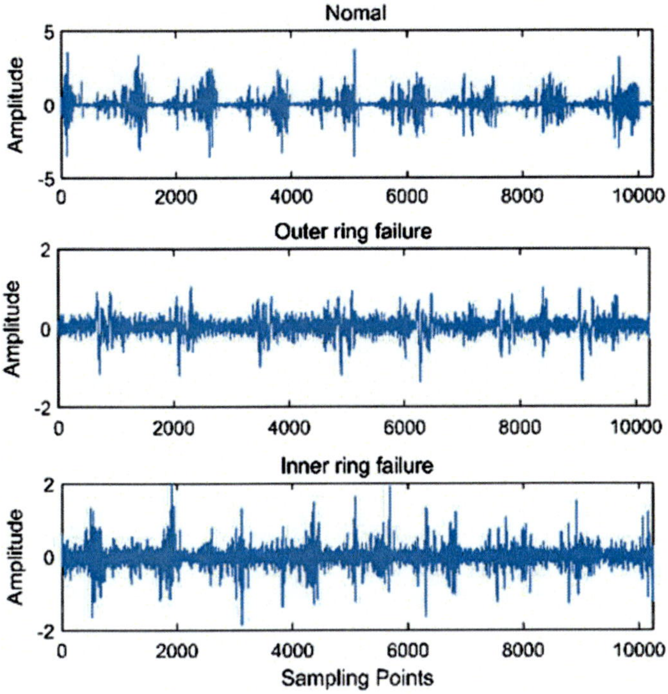

Fig. 2 Time-domain signal diagram

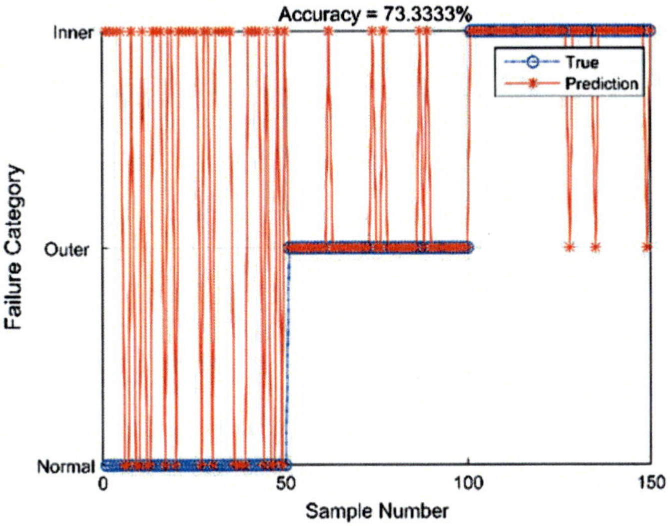

Fig. 3 Unreduced result

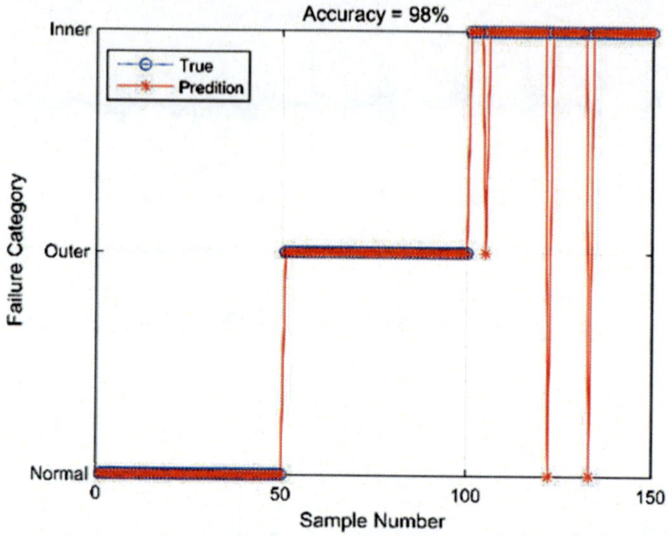

Fig. 4 Dimensionality reduction result

Table 3 Dimension reduction results of other methods

	KPCA	LLTSA	PCA	DM
Accuracy (%)	72.67	85.33	87.33	98

SVM is used to identify the feature set, the recognition rate without dimensionality reduction is 42% (63/150), and the recognition rate after dimensionality reduction is 78.67% (118/150); therefore, DM can effectively reduce the redundancy of high-dimensional feature species, and ELM can also provide more effective recognition accuracy.

Respectively use KPCA, LLTSA, and other methods to reduce the dimensionality of the signal, and the results are shown in Table 3.

7 Conclusion

Experiments show that DM can effectively reduce the redundancy of high-dimensional feature sets and improve the performance of the classifier compared with dimensionality reduction methods such as KPCA and PCA; ELM has better generalization performance than SVM and can improve higher accuracy rate.

References

1. Feng J, Yao Y, Lu S et al (2021) Domain knowledge-based deep-broad learning framework for fault diagnosis. IEEE Trans Industr Electron 68(4):3454–3464
2. Mauricio AMR, Gryllias K (2020) Cyclostationary-based multiband envelope spectra extraction for bearing diagnostics: the combined improved envelope spectrum. Mech Syst Signal Process 149:107–150
3. Xu L, Chatterton S, Pennacchi P (2021) Rolling element bearing diagnosis based on singular value decomposition and composite squared envelope spectrum. Mech Syst Signal Process 148:107–174
4. Rezaee M, Taraghi Osguei A (2020) Improving empirical mode decomposition for vibration signal analysis. Arch Proc Inst Mech Eng Part C J Mech Eng Sci 231(12):2223–2234
5. Yang H, Ning T, Zhang B, et al (2017) An adaptive denoising fault feature extraction method based on ensemble empirical mode decomposition and the correlation coefficient. Adv Mech Eng 9(4)
6. Li R, Ran C, Zhang B et al (2020) Rolling bearings fault diagnosis based on improved complete ensemble empirical mode decomposition with adaptive noise, nonlinear entropy, and ensemble SVM. Appl Sci 10(16):18
7. Yan X, Jia M (2018) A novel optimized SVM classification algorithm with multi-domain feature and its application to fault diagnosis of rolling bearing. Neurocomputing 313(3):47–64
8. Wang X, Lu Z, Wei J et al (2019) Fault diagnosis for rail vehicle axle-box bearings based on energy feature reconstruction and composite multiscale permutation entropy. Entropy 21(9):865
9. Jiao J, Yue JH, Pei D, et al (2020) Application of feature fusion using coaxial vibration signal for diagnosis of rolling element bearings. Shock Vibr
10. Damelin SB, Gu Y, Wunsch DC, et al (2014) Fuzzy adaptive resonance theory, diffusion maps and their applications to clustering and biclustering. Math Model Nat Phenom 10(3)
11. Tratanova Z, Leimkuhler B, Lelievre T (2019) Local and global perspectives on diffusion maps in the analysis of molecular systems. Proc Royal Soc A-Math Phys Eng Sci 476(2233)
12. Huang GB, Zhou H, Ding X et al (2012) Extreme learning machine for regression and multiclass classification. IEEE Trans Syst Man Cybern Part B 42(2):513–529
13. Porte JDL, Herbst BM, Hereman W, et al (2008) An introduction to diffusion maps. In: Proceedings of the 19th Symposium of the Pattern Recognition Association of South Africa (PRASA 2008)
14. Lessmeier C et al (2013) Chair of Design and Drive Technology, Paderborn University. KAt-DataCenter:mb.uni-paderborn.de/kat/datacenter

Design and Analysis of High-Impedance Integral Transformer

Jinmei Zhang, Wei Liu, and Yonghang Tai

Abstract Transformer is a static electrical equipment that uses the principle of electromagnetic induction to achieve voltage change. Its main function is to realize voltage conversion, current conversion, impedance conversion, isolation, voltage regulation, etc. In order to limit the short-circuit current of the transformer, high-impedance transformers are used more and more widely. Traditional transformer series reactor can solve the problem of transformer high impedance, but it has certain limitations in terms of limited space and economic cost. In this paper, we analyze three ways to increase transformer impedance through principle analysis, calculation simulation, and test verification results by using the scheme of placing reactors in transformers to form high-impedance all-in-one transformers.

Keywords Transformer · Reactor · High impedance · Integral transformer

1 Introduction

At present, China's transformer market high-end product capacity shortage, low-end product overcapacity is quite obvious, so many companies began to invest heavily in energy saving, environmental protection, miniaturization, low noise, high impedance and other high-tech new product development [1]. Under the guidance of national policies, the current rapid growth of new energy inverter products, the traditional transformer series reactor to increase impedance applied to the topology circuit method gradually applied [2]. However, the traditional transformer series reactor method has certain problems in terms of economy and volume. Therefore, we introduce the design of high-impedance integrated transformer with reactor in the transformer to solve the defects of transformer series reactor or traditional high-impedance design and realize the purpose of transformer high impedance, small size, and energy saving.

J. Zhang · W. Liu · Y. Tai (✉)
School of Physics and Electronic Information, Yunnan Normal University, Kunming, China
e-mail: taiyonghang@126.com

Z. Yu et al. (eds.), *Advancements in Mechatronics and Intelligent Robotics*,
Advances in Intelligent Systems and Computing 1220,
https://doi.org/10.1007/978-981-16-1843-7_50

2 Transformer Short-Circuit Impedance and Core Reactor

2.1 Transformer Short-Circuit Impedance

Short-circuit impedance is an important technical parameter of dry-type transformer, and the impedance of the winding is determined by the number of turns and size of the winding and therefore also determines the technical and economic index of the dry-type transformer. Short-circuit impedance calculation will be a side of the short-circuit impedance through the conversion of the way to the other side, the use of the standard value of the way to reflect, so in the actual test verification process, according to the test equipment testing capacity, from any side can get the same short-circuit impedance value [3].

The leakage reactance of the winding is the energy stored in the winding by the leakage magnetic field of the dry-type transformer, which is concentrated on the leakage reactance of the winding. Therefore, the resistance part of the winding is called the active component of the winding, and the reactance part of the winding is called the reactive component [4]. The relationship between impedance voltage $(U_{ZK}, \%)$, resistance voltage $(U_r, \%)$, and reactance voltage $(U_X, \%)$ is

$$(U_{ZK}, \%) = \sqrt{U_r^2 + U_X^2} \tag{1}$$

The calculation formula of resistance voltage in engineering calculation is

$$U_r = \frac{P_f}{10S_n} \tag{2}$$

The formula for calculating reactance voltage is

$$U_X = \frac{24.8IN\sum D}{e_z H_x 10^4} \tag{3}$$

where I is rated current, unit is A, N is total number of rated tap turns, $\sum D$ is equivalent magnetic leakage channel area, unit is cm^2, e_z is turn potential, H_x is average reactance height of winding, unit is cm, P_f is load loss, unit is W, and S_n is rated capacity, unit is kVA.

2.2 Iron Core Reactor

One of the uses is to limit the short-circuit current of the system, which has the same effect as the short-circuit impedance of a transformer. Usually, the magnetic field of the conductor itself is not strong, so it is manufactured as a bolt-type winding, which is an air-core reactor. In order to obtain a larger inductance value during actual use, an iron core is inserted in the center to produce a larger inductance, which is an iron core reactor [5].

Engineering calculation of reactance voltage of reactor

$$U_{Ln} = 2\pi f L I_n * 10^{-3} \tag{5}$$

where U_{Ln} is reactance voltage of reactor, f is rated frequency, L is inductance, and I_n is rated current.

3 Theoretical and Structural Characteristics of High-Impedance Integral Transformers

3.1 High-Impedance Integrated Transformer

By adding an auxiliary core to the primary winding of the transformer, a larger inductance is created on the primary side, resulting in a larger reactance voltage drop. It is equivalent to the use of the primary side of the transformer winding in series into a reactor. Solve the problem of increasing the number of winding turns and increasing the thickness of the winding when the traditional transformer requires high impedance, reduce the winding reactance height and other methods to achieve high-impedance voltage drop, while solving the problem of space constraints caused by the use of transformers, reactors separated in series combination [6].

3.2 The Engineering Calculation Method

Calculate the resistance voltage of the active component of the transformer, and the calculation method is shown in formula (2); calculate the reactance voltage of the reactive component part of the transformer, see formula (3); calculate the reactance voltage of the reactor, see formula (5) for the calculation method; calculate the equivalent impedance voltage of the integrated reactor, see formula (6); calculate the combined impedance voltage of the high-impedance integrated transformer, and the calculation method is shown in formula (7).

Engineering calculation of equivalent impedance voltage of reactor

$$U_{Lk} = \frac{U_{Ln}}{10U_{1n}} \times 100\% \tag{6}$$

where U_{Lk} is equivalent impedance voltage of reactor.

3.3 Engineering Calculation of Composite Impedance Voltage of High-Impedance Integrated Transformer

$$U_{\mathrm{k}} = \sqrt{U_{\mathrm{r}}^2 + (U_{\mathrm{x}} + U_{\mathrm{Lk}})^2} \tag{7}$$

where U_{k} is resultant impedance voltage of high-impedance integrated transformer.

3.4 Three Ways to Achieve High Impedance and Increase Impedance

Traditional high-impedance transformer: Reduce the core magnetic density, increase the number of winding turns, increase the thickness of the winding, increase the distance of the main channel, and reduce the height of the winding reactance to achieve high impedance.

Separate series connection of transformer and reactor: High impedance is achieved by the combination of reactors in series at the input end of the transformer.

High-impedance integrated transformer: The auxiliary core is added to the primary winding of the transformer to increase the leakage of the primary winding, provide the required leakage reactance, and achieve the purpose of high-impedance voltage of the system.

4 Engineering Simulation Analysis of Practical Application of High-Impedance Integrated Transformer

4.1 Economic Analysis Results

Based on the simulation results, output shape, cost, etc., perform overall analysis to make economic analysis and comparison results, which are shown in Figs. 1 and 2.

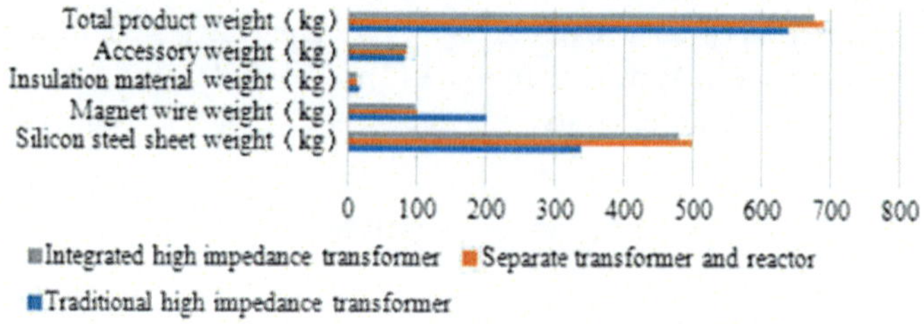

Fig. 1 Overall weight and material consumption analysis

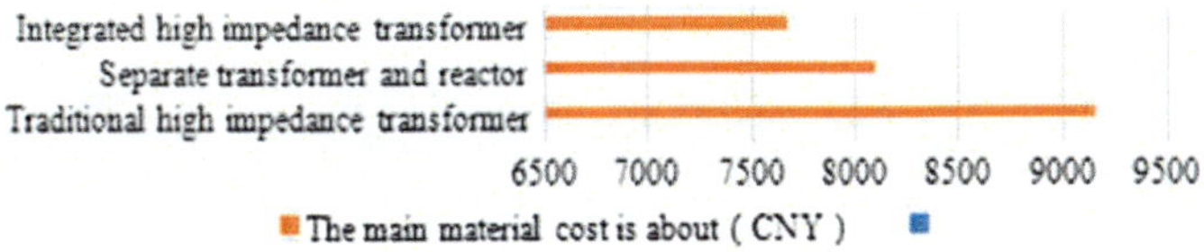

Fig. 2 Cost analysis

4.2 Traditional High-Impedance Transformer

In order to achieve high impedance, reduce the use of silicon steel sheets and increase the number of winding turns. Since the transformer uses aluminum windings, the aluminum density is only about 1/3 of the silicon steel sheet density. Therefore, the overall transformer is light in weight; increasing the main airway to increase the impedance leads to larger dimensions; the traditional high-impedance transformer has the highest cost among the three methods. Compared with the other two methods, the increase in cost is approximately 13 and 19%, respectively; the overall cost performance is poor.

4.3 Series Separation of Transformer and Reactor

Independent design of transformer and reactor, series use, maximum overall weight; the overall dimensions of the transformer and reactor become larger after installation; the cost is somewhere in between; the overall cost performance is moderate.

4.4 Integrated High-Impedance Transformer

Due to the existence of the auxiliary core and the transformer and reactor sharing the primary winding, the overall weight is between the two; the auxiliary core is built in between the primary winding and the secondary winding main channel, and the corresponding volume is the smallest; the main material cost is only 83.7% of the traditional type and 94.7% of the separated structure, and the overall cost is the lowest; overall it is cost effective.

4.5 Product Test Results

According to the simulation design, the electromagnetic design is output, and the product scheme structure design is carried out according to the integrated structure design scheme. After the finished product is produced.

Table 1 High-impedance integrated transformer actual test result 1

Rated capacity (kVA)	Primary side voltage (V)	Secondary side voltage (V)	Rated current of primary side (A)	Rated current of secondary side (A)	Connection group	Rated frequency (Hz)
160	270	400	342.1	230.9	Dyn11	50

Table 2 High-impedance integrated transformer actual test result 2

Total impedance voltage (%)	Effectiveness (%)	Temperature rise of primary winding (K)	Temperature rise of secondary winding (K)	No-load current (%)	No-load loss (W)	Load loss (W)
10.76	98.13	88.7	91.1	2.2	909	2,436

The test results are shown in Tables 1 and 2, and the test results meet the requirements of basic performance parameters.

5 Conclusion

The integrated structure design scheme of transformer and reactor can solve the problem of high impedance, large size, and high cost of transformer from the aspects of physical space size and comprehensive cost performance.

Photovoltaic and wind power new energy projects are mainly installed and used in special environments such as deserts, mountains, and oceans, and the traditional method of using a transformer and a reactor to increase the impedance increases the volume and weight of the inverter and increases the difficulty of transportation and installation [7]. The integrated high-impedance structure design can effectively reduce the volume and weight of the inverter, which is more conducive to the installation and transportation of inverter products.

In short, in terms of energy saving, miniaturization, and high impedance of transformers, the application of integrated design has more economic research value that can be used and continuously improved.

References

1. Ieee, Bwmeta. Element (2008) IEEE recommended practice for establishing liquid-filled and dry-type power and distribution transformer capability when supplying nonsinusoidal load currents—redline
2. Baoding Tianwei Baobian Electric Co., Ltd. (2000) Transformer test technology. Machinery Industry Press

3. Cui L (1996) Theory and design of special transformers. Science and Technology Literature Publishing House
4. Shenyang Transformer Research Institute (1985) Transformer design manual (electromagnetic calculation part). Shenyang Transformer Research Institute
5. Tang Y, Shi N, Yao S, Hu S (1993) Electrical engineering. Xi'an Jiaotong University, Xi'an
6. Qianzhan Industry Research Institute (2018) Analysis of the development status and future trends of the power transformer industry in 2018, exploring new processes and new materials. 2018–2023 China's power transformer industry market demand forecast and investment strategic planning analysis report
7. Liu G (2015) Design ideas of high impedance power transformers. Transformers

Design and Manufacture of a Pulse Driving Circuit for Semiconductor Laser

Yuming Liu, Tao Jiang, Zhikun Yang, Yonghang Tai, and Chao Zhang

Abstract This paper designs a drive circuit for semiconductor lasers that can generate high-speed pulses. The system uses the STM32F429VGT6 chip with an ARM Cortex-M4 core as the controller uses a single-channel, high-speed, and low-side GaN driver UCC27611 to drive high-speed MOSFET field, which effects transistor RD07MUS2B to generate high-speed pulses by using the HMI display or keys to adjust the pulse width and frequency of the signal. The semiconductor laser is provided with a pulse driving current with a short period and a narrow pulse width.

Keywords Semiconductor lasers · High-speed pulses · Narrow pulse width

1 Introduction

The laser is one of the major human inventions in the twentieth century. The birth of laser means that optical science will enter a whole new field. The first laser in the world was born in the twentieth century. Since then, laser theory and laser technology have been gradually popularized and applied, followed by various lasers [1]. Nowadays, there are many types of lasers. According to the laser working material, it is divided into solid laser, gas laser, semiconductor laser, and dye laser [2]. Among them, semiconductor lasers have developed rapidly, have become the leader among many lasers, and have been widely used in laser communications, laser printing, laser pointers, and radar.

With the continuous development and progress of semiconductor laser technology, laser drive power supplies are also constantly developing, with continuous improvement in performance, which has laid a foundation for the better development of laser devices. High-power narrow pulse semiconductor laser technology was first proposed and researched by some developed countries such as the USA,

Y. Liu · T. Jiang · Z. Yang · Y. Tai (✉) · C. Zhang
School of Physics and Electronic Information, Yunnan Normal University, Kunming, China
e-mail: taiyonghang@126.com; zhchynnu@foxmail.com

Z. Yu et al. (eds.), *Advancements in Mechatronics and Intelligent Robotics*,
Advances in Intelligent Systems and Computing 1220,
https://doi.org/10.1007/978-981-16-1843-7_51

Germany, Japan, and other countries. Now their research in this area has reached a very high level. Power supplies developed by many companies for driving semiconductor lasers have gradually entered the market. For example, the multinational company AVTECH, which is famous for producing portable high-speed pulse signal sources, has launched pulse power supplies with models AVOZ-A1A-B and AV-1011-B [3]. The highest pulse peak current can reach 2 A, the repetition frequency can reach 1 MHz, and the pulse width can reach 100 ns. It has strong versatility and can be applied to a variety of semiconductor lasers. In addition, the PCO-7219 pulse power supply produced by another company DEI is not to be outdone. Its pulse width can reach about 50 ns, but their price requires 10,000 or 20,000 coins, which is extremely expensive. Although there is a lot of research on this aspect in China, most of them are used in optical fiber communication. Although this type of driving power has a very narrow pulse width, it generally has the characteristics of small driving current. Therefore, this article will design a drive circuit that can generate high-speed pulses with a pulse width of 100 ns, a repetition frequency of 2 MHz, and a 2.5 A peak current output.

2 System Solution Analysis

In order to realize a driving circuit with a pulse width of 100 ns, an electric pulse peak of 2.5 A, and a repetition frequency of 2 MHz, which can generate high-speed pulses, this article analyzes the design requirements and finds that the difficulty lies in how to generate high-speed pulses with a pulse width of 100 ns. And looking for a switch tube that can respond to high-speed pulses. Therefore, we want to use a high-clock frequency single-chip microcomputer as the control center and choose a high-speed switch tube used in radio frequency to control the high-speed switching tube to generate high-speed pulses. Aiming at the weak drive capability of the I/O port of the microcontroller, a high-speed drive chip for MOSFET is proposed to provide sufficient drive capability for the switch tube. Taking into account the convenience of human–computer interaction, the HMI display screen is used to display and adjust the pulse width and frequency of the output pulse. And the addition of a key function module makes it more convenient to adjust the pulse width and frequency of the output pulse.

2.1 Controller Selection

The STM32F429VGT6 chip is based on the high-performance Arm Cortex-M4 core, and the core operating clock frequency can reach 180 MHz. The Cortex-M4 core has a single-precision floating-point calculation unit (FPU) [4]. It supports all arm single-precision floating-point data processing and data type conversion. It also has a complete set of DSP instructions and a memory protection unit (MPU) to improve the security of the application. It has a high-speed flash memory of 1 MB flash, which supports reading and writing. With up to 260 KB of SRAM, which contains 64 KB of core coupled memory (CCM). Up to 17 timers, up to 180 MHz, each timer has up to 4 PWM or pulse counter inputs [5].

2.2 *Driver Chip Selection*

UCC27611 is a single-channel, high-speed, and gate driver with 5 V drive [6], which is specifically used to drive enhancement mode GaN FETs. UCC27611 has a very strong peak current drive capability and can provide a source current of close to 4 A and a sink current of 6 A. With its dual output configuration, the turn-on and turn-off time of the MOSFET can be optimized. In high-frequency applications, UC27611 can handle input pulse frequencies up to 5 MHz. Its TTL and CMOS low-voltage logic is not affected by VDD input voltage and has high compatibility. And UCC27611 has a unique safety protection design. When the input pin is floating, the internally designed pull-up and pull-down resistors will keep the output low. The small 2 × 2mm SON-6 package with external heat dissipation pad and ground pad is used to reduce the volume to the extreme while ensuring heat dissipation. UCC27611 is also widely used in the design of synchronous rectification, switching power supply, etc. The parameter characteristics are shown in Table 1.

2.3 *Transistor Selection*

For the selection of transistors, its maximum rated parameters, VDSS maximum drain-source voltage, VGS maximum gate-source voltage, ID continuous drain current, IDM pulse drain current, PD allowable channel total power consumption, TJ operating temperature and storage environment temperature range, EAS single pulse avalanche breakdown energy, EAR repeated avalanche energy, IAR

Table 1 Parameter characteristics of UCC27611

Parameter	Minimum	Standard value	Maximum	Unit
VDD	4	12	18	V
OUTH, OUTL	−0.3	VREF	VREF + 0.3	V
VREF	0	5	6	V
IN+, IN-	−0.3	–	18	V
Temperature	−40	–	150	°C
Rise/fall time	–	5	–	ns
Turn on/off delay	–	14	25	ns

avalanche breakdown current, V(BR)DSS drain-source breakdown voltage, VGS (th), VGS(off) threshold voltage, RDS(on) on-resistance, IDSS zero gate voltage drain current, IGSS gate-source leakage current, Ciss input capacitance, Coss output capacitance, Crss reverse transfer capacitance, and other parameters are used to measure whether it can be applied in the design. The parameters that need to be paid attention to are different for different applications. For this design, the switching frequency is as high as 2 MHz, which places high requirements on Ciss input capacitance, Coss output capacitance, and Crss reverse transmission capacitance.

RD07MUS2B is a RoHS-compliant, high-speed MOSFET transistor, which is dedicated to 870 MHz RF power amplifier applications. Therefore, it has a very high-frequency response. Its working characteristics are shown in Table 2.

Refer to the chip data sheet to know that the Ciss input capacitance, Coss output capacitance, Crss reverse transmission capacitance, and Vds drain-source voltage corresponding relationship at 1 MHz input frequency are shown in Table 3. According to the table data, when the drain-source voltage is equal to 5 V, the capacitance of about 50 pF can shorten the response time.

The extremely low gate-source voltage VGS of RD07MUS2B reduces the driving difficulty of the driver chip. For the control of the MOSFET power transistor, only when the applied gate-source control voltage VGS is greater than the threshold voltage VTH, the MOSFET power transistor can be turned on to achieve the switching effect. Refer to the chip data sheet to know that when the gate-source voltage VGS is equal to 2.5 V, the drain current ID can reach 3 A, which fully meets the requirements.

Table 2 Parameter characteristics of RD07MUS2B

Parameter	Condition	Typical value	Unit
VDSS	Vgs = 0 V	25	V
VGSS	Vds = 0 V	−5/+10	V
ID	–	3	A
TCH	–	150	°C
TSTG	–	−40/+125	°C
VTH	Vds = 7.2 V	1.5	V
POUT	F = 175 MHz	7.2	W

Table 3 Capacitance characteristics

Parameter	Vds = 0v	Vds = 5v	Vds = 10v	Unit
Ciss	75	70	70	pF
Coss	100	50	40	pF
Crss	8	4	4	pF

3 Overall Design

Supply 5 V power supply voltage, adjust the voltage to 3.3 V through linear regulator for STM32F429VGT6 use, and 5 V power supply voltage is directly connected to UCC27611 drive module and RD07MUS2B MOS tube module. The STM32F429VGT6 microcontroller generates an adjustable pulse width and frequency PWM input to the UCC27611 drive module, and then controls the RD07MUS2B MOS tube module to generate the desired pulse current. In this design, there are two ways to adjust the frequency and pulse width: one is to adjust through the HMI display, and the other is to adjust by pressing the button [6].

4 Test and Result Analysis

Test the system, fix the pulse width at 100 ns, and adjust the frequency in the range of 1–2 MHz, you can see that the generated pulse is basically without distortion, as shown in Figs. 1 and 2.

Carry on the electric current and power test to the system, the test result is shown in Table 4.

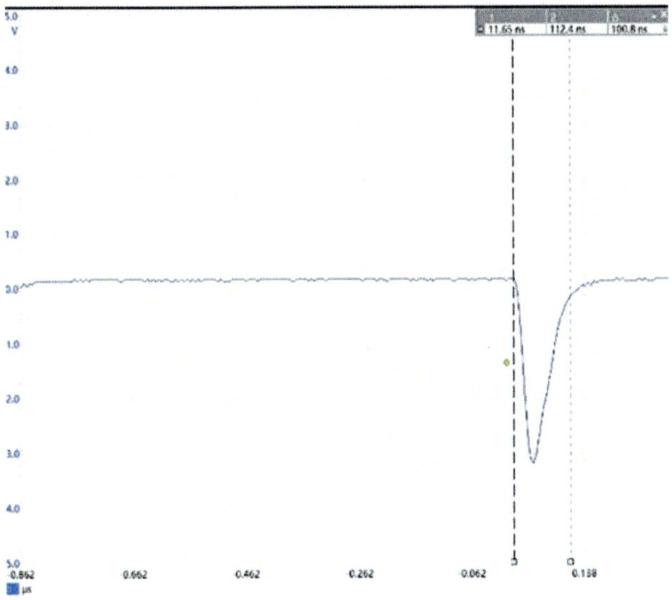

Fig. 1 Impulse waveform of 1 MHz and 100 ns

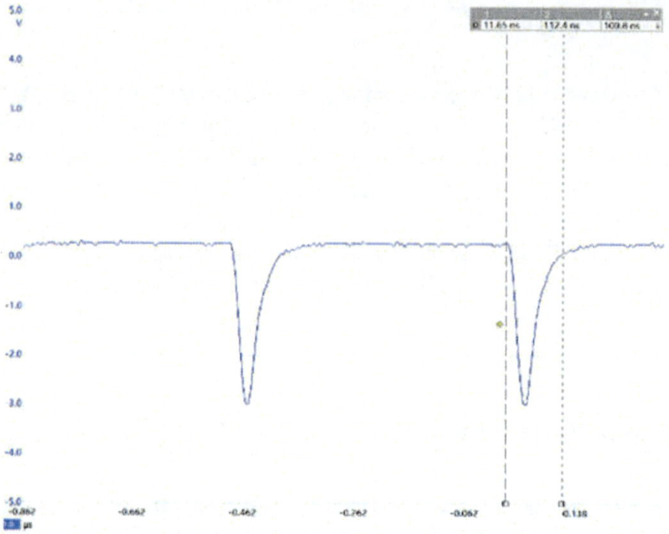

Fig. 2 Impulse waveform of 2 MHz and 100 ns

Table 4 Correspondence table of current and power

Pulse width (ns)	Average current (mA)	Peak current (A)	Average power (W)	Peak power (W)
70	200	1	0.72	3.6
100	300	1.2	1	4.32
150	400	1.32	1.4	4.75
200	500	1.6	1.8	5.5

5 Conclusion

This paper has completed the design and production of a pulse drive circuit for semiconductor lasers from the aspects of system scheme analysis, overall scheme design, hardware design, software design, and system testing. Using STM32F429VGT6 as the controller to drive the high-speed MOSFET transistor through the driver chip UCC27611 provides a short period of short pulse width driving current for the semiconductor laser. The pulse width and frequency of the pulse can be adjusted through the button or the HMI display, and the pulse current frequency can reach 2 MHz, and the pulse width can reach 70 ns when the test is stable. We will continue to study how to further increase the frequency to generate a narrower pulse current.

References

1. Yao P, Zhang Y, Tu B, Wang X, Zhao P, Zhao X (2011) Design and application of pulsed semiconductor laser power supply for mortar laser fuze. Infrared Laser Eng 40(09):1696–1700
2. Xing S, Wang R, Guo R, Cui W (2019) Constant current drive design and realization of butterfly diode laser. Laser Infrared 49(05):553–558
3. Gao M, Bu X, Zhang Y, Yang H (2019) Design of tunable semiconductor laser drive circuit based on FPGA. Laser J 40(01):115–119
4. Chen H, Lu W, Wang K (2019) The design of high-power semiconductor laser pulsed constant current drive circuit. Electr Measur Instrum 56(02):129–133
5. Luo S, Li W (2019) SPL_PL90_3 semiconductor laser drive circuit design. J Jiangsu Univ Sci Technol (Nat Sci Ed) 33(01):66–72
6. Qiu X, Li N, Sun D, Li C, Jiang L, Wei J, Wang G (2018) Research on miniaturization and high stability semiconductor laser driving circuit. Laser Infrared 48(04):469–475

The Efficiency of Vulnerability Detection Based on Deep Learning

Xue Yuan, Peng Zeng, Yonghang Tai, and Feiyan Cheng

Abstract Computer software has been widely used in various industries, and many software cannot avoid receiving network attacks. The phenomena suggest that existing solutions for vulnerability detection demand improvement. This has motivated researchers to find and fix these vulnerabilities early. Deep learning is now widely used for vulnerability detection. Aiming at investigating the training efficiency of distinct neural network models, we leverage three datasets, covering 126 types of vulnerabilities. Each dataset is partitioned into three sets with a ratio of 6:2:2. Word2vec has been applied to transfer program code to vectors. Experiments results have shown that the DNN network achieved maximum efficiency.

Keywords Deep learning · Vulnerability detection

1 Introduction

Vulnerabilities in software undermine the security of computer systems. Companies may suffer from a cybersecurity crisis. This has been proved by the 'Shellshock' vulnerability. Vulnerability [5] refers to specific defects in software that allow attackers to harm the system. Despite plenty of studies that have been proposed to aid vulnerability detection, the number of vulnerabilities released in the common vulnerabilities and exposures. Traditional vulnerability detection can be categorized into three categories as follows: static methods such as data flow analysis, dynamic methods, and hybrid methods. Source code is analyzed when using static methods. Compared with static methods, the dynamic method is designed for detecting the error of the program at run time [4]. The mixed-method combines the two methods. The vulnerability detection method based on machine learning improves the efficiency of vulnerability detection, but this method relies on experts to define features, which is tedious work for experts, and there is a margin for error. The

X. Yuan · P. Zeng · Y. Tai (✉) · F. Cheng
School of Physics and Electronic Information, Yunnan Normal University, Kunming, China
e-mail: taiyonghang@126.com; chengfy@ynnu.edu.cn

Z. Yu et al. (eds.), *Advancements in Mechatronics and Intelligent Robotics*,
Advances in Intelligent Systems and Computing 1220,
https://doi.org/10.1007/978-981-16-1843-7_52

performance of the vulnerability detection system is also strongly related to expert-defined features. Existing machine learning-based vulnerability detection systems [2] may have problems like high negative rates. This has been proved by VUDDY [12] and VulPecker [13], when detecting vulnerabilities of Apache HTTPP2.4.23, VUDDY showed a false negative rate of 18.2%. Li et al. [2] conducted experiments on the Be-Sel-NVD dataset, the false negative rate of VUDDY reached 95.1%. This indicates the VUDDY and VulPecker [3] focus on low false positive rates. In practice, however, such high false positive rates vulnerability detection systems are inefficient to operate. Hence, this paper focusses on the techniques using deep learning for the detection of vulnerabilities.

2 Related Work

In the field of software security, various surveys have been published. Liu et al. [10] provided reviews of literature using code analysis and vulnerability detection using ML. Shahriari [6] reviews the studies using conventional ML techniques for vulnerability detection. Singh et al. [14] make a brief analysis of deep learning vulnerability detection, and the proposed study has demonstrated the advantages of vulnerability detection using deep learning whether it is feasible of using deep learning for vulnerability detection. An extensive review [1] has provided a systematic study of how to capture the characteristics of vulnerable codes.

3 Datasets

The vulnerable data is extracted from open-source projects by Lin et al. [1]. According to the description of vulnerabilities in the National Vulnerability Database (NVD) [11], vulnerable files and functions were labeled. The dataset containing real-world vulnerabilities was constructed [7–9]. We evaluate the performance of four neural networks on three distinct datasets. LibPNG has the capability of processing PNG files. LibPNG can be used to write each line of pixels in a PNG file. Pidgin is a software used for chatting that supports the IRC protocol on the

Table 1 Datasets involved in experiments, datasets are extracted from three open-source projects. There are dual-granularity samples

	File-level		Function-level	
	Non-vulnerable	Vulnerable	Non-vulnerable	Vulnerable
LibPNG	34	44	577	45
VLC player	616	45	6115	44
Pidgin	448	42	8626	29

Linux platform. VLC player is a video player. Table 1 offers a summary of the project involved in experiments. These projects are open source, mainstream, and commonly used. For this reason, these projects are utilized in various studies. Each data is divided into a training set, validation set, and test set according to the ratio of 6:2:2.

4 Experiments and Results

In this section, we present four neural networks, and the efficiency has been compared. A framework for vulnerability detection has been proposed by Lin et al. [1], which is used for the experiment in this paper. Figure 1 shows the process of data in the embedding module.

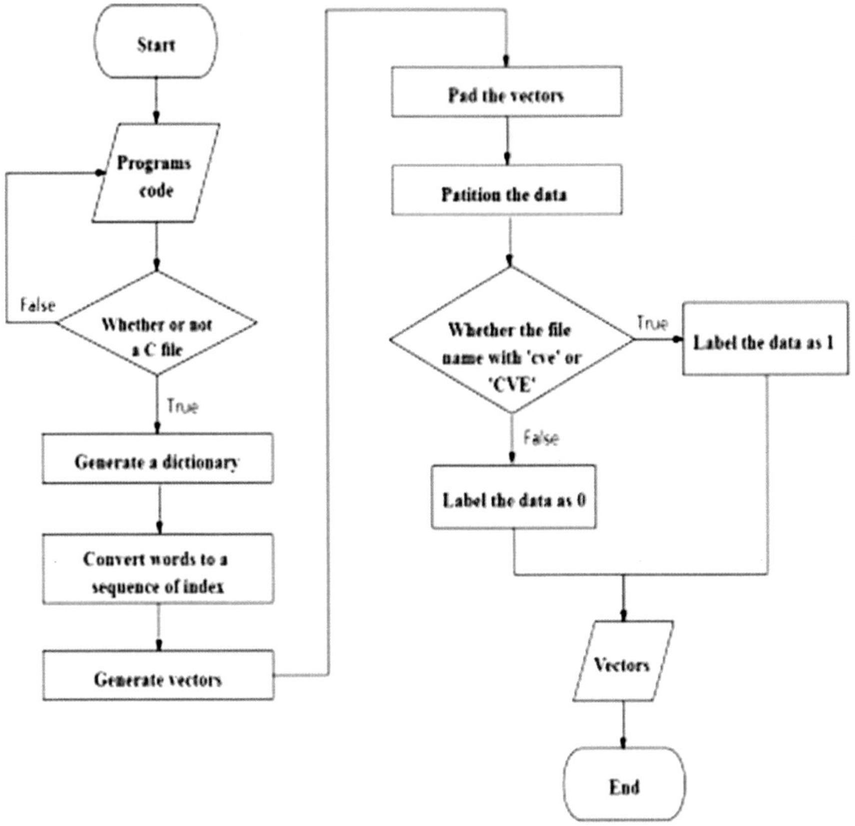

Fig. 1 Process of the training program is converted into vectors. Generating vectors that are used to the input of the neural network

The implementation of four network used Keras (version 2.2.5) with TensorFlow (version 1.14.0) backend. The word2vec is provided by the genism package (version 3.8.3). The framework proposed by Lin et al. [1] provides six available neural networks and three embedding modes. In this paper, four neural networks and word2vec embedding solutions are selected to carry out experiments.

As depicted in Fig. 1, the training programs code is represented in vectors that are suitable for the input of neural networks, which is described as follows:

Step I: After obtaining the programs, generating a dictionary, then the word is converted into an index sequence.
Step II: Training word2vec, generating vectors for neural network input using trained word2vec.
Step III: The vectors are padded with a uniform length.
Step IV: Input data is divided into a training set, verification set, and test set according to the ratio of 6:2:2.
Step V: The file's name with 'cve' or 'CVE' is labeled as 1, the rest are labeled as 0

Four neural networks were experimented on three distinct datasets, and the results are shown in Table 2.

Table 3 shows the comparative result of training time. When applying the DNN for learning, the training process on LibPNG took up 18 s. The test results of DNN

Table 2 Summary statistics for the performance of four neural networks on three datasets

		LibPNG	VLC player	Pidgin
GRU	Accuracy	0.934	0.993	0.997
	Val_acc	0.900	0.995	0.998
	Total accuracy	0.937	0.991	0.996
	Val_loss	0.341	0.013	0.007
BiGRU	Accuracy	0.932	0.992	0.996
	Val_acc	0.900	0.995	0.998
	Total accuracy	0.936	0.991	0.996
	Val_loss	0.325	0.015	0.008
DNN	Accuracy	0.995	0.999	0.999
	Val_acc	0.950	0.997	0.998
	Total accuracy	0.960	0.991	0.996
	Val_loss	0.246	0.017	0.011
LSTM	Accuracy	0.954	0.995	0.996
	Val_acc	0.950	0.995	0.997
	Total accuracy	0.960	0.959	0.996
	Val_loss	0.170	0.006	0.008

Table 3 Comparative result of training time

	LibPNG(s)	VLC player(s)	Pidgin(s)
DNN	18	138	117
LSTM	296	2850	3900
GRU	206	2603	4050
Bi-GRU	178	4500	6300

on the three datasets are shown in Fig. 2. The test on LibPNG shows that FPR = 0.38, as shown in Fig. 2a. Figure 2b, c illustrates the result of VLC player and Pidgin, FPR = 1. The data imbalance has contributed to this phenomenon.

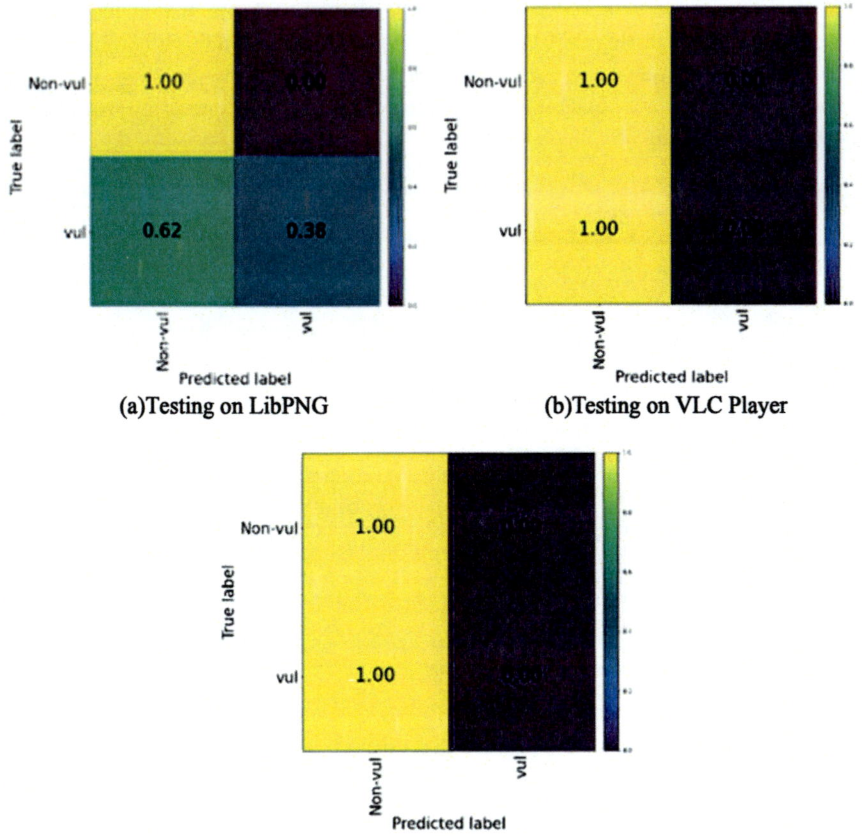

(a)Testing on LibPNG

(b)Testing on VLC Player

(c)Testing on Pidgin

Fig. 2 Results of testing DNN models on three distinct datasets. The first dataset is LibPNG, the ratio of non-vulnerable functions to vulnerable functions is 577:45, the second dataset is VLC player, the ratio is 6115:44, the second dataset is VLC player, the last dataset is Pidgin, and the ratio is 8626:29

Compared with F1-score and precision, accuracy reveal different information, $\text{Accuracy} = \frac{TP+TN}{TP+TN+FP+FN}$. Accuracy refers to the samples that are correctly classified and divided by the number of total samples. Accuracy is an intuitive indicator. Due to the data imbalance, however, high accuracy is not considered that the classifier performs well. For instance, there are a hundred functions, 99 functions are non-vulnerable, and 1 is vulnerable. If the test results are all non-vulnerable, accuracy = 99%. Although the accuracy is high, this classifier can cause a lot of damage if applied. Therefore, the evaluation of a model by accuracy alone is not accurate enough. For this reason, many studies utilized F1-score, precision, and recall. The datasets are designed for simulating real situations by Lin et al. [1], when retrieving less than 1, 10, 20, and 50%, respectively, the precision of DNN is 44, 15, 10, and 5%. Due to the synthetic datasets are constructed artificially, it follows a template-like encoding format. Compared with the performance on SARD datasets, the detector on the synthetic dataset performs better. Neural networks can easily distinguish between vulnerabilities and non-vulnerabilities. If accuracy is used to evaluate the performance of a neural network alone, it may be unscientific. In the next experiment, we should focus on like precision, recall such evaluation indicators.

5 Discussion

The application of deep learning for vulnerability detection has many problems unsolved, which motivates more researchers to contribute to solving these problems. Our experiments select word2vec as the embedding solution. Most recently, ELMo [15] model has been applied for embedding, indicating a promising research direction for vulnerability detection.

References

1. Lin G, Xiao W, Zhang J et al (2020) Deep learning-based vulnerable function detection: a benchmark. In: Information and communications security
2. Li Z, Zou D, Xu S et al (2018) Vuldeepecker: A deep learning-based system for vulnerability detection. arXiv preprint arXiv:1801.01681
3. Lin G, Zhang J, Luo W et al (2018) Cross-project transfer representation learning for vulnerable function discovery. IEEE Trans Industr Inf 14(7):3289–3297
4. Harer JA, Kim LY, Russell RL et al (2018) Automated software vulnerability detection with machine learning. arXiv preprint arXiv:1803.04497
5. Lin G, Wen S, Han QL et al (2020) Software vulnerability detection using deep neural networks: a survey. Proc IEEE PP(99):1–24
6. Ghaffarian SM, Shahriari HR (2017) Software vulnerability analysis and discovery using machine-learning and data-mining techniques: a survey. Comput Surv (CSUR) 50(4):56
7. Zeng P, Lin G, Pan L et al (2020) Software vulnerability analysis and discovery using deep learning techniques: a survey. IEEE Access

8. Chen X, Li C, Wang D et al (2019) Android HIV: a study of repackaging malware for evading machine-learning detection. IEEE Trans Inf Forensics Secur 15:987–1001
9. Zhang J, Xiang Y, Wang Y et al (2012) Network traffic classification using correlation information. IEEE Trans Parallel Distrib Syst 24(1):104–117
10. Liu L, De Vel O, Han QL et al (2018) Detecting and preventing cyber insider threats: a survey. IEEE Commun Surv Tutorials 20(2):1397–1417
11. NVD, https://nvd.nist.gov/
12. Kim S, Woo S, Lee H et al (2017) VUDDY: a scalable approach for vulnerable code clone discovery. In: 2017 IEEE Symposium on Security and Privacy (SP)
13. Li Z, Zou D, Xu S et al (2016) VulPecker: an automated vulnerability detection system based on code similarity analysis. In: Proceedings of the 32nd annual conference on computer security applications, pp 201–213
14. Rattan D, Bhatia R, Singh M (2013) Software clone detection: a systematic review. Inf Softw Technol 55(7):1165–1199
15. Peters M, Neumann M, Iyyer M et al (2018) Deep contextualized word representations

Automation Control

The Design and Simulation of Sit-Up Counter Based on MCU

Hong Wang, Gongping Chen, Fan Yang, and Shuhao Yu

Abstract Our country's economy and people's material life level rapid ascension, people also pay more and more attention to personal health in today's busy society. We do not have the time and space to do strenuous exercise. You can squeeze in some sit-ups during the AD breaks on TV in the evenings. To get a better count, this work designs a sit-up counter based on MCU. This design put computer software and hardware technology as one solution, using 51 SCM as the control core, achieve the automatic counting crunches. Specific processes include start, end control, real-time display the count value, tip end of the test, etc. This design has a simple structure, functional, practical, and high reliability.

Keywords Buzzer · Single chip · Nixie tube · Photoelectric sensor

1 Introduction

Exercise is of a great help and necessity to our bodies. We can do some exercises like sit-ups at home after we get up early and work late. In addition, students also have to do sit-ups, but the manual counting method is still used in the physical fitness test. There are fewer sit-up counters in using today, and only some professional high-end gym has such equipment. Because of the low penetration, it seriously affects the daily exercise effect of supine exercise lovers and it also wastes a lot of manpower and time on fitness tests [1]. Therefore, it has great significance for us to design a convenient and efficient sit-up counter by using the single-chip microcomputer, photoelectric sensor, buzzer, and digital tube [2].

H. Wang · G. Chen
College of Information and Electronic Engineering, Lu'an Vocational Technical College, Lu'an 237158, China

F. Yang · S. Yu (✉)
College of Electronic and Information Engineering, West Anhui University, Lu'an 237012, China
e-mail: 1956345025@qq.com

Z. Yu et al. (eds.), *Advancements in Mechatronics and Intelligent Robotics*, Advances in Intelligent Systems and Computing 1220,
https://doi.org/10.1007/978-981-16-1843-7_53

2 The Design for Sit-Up Counter Based on Single Chip

The design mainly includes hardware and software. The system module includes SCM module, human body infrared sensor module, beeping prompt module, digital tube display module. The system hardware circuit design mainly has the single-chip microcomputer control circuit, the human body induction circuit, the digital tube display circuit, and so on [3]. The sit-up counter is controlled by the software which the single-chip microcomputer can operate the command to control the various hardware modules and realize the function of counting. The system software design mainly has the circulation detection, the count, the terminal buzzer prompt, and so on function.

2.1 The Design of Hardware System

The hardware components include AT89S52 single chip, human body sensor module, digital tube and buzzer, etc. Here is an introduction one by one.

The design of SCM system. In the design, we use AT89S52 single chip. It is a low-power consumption and high-performance 8-bit microcontroller. It has 8 K system programmable flash memory, and it allows program memory to program within the system. AT89S52 has the advantages of simple interface, easy control, and powerful performance, so it is widely used in many embedded control systems. Furthermore, it can be reduced to 0 Hz static logic operation and it supports two kinds of software which can choose power-saving mode. In idle mode, CPU stops working and can support some components such as RAM, timer/counter, serial port and interrupt to continue working. In power down protection mode, the RAM content is saved and the oscillator is frozen. MCU all work stop until the next interrupt or hardware reset.

The design of human body sensor module. Firstly, we use the photoelectric sensor to detect the human body information in the two standard postures. The system determines whether the campaign is a standard. Then the information will be uploaded to the system if it meets the specification. The photoelectric sensor is a device that converts optical signal into electrical signal, and it works based on the photoelectric effect. The photoelectric effect refers to the phenomenon that the electrons of the materials absorb the energy of photons, and the corresponding electrical effect occurs when light is irradiated on some materials. The photoelectric sensor is a kind of small electronic equipment, and it is the key element to realize photoelectric conversion in various photoelectric detection systems. It is a kind of sensor which can detect the existence of objects and the change of the surface state by using various properties of light.

The photoelectric sensor is mainly composed of a luminous part and a receiving part. If the projected light is obscured or reflected by the object being detected, the amount of light reaching the receiver will vary. The receiver will detect this change and converts it into an electrical signal and output it. In our work, we use the optoelectronic switch of the counter—firing type [4]. The optoelectronic switch consists of a transmitter and a receiver that are structurally separated from each other and whose optical axis is opposite. The light from the transmitter goes directly into the receiver [5]. The photoelectric switch outputs a switching signal when the detected object passes between the transmitter and the receiver and blocks the light. The schematic diagram of photoelectric switch shows in Fig. 1.

The design of other module. These modules include system control module, digital tube counting and timing display module, and buzzer prompt module because they are not the focus on this paper and will not be repeated. The circuit diagram of beeping prompt shows in Fig. 2.

2.2 The Design of System Software

The System software was written with Keil 5. Keil C51 is a 51 series MCU C language Software development system produced by Keil Software. It includes a C compiler, macro-assembler, linker, library management and a powerful simulation

Fig. 1 Schematic diagram of photoelectric switch

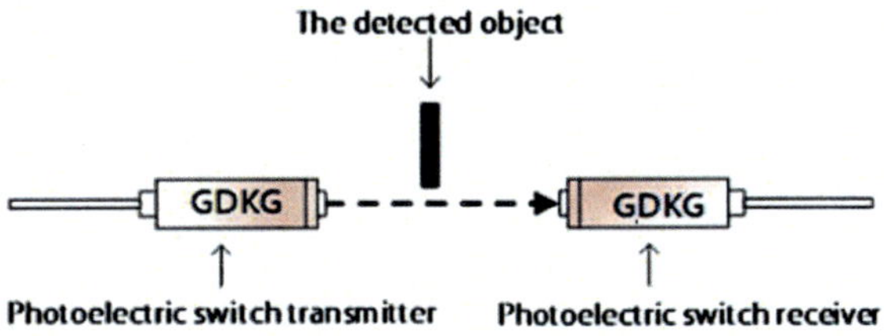

Fig. 2 Circuit diagram of beeping prompt

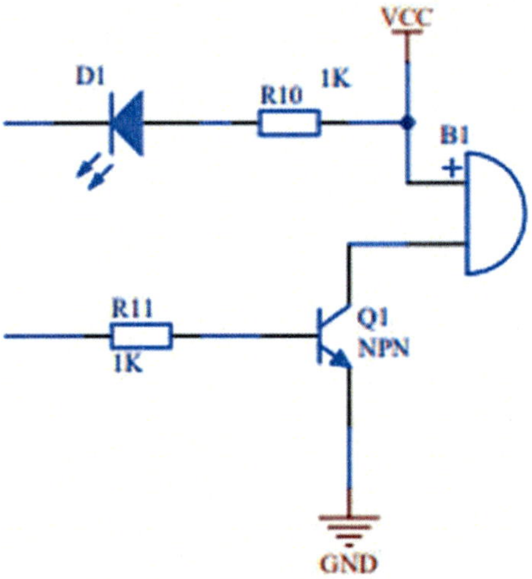

Fig. 3 System flowchart

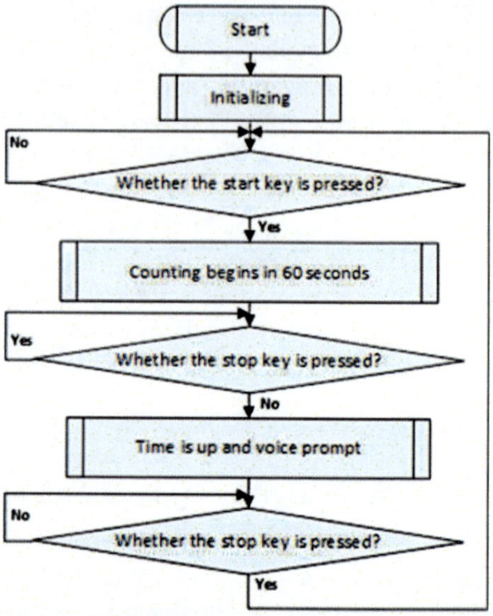

debugger, etc. [6]. These pieces are put together through an integrated development environment.

The system is initialized first. Then the system determines if there is a start button pressed. When a start button is pressed, the system starts working and the two digital tubes will count and 60 s countdown, respectively. The system then determines if there is a stop button pressed. If a stop button is pressed, the system stops working. The buzzer will sound when the 60 s countdown is over. Finally, the system determines whether there is a reset key pressed. If the reset key is pressed, the system is automatically initialized. The main program flowchart of the system is shown in Fig. 3.

3 System Simulation

By the use of software Proteus, the sit-up system is modeled and simulated. Proteus is an EDA tool software published by Lab Center Electronics. It not only has the simulation function of other EDA tool software, but also can simulate the single-chip microcomputer and peripheral devices. It is a better simulation of the SCM and peripheral devices tools [7]. From schematic layout, code debugging to SCM and peripheral circuit co-simulation, one-key switch to PCB design, realizing the complete design from concept to product [8]. On the compilation side, it also supports a variety of compilers such as IAR, Keil, and MATLAB. The system simulation circuit diagram is shown in Fig. 4.

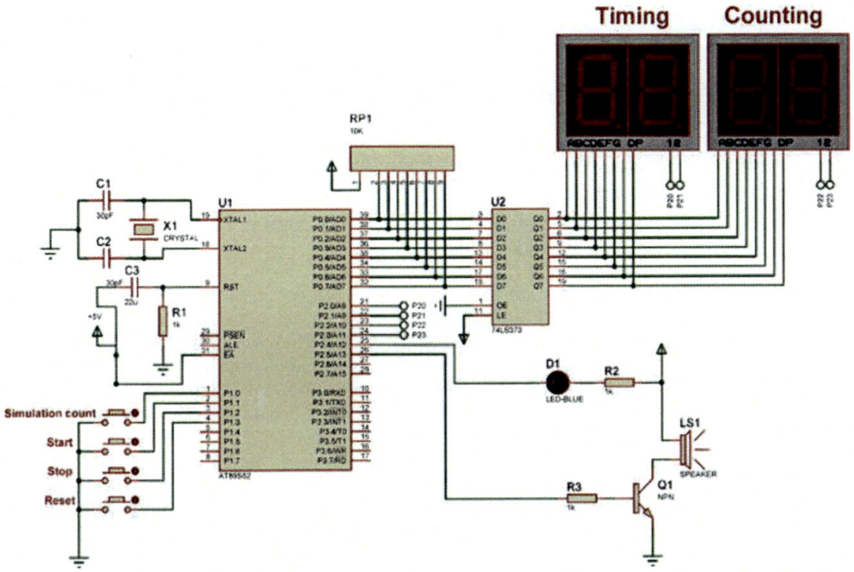

Fig. 4 Circuit diagram of system simulation

The system simulation process is as follows.

Firstly, press the system in the lower left corner of the software to start preparing the simulation button. The system waits for one second, all the components of the system controlled by the single-chip microcomputer are energized and lit up. The two digital tubes display 0, respectively. At this time, the system is waiting for work.

Secondly, press the start button and the analog count button. The whole system starts to work, and the 60 s countdown begins. The digital tube on the left shows the real-time countdown. The system start counting by pressing the analog count button. Each press of the analog count button is equivalent to doing a sit-up. The MCU receives the signal and then displays the number of digital tubes on the right side in real time. When the countdown time of 60 s arrives, the system will sound a buzzer to prompt the end of the system simulation. The counting action will stop, and the counting number of the digital counting tube will remain unchanged. This one-minute sit-up exercise is completed. The exercisers will know their own movements by counting the number of movements displayed by the digital counting tube.

Thirdly, after pressing the reset button, the MCU starts up again and number displayed by the two digital tubes is set to zero, respectively. When a complete 60 s movement simulation is completed, the reset key can be pressed to prepare for the next movement.

Finally, when the stop button is pressed, the system will stop working, the countdown of the timing digital tube will stop and the counting digital tube will stop. This function can be used by our exercisers when they want to stop in the middle of their exercise. Press the reset button again to reset the processing.

4 Conclusion

The sit-up counter is a tool that gives exercisers an intuitive sense of how much they are doing in a given amount of time. It can help athletes count and analyzes their exercise quality. The scheme we designed USES is the single-chip microcomputer as the control center. The infrared photoelectric sensor is used as a device to detect the movement of the movers. The two digital tubes count and time, respectively. The timer ends, and it sounds through the buzzer. The control area has start, stop, and reset button.

The design of the sit-up counter is not only high sensitivity, accuracy, but also greatly improves the convenience and reliability. It is small and easy to operate. It will make it easier for people to buy the sit-up counters we want. It has a very good application prospect which is very beneficial to our vast student community and sports lovers. In the future, we will make physical products according to this design.

Acknowledgements This research was financially supported by the top academic aid project for academic (professional) talents in colleges and universities of Anhui Province (gxbjZD74), the Key project of natural science research in colleges and universities of Anhui Province (KJ2019A1065) and the Quality engineering project of Anhui University (2018mooc340, 2019xfzx04).

References

1. Zhang J, Li Y (2019) Combined with the "principle of single chip microcomputer" and "sensor technology" curriculum design comprehensive case base research. Chin J ICT Educ 14:51–53
2. Bao S, Yang S (2019) Design and case study of single chip microcomputer principle and application. Ind Control Comput 32(06):134–137
3. Xiao Ningbo (2014) Sit-ups counter based on single chip microcomputer. Electron Des Eng 22(18):113–116
4. Han L (2020) Principle and characteristics of photoelectric sensor. Technol Innov Appl (10):77–78
5. Bai S (2020) Design and implementation of a new photoelectric sensor detection method. Electron world (06):137–140
6. Gao Z (2020) The design, simulation and production of electronic clock based on MCU. Electron Test (05):13–15
7. Yu D (2019) Design and realization of software simulation of single chip microcomputer system. Hubei Agric Mechanization 24:152
8. Shi H, Yang Y (2020) Design and implementation of clock based on STC89C52 single chip microcomputer. Pract Electron (Z1):96–98

Adaptive Interference Simulation of Environment Noise

Chun Xia Meng, Xiao Yuan Li, and Liang Zhang

Abstract The sonar equation satisfied by the acoustic signal of the moving sound source is derived in the presence of artificial interference noise. Then, under the typical sound velocity conditions of shallow sea, considering the sound absorption of seawater, the propagation loss value of each frequency in the frequency band of 20–40 kHz is calculated, and the noise intensity of the marine environment is quantitatively analyzed from the sonar equation according to the speed of sound source motion. Change over time. Finally, in different time periods, the simulation generates artificial intervention noise. The spectrum of this noise satisfies the spectral statistical characteristics of the measured ambient noise, and its intensity enables the sonar equation to be established all the time. The simulation results show that when using the transducer to transmit human intervention noise, we can change the ocean environment noise level and significantly reduce the acoustic performance of the moving sound source.

Keywords Marine environment noise · Moving sound source · Adaptive simulation · Underwater acoustics

1 Introduction

After the Second World War, along with the development and technical progress of subjects such as underwater acoustics, signal processing, and computer science, the rapid development of oceanography has led to a leap in marine technology research [1, 2]. The research and utilization of marine environmental characteristics have

C. X. Meng (✉)
Science and Technology on Underwater Test and Control Laboratory,
16, Binhai Street, DaLian 116013, China
e-mail: lxy319@163.com

X. Y. Li · L. Zhang
Dalian Scientific Test and Control Technology Institute,
16, Binhai Street, DaLian 116013, China

Z. Yu et al. (eds.), *Advancements in Mechatronics and Intelligent Robotics*,
Advances in Intelligent Systems and Computing 1220,
https://doi.org/10.1007/978-981-16-1843-7_54

important practical significance [3]. In order to reveal the impact of marine environmental noise on acoustic dynamic testing, this paper starts from the acoustic energy equation under the working mode of noise [4, 5], quantitatively analyzes the change in the intensity of marine environmental noise over time according to the speed of sound source movement, and performs adaptive simulation of marine environmental noise Research.

2 Acoustic Energy Equation in the Presence of Artificial Interference

The acoustic energy equation comprehensively considers the equipment performance, the influence of the acoustic channel, and the characteristics of the sound source. It is a basic equation established according to the energy criterion. It quantitatively reflects the quantitative relationship between the three. The active acoustic energy equation can be divided into two cases, working in the background of noise and working in the background of reverberation. When the reverberation is the main interference of the active system, the array signal system works in the background of the reverberation. When noise is the main interference of an array system, the array signal system is said to work in a noise background. The active acoustic energy equation working in a noisy background is as follows:

$$\mathrm{SL}_D - 2\mathrm{TL} + \mathrm{TS} - (\mathrm{NL} - \mathrm{DI}) - P = \mathrm{DT} \tag{1}$$

where SL_D is the emission source level, DT is the detection threshold, DI is the directivity index of the array, P is the scanning loss, TS is the energy intensity of underwater sound source, NL is the noise level, it is the sum of the interference caused by the self-noise of the moving sound source and the ambient noise. Generally, the self-noise of the moving sound source is much larger than ambient noise, so self-noise is the main noise interference of the receiving array. In the active acoustic energy equation, SL_D, DT, DI, P are all determined. TS is a function of the target chord angle, and the values corresponding to the different chord angles are different. TL is propagation loss.

The propagation loss corresponding to the frequency f_i at distance R_i is $\mathrm{TL}_{ij}(R_i)$. At this time, the noise level received by the moving sound source signal system can be expressed as:

$$\mathrm{NL}_{ij}(R_i) = \mathrm{SL}_D - 2\mathrm{TL}_{ij}(R_i) + \mathrm{TS} - P - \mathrm{DT} + \mathrm{DI} \tag{2}$$

The source level of the artificial simulation signal can be expressed as:

$$\mathrm{SNL}_{ij} = \mathrm{NL}_{ij}(R_i) + \mathrm{TL}_{ij}(R_i) \tag{3}$$

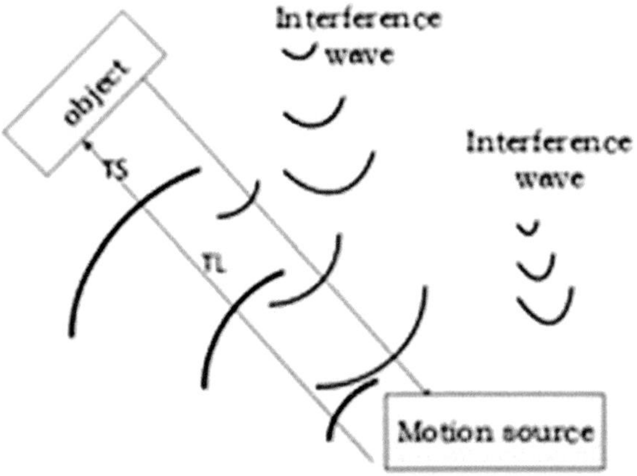

Fig. 1 Schematic of environmental adaptive simulation

According to the movement speed of the moving sound source, the source level of the simulation signal is calculated one by one at a distance and one by one at a frequency, so as to achieve adaptive interference with the acoustic environment. Figure 1 shows the principle of environmental simulation.

3 Simulation of High-Frequency Propagation Loss

When a sound wave propagates in seawater, its energy gradually decreases as the propagation distance increases. The absorption loss of sound waves in seawater is related to frequency, especially in high-frequency bands. TL determines the distance of the sound waves travel. If I_0 represents the sound intensity at 1 m from the center of the sound source, and Ir is the sound intensity at the sound receiver, then the propagation loss from the moving sound source to the sound receiver is:

$$\mathrm{TL} = 10\lg\left(\frac{I_0}{I_r}\right) \tag{4}$$

For simplicity, suppose there is a Pekeris waveguide, it is assumed that the ocean waveguide model is composed of a seawater layer and a liquid semi-infinite space. The parameters of the seawater layer include sound velocity, density, and attenuation coefficient, wherein the sound velocity is represented by the measured seawater sound velocity profile. In the simulation calculation described below, the density is taken as 1000 kg/m^3 and the attenuation coefficient is taken as 0.04 dB/wavelength. The sound velocity of the seabed is 1732 m/s, the density is 2008 kg/m^3, and the attenuation coefficient is 0.866 dB/wavelength. The depth of a moving sound source is that the sound source depth is 50 m and the receiving depth is 50 m.

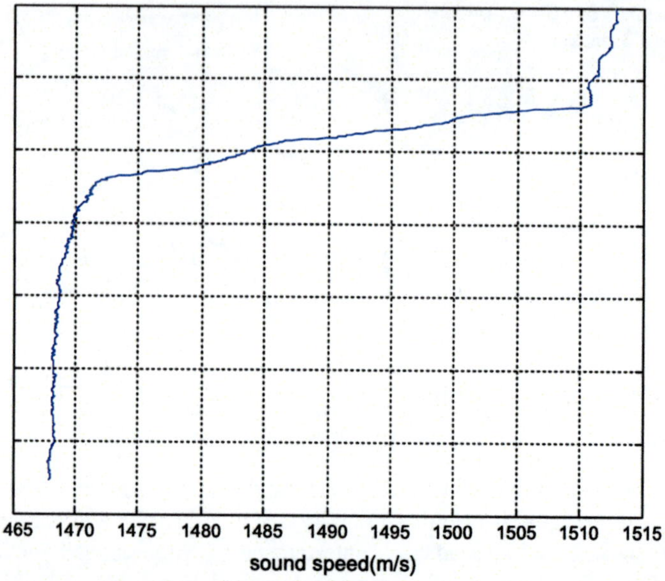

Fig. 2 Sound velocity of seawater

Figure 2 shows the seawater sound velocity profile measured in a shallow sea in summer, and the vertical coordinate is the depth of the seawater. According to the waveguide environmental parameters, the ray acoustic theory is used to simulate the propagation loss of high frequency in the sound waveguide. The simulation calculations are carried out for each frequency, and the results show that the acoustic propagation loss of different frequencies is not much different. For the sake of simplicity, in the subsequent studies, the propagation loss of each frequency in the range of 20–40 kHz takes the calculation result of the propagation loss corresponding to 30 kHz. Figure 3 shows the variation of the acoustic signal propagation loss with distance at a frequency of 30 kHz.

It can be seen from Fig. 3 that with time, the distance between the moving sound source and the underwater object decreases, and the propagation loss of the sound signal gradually decreases. According to the active acoustic energy equation of a moving sound source working under a noise background, it can be known that if no artificial noise simulation signal is applied, the echo of the underwater object received by the moving sound source signal processing system will become larger and larger. In order to ensure that the left and right sides of the active acoustic energy equation are equal at any time, the noise simulation signal must be artificially applied, and the noise simulation signal source level must be increased as the distance between the moving sound source and the underwater object decreases.

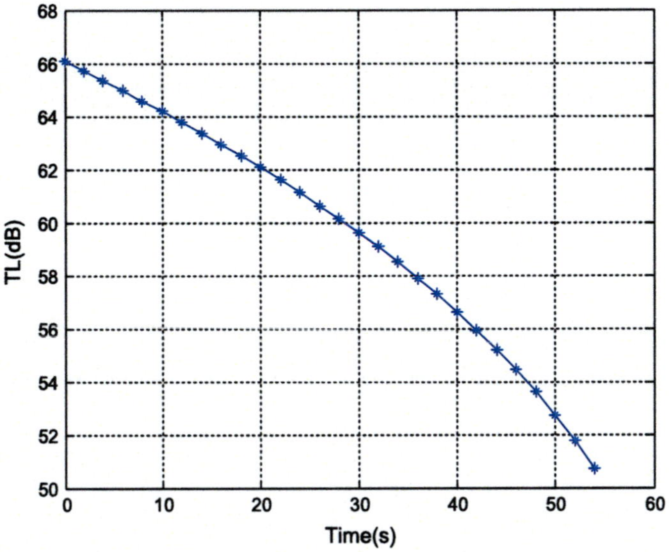

Fig. 3 Variation of propagation loss with time at 30 kHz

4 Adaptive Simulation of Marine Environmental Noise for Different Moving Sound Source Parameters

4.1 Simulation of Moving Sound Source A as an Example

Assume the main parameters of moving sound source A and its signal processing system: The sound source level is 210 dB, the noise level is 59 dB, the reception directivity is 20 dB, the scanning loss is 5 dB, and the detection threshold is 38.2 dB. The energy intensity of underwater objects is 10 dB, and the moving speed of source is 35 knots. Figure 4 shows the change in noise level over time near a moving sound source, and Fig. 5 shows the change in distance between a moving sound source and an underwater object, and the change in the level of a simulation signal source.

Assume the main parameters of moving sound source B and its signal processing system: The sound source level is 210 dB, working frequency is 20–40 kHz, the noise level is 55 dB, the reception directivity is 21.48 dB, the scanning loss is 5 dB, and the detection threshold is 10 dB. The energy intensity is 10 dB, and the moving speed of source is 55 knots. Figure 6 shows the change in noise level, and Fig. 7 shows the change in distance between a moving sound source and an underwater object over time, and the change in the level of a marine environment noise simulation signal source.

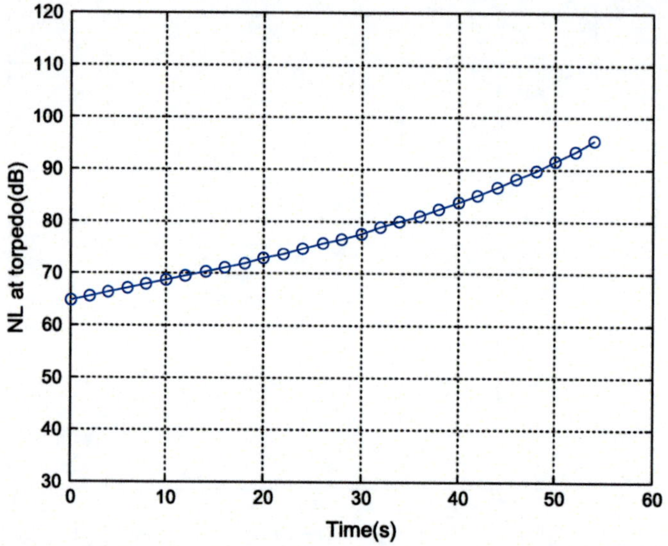

Fig. 4 Noise level at moving sound source A

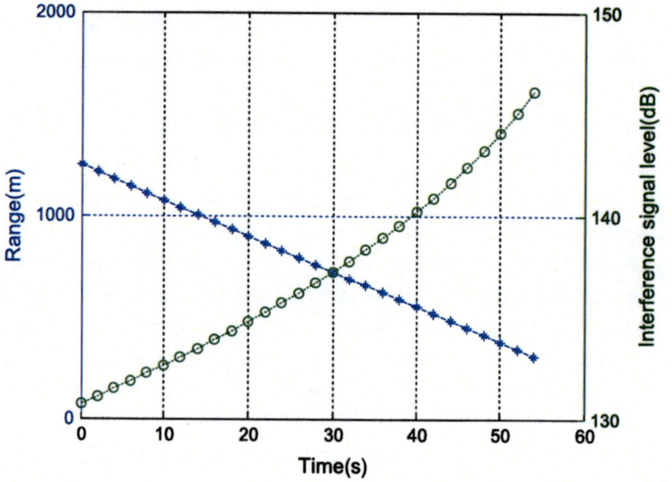

Fig. 5 Moving sound source A distance and simulated signal source level change with time

Comparing Figs. 4 and 6, it can be seen that the moving sound source B has a lower detection threshold. In order to keep the left and right sides of the acoustic energy equation equal, the noise level received by the moving sound source B is large at any time. In order to interfere with the signal processing system of the moving sound source B, a larger simulated signal source level needs to be

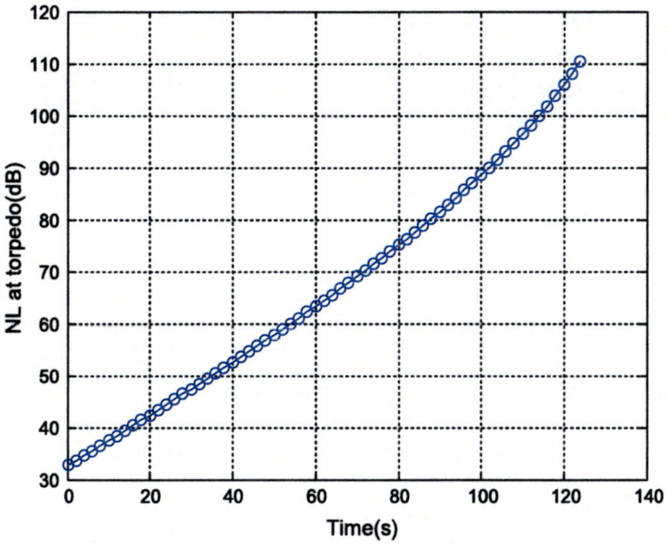

Fig. 6 Noise level at moving sound source B

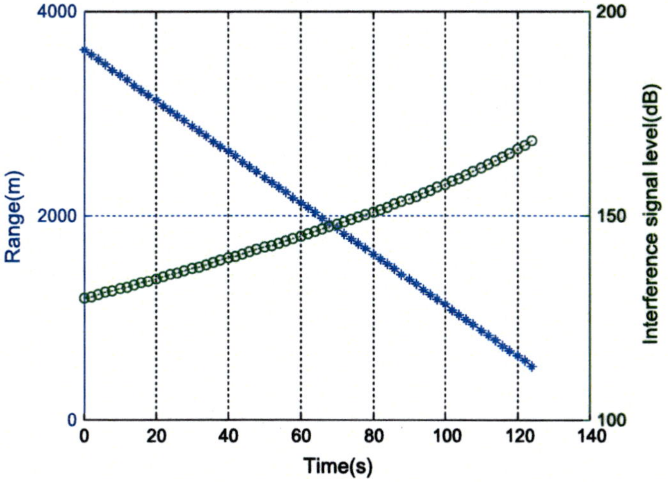

Fig. 7 Moving sound source B distance and simulated signal source level change with time

generated. Comparing Figs. 5 and 7, it can be seen that under the same background noise condition, the moving sound source B has a low detection threshold, so the effective range of the sound signal is also large. At the same time, a larger simulation signal source level is required to interfere with the moving sound source B.

5 Realization of Artificial Simulation of the Marine Environment

Analyze and process the marine environmental noise data from 20–40 kHz measured at sea, and obtain the wideband environmental noise spectral characteristics in the selected marine channel. Amplify the marine environmental noise according to the above-mentioned simulation calculation method and use several separate high-frequency transducers to transmit signals in a large sea area. Figure 8 shows the comparison between the noise simulated interference signal and the original ambient noise spectrum, where the curve with "blue asterisk" is the spectrum of the noise simulated interference signal, and the curve with "red inverted triangle" is the spectrum of the measured marine environmental noise.

The simulated interference signal of the marine environmental noise generated by the above method has the same spectral characteristics as the actual marine environmental noise, which makes the signal processing system of the moving sound source unable to detect the artificial simulated interference signal from the background. The source level of the signal is different, which can ensure that the left and right sides of the acoustic energy equation are equal at any time, which reduces the performance of the signal processing system.

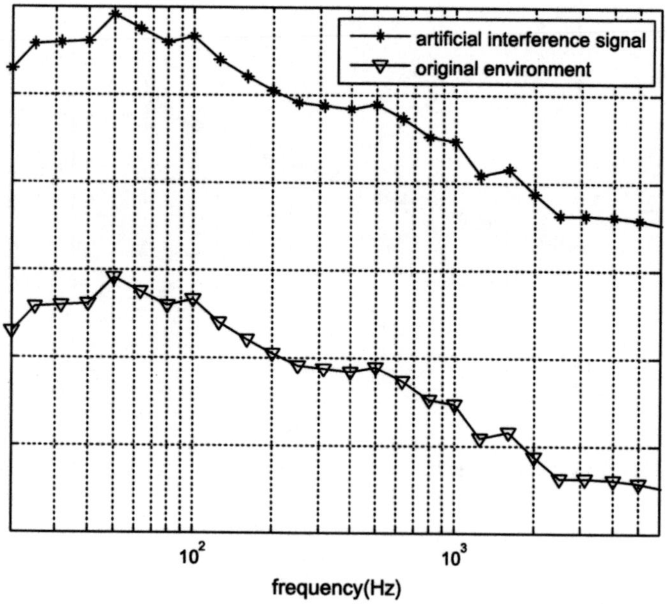

Fig. 8 Comparison of noise simulation signal and original ambient noise spectrum

6 Summary

In this paper, a high-frequency moving sound source at a stable moving speed is taken as an example, and the high-frequency propagation loss under shallow ocean channel conditions is calculated using ray theory. Based on the active acoustic energy equation working under noise background, a method of artificially interfering with the working environment of a moving sound source signal processing system using marine environmental noise was studied. Simulation calculations are made for moving sound sources with different parameters. It can be seen that in the same marine environment, the source level of artificial interference noise changes with the movement speed of the moving sound source and the detection threshold of the signal processing system. At different times, the measured intensity of the marine environmental noise is amplified by different multiples, and several high-frequency transducers are used for transmission, which can adaptively change the marine noise level near the moving sound source. The research method in this paper provides a reference for the performance evaluation of signal processing systems under different SNR conditions in a laboratory environment.

Acknowledgements This study was funded by the foundation research project (No. JCKY2016207A037). The authors would like to thank colleagues for technical assistance with providing marine environmental parameters, and the organizers are thanked very much by the author for providing the international academic exchange opportunity.

References

1. Robert J (1984) Uric.: principles of underwater sound. McGraw-Hill, New York
2. Pau C (2010) Etter.: underwater acoustic modelling and simulation. Publishing House of Electronics Industry, Beiji
3. Yang S (1994) e.: principles of underwater sound propagation. Harbin Engineering University Publishing Company, Harbin
4. Kuperman WA, Ingenito F, Fialkowski LT (1993) Modelling ambient noise in three-dimensional ocean environments. J Acoust Soc Am 108:739–752
5. Fried SE, Walke SC, Hodgkiss WS, Kuperman WA (2013) Measuring the effect of ambient noise directionality and split-beam processing on the convergence of the cross-correlation function. J Acoust Soc Am 121:1824–1832

Comprehensive Evaluation of Real Estate Development Based on Factor Analysis and Cluster Analysis: A Case of Hubei Province

Mengyan Zhao, Lili Meng, Hongxing Liu, Yazhou Xiong, and Jing Mo

Abstract At present, China's real estate and related industries account for a large proportion of the GDP, which has played an important role in the rapid development of China's economy. This paper first constructs an evaluation index system for the development of real estate and then uses factor analysis to analyze the data of the real estate development in Hubei Province in the past 19 years, analyzes the trend of real estate development in Hubei Province, and important factors affecting the development of real estate, and uses *K*-Means cluster analysis to perform cluster analysis on 19 years from 2000 to 2018. Finally, the government can take measures from three aspects to promote the healthy development of the real estate industry.

Keywords Hubei Province · Real estate development · Factor analysis · Variance analysis · Influencing factors

1 Introduction

China's economic development achievements have already created many miracles, and the development of real estate has an indelible important position in China's economic development. Since 1998, China's real estate market has entered a period of rapid development. Since then, housing prices have risen all the way. Real estate investment not only promotes the development of basic urban construction in cities, but also improves people's living standards and increases the rate of urbanization. The development of real estate is related to a series of economic developments and to the development of real estate.

At this stage, the development of the regional real estate industry in Hubei Province is relatively lagging, and development investment is insufficient. And the development of the real estate industry in Hubei Province needs to be improved.

M. Zhao · L. Meng · H. Liu · Y. Xiong · J. Mo (✉)
School of Economics and Management, Hubei Polytechnic University, Huangshi 435003, Hubei, China

Z. Yu et al. (eds.), *Advancements in Mechatronics and Intelligent Robotics*,
Advances in Intelligent Systems and Computing 1220,
https://doi.org/10.1007/978-981-16-1843-7_55

If the data analysis on real estate development can be carried out, the overall economic development of Hubei Province will be improved. This article mainly explores the real estate development status of Hubei Province in the past 19 years, so that Hubei Province can achieve more stable and sustainable development in the future and accelerate the development of a well-off society.

Under the above research background, this study focuses on the real estate development of Hubei Province from 2000 to 2018. It analyzes the development of real estate from multiple angles, including national economy, population, employment and wages, fixed asset investment, etc. And this study finds out the influencing factors of the province's real estate development, and evaluates and analyzes the development of the past 19 years [1].

2 Theory on Feature Extraction Based on Factor Analysis

2.1 Overview and Basic Ideas of Factor Analysis

Factor analysis is a method that can greatly reduce the number of variables involved in data modeling without causing a large amount of loss of information. It is a method that can effectively reduce the dimensionality of variables and is widely promoted.

It is a multivariate statistical analysis method, whose goal is to condense many original variables into a few factors with the least loss of information [2].

2.2 Mathematical Model of Factor Analysis

The core of factor analysis is to use fewer independent factors to reflect most of the information of the original variables. There are p original variables, which are $x_1, x_2, \ldots, x_p$, and the mean of each variable (after standardization) is 0, and the standard deviation is 1. Now that each original variable can be represented by a linear combination of $k(k<p)$ factors, which are $f_1, f_2, \ldots, f_k$, then there is following formula (1):

$$\begin{cases} x_1 = a_{11}f_1 + a_{12}f_2 + \cdots + a_{1k}f_k + \varepsilon_1 \\ x_2 = a_{21}f_1 + a_{22}f_2 + \cdots + a_{2k}f_k + \varepsilon_2 \\ \vdots \\ x_i = a_{p1}f_1 + a_{p2}f_2 + \cdots + a_{pk}f_k + \varepsilon_p \end{cases} \tag{1}$$

The above formula (1) is the mathematical model of factor analysis.

3 Theory on Cluster Analysis

3.1 Overview and Basic Principles of Cluster Analysis

Cluster analysis is to classify things based on the characteristics of the value of objects. It is the result of the combination of numerical taxonomy and multivariate statistical techniques. In the process of classification based on the cluster analysis, people do not need to give a classification standard in advance. Cluster analysis can start from sample data and automatically perform classification analysis. The number of clusters obtained may not be consistent, so we may choose the appropriate number of categories according to the needs.

The basic idea of analysis is to observe the close relationship between samples according to the numerical characteristics of things. And the close relationship is measured by the distance between the samples. Once the distance between the samples is defined, the close samples can be classified into one category [3–5].

3.2 Hierarchical Clustering

There are two types of hierarchical clustering: one is *R*-type clustering, and the other is *Q*-type clustering [2].

R-type clustering: It clusters the samples, so that the samples with similar characteristics are clustered together, and finally the samples with large differences are separated.

Q-type clustering: It clusters the variables, so that the variables with large differences are separated, and finally, the variables with similarity are gathered together.

3.3 K-*Means Clustering*

K-Means clustering method, also known as fast clustering method, is a stepwise clustering analysis method for large sample data after the user first specifies the number of categories. Compared with hierarchical clustering, its execution efficiency is higher and it takes up less memory. But the difference between these two clustering methods is that *K*-Means clustering needs to achieve a specified number of categories.

4 Empirical Analysis

4.1 Index System Construction, Data Collection, and Processing

When selecting real estate data indicators, this paper mainly follows two principles. One is that the indicators cover a wide range of real estate industries; the other is that they are representative and influential real estate development indicators. To this end, this paper selected the following 12 indicators to fully reflect the development of the real estate industry in Hubei Province. The corresponding relationship between specific indicators and corresponding variable codes is shown in Table 1.

And there are large differences in the dimensions of the selected indicator data, it is difficult to observe and analyze data, so the data must be standardized. There are many standardization methods. The default standardization by SPSS software is z-score standardization. And the standardized formula is as following (2):

$$x^* = \frac{x - u}{\sigma} \tag{2}$$

The following correlation and cluster analysis are calculated with standardized data.

Table 1 Index system and variable code table

Indicator name	Unit	Variable code
Resident consumption level	Yuan	x_1
Gross domestic product (GDP)	100 million yuan	x_2
The added value of primary industry	100 million yuan	x_3
The added value of secondary industry	100 million yuan	x_4
Permanent population at the end of the year	10,000 people	x_5
Natural population growth rate	%	x_6
Investment in fixed assets of the whole society	100 million yuan	x_7
Real estate development investment	100 million yuan	x_8
Building area under construction by real estate development enterprises	10,000 m^2	x_9
Commercial housing sales area	10,000 m^2	x_{10}
Number of real estate development enterprises	Piece	x_{11}
Total retail sales of consumer goods	100 million yuan	x_{12}

4.2 Correlation Analysis of Real Estate Development Indexes

Correlation coefficient matrix. Pearson method is used for correlation analysis, and the correlation coefficient matrix of 12 indicators is calculated.

And all the correlation coefficients are all greater than 0.85, and the correlation coefficients are very high, indicating that the 12 variables present a strong linear relationship, and common factors can be extracted from them. Therefore, it is necessary to perform factor analysis on the 12 indicators.

Bartlett sphericity test and KMO test. The specific test results are shown in Table 2.

It can be seen that in Bartlett's sphericity test, the observed value of the test statistic is 765.125, and *P* value is close to 0, which means the test statistic is very significant. The null hypothesis is rejected, and the correlation coefficient matrix is considered unlikely to be a unit matrix, so it is suitable for factor analysis.

The KMO test has a value greater than 0.7. According to the KMO metric given by Kaiser, it is suitable for factor analysis.

In summary, the factor analysis of 12 indicators is appropriate and necessary.

4.3 Factor Analysis of Real Estate Development Index

According to the correlation coefficient matrix *R*, this paper uses the principal component analysis method to extract the main factors to find the eigenvalues and the variance contribution rate, and the cumulative variance contribution rate. It can be seen from Table 3 that the first three factors can explain 99.437% of the vast majority of the information of the original variables, and the remaining factors contribute very little to the explanation of the original variables. Therefore, the original 12 indicators can be successfully compressed into 3 indicators, and the extraction of 3 indicators is appropriate and effective.

Table 2 KMO and Bartlett's test

The Kaiser-Meyer-Olkin measure of sample adequacy		0.873
Bartlett's sphericity test	Approximate chi-square	765.125
	Degrees of freedom	66
	Significance	0.000

Table 3 Explained total variance

Components	Initial eigenvalue			Extraction sums of squared loadings			Rotation sums of squared loadings		
	Total	% of variance	Cumulative (%)	Total	% of variance	Cumulative (%)	Total	% of variance	Cumulative (%)
1	11.481	95.679	95.679	11.481	95.679	95.679	6.607	55.054	55.054
2	0.291	2.424	98.103	0.291	2.424	98.103	2.852	23.764	78.818
3	0.160	1.335	99.437	0.160	1.335	99.437	2.474	20.619	99.437
4	0.034	0.284	99.721						
5	0.014	0.116	99.837						
6	0.011	0.091	99.928						
7	0.005	0.040	99.967						
8	0.002	0.021	99.988						
9	0.001	0.008	99.996						
10	0.000	0.003	99.999						
11	0.000	0.001	100.000						
12	4.991E-5	0.000	100.000						

4.4 The Naming Explanation of the Main Factor

In order to make the practical significance of the extracted principal factors more clear, we use the maximum variance method to perform orthogonal rotation on the factor-loading matrix, so as to make the factor naming explanatory. The rotated factor-loading matrix is shown in Table 4.

It can be seen from the table that the first factor has a higher load on *x*1, *x*5, *x*7, *x*10, and *x*12. The first factor mainly explains the above variables; the second factor has a higher load on *x*11; the third factor has a higher load on *x*6. Therefore, the first factor can be interpreted as the level of national economy, the second factor can be interpreted as the level of real estate development, and the third factor can be interpreted as the level of population growth.

4.5 Comprehensive Evaluation of Real Estate Throughout the Year

A comprehensive evaluation of the real estate development is conducted in Hubei Province from 2000 to 2018. This paper considers from the simple quantity, the variance contribution rate after rotating by two factors is the weight.

$$F = 0.55054F_1 + 0.23764F_2 + 0.20619F_3 \tag{3}$$

We calculated comprehensive evaluation scores from 2000 to 2018 according to formula (3) and sorted them in time series. As shown in Fig. 1, the factors F_1 and F_2 have a stronger influence than factor F_3.

Table 4 Rotated factor-loading matrix

Component	Main component		
	1	2	3
*x*1	0.818	0.430	0.377
*x*2	0.777	0.469	0.417
*x*3	0.663	0.584	0.462
*x*4	0.724	0.513	0.455
*x*5	0.846	0.402	0.335
*x*6	0.478	0.412	0.775
*x*7	0.816	0.393	0.418
*x*8	0.775	0.431	0.453
*x*9	0.786	0.433	0.431
*x*10	0.812	0.417	0.393
*x*11	0.474	0.794	0.379
*x*12	0.810	0.423	0.402

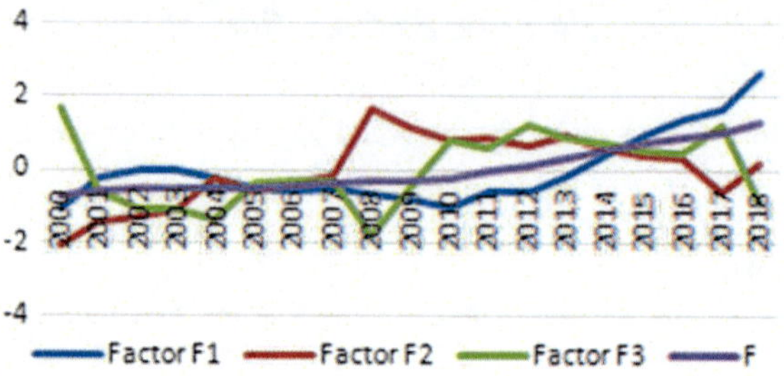

Fig. 1 Factor and score line chart

4.6 K-*Means Clustering*

First select the number of clusters, which is represented by the symbol k:

(1) when $k = 3$;

The first category: 2018, 2017, 2016, 2015;
The second category: 2014, 2013, 2012, 2011, 2010;
The third category: 2009, 2008, 2007, 2006, 2005, 2004, 2003, 2002, 2001, 2000;

(2) when $k = 4$;

The first category: 2018, 2017, 2016;
The second category: 2015, 2014, 2013;
The third category: 2012, 2011, 2010;
The fourth category: 2009, 2008, 2007, 2006, 2005, 2004, 2003, 2002, 2001, 2000;

From the above clustering results, we can know that when the number of clusters is three types, 2009–2010 and 2014–2015 are a transition period for real estate development; when the number of clusters is four types, 2009–2010, 2012–2013, 2015–2016 are a turning period for real estate development.

It can be seen that in recent years, the speed of real estate development has become faster and faster, and it also reflects the gradual increase in fluctuations in real estate development, which has to arouse people's attention.

5 Some Suggestions

5.1 Making Reasonable Positioning and Coordinate

The sustainable development of real estate focuses on the national economy, real estate development, and population growth.

5.2 Reasonably Development and Utilization

We must pay attention to environmental protection, look at problems from a development perspective.

5.3 Guiding the Correct Housing Needs and Control the Population in the Area

The house is for living, not for speculation. We must control regional population growth and stabilize residents' potential real estate needs.

Acknowledgements The study is supported by the college students' innovation and entrepreneurship training projects of Hubei Province Grant No. S202010920070.

References

1. Xiao Y (2017) Research on the association between new urbanization and real estate market in Hubei Province. Master's Thesis, Central China Normal University
2. Xue W (2011) Analysis based on SPSS, 2nd edn. Renmin University of China Press, Beijing
3. Chen C, Qiu Y (2009) Common functions and application examples of SPSS15.0. Publishing House of Electronics Industry, Beijing
4. Xiong Y, Zhang S, Lan J, Chen F (2019) Efficiency evaluation of regional economic development of mining and metallurgy city based on DEA model. Int J Appl Decis Sci 12(3): 242–256
5. He L (2011) Research on sustainable development index system of real estate industry based on factor analysis. Prod Res 08:102–105

Design of a Portable 3D Scanning Device

Kangwei Chang, Penghui Ding, Shixun Luan, Kaikai Han, and Jianyong Shi

Abstract With the development of point cloud data acquisition technology, 3D scanners are becoming more and more widely used. The 3D scanner collects data such as the shape and surface characteristics of specific target in reality and processes it to obtain a three-dimensional model of the target, which can be used in many fields, such as industrial design, biomedicine, and film production. This article introduces the design and working principle of a low-cost portable 3D scanning device. The equipment uses laser scanning target surface. Compared with existing products on the market, it has a lower cost and smaller size. It is suitable for mass production and can be widely used in engineering fields.

Keywords 3D scanning · Portable · Low cost · Laser

1 Introduction

3D scanning devices are very popular these days. There are many good products on the market.

Considering detection method, there are roughly two types: contact and non-contact device.

The contact-type 3D scanner directly touches the target through a probe and moves the probe for multipoint measurement. This method is intuitive and accurate and has been put into practical application for a long history and is often used in manufacturing industry. However, the measurement probe may damage the surface of the measured object, the measurement process takes a long time, and the probe itself is extremely easy to damage, resulting in high cost. The typical application of contact measurement is a CNC three-coordinate measuring machine.

Non-contact measurement uses the principle of reflection, and it projects laser, ultra-sound wave, X-ray, etc., onto target surface to collect feedback information.

K. Chang (✉) · P. Ding · S. Luan · K. Han · J. Shi
Shanghai Aerospace Control Technology Institute, Shanghai 201109, China
e-mail: ccckkkwww1@163.com

Z. Yu et al. (eds.), *Advancements in Mechatronics and Intelligent Robotics*,
Advances in Intelligent Systems and Computing 1220,
https://doi.org/10.1007/978-981-16-1843-7_56

This method will not damage the surface of target and also has a high resolution and can measure a bigger object. It easily shows its superiority in the measurement of irregular curved surfaces.

The scanner uses a input recorder to obtain the raw three-dimensional information and uses a reverse 3D software to reconstruct the surface. This is commonly referred as reverse engineering technology. The 3D scanning technology provides a new and efficient route for the manufacturing industry.

Non-contact 3D scanning device currently on the market can be roughly divided into three categories: one is a radial 3D scanning system based on the principle of pulse ranging; the other is a phase interference scanning system based on the principle of optical interference; the third is based on a 3D projection triangulation scanning system constructed by a camera and a structured light source.

2 Principle of Design

This scanner is used for establishing 3D digital models of small or medium-sized objects, such as mechanical parts, electronic components, and structural parts. To be portable, the scanner must be miniaturized.

Our device focuses on small- or medium-sized targets. Nowadays, most existing products on the market focus on large targets (such as human bodies and large structural parts) and use multicamera system combined with depth information to generate 3D models. The total cost is relatively high. They also require high brightness of light, because blue light scanning system needs all reflection information on target surface during the whole scanning process, and environmental factors such as temperature, humidity, and even light brightness will affect it. Generally speaking, the temperature should usually be controlled at −10 to 50 °C, and the relative humidity should be less than 65%. When scanning, this method usually scans under fluorescent lamps or natural light sources and try to avoid scanning under strong sunlight. The shape, material, color, and roughness of the product itself results in quite different reflection and absorption behavior of light, especially the surface roughness and refractive index, for example, scanning a very reflective surface without spraying before will often cause errors.

Moreover, the resolution of millimeter-level target details is not high, thus most existing products function poorly on very small surface. Our design uses a different method from them, by utilizing the good focus character of laser, our scanner uses focused laser beam scanning the target surface, thus can get better target details.

Our design is very simple, eliminating all unnecessary parts, so it can reduce production costs and product size (improve portability). On the other hand, most of the products on the market are in the conceptual stage and have complex structures, which are expensive.

Our main application areas are 3D printing, machining, electronics manufacturing, medical equipment, etc.

Our product category: A 3D digital modeling device, low-cost, and portable. Our product will become a powerful tool in the hands of front-line industry engineers.

3 Scan Coverage Rate Calculation

The equipment consists of a turntable, a laser rangefinder, a controller, and some other accessories. Among them, the turntable is responsible for rotating the target, the laser rangefinder projects leaser beam onto the target surface, and the feedback data is converted to the distance between the measured surface and the center of the turntable. In this way, the overall outer surface information of the target can be obtained by rotation. After the controller processes the information, we can use software to synthesize a target 3D model.

Calculation of the distance between the outer surface of the target and the center of the turntable ($X2$) relays on the value ($X1$) measured by the laser rangefinder ($X2 = \mathrm{H} - X1$). One rotation of the turntable can get a "slice" of the target at this current height, which can be integrated on different heights, thus obtaining complete target surface details.

According to different scanning paths, two scanning modes are designed:

In the fast scan mode, the turntable rises and rotates at the same time, and the trajectory of the laser spot is a spiral, and the pitch of rising table is d;

In the slow scanning mode, after the turntable has finished rotating 360°, it then raises a height of d, and the scan trajectory is a series of concentric circles with a distance of d.

Theoretical accuracy calculation process: Assuming that the target is a cylinder, the radius of the bottom surface is R, the height of the cylinder is L, and the radius of the laser beam spot is r, the proportion of the area that the spot sweeps across the outer surface of the target is calculated as the accuracy λ:

In the fast mode, the length of one circle of the spiral is $\mathrm{sqrt}\left((2\pi R)^2 + d^2\right)$, the line width is $2r$, the number of cycles is converted to L/d, and the cylindrical side surface area is $2\pi RL$, the scan coverage rate is $\lambda 1$. The calculation formula is:

$$\lambda 1 = (L/d) * \mathrm{sqrt}\left((2\pi R)^2 + d^2\right) * \frac{2r}{2\pi RL}$$
$$= r * \frac{\mathrm{sqrt}\left((2\pi R)^2 + d^2\right)}{\pi R d}$$

In slow mode: the length of the concentric circumference is $2\pi R$, the thickness of the round belt is $2r$, and the number of concentric circles is converted to L/d, the calculation formula of the scanning coverage $\lambda 2$ is:

Fig. 1 Laser trajectory of two scan modes

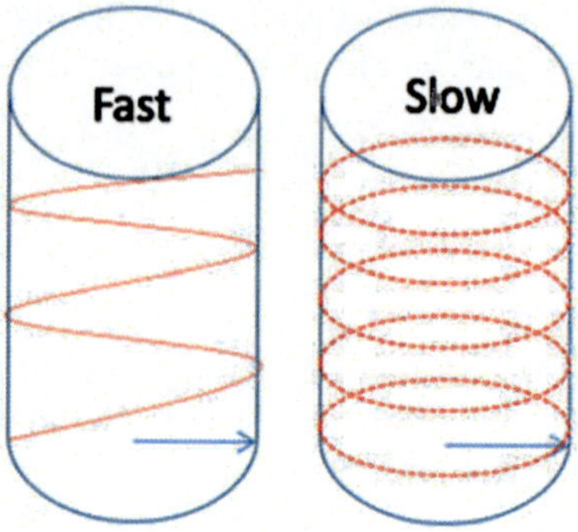

$$\lambda 2 = (\mathrm{L}/\mathrm{d}) * \frac{(2\pi\mathrm{R} \ * 2\mathrm{r})}{2\pi\mathrm{RL}}$$
$$= 2r/d$$

The final result is very simple, $2r/d$. It can be easily known from this result, that by adjusting the ratio of r to d to near 0.5, a very high scanning coverage rate can be achieved, up to 100%. Considering in reality, r is very small, due to the precision limitation of the mechanical structure of the turntable, d is usually greater than $2r$. So generally, the scan coverage is less than but near 100%.

In order to understand the scanning process, Fig. 1 simply demonstrates the trajectory of laser beam. The left figure is the fast mode, and the right figure is the slow mode.

4 3D Digital Model Reconstruction

In the model construction stage, we use several existing 3D design software on the market: SolidWorks, Pro/E, etc. The scanning raw data is used as input, using the form of coordinate point cloud data. By utilizing scripts on PC, we can easily process a large amount of data and generate the raw digital model. The main function of the PC is to filter and reduce the noise of the returned data and reorganization. We can use MATLAB to edit the obtained data, set the zero point to draw a three-dimensional image, and filter the obtained two-dimensional array out as needed. The data is processed to synthesize a complete final three-dimensional digital model and the data is saved in -ply format file.

The data processing unit requires point cloud preprocessed such as down-sampling and smoothing, while simplifying massive data, making the point cloud data coverage high, the data is the sparsest, and no details are lost. By using edge positioning and contour extraction technology, the unit can process three-dimensional information on the target area, data acquisition, analysis, comparison, and evaluation.

Closed loop and feedback are the result to the host computer.

5 Additional Modification

In order to further improve, the performance and ease of use of the equipment, double probes, fixed splints, etc., can be added.

5.1 *Hall Effect Sensor and Magnetic Levitation*

Hall effect sensor can be used to check the object rotation status, and we can add a magnetic levitation system to the turntable instead of traditional servo motor mechanism.

Magnetic levitation technology is becoming very mature and widely used. According to the control method, the levitation devices consist of static magnetic levitation and dynamic magnetic levitation. Static magnetic levitation only relies on the force between the magnets to achieve levitation, but gravitational and static electric and static magnetic fields will make the levitation unstable. Active dynamic magnetic levitation system detects the position of target object and changes the electromagnetic field in real time to achieve system stable. Our design uses dynamic magnetic levitation.

When the target rotates on the magnetic levitation turntable, the change of the magnetic field is determined by the STM32 embedded system. After reading the ADC, the angle information of the object can be obtained.

The controller program utilizes PID control to make the suspension more stable and output PWM pulse. The duty cycle changes with the output value of the linear Hall effect sensor and then controls the coil in which magnetic force keeps the object in suspension and stable rotation. Using magnetic force levitating objects in mid-air can reduce the interference of other objects around, which is convenient. After the scanning starts, the PWM duty cycle can be put into the STM32 embedded system to control the rotation speed.

5.2 *Double Probe*

It is possible to add a laser distance measuring head at the symmetrical position of the original probe, so that it only needs to rotate 180° to complete a scan, which can shorten the scan time by half, and the data of two probes can be obtained from the same feature point. Taking the average of the probe data can further improve the scanning accuracy (turn 360°);

Fixed splint: the turntable can be equipped with splints, the target is clamped between the turntable and the splint, and the screw is fixed to prevent the target from slipping due to the high speed of the turntable, causing measurement failure.

6 3D Scanning System Schematic Diagram

The following shows schematic diagram of an example of "slicing" the target at a certain height (Figs. 2, 3 and 4).

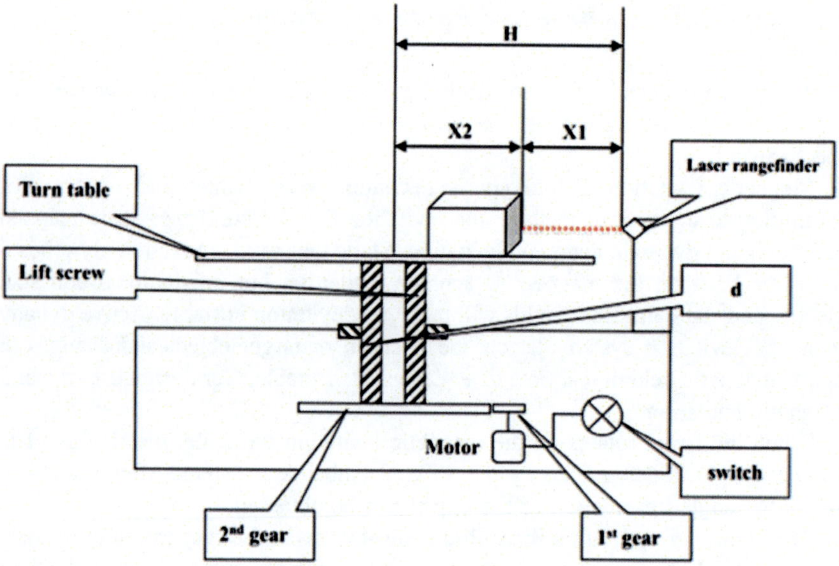

Fig. 2 Schematic diagram of our device

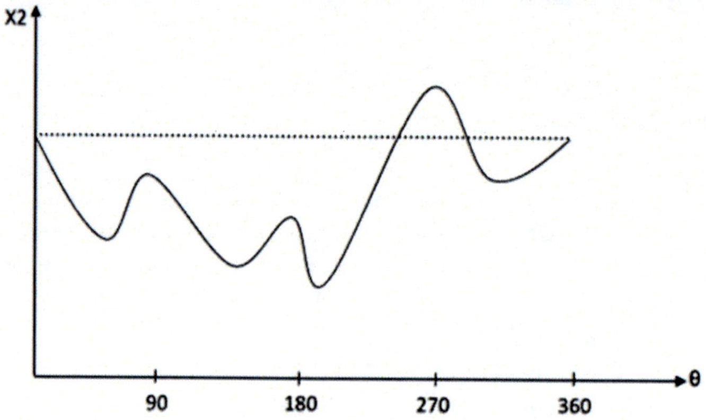

Fig. 3 Diagram of *X2/θ*

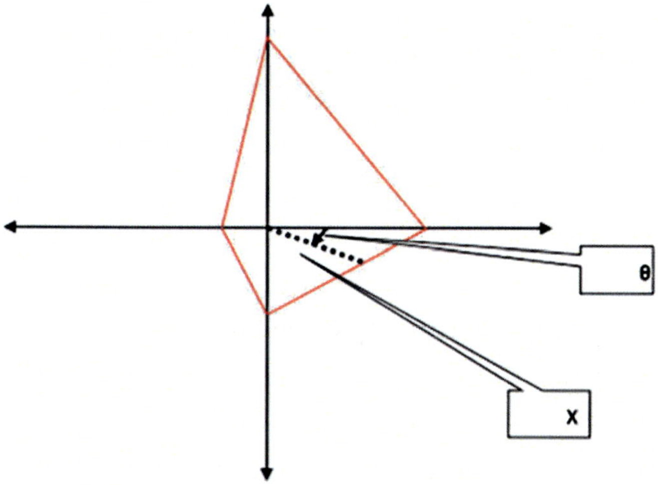

Fig. 4 Corresponding outline of target at certain height

From the functional relationship between $X2$ and θ, the contour slice of the target at certain height is obtained, and then the 3D model can be established by stacking layer by layer.

Bibliography

1. Xiujun D (2007) Research on 3D laser scanning technology and its engineering application. Chengdu University of technology, Chengdu
2. Yunlan G (2008) Research on some problems in 3D laser scanning data processing, D. Tongji University, Shanghai
3. Qun L (2020) Application of 3D laser scanning technology in engineering survey. J Build Mater Decoration 01:230–231
4. Siyuan J, Rong T, Huajun P (2019) 3D laser scanning in space surveying and mapping. J Urban Surv 06:64–67
5. Zunyan Q, Zhonglei S (2019) 3D laser scanning technology in space modeling. J Beijing Surv Mapp 33(11):1340–1343
6. Jiang W (2019) Application of 3D laser scanning technology in measurement. J Beijing Surv Mapp 33(11):1344–1347
7. Ming F, Minglong Y, Yonghua X et al (2020) 3D laser scanning and tilt photogrammetry. J Surv Mapp Sci 45(01):99–107
8. Jian H (2019) Application and analysis of 3D laser scanning measurement system. J New Technol China, New Prod (21):15–16

Application of Electronic Wiring Board in Astronautic Field

Wang Shao, Kaikai Han, Li Ma, and Fan Guo

Abstract Wiring is an important link in the production process of cable harness. The quality of wiring is related to the final compliance and external structure of cable harness. Cables with fewer branches and large dimensional tolerance are generally directly wired manually, while cables with complex structure and high-dimensional accuracy requirements are generally wired by wiring boards. Wiring boards can be divided into traditional templates [Shen and Zhang in Technical requirements for wire binding manufacturing process, No. 200 Factory. Ninth Research Institute of China Aerospace Science and Technology Corporation, pp 4–7, 2014, 1] and modern electronic wiring boards. The traditional pattern wiring method is complicated and time-consuming from the printing of wiring drawings, blueprint, drawing distribution, drawing fixing, and tooling positioning to the start of wiring. The pattern wiring can no longer meet the requirements of mass production and is basically no longer used. Compared with the traditional model wiring process, the electronic wiring board has simpler process and faster wire changing speed, and the electronic wiring board meets the core requirements of Industry 4.0, which realizes synchronization of design and manufacturing, and intelligent and digital production. The electronic wiring board has been widely used in the field of automobiles. At present, it has also been applied to some extent in the astronautic field.

Keywords Electronic wiring board · Wire changing speed · Industry 4.0 · Intelligence · Digitalization

1 Cable Production Process

For different cable network products in the astronautic field, the general flow of production process [2] is shown (See Fig. 1).

W. Shao (✉) · K. Han · L. Ma · F. Guo
Shanghai Aerospace Control Technology Institute, Shanghai 201109, China

Z. Yu et al. (eds.), *Advancements in Mechatronics and Intelligent Robotics*,
Advances in Intelligent Systems and Computing 1220,
https://doi.org/10.1007/978-981-16-1843-7_57

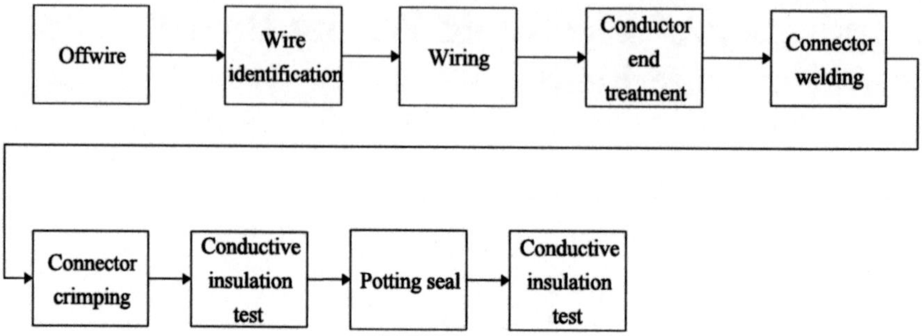

Fig. 1 General process of production of cable products

As the "nerve" of many large aircraft equipment (aircraft, missile, spaceship, etc.), cables are responsible for transmitting sensing and control signals. Conductors of different specifications and different working purposes are wrapped into bundles, which are called wire bundles. Different cable bundles are assembled together to form a cable net. Wiring occupies an important link in the cable manufacturing process and is a necessary process. When the cable laying astronautic is open, the cable structure is usually relatively simple and the size and tolerance are large. The wiring process of such cables is placed at the front end of the cable production process. Due to the large tolerance, the size is easy to control, and the efficiency of wiring before installing connectors is high. In the case of narrow cable laying astronautic, cable structures are often complex and cable sizes and tolerances are small, which is easy to interfere with other structural parts. In local positions prone to interference, the requirements for dimensional accuracy and cable modeling are relatively high. The wiring process of such cables is placed in the middle of the cable production process, i.e., the branches with high-dimensional accuracy requirements are first manufactured, then necessary forming is carried out, and the remaining wires are continuously inserted and bound into bundles to assemble the remaining connectors.

In the process of wiring, workers need to arrange hundreds of wiring harnesses and nuclear contact information under the guidance of process documents. Workers are easily fatigued and suffer from cloth misalignment and cloth leakage. After the arrangement is completed, manual verification is still required and a lot of time is required.

2 Electronic Wiring Board Function

2.1 Electronic Wiring Board Composition

Electronic wiring board is a set of equipment with digital wiring function, which consists of LCD screen, industrial personal computer, wiring software developed on the basis of drawing software and wiring fixture. Wiring software is the most important part of electronic wiring board. Users only need to import the design drawings into the system and put them on the screen; place the wiring fixture at each branch point on the drawing, then scan the code with a code scanning gun according to the QR code label on the wire and wiring according to the path displayed by the system on the LCD screen (usually displayed in highlighted or prominent colors). During the wiring process, suggestive information such as text, pictures, videos, and the like imported into the system in advance can be retrieved. Through human–computer interaction, not only can the specific information, connector information, tooling information, and the like of each wire be displayed digitally quickly, but also certain errors can be automatically identified, such as wire duplication, missing and errors. Compared with the traditional template wiring method, the efficiency is higher and the process is easier to control. The general flow of electronic wiring is shown (See Fig. 2).

2.2 User Management

Relevant personnel using the application program are preregistered and managed and stored in the application server, and the administrator sets the user name, user rights, and password of the relevant personnel. It is mainly used for sorting out project data, processing time, and MES data. Users must enter the assigned user name and password before opening the application, otherwise they will be alerted. The login time, exit time, operation items, wiring time, and other data of all users can be viewed by the administrator. Only administrators can use the "Manage User" module. This module can add and delete user-related information.

2.3 Data Import and Analysis

Managers who normally log into the application program can open the design drawings and analyze the graphic objects in the original file through the functions of the application to form branches, connectors, etc. By importing the wiring table corresponding to the file, analyzing the information of wires, connectors and the like in the table, automatically matching the graphic object data with the relevant data in the Excel file through the automatic matching function of the application program, and feeding back the matching results.

Most of the drawing software used in the astronautic field is AutoCAD. CAD files generally include frame information and wiring harness information. Some drawing habits of designers will lead to non-standard drawings, such as not marking according to actual dimensions when marking dimensions, no continuous break-points between wire segments, and the format of wiring tables is not Excel. Non-standard drawings directly affect the use of the screen. According to

Fig. 2 General flow of electronic wiring

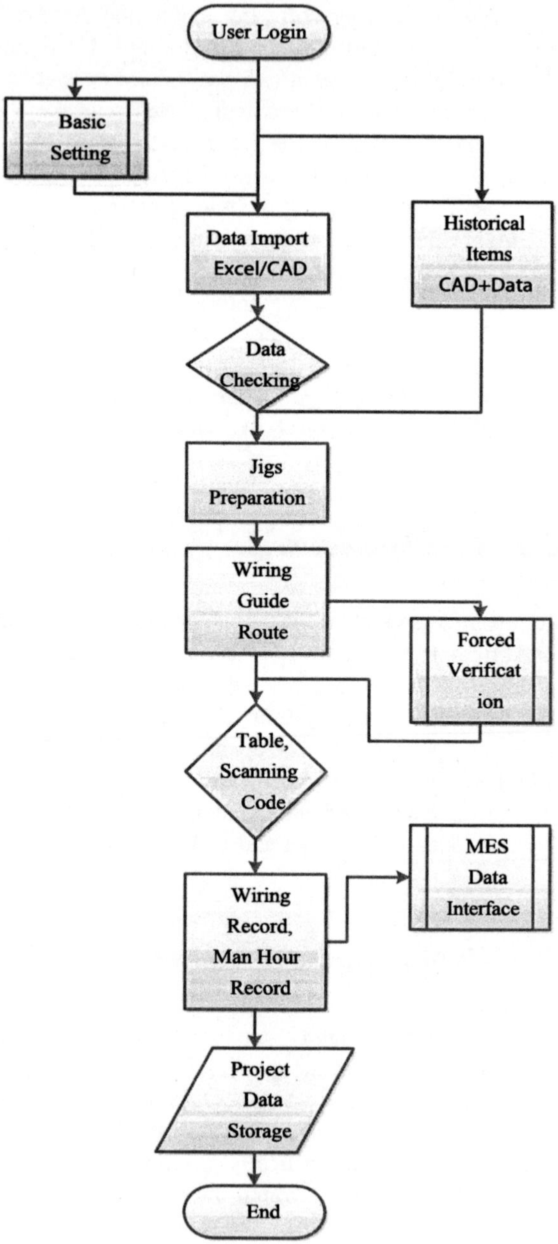

predefined data recognition rules, the technician can edit the non-standard drawing twice and convert all kinds of information in CAD files, such as layers, text blocks, blocks and wires (traces, branches and paths). The format can be analyzed by the application program to form various objects corresponding to the wire harness cable and stored in an independent directory for management. All data are bundled and stored together according to the picture number information of the picture frame and the user name to ensure the validity of the data when opened in the application. The tooling defined in the drawing is also managed as a separate object, which only needs to be marked in the CAD file, and some structural management is carried out on the tooling in the process of data analysis, and the tooling drawing number is automatically highlighted on the wiring board through scanning QR code.

2.4 *Wiring Module*

This module is mainly used by operators and inspectors. Electronic wiring in the astronautic field generally adopts the "code scanning wiring" method. The wire identification is a QR code, which is completed by the upstream wire label automatic pasting equipment. The QR code contains information such as product code and node definition of the wire. This information determines the uniqueness of the wire, and the material information of each wire can be tracked in MES system. Before formal wiring, select a wire and scan the QR code on the wire through the code scanning gun. The system will automatically identify the information in the QR code and highlight the wiring path of the wire. After the wiring according to the path is completed, the operator can start to arrange the next wire after the confirmation is completed in the confirmation interface. After all wires are arranged, the operator confirms the completion and the inspector reconfirms all wires.

3 Application of Electronic Wiring Board

3.1 *User Login and Authority Management*

Users can be added or deleted in the user management interface after logging in as an administrator. By this function, the configuration management of personnel in the production process can be realized, and the login time, logout time, operation items, wiring time, and other data of all users can be viewed.

3.2 *Data Import and Screen Projection*

Taking a cable as an object, the drawing and wiring table are imported into the system. The drawing is in DWG format and the wiring table is in Excel format. After importing, the system associates the drawing with the wiring table with the connector number as the identification target and automatically completes the matching between the drawing and the wiring table. After the matching is successful, the drawing can be normally displayed on the screen. The wiring fixture is placed at each cable branch as the branch point of the cable, and different branch situations are met through the special design of the fixture, thus achieving accurate control of the cable size.

The QR code on the conductor identification is scanned with the code scanning gun configured by the wiring system. The QR code information on each conductor is unique. When the selected conductor does not belong to the current cable product, the code scanning will pop up a prompt box to report an error. When the selected wire belongs to the current cable product, the path will be highlighted on the LCD screen and wiring will be carried out according to the prompt path (See Fig. 3).

3.3 *Wiring*

For cables with simple structure and no need for 3D forming, that is, the finished cable is two-dimensional and can be directly wired. As a result of that diversity and complexity of product in the astronautic field, tactical cabin cables are located in the

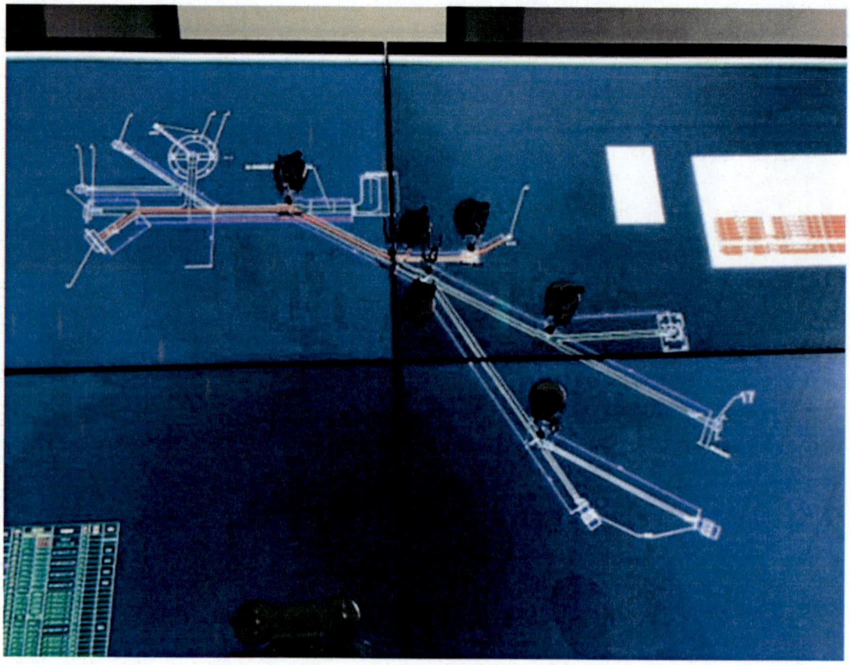

Fig. 3 Drawing screen projection status

assembly cabin with limited astronautic and complex internal structure, which is easy to interfere with other structural parts in astronautic. Therefore, strict requirements are imposed on cable size and even the appearance of cable finished products, cables need to be preformed before assembly, and are generally formed by using manufacturing tooling during production. In response to this situation, when using electronic wiring board wiring, 3D wiring is realized through a special design that combines 3D forming tooling with wiring fixture. That is, the branches with the most difficult size and shape to control are manufactured first, and then one end is fixed on the wiring board and formed. After one end is arranged, the operator can wire the other end or the remaining wires, thus realizing the simultaneous wiring and forming. The operator can only lay one wire after each wire is laid for confirmation. The current process can be saved during the wiring. After all wires are laid, click Finish. The inspector scans all wires to confirm whether there are missing wires or wrong wires. The final effect of 3D wiring is shown (See Fig. 4). Different operators have different dimensional tolerances when manually wiring. Although the dimensions of the final finished cables are all within the tolerance range, they are difficult to assemble. The adoption of electronic wiring improves the accuracy of cable size, avoids the size difference caused by tolerance superposition under the condition of more branches, and improves the consistency of cable size. The human–computer interaction system makes the operator break away from the paper process when wiring, and does not need to identify hundreds of wire marks with naked eyes, so that wiring is easier and cloth misalignment and cloth leakage caused by long-term fatigue work are avoided. Manual wiring needs to classify wires during the preparation work, which is time-consuming and laborious, while electronic wiring can randomly take one wire to scan the code, thus reducing the preparation time and improving the wiring efficiency.

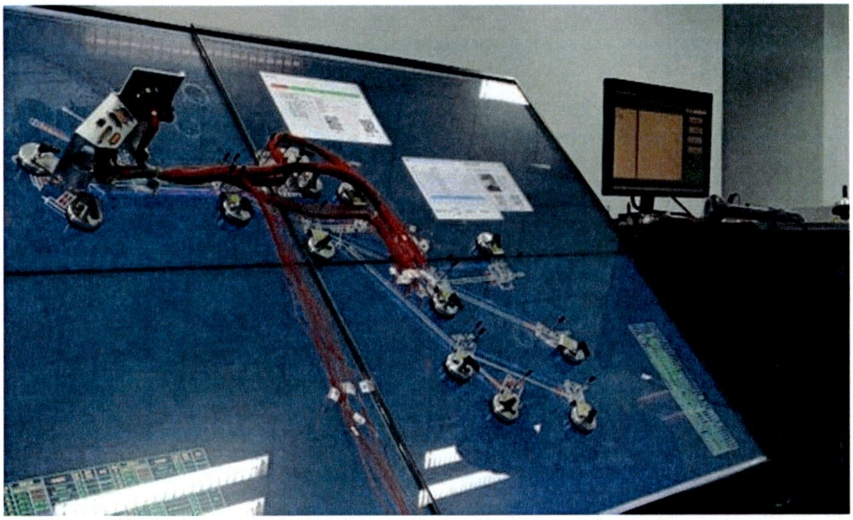

Fig. 4 3D wiring

In addition, you can also view suggestive information in JPG, word, PDF, video, and other formats through the auxiliary functions of the system, which can be the model and specification of the connector, the lateral view, and the actual installation of the cable. The system can reserve a data interface for data interaction with MES system.

Through human–computer interaction, the system improves the digital level of cable wiring, frees laying personnel from consulting a large number of relevant technical data, provides technical basis for paperless and automation of the system, improves manufacturing efficiency, reduces wiring time and reduces labor intensity of workers.

4 Conclusion

Electronic wiring board has been applied in the astronautic field to some extent. Wiring software associates drawings with wiring tables and analyzes data to realize screen projection. Through the design of wiring fixture, the size of cables of different branch types is accurately controlled, and the cable forming tooling and wiring fixture are combined to realize 3D wiring, and suggestive information can be viewed through auxiliary functions, so that the operator can understand the product more intuitively. The application of electronic wiring board not only greatly improves the wiring efficiency, but also controls the cable size accurately. All data generated in the wiring process are saved as XML format files for data interaction with MES system. Intelligent and digital production is easier to control.

References

1. Shen P, Zhang S (2014) Technical requirements for wire binding manufacturing process, No. 200. Factory, Ninth Research Institute of China Aerospace Science and Technology Corporation, pp 4–7
2. Ma L (2016) Wire binding and cable manufacturing process specifications, No. 803. Research Institute of Eighth Research Institute of China Aerospace Science and Technology Corporation, pp 2–3

Research on Capacity of Mixed Vessels Traffic Flow Based on Vessel-Following Theory

Yan Huaran, Zhou Guoxiang, Liu Tao, and Zhao Chunbo

Abstract In order to study the characteristics of mixed vessel traffic flow, based on classical head distance model and probability analysis, by studying the combination time head way of different vessel-following sequences, the capacity model of mixed vessels traffic flow was established. Through analyzing two representative types of vessels, research results indicate that the capacity of mixed traffic increases with the traffic flow speed in a certain speed range, but the increasing trend slows down. The closer length and inertial stopping distance of different kind vessels are the more capacity of mixed traffic increase. And the influence of reaction time on the capacity is related to proportion of different kinds of vessels.

Keywords Mixed vessels traffic flow · Vessels traffic capacity · Vessel-following theory · Time headway model

1 Introduction

In the maritime traffic engineering, traffic capacity refers to the capacity of a channel to manage vessel, which is measured by the maximum number of vessels passing through in a certain time [1]. At present, the formulas for calculating the passage capacity mainly include the West German formula, the Polish formula, the Yangtze River formula, and the Changjiang formula. Its common characteristic is that a series of parameters need to be analyzed and determined according to the actual situation and data of the channel. The value of parameters varies from person to person and is highly subjective, which leads to the non-standard calculation of

Y. Huaran · Z. Chunbo
Merchant Marine College, Shanghai Maritime University, Shanghai 201306, China

Z. Guoxiang (✉)
Wusong Maritime Administration, Shanghai 201908, China

L. Tao
Communication and Transportation College, Shanghai Maritime University,
Shanghai 201306, China

Z. Yu et al. (eds.), *Advancements in Mechatronics and Intelligent Robotics*,
Advances in Intelligent Systems and Computing 1220,
https://doi.org/10.1007/978-981-16-1843-7_58

channel passing capacity. Moreover, existing studies on channel passing capacity often take traffic flow velocity and vessel density as fixed values without considering their mutual influence and lack of further study on their internal traffic characteristics and mechanism [2, 3].

The study of traffic flow theory shows that flow rate, velocity, and density are a kind of dynamic equilibrium relation vessel. In recent years, some scholars have been aware of the deficiencies in the previous studies on channel passage ability and have begun to make a preliminary discussion on channel passage ability using the macroscopic traffic flow theory. Shao Changfeng made a preliminary dynamic exploration and analysis of vessel traffic flow by applying fluid model [4]. Using the research method of highway traffic for reference, He Liangde and Zhu Jun established the direct functional relation vessel between vessel density and vessel speed by using the following theory and strengthened the analysis of vessel traffic mechanism [5, 6]. However, the above researches are limited to the analysis of a single vessel type in the waterway. In practice, due to the different sizes and types of vessels in the channel, the sequence composition of vessel following in the mixed vessel flow is also random, resulting in different vessel spacing, which has a great impact on the channel passage capacity. Therefore, it is significant to study the passage capacity in the case of mixed traffic flow.

2 Theoretical Model of Vessel Following

The car-following theory is a dynamic study model of the following vehicle's corresponding behavior caused by the change of the leading vehicle's motion. The characteristics of single-lane traffic flow are understood by analyzing each vehicle following each other so as to connect the microscopic behavior of vehicles with the macroscopic phenomenon of traffic flow. This model has been widely applied in microtraffic simulation, self-cruise control, capability analysis, traffic safety evaluation, and other fields [7].

2.1 Fundamental Assumption

The vessel-following model in the channel is based on the following hypothesis: a group of vessels navigate one by one, the officer only responds to the action of the front vessel. The movement of the response is accelerating or decelerating, and there is no overtaking. Channel width and lateral interference are ignored.

In addition, from the perspective of safety, the following vessel should meet two requirements. Firstly, the speed of the behind vessel should fluctuate around the speed of the front vessel, and not greater than that of the front vessel for a long time. Secondly, a safe distance must be kept between the front and behind vessel to make sure that there is enough time for the behind vessel officer take action to responding [8].

There are two means for vessel to brake, stop engine and reverse engine. Generally speaking, the reverse engine will cause the vessel's bow to turn uncontrollably under the influence of the deep transverse force, the discharge transverse force, the wind and current force, the shallow water effect or the bank effect when the vessel navigating on the channel. In addition, emergency reverse engine will cause the main engine rotating parts stress too much. Therefore, vessels usually braked by stop engine.

2.2 Prow Time–Distance Model Based on STOP Engine

The prow time–distance model that applied to analyze the following up of vessels as shown in Figs. [9, 10]. n is the front vessel, $n + 1$ is the behind vessel. $x_{n(t)}$ and $x_{n+1}(t)$ is the position of two vessels. t_m is the responding time, which include the officer responding and action time and the engine receiving command time. d_1 is the distance that vessel navigating with the original speed in a certain time tm. d_2 is stop stroke of the behind vessel, d_3 is stop stroke of the front vessel, ε is the lee-way factor, m_0 is the safety margin after the two vessels stop [11].

From Fig. 1, to ensure the safety of the two vessels after stop engine, it should meet the requirement:

$$d_1 + d_2 - d_3 \geq \varepsilon m_0 \tag{1}$$

In critical condition:

$$x_n(t) - x_{n+1}(t) = d_1 + d_2 + \varepsilon m_0 - d_3 \tag{2}$$

$$d_1 = t_0 v \text{ (} v \text{ is the speed before decelerate)} \tag{3}$$

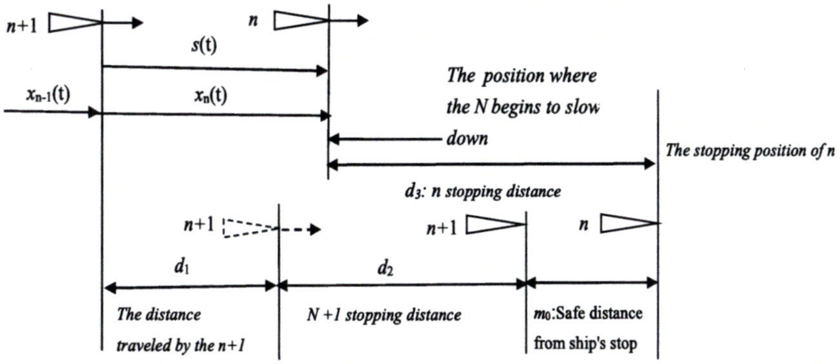

Fig. 1 Follow theoretical analysis diagram

So:

$$x_n(t) - x_{n+1}(t) = t_m v + \varepsilon m_0 + d_2 - d_3 \tag{4}$$

Corresponding distance of the bow is:

$$h_t = \frac{[x_n(t) - x_{n+1}(t)]}{v} \tag{5}$$

3 Mixed Traffic Flow Capacity Model

3.1 Mixed Traffic Flow Capacity Calculation Model

Let $h_{i,j}$ is the bow distance of the combined two vessels. i is the front vessel, j is the behind vessel, $i, j = 1, 2, \ldots, r$. Because the vessel type of two adjacent vessels in the traffic flow is random, the probability of the front vessel i and the behind vessel j is $p_i p_j$, and:

$$\sum_{i=1}^{r} \sum_{j}^{r} p_i p_j = (p_1 + p_2 + \cdots + p_r)^2 = 1 \tag{6}$$

Take the combination bow time interval h_t to represent the average minimum bow time interval under different tracking sequences of the mixed traffic flow consisting of r vessel types, according to the theory of probability:

$$h_t = \sum_{i=1}^{r} \sum_{j=1}^{r} p_i p_j h_{i,j} \tag{7}$$

So, it can be seen that in the calculation of mixed traffic passing capacity, it is very important to solve the interval between different vessel types.

3.2 Bow Time Interval Analysis of Mixed Traffic Following

Assume that the vessel traffic flow is composed of a mixture of large and small vessel types, the length of little vessel is l, the proportion is p, and braking reaction time is t. The length of large vessel is L, the proportion is $1 - p$, and braking reaction time is T. Same speed before emergency braking. The braking distance of little vessel is $d(v)$, and that of large vessel is $D(v)$, the difference of them is $S(v)$. If there is a large vessel of the combination, the safety margin is M, otherwise, the safety margin is m. The safe distance between two vessels at rest is often expressed as a multiple of the length. Referring to the anchor distance, $M = 3L$, $m = 3l$ [12]. The corresponding situation as follows:

Both of them are large vessel.

$$x_n(t) - x_{n+1}(t) = Tv + \varepsilon M + [D(v) - D(v)] \tag{8}$$

$$h_{2,2} = \frac{(x_n(t) - x_{n+1}(t))}{v} = T + \frac{\varepsilon M}{v} \tag{9}$$

Both of them are little vessel.

$$x_n(t) - x_{n+1}(t) = tv + \varepsilon m + [d(v) - d(v)] \tag{10}$$

$$h_{1,1} = \frac{(x_n(t) - x_{n+1}(t))}{v} = T + \frac{\varepsilon m}{v} \tag{11}$$

A small vessel in front and a large vessel behind.

$$x_n(t) - x_{n+1}(t) = Tv + \varepsilon M + [D(v) - d(v)] \tag{12}$$

$$h_{1,2} = \frac{(x_n(t) - x_{n+1}(t))}{v} = T + \frac{\varepsilon M}{v} + \frac{s(v)}{v} \tag{13}$$

(4) A small vessel in front and a little vessel behind. Since the large vessel has a larger inertia, it is assumed that when the large vessel is braking, the little vessel will be reducing with a same accelerated speed that belong to the large vessel.

$$x_n(t) - x_{n+1}(t) = tv + \varepsilon M + [D(v) - D(v)] \tag{14}$$

$$h_{2,1} = (x_n(t) - x_{n+1}(t)) = t + \varepsilon M/v \tag{15}$$

The average bow time interval of the traffic flow is:

$$h = p^2 \times h_{1,1} + (1-p)^2 \times h_{2,2} + p(1-p) \times h_{1,2} + (1-p)p \times h_{2,1} \tag{16}$$

$$C = \frac{24 \times 3600v}{[T(1-p) + tp]v} + 3(1-p^2)L + p^2 l + p(1-p)s(v) \tag{17}$$

4 Model Parameter Analysis

The parameters in the above equation can be divided into two categories according to their respective ownership relationship. The one is macro traffic flow characteristics, v and p. The other one is vessel characteristics, t_m (t, T), $l_m(l, L)$, $S(v)$. From the point of view of water traffic control management, the first type of parameters is more important, which will be discussed in detail below.

4.1 *The Relationship of* C-P

To confirm the influence of the first type parameters, we should first confirm the value of the second type parameters. According to China's "Design Code of General Layout for Sea Ports" and the actual situation of small and medium-sized coastal ports in China, GT3000 with a total length of 96 m and GT10,000 with a total length of 135 m are taken as the representative ship type of bulk carrier [13]. The response time of little vessel is 10 s, and that of large vessel is 20 s [14]. $\varepsilon = 1$. For ships, the speed of 5 kn is generally the minimum speed to maintain rudder effect, and the ship braking distance is mainly determined based on this. According to the latest research results of the China transportation planning and design institute on braking distance (see Table 1), when the speed before parking is 4, 6, 8, 10 kn, the ship's stopping braking distance is 2 lm (length), 5, 9, 16 lm [15].

When the value of p is 0.3–0.9, we can obtain the relationship of C-p-v. From Fig. 2, we can see that as p goes up, C goes up faster and faster. That is to say, under the same vessel traffic flow speed, the larger the proportion of small vessels, the greater the channel traffic capacity.

4.2 *The Relationship of* C-V

From Fig. 2, we can see that when p is constant, if v is smaller, C is also smaller. With the increase of v, C first grows rapidly and then slows down, and the smaller the probability is, the larger the growth rate decreases.

If the ship's stop engine brake is regarded as uniform deceleration motion, acc according to,

$$s(v) = D(v) - d(v) = \frac{v^2}{2A} - \frac{v^2}{2a} = \frac{v^2(a-A)}{2aA} \tag{18}$$

Table 1 Stop engine stroke

Speed before stop engine	4–6 kn	6–8 kn	8–10 kn
Straight-line distance (majority concentration)	2–5 lm	4–7 lm	5–15 lm
Mean straight-line distance	5.3 lm	9 lm	14.1 lm
Actual distance (majority concentration)	2–6 lm	4–10 lm	7–17 lm
Mean straight-line distance	5.9 lm	10 lm	15.8 lm

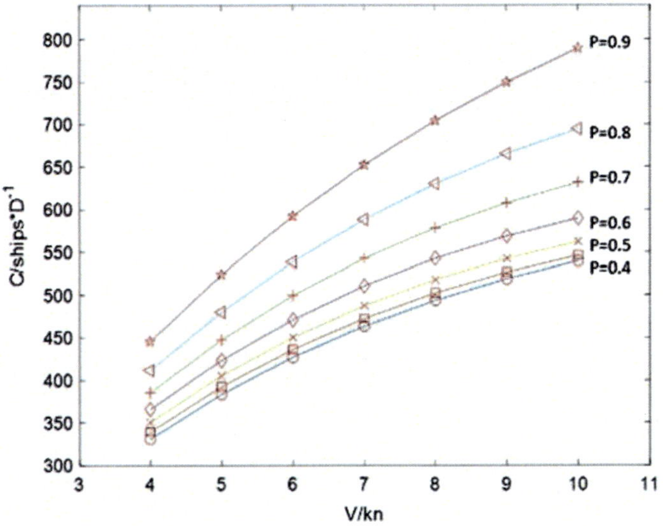

Fig. 2 Relationship of *C-p-v*

a is the acceleration of smaller vessel, A is the acceleration of larger vessel. Deriving from A, when $\mathrm{d}c/\mathrm{d}v = 0$,

$$v_{\max}^2 = \frac{2aA[(1-p^2)\varepsilon M + p^2 \varepsilon m]}{p(1-p)(a-A)} \tag{19}$$

From this, we can see that C is not increased by the increase of v, when, $v = v_{\max}$, C is maximum. And from (19), $v_{\max}$ has nothing to do with the two vessels' reaction time T and t.

Under normal circumstances, the average speed of small- and medium-sized vessels in channel is 8 kn, the stop stroke: $m_0(v = 8\ \text{kn}) = 9$ lm.

The average acceleration of ship stop brake, $a = 9.3 \times 10^{-3}$ m/s^2, $A = 6.6 \times 10^{-3}$ m/s^2.

As it is difficult for the traffic flow in the channel to reach this speed, in general, with the increase of the ship traffic flow speed, it gradually increases, but the increase trend gradually slows down.

4.3 *The Relationship of* C-L$_m$, S(V) *and* T$_m$

As for the influence of the second type of parameters, it can be seen from the 2.2 conclusion formula that with the increase of reaction time, ship length, and stopping stroke, the passage capacity of mixed traffic decreases gradually. Then, the

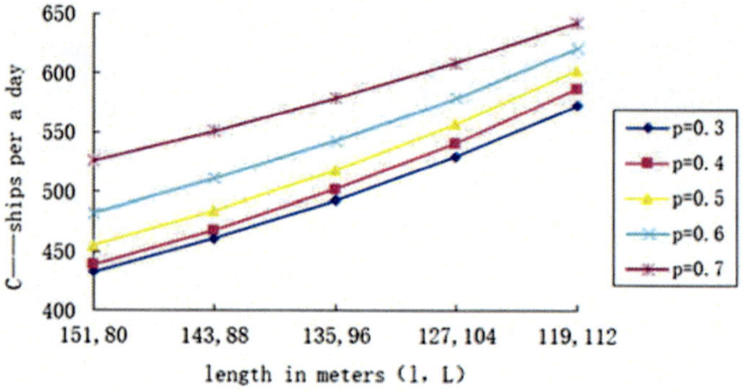

Fig. 3 Relationship of *C*-lm

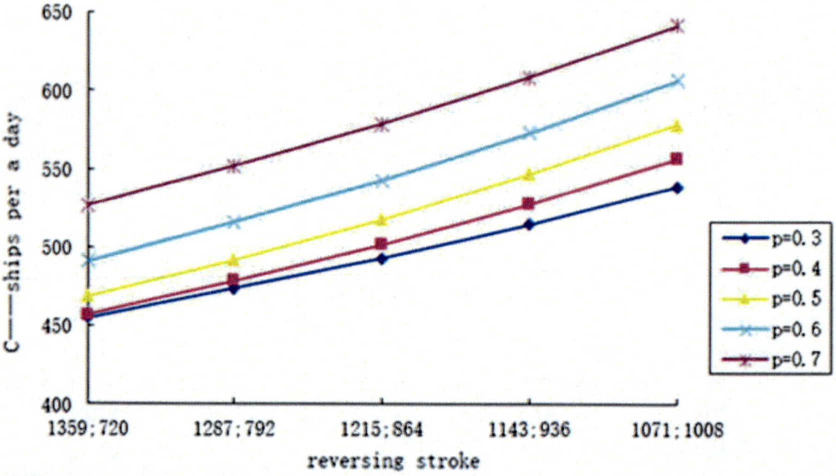

Fig. 4 Relationship of *C*-*S*(*v*)

influence of the similarity degree of different types of ships on the passage capacity is analyzed.

Assuming the vessel traffic flow speed is 8 kn, the original value in 3.1 is taken as the intermediate value to gradually expand and reduce the vessel-type parameter interval. The results are shown as follows (Fig. 3).

It can be seen from Figs. 4 and 5 that, with the narrowing of the value interval of the two ship types' length and stop stroke, the channel traffic passing capacity gradually increased. In other words, the more similar the length and stop stroke of different ship types in the channel, the greater the passing capacity.

It can be seen from Fig. 5 that, the effect of response time of different types of ships on passing capacity is related to probability (p). If $p < 0.5$, with the decrease

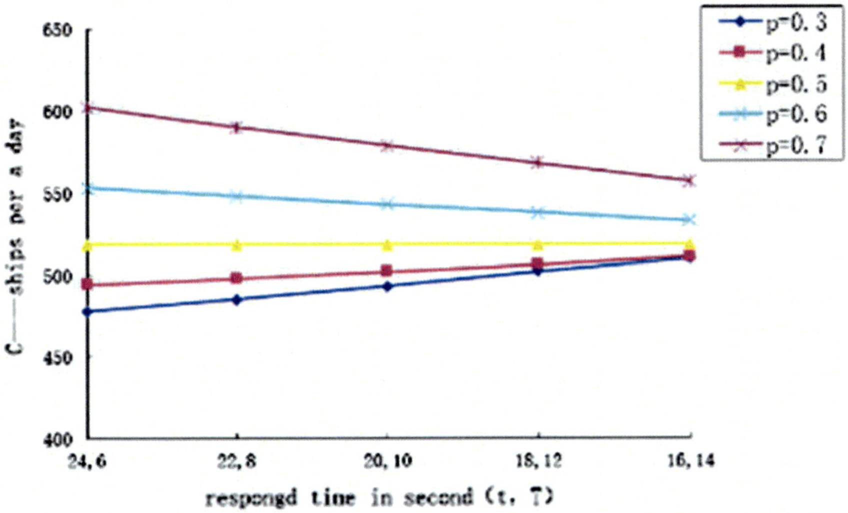

Fig. 5 Relationship of $C\text{-}t_m$

of the reaction time interval between the two types, the passing capacity shows a trend of gradual increase. It is reversed when $p > 0.5$.

5 Conclusion

The above studies show that the mixed traffic flow passing capacity in channel is not only related to the vessel traffic flow speed and vessel-type combination, but also related to the reaction time, vessel length, and vessel stopping performance. Within a certain range, C is increasing with the increase of v, but the trend of increase gradually slows down. The closer the ship length and stopping stroke of different ship types are, the greater the capacity of mixed traffic to passing. At the same time, the influence of response time of different types of ships on the passing capacity (C) is related to probability (p), and the variation trend of the passing capacity (C) of mixed traffic is different with different probability (p).

References

1. Zhaolin W, Jun Z (2004) Marine traffic engineering. Dalian Maritime University Press, Dalian
2. Xiaoyu D, Yinzhen L, Yaling Z (2011) Overview of research on traffic capacity of harbor channel. Port Waterway Eng 3:10–15
3. Yu D, Ye J, Liangde H (2007) Calculation method of inland waterway's throughput capacity. Port Waterway Eng 1:59–65

4. Changfeng S, Xianglin F (2012) Fluid mechanics model for vessel traffic flow. J Dalian Marit Univ 28(1):52–55
5. Liangde H, Ye J, Zhaojin Y (2012) Following distance model of inland ship. J Dalian Marit Univ 12(1):55–62
6. Jun Z, Wei Z (2009) Calculation model of inland waterway transit capacity based on ship-following theory. J Traffic Transp Eng 9(5):83–87
7. Hong H, Yong C (2019) Research on high speed car-following traffic flow considering dynamic effect of preceding vehicle. Comput Eng Appl 55(14):209–214
8. Yi L (2017) The application research of control in car-following model. Ningbo University, Ningbo
9. Helbinu D (2001) Traffic and related self-driven many particle systems. Rev Mod Phys 73(4): 1067–1141
10. Toledo T (2007) Driving behavior: models and challenges. Transport Rev 27(1):65–84
11. Qi C (2011) Influence of safety interval between ships on throughput capacity of costal fairway. Dalian University of Technology, Dalian
12. Fei G (2018) The study of single direction ships formation based on ship-following theory. Dalian Maritime University, Dalian
13. Ministry of Transport of the People's Republic of China (2014) Design code of general layout for sea ports (JST165-2013). China Communications Press, Beijing
14. Zhouhua X, Junmin M, Yongqing J (2004) A study of 3D model of ship domain for inland waterway. J Wuhan Univ Technol (Transp Sci Eng) 28(3):380–383
15. Zhipeng Z, Xin X (2011) On braking distance of large vessels based on vessel observation. Port Waterway Eng 11:6–12

Agent-Based Simulation of Speculation in China's Refined Oil Market

Qing Zhou, Qinlan Yuan, and Yanli Li

Abstract The refined oil market of China is gradually becoming market-oriented, and the oil pricing mechanism has experienced a series of reforms. Using agent-based simulation method, this paper analyzes the speculative phenomenon of hoarding oil. The result shows that the speculative phenomenon of hoarding oil is more common when the price of refined oil is lower. When the price of refined oil is higher, the speculators will reducing the amount of refined oil that has been hoarded, and even there is no oil to be hoarded. Besides, the government price adjustment policy will influence the speculators' strategy of hoarding oil.

Keywords Refined oil pricing mechanism · Agent-based simulation · China's refined oil market

1 Introduction

After the China entered the WTO, the refined oil market of China is gradually becoming market-oriented; China opened its retail market of refined oil in 2004 and then opened its wholesale market of refined oil at the beginning of 2007. The entrance of these market participants breaks up the monopoly of the two big oil giants and forms a pluralistic competition market situation in Chinese refined oil market.

In recent years, scholars have tried to figure out the solutions of some problems in refined oil market by giving detailed studies of the refined oil market. In order to integrate to the international refined oil market as well as to achieve the goal of market-oriented, Chinese refined oil market has experience a series of reforms since 1998. According to the regulation of 2009 [1], when the moving average prices of

Q. Zhou (✉)
Faculty of Petroleum, China University of Petroleum, Beijing 834000, At Karamay, China
e-mail: zhouqing@cup.edu.cn

Q. Yuan · Y. Li
School of Business Administration, China University of Petroleum, Beijing 102249, China

Z. Yu et al. (eds.), *Advancements in Mechatronics and Intelligent Robotics*,
Advances in Intelligent Systems and Computing 1220,
https://doi.org/10.1007/978-981-16-1843-7_59

international crude oil change more than 4% within 22 working days in a row, the domestic gasoline prices and diesel prices can adjust accordingly. In March 26, 2013, a new pricing mechanism was introduced, according to the "Regulations on Administration of oil prices (trial implementation)," the accounting and adjusting cycle of gasoline prices and diesel prices changed to 10 days instead of 22 days and canceled the restriction of the fluctuation range of 4%. [2] And to save social cost, when the prices fluctuation of gasoline and diesel is lower than 50 RMB per ton, the prices of gasoline and diesel do not adjust immediately. Instead, the prices fluctuation is going to be taken into account in the next price adjustment cycle.

Since 2007, the "diesel shortage" phenomenon happened frequently in Chinese refined oil market, and the market players had a strong desire to predict the ups and downs of the oil prices. The speculative activities of "hoarding oil and arbitraging" became a barrier to the standardization and the well-order of the market and blocked the healthy development of the refined oil market. However, the studies about the underlying cause of the phenomenon are limited, and the researches about the factors that can influence the hoarding oil activities are also scarce. Using agent-based simulation method, this paper gives a comparison between the old and new pricing mechanism and also gives an analysis about the speculative activities of the refined oil wholesalers.

In recent years, with the development of simulation method, thanks to its strong reproduction ability of the real world, there are more and more achievements about refined oil market by using the simulation method, and the achievements are mainly in the following aspects:

The studies of the Chinese refined oil market using system dynamics simulation. An and Zhang [3] have used this method to build a SD model about the coordinated development of the oil and gas resources and socio-economic system, and the annual oil production curve and oil resource shortages curve were simulated in this study, and this paper also predicted the change of the domestic oil supply gap under different policies. Yuan [4] gave a study of Chinese diesel demand market using system dynamics simulation, built a system dynamic model of diesel demand market and simulate the domestic diesel demand under two scenarios; the result showed that the reform of the diesel pricing mechanism did not cause dramatic changes in the diesel market.

The simulation researches on the forecasting of the oil prices. Jia [6] forecasted the Chinese oil price using principal component analysis method and the BP neural network method. Zhu and Guo [7] built a model to predict the oil price using the wavelet network method; the method integrates the advantages of the wavelet analysis "multi-scale transform analysis and the neural network method" nonlinear prediction ability, which helped to improve the prediction accuracy and reduce errors.

Shi [9] simulated the influence of refined petroleum's tax and fees on Chinese oil market. This paper analyzed the properties and behavior of the market participants, including oil producers, dealers and petrochemical corporations as the agents, how to affect by adjusting the tax and fees.

2 Refined Oil Market Model

In Chinese oil market, PetroChina, Sinopec and CNOOC have strong monopoly power, and these three companies (especially the PetroChina and Sinopec) have strong internal integration; we divide the whole market into two systems: One is consisted of the three state-owned oil companies, including refineries, oil depots, wholesalers, oil distribution centers and gas stations that owned by them; another is consisted of the private enterprises, joint ventures and foreign-invested enterprises, including their refineries, wholesalers and gas stations. The composition of the agents is shown in Fig. 1:

2.1 The Behavior of the Wholesalers

In order to simulate the speculative behavior of the market participants, the model gives a detailed analysis about the decision-making behavior of different market participants, including the behavior rules of the wholesalers and gas stations and so on. This paper focuses on the behavior rules of the wholesalers, and the oil price forecasting behaviors of market participants and the price regulation behaviors of government will be analyzed in detail in other paper, and other behaviors rules of

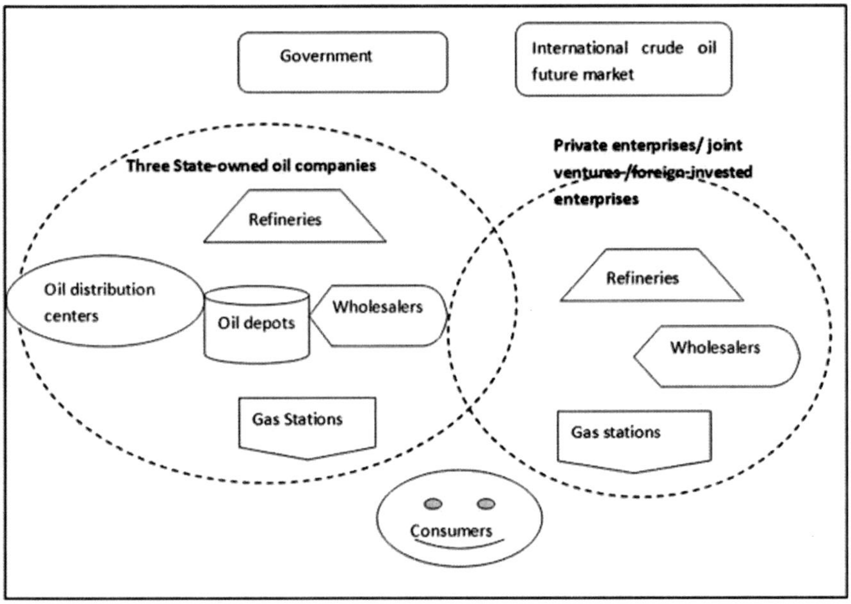

Fig. 1 Structure of petroleum market agent in China

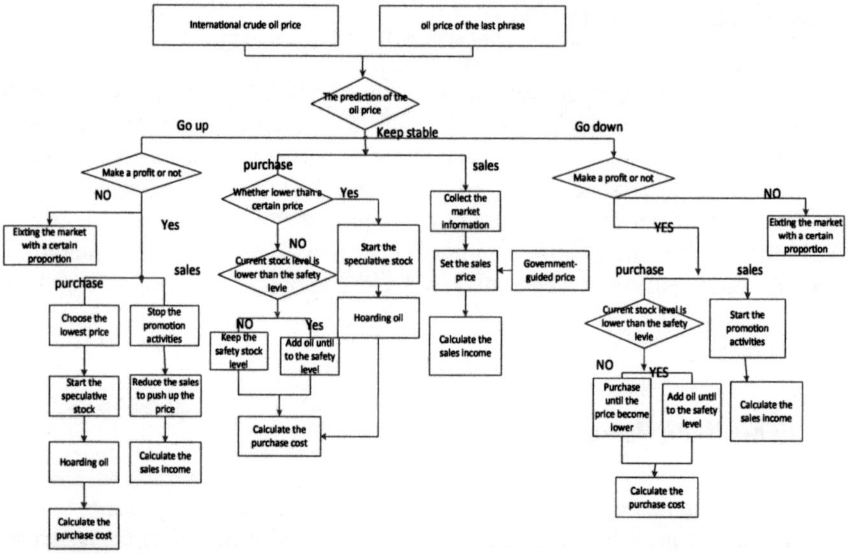

Fig. 2 Executive-lever model of petroleum wholesaler—the first phase

the market participants will be omitted. The behavior rules of the wholesalers will be shown in Fig. 2.

The purchase behavior of wholesalers: The purchase cost is calculated based on the purchase price and the purchase quantity.

The sales behavior of wholesalers: The sales income is calculated based on the sales price and sales volume.

The wholesalers' decision-making behavior in the next phase: The wholesalers calculate the profit of the last phase firstly, if the profit is positive, which means that the decision-making of the last phase is correct, and the next phase will use the same purchase and sales rules as the last phase; if the profit is negative, which means that the decision-making of the last phase is not correct, the wholesalers should set up a punishment mechanism and a learning mechanism for themselves.

2.2 *The Behavior of the Wholesalers*

The variables of wholesalers

$S_t^{w_j}$: The stock level of wholesaler j in phase t, $j = 1, 2, 3, \ldots, n$;

$SL_t^{w_j}$: The upper limit of the stock level of wholesaler j in phase t, which is set to a fixed value of 500, $j = 1, 2, 3, \ldots, n$;

$SS_t^{w_j}$: The speculative stock level of wholesaler j in phase t, with the initial value being set to 200, $j = 1, 2, 3, \ldots, n$;

The parameters of wholesalers

GWR: The market share of the state-owned wholesalers;
MWR: The market share of the private wholesalers;
r^{w_j}: The risk preference of wholesaler j;
δ^{w_j}: The probability of learning for wholesaler j, with the initial value being set to different value for different wholesalers according to different risk preference value.
θ^{w_j}: The speculative stock change rate for wholesaler j, with the initial value being set to different value for different wholesalers according to different risk preference value

The behavior rules of wholesalers

The wholesalers are not only the refined oil demanders but also the refined oil suppliers, and they are active in the oil market. The main behaviors of wholesalers include their purchase behaviors, their sales behaviors and their decision-making for the purchase and sales behaviors in the next phase based on the profit level of the last phase.

The decision-making process of purchase behavior:

The wholesalers will make a decision about whether to purchase based on the current stock level first; the purchase behavior rules are as follow:

If $S_t^i \geq 750$, the wholesaler will not purchase;

If $S_t^i < 750$, the wholesaler will purchase;

When deciding to purchase, the wholesaler will purchase from nearby supplier who has the lowest price, the suppliers include state-owned refineries, private refineries and oil depots, and the purchase price should meet the following function:

$$\mathrm{PP}_t^{w_j} = \min\left(\mathrm{SP}_t^{p_1}, \mathrm{SP}_t^{p_2}, \ldots, \mathrm{SP}_t^{p_i}\right), j = 1, 2, 3, \ldots, n; \tag{1}$$

After finding the best supplier with the lowest price, the wholesaler will predict the future trend of the oil price, and then the wholesaler should decide the purchase volume and make a decision about whether to hoard oil as well as the oil volume that should be hoarded. Finally, the wholesaler should calculate the purchase cost and update the stock level.

The purchase volume should be decided as follows:

$$\mathrm{PA}_t^{w_j} = \left\{ \begin{array}{ll} \mathrm{SL}_t^i + \mathrm{SS}_t^i - S_t^i, & \mathrm{PF}_t^i = \text{rise}; \\ \mathrm{SL}_t^i * 0.65 - S_t^i, & \mathrm{PF}_t^i = \text{down} \ \& \ S_t^i < 550; \\ 0, & \mathrm{PF}_t^i = \text{down} \ \& \ S_t^i \geq 550; \\ \mathrm{SL}_t^i * 0.85 - S_t^i, & \mathrm{PF}_t^i = \text{balance} \ \& \ S_t^i < 550; \\ 0, & \mathrm{PF}_t^i = \text{balance} \ \& \ S_t^i \geq 550. \end{array} \right\} \tag{2}$$

The purchase cost is:

$$\mathrm{PC}_t^{w_j} = \mathrm{PA}_t^{w_j} * \mathrm{PP}_t^{w_j} \tag{3}$$

The decision-making process of sales behavior

Another important behavior of the wholesalers is their sales behaviors, including determining the sales volume and determining the sales price. The sales volume should be:

$$\mathrm{SA}_t^{w_j} = \mathrm{PA}_t^{s_m} \tag{4}$$

The sales price should be set according to the following rules:

$$\mathrm{SP}_t^{w_j} = \left\{ \begin{array}{ll} \mathrm{GMP}_t, & \mathrm{PF}_t^i = \text{rise}; \\ P_t, & \mathrm{PF}_t^i = \text{down} \,\&\, S_t^i < 550; \\ P_t^i * 0.75, & \mathrm{PF}_t^i = \text{down} \,\&\, S_t^i \geq 550; \\ \mathrm{GMP}_t * 0.85, & \mathrm{PF}_t^i = \text{balance}. \end{array} \right\} \tag{5}$$

The sales income should be calculated as follows:

$$\mathrm{SI}_t^{w_j} = \mathrm{SA}_t^{w_j} * \mathrm{SP}_t^{w_j} \tag{6}$$

The decision-making process of the next phase

At the end of each phase, after finishing the purchase behavior and the sales behavior, a wholesaler should calculate its profit based on its stock-holding cost, sales income and purchase cost firstly, and the behavior rule of the next phase should be adjusted accordingly, including the adjustment of 'the speculative stock change rate' and "the probability of learning."

The speculative stock change rate and the probability of learning are set as follows in Table 1.

The wholesaler will behave according to the rules; if the wholesaler wants to learn, it will learn from the competitor who has the maximum profit, and the wholesaler will behave like its learning object in the next phase; if the wholesaler does not want to learn temporarily, he will adjust its speculative stock of the next phase according to the adjustment rules. In a word, every wholesaler tries its best to maximize its profit and make profit as more as possible, and adjusts its behavior parameters to avoid losses.

Table 1 Change rate of speculate stock and learn rate of wholesaler

Risk preference	Risk preference type	The speculative stock change rate	The probability of learning
$r^{w_j} = 2$	Risk lover	$\theta^{w_j} = 0.07$	$\delta^{w_j} = 0.85$
$r^{w_j} = 1$	Risk neutral	$\theta^{w_j} = 0.05$	$\delta^{w_j} = 0.5$
$r^{w_j} = 0$	Risk averter	$\theta^{w_j} = 0.03$	$\delta^{w_j} = 0.15$

3 The Simulation Result Analysis

In order to simulate the phenomenon of "hoarding oil" in the refined oil market, this paper adopts an agent-based simulation method, on the basis of the detailed design of the agents' behavior rules; this paper uses Starlogo to finish the simulation process. The simulation model sets different speculative stock according to different risk preferences of different wholesalers, and the behavior rules in the next phase are determined by the current profit. If the wholesaler can make a profit in current phase, it will promote the stock level ("hoarding oil"); if the wholesaler makes losses, it will learn from other or reduce its stock level.

Figure 3 shows the scatter diagram of the speculative stocks after 1200 simulation steps. From the figure, we can see that there is a significant nonlinear relationship between the speculative stock level and the refined oil price. The wholesalers' speculative "hoarding oil" behaviors have some rules under the influence of oil price; when the predicted oil price will be lower, the wholesalers generally hoard the oil and earn the spread; so the speculative "hoarding oil" behaviors always happen in a certain price range with lower price which is between 5 and 8. And we can also from the Fig. 4 that when the oil price is lower, more wholesalers will hoard oil, and the volume of hoarding mainly distributed in the range from 100 to 500. When the oil price is higher, the wholesalers who will hoard oil will be decrease, and when the price is higher than 8, there is almost no wholesaler who will hoard oil.

Figures 4 and 5, respectively, show wholesalers' speculative stock (SP) level under the old and new pricing mechanisms after 1200 simulation steps. From the figures, we can see that the refined oil market have the "hoarding oil" phenomenon,

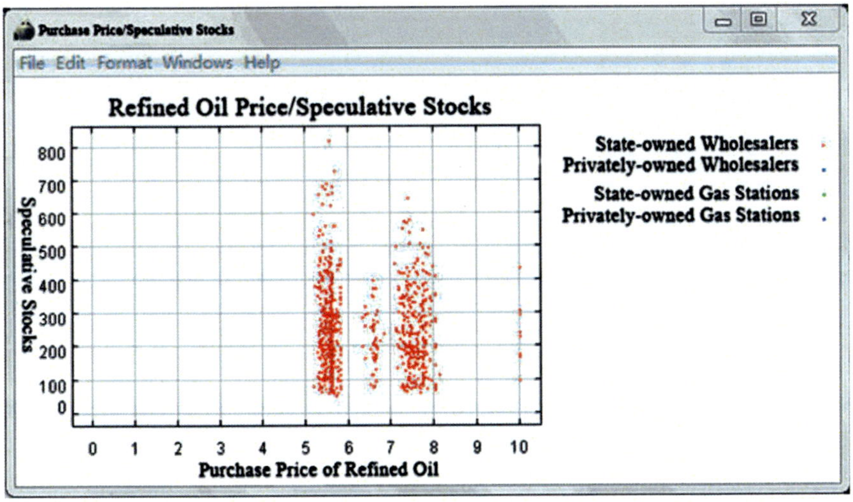

Fig. 3 Scatter diagram of the speculative stocks

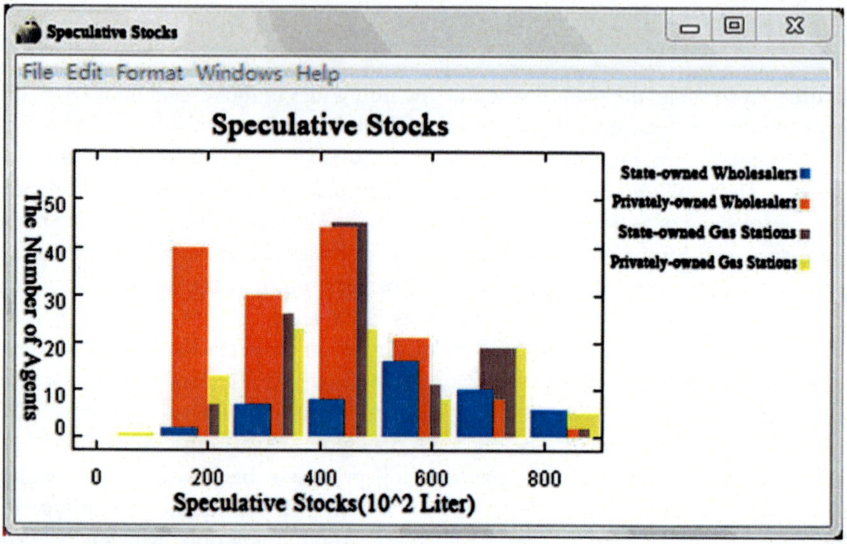

Fig. 4 SP in old mechanism

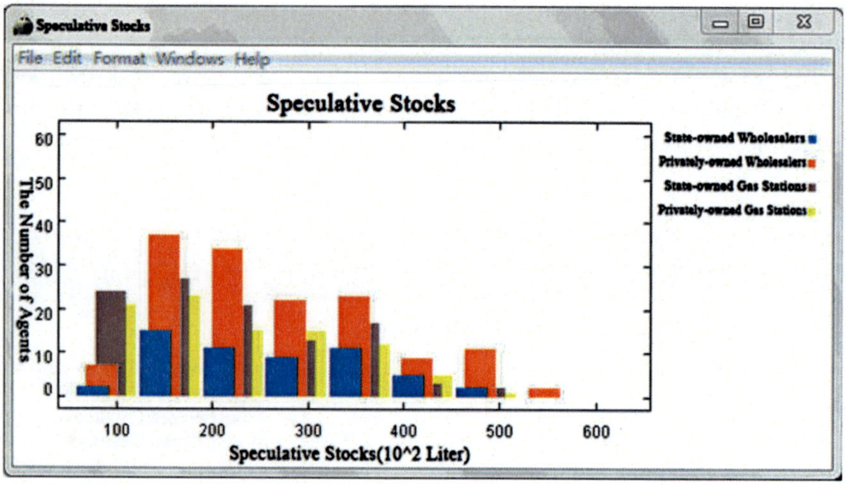

Fig. 5 SP in new mechanism

and the distributions of hoarding volume of state-owned wholesalers and private wholesalers are almost same. Comparing these two figures, we can see that the hoarding volume under the old pricing mechanism is distributed from 200 to 800, while the hoarding volume under the new pricing mechanism is distributed from 100 to 500, which means that the new pricing mechanism reduces the "hoarding oil' phenomenon to some extent.

4 Conclusion

From the simulation result, we can see that the speculative activities of the wholesalers are affected by the market oil price and government price policy. Under the new pricing mechanism, the speculative activities still exist, but the government can reduce the wholesalers' speculative stocks and speculative risk preference by setting the appropriate probability of price adjusting. Due to the special nature of the refined oil market, the wholesalers can predict the government's price adjustment decision effectively, according to the prediction of the oil price and the current pricing mechanism. So in the refined oil market, there is a general phenomenon that the wholesalers will hoard the oil before the price is going up and then gain the spread. And at the same time, the government can guide the market behaviors by setting the appropriate probability of price adjusting.

The agent-based simulation method is an effective method for some complexity research. Starting from the analysis of the micro-agents, using the bottom-up analysis pattern, the agent-based simulation method can simulate the macroscopic complex phenomenon by setting the simple behavior rules of micro-agents, using the multi-agent technology and the complexity science knowledge to discover and explore the problem. So far, the agent-based simulation studies about the refined oil market are still relatively limited; this paper applies the agent-based simulation method to the analysis of the refined oil market and studies the speculative activities of the agents in the refined oil market. The follow-up research will improve the model and the behavior rules of the agents based on the existing research.

Acknowledgements This work is supported by Research Foundation of China University of Petroleum-Beijing at Karamay (RCYJ2017B-03-002) to Q. Zhou.

References

1. Regulations on Administration of Oil Prices (Trial Implementation) (2009) No. 1198. http://bgt.ndrc.gov.cn/zcfb/200905/t20090508_498927.html
2. The PRC Further Improved the Pricing Mechanism for Refined Oil Products (2013) No. 624. http://jgs.ndrc.gov.cn/jggs/sytrqjg/201303/t20130326_534087.html
3. An G, Zhang Z, Zhang X (2010) Policy simulation of coordinate development for China's oil-gas resources and social economy system based on SD. Sci Technol Manage Res 30(17): 231–237
4. Yuan L, Zhang B-s, Jiang W-r (2009) Dynamic simulation system of Chinese diesel fuel demand and its simulation. Stat Decis (01):54–57
5. Yuan L, Zhang B-s, Wang Z (2010) Dynamic simulation system of Chinese oil products demand and its simulation. Syst Eng Theory Pract 30(4):654–666
6. Jia Z-h, Siqingbala, Chen Y-j (2011) Research on petroleum price prediction based on neural network. Comput Simul 28(11):354–357
7. Zhu X-m, Guo Z-g (2011) Simulation study on forecasting method of oil price forecasting. Comput Simul 28(6):361–364

8. Wang L, Fan Y, Wei Y-m (2007) Agent-based model for Chinese refined oil market and its application. J Manage Sci 20(5):76–82
9. Shi Y-l (2011) China's oil development policy modeling and simulation based on complex adaptive system. China University of Geosciences, Beijing
10. Lin B, Liu X (2013) Reform of refined oil product pricing mechanism and energy rebound effect for passenger transportation in China. Energy Policy (57):329–337

Research on Fitness Effect Evaluation of Elderly Based on Data Mining Technology

Donghua Zhou

Abstract Based on the concept of health, physical activity, and fitness risk evaluation of the elderly, this paper puts forward the feasibility of constructing the evaluation index system of physical fitness ability of the elderly from the theoretical and practical aspects. By using the methods of literature review and expert interview, this paper studies the evaluation index of physical fitness ability of the elderly. This paper constructs the fitness evaluation model of the elderly based on data mining and carries out the case analysis of the elderly fitness effect evaluation.

Keywords Elderly · Fitness · Evaluation · Data mining

1 Introduction

The aging of population is an inevitable trend that the age structure of population will change when the society develops to a certain stage. However, due to the full implementation of the family planning policy in China in the 1970s, the aging process of China's population is particularly rapid, and the number of elderly population is increasing. Therefore, how to alleviate the problem of population aging on the basis of social sustainable development has attracted the attention of all walks of life. Healthy aging is an important strategy to deal with the aging of the population and is an effective way to solve the problem of population aging. Physical fitness of the elderly has a good effect on promoting health and strengthening physical fitness.

Since 2000, many areas in China have gradually entered the aging society, and "healthy aging" has been rapidly implemented in the country. The research on the fitness theory of the elderly began to focus on the practical significance and social value The research scope mainly includes the investigation and analysis of the current situation of elderly people's physical fitness under different cultural envi-

D. Zhou (✉)
Research Center of Sports and Health of Wuhan Business University, Wuhan, China
e-mail: zhoudh2003@163.com

Z. Yu et al. (eds.), *Advancements in Mechatronics and Intelligent Robotics*,
Advances in Intelligent Systems and Computing 1220,
https://doi.org/10.1007/978-981-16-1843-7_60

ronment background, the monitoring of elderly physical fitness from the perspective of single discipline, the research of elderly sports fitness service system and the significance of elderly physical fitness. This study will be based on interdisciplinary theory, to the evaluation of fitness effect of the elderly before physical fitness as the starting point, to build an evaluation model of elderly physical fitness effect based on data mining technology. Its theoretical significance lies in the further discussion of the general theory of the elderly physical fitness, which will play a certain complementary role in the current related research of the elderly physical fitness, and further improve and innovate the theoretical research on elderly physical fitness; its practical significance lies in the realization of the physical fitness effect evaluation model of the elderly based on data mining technology The comprehensive evaluation of the effect provides a theoretical basis for better serving the elderly's physical fitness practice and improving the scientific level of physical fitness [1].

2 Physiological Function Characteristics of the Elderly

With the increase of age, the function of the organs and tissues of the human body gradually degenerates. In the elderly, the body water is reduced, the fat is increased, and the cell number is reduced. The function of glucose metabolism in the elderly is decreased, and they are prone to diabetes. With the aging of the human body, the quality metabolism of the elderly changes, and the lipoprotein in the snow increases significantly, which makes them prone to hyperlipidemia, atherosclerosis, hypertension, and brain diseases. With the increase of age, the decomposition of protein metabolism is greater than synthesis, and its digestion and absorption function decreases. Mild protein deficiency may lead to fatigue, weight loss, and resistance. Severe deficiency can lead to malnutrition, edema, hypoproteinemia, and decreased liver and kidney functions. However, excessive high protein diet can increase the burden on liver, kidney, and other organs [2].

The characteristics of the cardiovascular system in the elderly are: myocardial degeneration, fibrosis, and atrophy, as well as the decline of muscle strength and the weakening of repairability. The results showed that the heart valve became hard and the opening and closing were inflexible; the whole body blood vessels, especially the coronary artery, had calcification, decreased elasticity, lipid infiltration, rough intima, stenosis of lumen, increased blood pressure, resulting in insufficient blood supply. The above two changes are the main reasons for cardiac vascular insufficiency in the elderly. The total muscle weight of the elderly can be reduced to 25% of their body weight. The muscle and its ligament atrophy, the oxygen consumption and muscle contractility decrease. The muscle strength of the elderly hand decreases most significantly. After middle age, the density of capillaries in muscle tissue will also decrease. On the one hand, it reduces the supply of oxygen and nutrients in muscle tissue. On the other hand, many metabolites produced by muscle cells in muscle work can not be transported out in time, so the intensity of muscle work decreases, and the duration of muscle work is significantly shortened.

3 The Influence of Old People's Fitness Exercise on Fitness Effect

The influence of fitness exercise on body shape of the elderly
The influence of physical exercise on the body shape is mainly concentrated in walking, aerobics, gas volleyball, Taijiquan, and other different exercise methods on the body shape of the elderly. There are few studies on the influence of fitness paths on the body shape of the elderly. Aerobic exercise can improve the waist-hip ratio of middle-aged and elderly women, reduce the fat accumulation of waist and abdomen, enhance heart function and exercise ability. Sticking to the air volleyball has a positive effect on controlling body weight and reducing body fat content, which is helpful to improve the body shape and body composition of the elderly. Walking can effectively improve the body shape index of the elderly.

The influence of old people's fitness exercise on their physical fitness
The influence of fitness exercise on physical fitness is mainly concentrated in Yangge Dance exercise, gas volleyball exercise, fitness walking, and other scientific and reasonable physical exercise can comprehensively improve the strength, flexibility, agility, speed, and endurance of the elderly. Yangge Dance is a sport that plays an important role in the physical and mental health of middle-aged and elderly women. Yangge Dance exercise can improve the grip strength, sit forward, stand with eyes closed and stand on one foot. It can effectively improve and improve the physical quality and quality of life of middle-aged and elderly women. Fitness walking can significantly improve the lower limb muscle strength of the elderly, and delay the decline of muscle strength caused by age. Sticking to the practice of air volleyball can improve the grip strength, sitting forward flexion, and reaction time of the elderly subjects. It has a positive effect on improving the strength, flexibility, and agility of the elderly [3].

Effect of fitness exercise on cardiopulmonary function in the elderly
Long-term adherence to aerobic exercise, gas volleyball exercise, Taijiquan, Daoyin health exercise, and fitness exercise can improve the blood lipid status of middle-aged and elderly people, prevent atherosclerosis and hyperlipidemia, and improve health level.

Long-term aerobic exercise can improve the function of cardiopulmonary system, especially the vital capacity. Moreover, the ability of aerobic exercise group to complete the quantitative load is also stronger than that of the ordinary group. The heart rate immediately after aerobic quantitative load is significantly lower than that of the ordinary group. The heart rate also shows the characteristics of short recovery period and rapid recovery, which indicates that long-term aerobic exercise can maintain and improve the cardiovascular system The function of the system has obvious effect. After the experiment, the resting heart rate and blood pressure decreased, the step test index and lung capacity were increased. Different exercise intensity has different effects on cardiovascular function of the elderly. Jogging can reduce diastolic blood pressure, systolic blood pressure, and myocardial function.

Aerobic exercise is of great practical significance to maintain muscle strength of the elderly. Scientific and reasonable physical exercise can improve the strength, flexibility, agility, speed and endurance of the elderly, and has good effects on the common diseases of the elderly. More and more scholars began to pay attention to the effect of exercise on the muscle function of the elderly. The results showed that "aerobic exercise can increase the nerve stimulation of the elderly, mobilize more exercise units to participate in work, thus causing changes in muscle strength" [4].

4 Evaluation Index System of Physical Fitness Effect for the Elderly

Generally, most single indicators can be obtained directly, such as height and weight, and some need to be specially constructed, such as the mental health of the elderly. The construction process of any single indicator is a process of logical reasoning, including multiple steps, including three main aspects: first, clarify the theoretical significance of the indicators; secondly, define the operation of the selected indicators; finally, design the calculation content and index calculation method. Through the above steps to achieve the "operation" of indicators.

The physical fitness effect of the elderly is a complex system with multi-levels and multi-elements. Through the understanding of the relevant research results, the author thinks that the physical fitness effect of the elderly not only includes the static adaptation content of physical health, but also involves the dynamic adaptation content of physical activity, and also reflects the adaptive characteristics of the elderly to sports. This paper will divide the physical fitness effect according to the dynamic characteristics of the process of physical fitness and the static factors affecting physical fitness. At the same time, combined with the risk incentives of physical fitness, focusing on the goal of maximizing the benefits of physical fitness, and drawing on the excellent evaluation index system related to this research topic, this paper selects representative and authoritative important indicators to construct the evaluation index system.

The following principles will be followed in the construction of the evaluation index system of the elderly physical fitness effect.

The principle of comprehensiveness
The evaluation of the fitness ability of the elderly should be closely related to the fitness goal [5]. The evaluation index must comprehensively reflect the factors affecting the elderly's physical fitness, and comprehensively reflect the actual level of the elderly's physical fitness from different levels and levels. The selection of

indicators should follow the objective as the main factor and subjective as the auxiliary factor. In the evaluation, it is required to comprehensively evaluate the health adaptation, physical fitness adaptation, social adaptation, and psychological adaptation of the evaluation object, so as to obtain the evaluation information extensively, fully, and comprehensively. At the same time, we should pay attention to the diversity and diversity of evaluation objects, rather than one-sided pursuit of the unity of evaluation standards [6].

Scientific principle
In order to make the evaluation index system can accurately reflect the actual situation of the elderly's physical fitness ability, timely assess the risk and benefit of the elderly's physical fitness process and make clear the existing problems and weak links of their physical fitness, the whole index system must be scientifically selected from the elements to the structure, from each index content, and the calculation method must be accurate Form a scientific evaluation model.

Principle of representativeness
These indexes are not related to the essence of the elderly, but it is not related to the essence of fitness. At the same time, the refinement of indicators can reduce the time and cost of evaluation, which can facilitate the development of evaluation activities and ensure the economy in the operation process of elderly fitness effect evaluation.

The principle of combining qualitative with quantitative
The evaluation of the physical fitness ability of the elderly includes a lot of information about the health adaptation, physical activity adaptation, social adaptation, and psychological adaptation of the elderly. At the same time, complete qualitative analysis, because too much rely on experience, easy to make the understanding stay in the fuzzy stage, and complete quantitative analysis, can not fully include the evaluation of the main body of the effect index. In the construction of evaluation index system, it is necessary to combine quantitative evaluation with qualitative evaluation. Only by combining the two can we make an objective evaluation of the evaluation object efficiently and comprehensively.

Through literature review and expert consultation, this paper analyzes the physical fitness of the elderly from the static and dynamic adaptation and constructs the evaluation index system of the elderly sports fitness effect according to the construction principles of the index system. After listing a large number of indicators reflecting the effect of physical fitness of the elderly, in order to improve the evaluation index system of physical fitness ability of the elderly, this paper selects 15 experts and scholars who have made great achievements in the field of physical training, psychology, social sports and sports of the elderly as the expert consultation group, and conducted three rounds of expert questionnaire survey. The final evaluation index system is listed in Table 1.

Table 1 Evaluation index of physical fitness effect of the elderly

Object level	First-level index	Second-level index
Physical fitness effect of the elderly	Body function (I1)	Body mass index (I11)
		Quiet heart rate (I12)
		Waist hip rate (I13)
		Blood pressure (I14)
		Blood sugar (I15)
		Serum cholesterol (I16)
	Life quality (I2)	Sleep (I21)
		Diet (I22)
		Activity index (I23)
		Health index (I24)
	Physical quality (I3)	Aerobic endurance (I31)
		Muscle strength (I32)
		Muscle endurance (I33)
		Flexibility (I34)
		Balance (I35)
	Mentality (I4)	Anxious (I41)
		Depressed (I42)
		Force (I43)

5 Data Mining Algorithm

Data mining refers to the process of extracting potentially valuable information from the jumbled data after scientific processing. Data mining uses database artificial intelligence, machine learning, neural network, mathematical statistics, pattern recognition, high performance computing, and data visualization to find valuable hidden information from huge data sources, analyze and summarize them into structural patterns to help enterprises make scientific decisions.

Adopting factor analysis method can find out the public factors from the complex indicators to determine the correlation between the variables of the elderly fitness effect evaluation, and carry out comprehensive evaluation; at the same time, it can solve the problem of traditional multiple index equal weight or subjective determination of index weight, making the elderly fitness effect evaluation more objective and reasonable.

Suppose that there are p indexes in the evaluation index system of elderly fitness effect, $X_1, X_2, \ldots, X_p$, if the following expression is satisfied:

$$X_i = \mu_i + a_{i1}F_1 + a_{i2}F_2 \cdots + a_{im}F_m + \varepsilon_i, \quad i = 1, 2, \ldots, p \tag{1}$$

where $F_1, F_2, \ldots, F_m$ are the common factors, a_{ij} is the factor load. ε_i is the special factor, which obeys the normal distribution of mean value 0 and standard deviation σ_i. The calculation steps are listed as follows:

Step 1: The sample covariance matrix S and its eigenvalues are calculated. The eigenvalues are recorded as $\lambda_1, \lambda_2, \ldots, \lambda_p$, the corresponding unit orthogonal eigenvectors are $l_1, l_2, \ldots, l_p$, then S can be decomposed into:

$$S = \sum_{i=1}^{p} \lambda_i l_i l_i' \tag{2}$$

Step 2: Determine the number of factors m. Select m to satisfy

$$\frac{\lambda_1 + \lambda_2 + \cdots + \lambda_m}{\lambda_1 + \lambda_2 + \cdots + \lambda_m + \cdots + \lambda_p} \geq P_0 \tag{3}$$

where $0.7 \leq P_0 < 1$.
Step 3: Calculate load matrix A.

$$A = (l_1, l_2, \ldots, l_m) \tag{4}$$

On the basis of factor analysis of the samples, using the factor score as a variable for cluster analysis, all the elderly in the sample are classified according to the fitness effect evaluation results, which is convenient for us to understand the distribution of the elderly fitness effect evaluation grade, so as to effectively carry out the elderly health management. K-means clustering method is applied in this research. K-means clustering firstly divides the similar sample points by pre-set K value and the initial center of gravity of each class. The center of gravity of the initial class is the observation value of randomly selected samples.

6 Case Study

Using stratified random cluster sampling method, 300 aged men and 300 women aged 56–60 were selected in a city. Among them, 46% are mental workers, 44% are light manual workers, 10% are heavy manual workers, 12% are primary school culture, 61% are middle school culture, and 27% are university and above.

The factor analysis component matrix is calculated based on factor analysis method, and the calculating results are listed in Table 2.

On the basis of factor analysis, cluster analysis is carried out on the factor scores of the elderly health effect evaluation index to obtain a more scientific division of the elderly fitness affect grade. As the evaluation grade of the elderly fitness effect is divided into five grades: excellent, good, medium, qualified and unqualified, the

Table 2 Factor analysis component matrix

First-level index	F_1'	Second-level index	F_2'
I1	0.954	I11	0.923
		I12	0.954
		I13	0.889
		I14	0.916
		I15	0.972
		I16	0.983
I2	0.968	I21	0.890
		I22	0.923
		I23	0.948
		I24	0.965
I3	0.942	I31	0.986
		I32	0.965
		I33	0.938
		I34	0.927
		I35	0.965
I4	0.956	I41	0.986
		I42	0.934
		I43	0.896

Table 3 Clustering center of fitness effect of the elderly

Class	Factor score	Grade
1	0.09545	Good
2	−4.69533	Unqualified
3	0.88542	Excellent
4	−2.19433	Qualified
5	−0.91203	Medium

number of categories is determined as 5. When the number of classes is known, *K*-means clustering method with better speed and scalability is used for classification. After 22 iterations, the final clustering center of the fitness effect of the elderly is shown in Table 3.

Among all 300 elderly people, 36 of them were excellent, 57 were good, 105 were medium, 65 were qualified and 37 were unqualified. In the analysis of variance, *P* value is less than 0.001, the results show that the clustering results are effective.

7 Conclusions

In this paper, the data mining method is used to construct the evaluation model of the elderly fitness effect, and the value of the elderly fitness effect evaluation index is obtained by the factor analysis method. Based on the factor score of the comprehensive index, the elderly fitness effect is classified by cluster analysis method. The results of grade analysis show that the excellent and good rate of the elderly fitness effect is obviously low, which indicates that we should pay more attention to the elderly fitness.

Acknowledgements This paper is part of achievement of sports and health management program, the program number: 2019TD002. Thanks to the support of Academic Center of Sports and Health in Wuhan Business University.

References

1. He W, Peng N, Chen Q, Xiang T, Wang P, Pang J (2020) The relationships among the skeletal muscle mass index, cardiorespiratory fitness and the prevalence of coronary artery disease in the elderly population. Arch Gerontol Geriatr 90(9–10):
2. Ciprandi D, Zago M, Bertozzi F, Sforza C, Galvani C (2018) Influence of energy cost and physical fitness on the preferred walking speed and gait variability in elderly women. J Electromyogr Kinesiol 43(12):1–6
3. André H-, Carnide F, Moço A, Valamatos M-J, Ramalho F, Santos-Rocha R, Veloso A (2018) Can the calf-raise senior test predict functional fitness in elderly people? A validation study using electromyography, kinematics and strength tests. Phys Therapy Sport 32(7):252–259
4. Di Girolamo FG, Situlin R, Mazzucco S, Šimunič B, Pisot R, Biolo G (2016) OR22: Abdominal fat accumulation is a key determinant of muscle mass and strength as well as cardiovascular fitness in active elderly subjects. Clin Nutr 35(1):1–9
5. Jang SY, Park KH, Soc WY (2020) The association between breakfast consumption patterns and physical fitness Association entre les habitudes de consommation du petit déjeuner et les variables de condition physique dans un échantillon d'adultes et de personnes âgées en Corée. Sci Sports 35(10):1–7
6. McKune AJ, Peters B, Ramklass SS, van Heerden J, Roberts C (2017) Autonomic cardiac regulation, blood pressure and cardiorespiratory fitness responses to different training doses over a 12 week group program in the elderly. Arch Gerontol Geriatr 70(5–6):130–135

The Current Situation of Vocal Music Course' Assessment in Hubei Polytechnic University

Xuejing Han

Abstract The purpose of this research is to study the current situation of vocal music course's assessment in music education major at Hubei Polytechnic University, China: To improve the vocal music course assessment scientifically, the current vocal music course's assessment used regular method of evaluation by evaluated the result of learning after lesson by the teacher. After using scientifically variety of assessment such as student's self-assessment, assessment by teachers during learning process as well as final test, the students had higher ability of singing, the teacher could lay foundation for adjustment and establishment of the assessment system for the class.

Keywords Vocal music · Assessment · Current situation · Hubei polytechnic university

1 Introduction

With the deepening of education reform, people begin to pay attention to the evaluation system of education and teaching, which is not only the demand of the development of the times, but also an important guarantee to improve the quality of education and teaching for all the people. As a subject of music education major in higher education-vocal music, it is very important to carefully study the overall process of its classroom teaching assessment.

The rise of curriculum theory has a history of more than 100 years. The essence of the curriculum is still divided into different opinions. However, as universities gradually move from "social fringe" to the "social center" as well as the general development of curriculum and teaching reform in various universities, people have great influence on the quality and effectiveness of curriculum. We have basically reached a consensus that the improvement of curriculum quality and effectiveness is the quality of higher education (teaching). So, how do people know and guarantee

X. Han (✉)
School of Art, Hubei Polytechnic University, Huangshi 435003, Hubei, China

Z. Yu et al. (eds.), *Advancements in Mechatronics and Intelligent Robotics*,
Advances in Intelligent Systems and Computing 1220,
https://doi.org/10.1007/978-981-16-1843-7_61

the quality and effectiveness of courses? What are the characteristics of people's awareness and safeguard activities? How do people use the results of these activities? These problems lead to an extremely important field of curriculum research: curriculum assessment, that is, the curriculum assessment process itself how to works.

2 Concept Definition and Related Theoretical Models

2.1 The Concept of Vocal Music

Vocal music: Vocal music is probably the oldest form of music, since it does not require any instrument besides the human voice. All musical cultures have some form of vocal music. It is an important and direct way to achieve quality education in music art.

2.2 The Concept of Music Education

Music education is a field of study associated with the teaching and learning of music. It touches on all learning domains, including the psychomotor domain (the development of skills), the cognitive domain (the acquisition of knowledge) and in particular and significant ways, the affective domain (the learner's willingness to receive, internalize and share what is learned), including music appreciation and sensitivity.

2.3 The Concept of Assessment

Assessment is a central feature of teaching and the curriculum. It powerfully frames how students learn and what students achieve. It is one of the most Significant influences on explicit focus on students' experience of higher education and all that they gain from it. The reason for an explicit focus on improving assessment practice is the huge impact it has on the quality of learning.

3 The Development of Curriculum Assessment in China

3.1 Vocal Music Teaching in China

Since the 1980s, the reflection and re-understanding of China's higher education have been mainly focused on the macro-system of higher education. By 1994, the former State Education Commission had initiated "The reform of teaching content and curriculum system of higher education for the twenty-first century." At the same time, on the basis of summing up the experience of educational reform and in-depth investigation and research, it put forward the overall policy of higher education reform "Increasing investment is the premise, ideological change is the forerunner, system reform is the key, and teaching reform is the core." These four sentences have become one of the signs that the focus of higher education reform in our country has begun to move downward. The reform in the micro-field of higher education is in full swing across the country.

In 2000, the Ministry of Education implemented the "Higher Education Teaching Reform Project in the New Century" on the basis of the phased results of the 1994 reform. It decided to carry out comprehensive reform research and practice on the personnel training mode, teaching content, curriculum system and teaching methods of higher education, push forward the teaching reform to a deeper level.

After the founding of New China, it put great importance to the development of vocal music education. In 1949, the establishment of China's higher vocal art education system has raised to the undergraduate level. Since the late 1970s, the training of master's and doctoral students has begun in higher art colleges. So far, China has formed a complete vocal music education system based on undergraduate course, formed the middle school, and post graduating to postgraduate education. Now, China has more than 30 higher art colleges, 120 secondary art schools which carry out vocal music teaching. China's vocal music education has made great progress. However, we cannot be so optimistic. Examine the current situation of our vocal music education, and there are still many problems and some misunderstandings. Because our vocal art education and teaching system have been established for less than a hundred years, only by clearly recognizing our problems we can promote the development of our vocal music education.

3.2 Curriculum Assessment in China

University curriculum assessment is an academic and professional activity and a highly practical application. Research field, so in the theoretical discussion of curriculum assessment, we must try our best to demonstrate its practicality. It enables the theoretical exploration and practical activities of assessment to be interdependent, mutually penetrating, mutually premise and mutually intermediary. On the one hand, theoretical exploration is based on assessment practice activities. Through practical investigation and theoretical promotion, it forms a general theory about assessment structure, process, influencing factors and other relevant university curriculum assessment operation process, and actively uses assessment practice to test and revise its own theory. On the other hand, in the assessment practice, we should take these general theories as the premise to guide and promote the specific behavior of curriculum assessment to be scientific, standardized and orderly.

First, 1949–1976, is the centralized curriculum assessment. After the founding of the People's Republic of China in 1949 to the "cultural revolution" 17 years ago, China's higher education as the other social fields, to learn from the Soviet Union in an all-round way, the most prominent is to learn from the Soviet Union according to the "professional" training of talent mode.

Second, 1978–1980s, the assessment of courses with Chinese characteristics started. After 1978, China implemented the policy of reform and opening up. In the increasingly frequent exchanges with the international community, people's minds have become more active and liberated. Higher educators began to rethink its 27 years of experience and lessons, began to doubt the legitimacy of the Soviet union's "Talent cultivation model" of over-specialization, starting with criticism of its rigid teaching program.

In the early 1980s, the assessment system of higher education in the modern sense of China had appeared. It was the result of the combination of external factors and internal needs, and finally organized by the government. It came into being with the establishment of the degree system in China. Although the assessment system appears in the macro-higher education activities, the assessment system also partly involves the curriculum level, But it did not really get into the college curriculum. Therefore, from the end of 1970s to the end of 1980s, It can only be said that it is the beginning of the theoretical exploration of curriculum assessment in Chinese universities.

During this period, the development of course assessment depended on the government's support, but at the same time, it began to explore the college course assessment in line with China's national conditions.

Third, after the 1990s, the assessment of courses with Chinese characteristics gradually matures. In the 1990s, on college curriculum and curriculum assessment, Both in theoretical research and in practice have made some achievements. In 1998, the Chap. 4 of the higher education law stipulated for the first time in the form of law stated that: The higher institutions independently set up and adjust disciplines and majors in accordance with the law; make teaching plan independently according to teaching needs. To compile teaching materials and organize teaching activities. At this time, how to ensure the quality of education through the quality of courses, how to ensure the implementation of their school-running ideas, school-running characteristics and development strategy, how to ensure the flexibility of university curriculum plan to meet the market demand, how to reflect the requirement of the society on the innovation ability in the course setting, and some problems are related to the assessment of university courses all present.

For a long time in the past, we have always taken the course assessment as a management tool and think the assessment results as the basis of teacher assessment, so we pay more attention to the course authority and the judgment of assessment experts. About that how to encourage curriculum compilers and teachers (university teachers should also be the main members of the curriculum development team) to carry out curriculum self-assessment, both practical experience and theoretical research are relatively lack. The deep reasons are as follows: First, because of the managerialism tendency of curriculum assessment, university teachers are quite disgusted with assessment activities and unwilling to participate in them. Second, teachers have rich practical experience but lack of theoretical knowledge of curriculum research, cannot be a good theoretical sublimation of practical experience.

4 The Vocal Music Course' Assessment in Hubei Polytechnic University

4.1 The Vocal Music Course in Hubei Polytechnic University

Hubei Polytechnic University is a general higher education institution with rapid development focusing on cultivating application-oriented talent, with engineering being our strong point. It is located in Huangshi, Hubei, a cradle of modern Chinese national industry. The faculty of music in Hubei Polytechnic University was founded in 1992. After more than 20 years of construction, it has developed into a professional teaching institute with large scale, strong faculty, and perfect teaching and research conditions. It was working in this faculty as a vocal music teacher for 12 years. Now, it has 10 vocal music teachers, including 6 doctors, They all have rich experience in vocal music teaching.

In order to understand the status of the existing vocal music course's assessment of Hubei Polytechnic University, to master the views and opinions of students and teachers, to test the feasibility of the application of assessment in vocal music course of Hubei Polytechnic University, the author carried out the study from March 2018. When gathering and analyzing responses from the students, the writer employed both quantitative and qualitative research methods. Before the trial of assessment, plenty of preparations were done from January to October in 2018.

4.2 The Current Situation of Vocal Music Course' Assessment in This University

According to the collected data analysis, get the existing problems of the Hubei Polytechnic University's vocal music course's assessment. It just has a final exam in every semester, have not for the learning process or teaching process, other detail on the final performance. The content of assessment is too singular, focusing on the mastery and application of vocal skills, and lack of assessment of the overall knowledge structure and comprehensive ability of students.

Under the influence of the curriculum reform of basic education in China, the vocal music curriculum of music education major has been reformed in different degrees, from class form and curriculum setting which has played a good role in promoting the improvement of vocal music teaching and the improvement of students' comprehensive ability in music education major.

However, for a long time, there is no perfect vocal music assessment model suitable for the teaching objectives for music education major. In most cases, refer to the assessment model of conservatory of music which are cultivate actor (in the end of the semester, the students sing one to two songs that the teacher choose for them, after singing, all the vocal music teachers give the scores, average is the student's final exam score). These assessment forms seem to have the effect of the fair to judge student achievement, but it just present to students the quantitative score. It is difficult for students and teacher to know the problems from a score that cannot objective and comprehensive assessment "The future music teachers" vocal music knowledge and the skills to master and use of comprehensive ability.

The method of assessment is too monotonous, focusing on summative assessment and quantification. The assessment method (i.e., final assessment) lacks of formative assessment that comprehensively and objectively reflects the progress of students' vocal learning and promotes the overall development of students.

Due to the influence of narrow assessment objects and the bondage of traditional examinations, we often equate curriculum assessment with examination. Examination becomes the master key, and it seems that only examination can be objective, scientific, to identify students, to provide information on curriculum regulation. In fact, the examination is just a way to check the learning effect. We must not be superstitious about examinations. In the contemporary society pursuing democracy and equality, it is generally believed that the main reasonable and legal means to provide everyone with equal educational opportunities should be examinations, which can provide social justice and promote social mobility and pass examinations.

In fact, it is not really science and reasonable, it reflects to a large extent concept that the winner is the king, the loser is the aggressors. And also, to extent the performance of the "the fittest survival." From this, the rationality and scientific of the examination are open to discussion. In Hubei Polytechnic University, except for the students' exam results at the end of the semester, another part of the grade comes from the students' usual learning performance. This score is given by the student's instructor. However, this score does not clearly indicate a student's overall learning situation. Students cannot get their own shortcomings from this score, for their next step of learning, it is difficult to set the right goal.

This part of score given by some teachers even cannot objectively reflect the students' learning attitude that will lead to the elimination of the enthusiasm of the students who are active in learning, and make some lazy students think that vocal music is an easy subject to pass, thus does not invest more energy in vocal music study and practice.

The main body of assessment is teacher-based, ignoring the status of students in assessment and the importance of students' self-assessment, co-assessment and group-assessment in assessment. The assessment in our country, the subject of curriculum assessment is from top to bottom. Government assessment the schools, the schools assessment the teachers, then the teacher assessment the students. It is too difficult to meet the needs of different stages of curriculum development.

Because our country is in a model of the planned economy to market economy transformation period. Current education system in our country still is centralized, the characteristics of all kinds of schools at all levels in not much education has the initiative, so the assessment of power mainly lies in the national and provincial education administrative departments at various levels, and of the hands of the curriculum assessment main body of the distribution and quantity of how many just reflects the social reality of this situation, the distribution of power and therefore caused the main body in the curriculum assessment of single feature.

Teachers are an important factor that cannot be ignored in the course or curriculum implementation. Therefore, the teachers should be the core of the subject of curriculum assessment. But the students from different classes and collectives, also have different social needs. They more concerned about the curriculum assessment that suits their own interests. There for, they can make up for some mistakes and prejudices in the assessment, which is more conducive to the continuous improvement of the curriculum itself.

The teacher-centered assessment mode can only form a single mode that "The teacher says, the students passively accept," resulting in the students lack of initiative and innovative spirit in learning. They do not take the initiative to distinguish the sound of good or bad, the practice is right or wrong, just rely on the teacher to tell them. In terms of teachers' teaching, students cannot give the timely feedback of their own experience in learning, resulting in teachers' failure to timely adjust their teaching methods and objectives.

The assessment function is more inclined to identify students' singing skills, deviating from the assessment principle of music education, and the goal of all-round development of students, affecting the transformation of vocal music teaching concepts and the improvement of teaching methods.

For a long time, the assessment of vocal music in Hubei Polytechnic University is mainly based on the master of methods in singing, which leads students to pay attention to the singing performance, while the teaching focuses on skills training, ignoring the comprehensive assessment of the students' innovative spirit, practical ability, learning attitude, discipline theoretical knowledge and pedagogical knowledge.

5 Some Suggestions

5.1 Subjectivity in Assessment

Compared with the traditional objective paper-and-pencil test, the subjective problem is more prominent in the implementation of the assessment. In the assessment of students in vocal music course teaching, there are usually three aspects of factors that may lead to the subjectivity of the assessment, so as to draw an inaccurate conclusion.

5.2 *The Fairness of Assessment*

The fairness of assessment refers to the degree of consistency in the observation and assessment of students' specific behavior by assessors. At least two assessors should assess at the same time. Since evaluation rules are qualitative descriptions of each assessment standard, and assessors have certain subjectivity in their cognition of assessment rules, it is difficult to guarantee the fairness of assessment. In fact, the process of assessment is the use of assessment tools. The assessor has decided subjectivity when using all kinds of reference rules in the tools to carry out the assessment work. In this way, it is difficult to guarantee the fairness and justice of assessment.

5.3 *The Quality of Judge*

Teachers are the main executors of assessment and assume the most important roles and responsibilities. Although teachers cannot require every teacher to assess like an assessment expert, they must have certain professional qualities of assessment. We should not only understand the theoretical knowledge of assessment as the basis of our own professional behavior, but also improve the ability of developing, designing evaluation tools and organizing, guiding and implementing assessment.

Acknowledgements The study is supported by Scientific Research Project of Education Department of Hubei Province (Young and Middle-aged Talents Project) Grant No. Q20204504.

References

1. Bloom BS, Hastings JT, Madaus GF (1971) Formative and summative evaluation of student learning
2. Chen L (2005) Puts forward the content and requirements of vocal classroom teaching assessment in the article
3. Cao D (2005) Puts forward the overall conception of vocal music teaching assessment
4. Chen L (2005) Assessment of vocal music teaching in music education majors in university
5. Han X (2018) Breeding and starting, pioneering and innovating–an overview of the development of Chinese vocal art in the 1920s and 1940s
6. Kan J (2004) The use of performance assessment in english teaching
7. Li J (2018) The blend of tradition and modernity–The development path of Chinese vocal music
8. Li L (2015) Research on the development process of Bel Canto in China since the reform and opening up
9. Luo J (2009) Research on the representatives of the development stages of Chinese Bel Canto
10. Li Z (2018) New thoughts on the teaching model of vocal music in music education major of university
11. Liu H (2010) A probe into the reform of vocal teaching courses in university——comparative study before and after the introduction of the new "course plan" and "syllabus"

Overview of Chinese Text Classification

Xing Meng and Jian Fei Shao

Abstract [Goal] The goal is to sort out the key development nodes, algorithms and models in the field of Chinese text categorization, so as to facilitate researchers to sort out the development context and provide practical guidance. [Methods] Through CNKI database and VIP database, the related literatures of Chinese text classification were retrieved, and the related algorithms and models of Chinese text classification were classified and summarized. [Result] The advantages and disadvantages of the mainstream text classification algorithms, such as Naive Bayes algorithm, support vector machine algorithm, neural network algorithm and decision tree algorithm, are summarized; [Conclusion] Considerable quantities of machine learning and deep learning theory are used in text classification, which tends to focus on phrase- and sentence-based topic representation.

Keywords Text categorization · The algorithm of text classification · Category of model

1 Introduction

The technology, for instance, big data and cloud computing, has contributed to the modern information science. The traditional paper documents are rapidly changing to electronic and digital. In the face of large amounts of data and information, people are more and more inclined to use computers to process data and information, which can not only improve the efficiency of related operations, but also improve the accuracy of related operations to a certain extent. Information mining and retrieval, natural language processing are the key technologies of data management, and text classification is an important basis for the operation of these technologies, which is a hot spot and a difficult problem. Traditional text classification mainly depends on manual work, which is time-consuming and laborious.

X. Meng (✉) · J. F. Shao
School of Information Engineering and Automation, Kunming University of Science and Technology, Kunming 650000, Yunnan, China

Z. Yu et al. (eds.), *Advancements in Mechatronics and Intelligent Robotics*, Advances in Intelligent Systems and Computing 1220,
https://doi.org/10.1007/978-981-16-1843-7_62

It is convenient to use automatic text classification technology for enhancing the efficiency of text classification and reducing the cost. There is an upward tendency for automatic text classification technology in recent research interests.

2 The Concept and Process of Text Classification

2.1 The Concept of Text Classification

Text classification, which refers to the process of automatically text categorization, accords to certain classification system or rules. It is widely used in information index, digital library management, information filtering and other fields. Text classification generally includes text preprocessing, word segmentation, model building and classification. With the flood of massive data, text and vocabulary show diverse features and rapid update, which brings great challenges to text classification. In order to understand the development of text classification algorithm more clearly, this paper combs and analyzes the related technologies and classification methods in the process of text classification [1].

2.2 Text Classification Process

General flow of text classification. The process can be divided into five steps, as shown in Fig. 1.

Step 1: Preprocesses the text to remove the redundant parts of the text, such as punctuation, preposition, etc.
Step 2: The text is segmented, the preprocessed text is segmented, and the unknown words are identified.
Step 3: Feature extraction and feature selection, after getting the result of text segmentation, the text feature extraction method is selected, and the features are

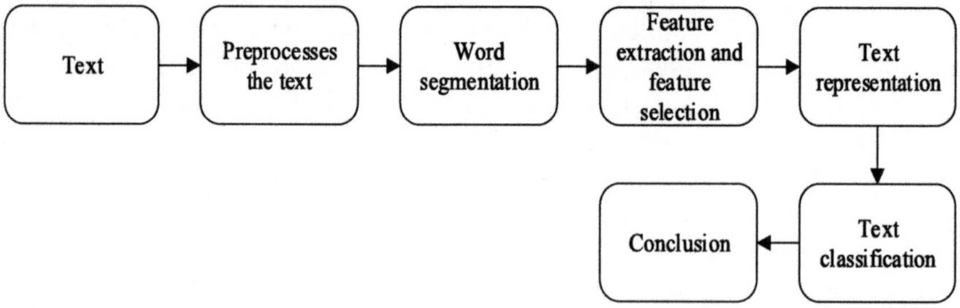

Fig. 1 General flow of text classification

selected to reduce the dimension as much as possible and reduce the amount of subsequent calculation.
Step 4: Text representation, select the appropriate method to represent the selected features, as the basis for classification.
Step 5: Text classification, select a reasonable classification method to classify the text, get the text category.

Among them, segmentation method, feature selection and classification algorithm are the key. Combined with the current research status of text classification, this paper mainly summarizes the text classification methods.

3 Text Classification

Text classification refers to the automatic classification of text according to a certain classification standard or system by computer [1]. It is not only a problem of natural language processing, but also a problem of pattern recognition. Therefore, the study of text classification can not only promote the development of natural language research, but also has great significance to the research of artificial intelligence technology.

3.1 General Methods of Text Classification

The knowledge engineering (KE) consists of expert experience and manual extraction, and the machine learning (ML) was the other category in text classification that is based on computer self-learning and extracting rules. KE and ML are disparate methods in text classification. Naive Bayes [2] is the earliest machine learning method, and then a majority of crucial machine learning algorithm is applied to the domain of text classification, for instance, support vector machine (SVM) [3], neural network [4] and decision tree [5].

3.2 Research Status of Text Classification

In the light of the defect of the traditional text classification methods has been found, which is revised and improved through many experts' experiment on the text classification methods. The literature [6] uses neural network algorithm that has many merits in natural language processing (NLP) field. KNN algorithm and SVM algorithm is used to classify Web text. The effect indicates that the accuracy on neural network algorithm is better than other algorithms. Compared with the traditional classification method which mainly adopts supervised method, depending on the existing natural language processing tools, it is easy to lead to the problem of error accumulation in the process of processing. In the literature [7], scholars project a text semantic feature learning theory based on convolution neural network. The convolution deep neural network is put on automatically learn the features of vocabulary, context features and entity locations of entity semantic relations. This method does not need NLP processing tools to extract features, which greatly improves the error accumulation problem caused by multiple processing links in the feature extraction process and improves the correctness of text classification. This paper [8] brings out the theory that rely on apparent semantics and ASLA. This paper works on the apparent semantics of Chinese vocabulary so as to feature them, next step is mine the latent semantics from PLSE and calculates the correlation between the apparent semantics and the latent semantics and the category of the document. This method [9] can well deal with the classification of irregular text such as Chinese network short text. Reference [10] proposed a short text classification model based on dense network so as to express our text directly. One hot coding was used to expand text feature selection by merging and random selection, which solved the problems of feature sparsity, dimension text data and feature representation. Literature [11] and document [12] respectively used the improved TF-IDF to modify the weight of word vector and manually build a dictionary to optimize the text classification algorithm. Finally, the convolution neural network is used to construct the classifier, which improves the accuracy of text classification. However, it does not deal with the high-order features reasonably, which leads to the time complexity of learning much higher than the traditional machine learning method, which needs further improvement. Reference [12] proposed a text classification method based on feature fusion model of deep learning, which uses convolutional neural network and bidirectional gated loop unit to extract the information of whole text and low-level information, this paper have made remarkable decline of impact of text expression on categorization.

Although the existing classification methods can meet the requirements of text classification in some aspects, there are still some problems, such as low efficiency of algorithm, poor domain pertinence and over fitting in the learning process. It is natural to reduce the learning time, improve the efficiency of classification, organically combine the advantages of different classification methods and achieve efficient and accurate text classification hot issues in the field of language processing.

4 Conclusion

This paper introduces the concept, process, key technologies and classification methods of text classification, summarizes the existing research and solution methods and summarizes the text classification in combination with the challenges in the process of text classification.

The development trend of related technologies is as follows.

(1) The representation of text features develops from discrete, high-dimensional to continuous and low-dimensional. When traditional text classification methods describe text, they are usually represented by words. With the continuous development of natural language processing methods, text representation tends to focus on phrases and sentences. This kind of method can effectively solve the problem of sparseness in the process of word representation.
(2) The learning of text develops from shallow to deep. With the continuous development of machine learning and deep learning, text-processing methods begin to change from traditional step-by-step learning to holistic learning, and the understanding of text develops from shallow analysis to deep understanding. Large amounts of machine learning and deep learning methods are applied to text classification process, such as fuzzy neural network, convolution neural network [13] and cyclic neural network [14] in text classification more and more widely used.,
(3) Text classification methods have developed from single to integrate. With the maturity of text classification technology, the advantages and disadvantages of various text classification methods are revealed. By reasonably integrating different classification methods, such as boosting improved algorithm, the text classification method can be further optimized.

References

1. Li S, Ma J, Zhao Y et al (2006) Automatic classification of digital science and technology papers [J]. J Shandong Univer (Science Edn) 2006(3):81–84
2. Cui W (2016) A Chinese text classification system based on Naive Bayes algorithm [C]. In: MATEC web of conferences, pp 1015
3. Zhang MY, Ai XB, Hu YZ (2019) Chinese text classification system on regulatory information based on SVM [C]. IOP Conf Ser: Earth Environ Sci:252
4. Saha D (2011) Web text classification using a neural network [C]. In: 2011 second international conference on emerging applications of information technology
5. Lei F (2018) Text research on text classification based on neural network and decision tree and its application [D]. University of Electronic Science and Technology of China, Chengdu
6. Zhou PX (2008) Study on web document classification based on neural network integration [J]. Lib Inf Serv 7:110–112
7. Zeng DJ, Liu K, Lai SW, et al (2014) Relation classification via convolutional deep neural network [C]. In: The 25th international computational linguistics, pp 2335–2344
8. Chen YW, Wang JL, Cai YQ et al (2015) A method for Chinese text classification based on apparent semantics and latent aspects [J]. J Ambient Intell Humaniz Comput 6(4):473–480
9. Li HM, Huang HN, Cao X et al (2018) Falcon: a novel Chinese short text classification method [J]. J Comput Commun 6:216–226
10. Wang GS, Huang XJ (2019) Convolutional neural network text classification model based on Word2vec and improved TF-IDF [J]. J Chin Comput Syst 40(5):1120–1126
11. Wang L (2019) Research on Chinese short text classification based on hybrid neural network [D]. Zhejiang University of Science and Technology, Hangzhou

12. Jin WZ, Zhu H, Yang GC (2019) An efficient character-level and word-level feature fusion method for Chinese text classification [C]. J Phys: Conf Ser: 12057
13. Wang P, Xu B, Xu JM et al (2016) Semantic expansion using word embedding clustering and convolutional neural network for improving short text classification [J]. Neurocomputing 174:806–814
14. Gong QJ (2016) Text classification based on recurrent neural network model [D]. Huazhong University of Science and Technology, Wuhan

Application Status and Prospect of Dot Matrix Digital Pen

Peng Guo, Jianfei Shao, Jianjie Ji, Lilin Pan, Hongfei Pu, and Rong Pu

Abstract The dot matrix digital pen has developed rapidly in recent years, and many products related to dot matrix digital pen have appeared in the market. This paper introduces the basic working mode and working principle of the dot matrix digital pen and shows the application of some dot matrix digital pen in the market in detail. Finally, it puts forward some ideas on the combination and development of dot matrix digital pen and artificial intelligence technology.

Keywords Dot matrix digital pen · Artificial intelligence

1 Introduction

With the development of science and technology, more and more science and technology products make our work and study more convenient, and education is no exception. Tablet computers with specific teaching software can make teaching work more efficient and interesting, but there are also difficulties and shortcomings in promoting the use of tablet computers in the classroom: the price of tablet computers is still relatively high, so it is unrealistic to equip all students. One disadvantage of using tablet computers is that they cannot produce paper documents. The above problems caused by using tablet can be solved by using dot matrix digital pen. The following will introduce some applications of dot matrix digital pen and dot matrix digital pen in education one by one.

P. Guo (✉) · J. Shao · J. Ji · L. Pan · H. Pu · R. Pu
Kunming University of Science and Technology, Kunming 650504, China

Z. Yu et al. (eds.), *Advancements in Mechatronics and Intelligent Robotics*,
Advances in Intelligent Systems and Computing 1220,
https://doi.org/10.1007/978-981-16-1843-7_63

2 Working Principle of Dot Matrix Digital Pen

2.1 Lattice Code

Dot matrix code is a special dot matrix printed on paper. It can mark different position information on paper by changing the position arrangement of points. In fact, the research of lattice code has been started a long time ago. QR (quick response) two-dimensional code [1], DataMatrix two-dimensional code [2] and Maxico de two-dimensional code are common coding methods.

2.2 Dot Matrix Digital Pen

The dot matrix digital pen has a pressure sensor, which can detect the pressing pressure value of the pen when writing. The dot matrix digital pen also has an infrared high-speed camera, which is combined with a special ink that is visible to the naked eye but invisible to the camera. When writing, the camera will not capture the writing track, but only the dot matrix code pattern that the pen passes through. The processor in the dot matrix digital pen decodes the dot matrix code pattern to get the position information of the pen passing by. The processor then after synthesizing the position information for a period of time, the writing track can be reproduced. The hardware of the dot matrix pen is shown in Fig. 1, and the effect of the dot matrix digital pen is shown in Fig. 2.

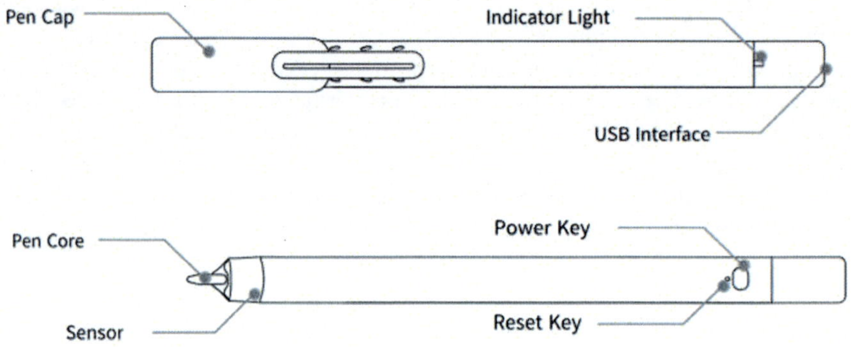

Fig. 1 Structure diagram of dot matrix digital pen

Fig. 2 Effect picture of dot matrix digital pen

3 The Application of Dot Matrix Digital Pen in the Field of Education

3.1 Electronic Notes

For example, it is impossible for students to lose their notes in class if they can't take notes from the notebook. There is also a problem with the traditional notebook note taking, that is, if students learn more subjects, if they use a notebook for each subject, it is not convenient to carry the notebook of all subjects. As like as two peas, the digital pen is used to solve this problem. By taking notes on special notebooks, a note of the same electronic version can be produced. The effect of electronic backup of notes is shown in Fig. 3.

Fig. 3 The electronic use effect of notes

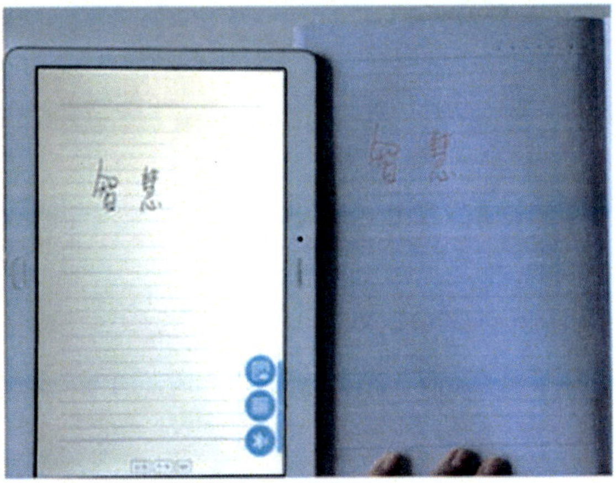

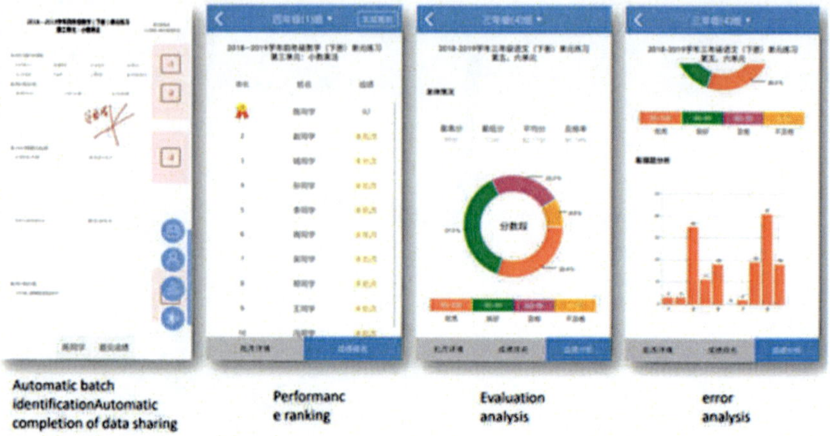

Fig. 4 The effect picture of using the system

3.2 *Intelligent Examination Paper Marking*

For subjective questions, using dot matrix digital pen with dot matrix code test paper can realize intelligent recognition of handwriting scores without changing teachers' handwriting marking habits. For objective questions, we can judge right and wrong directly and give scores. The system can count the score of the whole test paper and get the total score. In addition, it can also automatically generate evaluation report, grade ranking, wrong question situation, and can automatically record students' wrong questions, generate wrong question book and students' learning situation report. The effect of the system is shown in Fig. 4.

4 The Development of Dot Matrix Digital Pen

4.1 *Handwritten Information is Directly Input into Database*

Whether enterprises or banks treat customers to fill in forms and other information, if the traditional way to enter the database, customers or managers first fill in the information on the paper form, and then the special staff manually enter the database. If we use the dot matrix digital pen technology, we can avoid employing special staff to do heavy manual database entry work. Through the use of dot matrix digital pen technology to automatically complete the paper information written into the database, the idea is to lay the dot matrix code on the form paper in different areas, so that the system can recognize which specific form item is written in the form. For each table item, handwritten font is recognized by using handwriting recognition technology, and then the recognized content is automatically entered into the database by programming In this way, handwritten information can be directly inputed into the database.

4.2 *Monitor the Writer's Mental State*

Dot matrix digital pen has pressure sensor, high-speed infrared camera, and other hardware. Through these hardware, the speed of writing process and the pressure degree of writing can be recorded in detail when writing with dot matrix digital pen. Artificial intelligence technology can realize the state of the writer in the process of writing, for example, if a student is using the dot matrix digital pen to do a test paper The information collected by the dot matrix digital pen shows that his writing strength and speed are far faster than other students, so it can be judged that he is impatient to complete this test paper, and there are many similar applications.

5 Conclusion

In fact, the research of dot matrix digital pen began very early, but it seems that there has been no substantial application. In recent years, dot matrix digital pen has a rising momentum, and many applications in the field of education have appeared. Like the dot matrix pen, artificial intelligence is also a booming technology in recent years. The sensor of the dot matrix digital pen can record a lot of information. I believe that with the artificial intelligence technology, the dot matrix digital pen can achieve many unexpected functions.

References

1. Dong Q (2006) QR code recognition technology and its application in mobile phone [D]. Qingdao University, Qingdao, pp 66–72
2. Leng B (2007) A datamatrix-based mutant code design and recognition method research [C]. In: 2007 Fourth international conference on image and graphics, pp 570–574

Estimation of Knee Joint Motion Acceleration Using Mechanomyography Signals

Chenlei Xie, Daqing Wang, Dun Hu, and Lifu Gao

Abstract It has become a research hotspot to provide exercise assistance and rehabilitation training for the weakened limbs' elderly or disabled by wearable power-assisted robots' (WPAR) technology. How can the WPAR obtain effective human motion intentions is a bottleneck problem. We design a convolutional neural network (CNN)-long-short term memory neural network (LSTM) model, which adopts the MMG signals of lap muscles for estimating the knee joint movement acceleration. The estimated result can be used to calculate the angular acceleration of knee joint for the movement control of WPAR. We detect five muscles' MMG signals. The collected MMG data are divided by sliding window and stepping methods, and then, are extracted features by the CNN model. We input the features into the LSTM model. Finally, the acceleration values are estimated. It is proved by experiment that the mean correlation coefficient (R) is 97.5% for the acceleration estimation of the built model. This method will improve human-computer interaction of the WPAR.

Keywords Mechanomyography · Knee joint · CNN · Acceleration · LSTM

1 Introduction

As the elderly population was increased, the issue of elderly care has become the focus of attention from all walks of life. The partial body disability is particularly prominent in the elderly. In 2014, the disability rate of the elderly has reached more

C. Xie (✉) · D. Wang · D. Hu · L. Gao
Institute of Intelligent Machines, Hefei Institutes of Physical Science,
Chinese Academy of Sciences, Hefei 230031, China
e-mail: xiecl@mail.ustc.edu.cn

C. Xie · D. Hu
Department of Science Island, University of Science and Technology of China, Hefei 230026, China

C. Xie
Anhui Province Key Laboratory of Intelligent Building and Building Energy Saving,
Anhui Jianzhu University, Hefei 230022, China

Z. Yu et al. (eds.), *Advancements in Mechatronics and Intelligent Robotics*,
Advances in Intelligent Systems and Computing 1220,
https://doi.org/10.1007/978-981-16-1843-7_64

than 18.5% [1]. Meanwhile, the total number of disabled people in the country is about 85.02 million in the sixth national census, of which more than 50 million are limb disabled [2]. Therefore, solving the problems of life inconvenience caused by the weakened limbs of the above population has become an urgent task.

With the advancement of wearable sensors and human-computer interaction technology, many scholars and institutions have begun to use wearable power-assisted robots (WPAR) to provide exercise assistance and rehabilitation training for those elderly or physically disabled [3]. However, there are still some problems in how to continuously and reliably recognize the intention of the human body. As a result, the flexibility of the WPAR need to be further improved [4]. Modern human anatomy studies have shown that mechanomyography (MMG) signals [5, 6] are closely related to the state of muscle activity. Researchers have tried to use MMG signals as a signal source for the human-computer interaction [7].

Alves et al. used acceleration sensors to detect the flexor carpi radialis and extensor carpi's MMG signals, and constructed a wrist motion classifier [8]. The results showed that the average accuracy of recognition of the three repetitive wrist joint actions of flexion, extension, and rest was 89 ± 2%. Wu et al. [9] detected the MMG signals of the knee joint movement from the outer side of the experimenter's clothes and used the support vector machine model for recognizing the knee joint movement. Experimental results showed that the average accuracy of recognition was as high as 91%. Wang et al. collected the quadriceps muscle's MMG signals. The MMG signal-force relationship model was constructed by the support vector machine algorithm. The validity of the model used the MMG signals and the quadriceps contraction force was verified through experiments [10]. Lei et al. detected the MMG signals of the biceps brachii during the flexion of the elbow joint at different speeds, extracted the root mean square value and frequency variance as features. They inputted the features into the artificial neural network regression model, and then, estimated torque value [11]. Wilson et al. fused the inertial measurement unit and the MMG sensors information to sense the intention of the movement of the body, and then controlled the bionic hand prosthesis [12]. The results showed that the average accuracy of recognition of the gesture control of gripping and extending was as high as 93.3%.

We design the convolutional neural network (CNN)-long-short term memory neural network (LSTM) model (CNN-LSTM model) to estimate knee joint movement acceleration using MMG signals. We estimate the linear acceleration, and use the human stick model to further calculate the angular acceleration of knee joint for the movement control of WPAR [13]. The acceleration sensors are used as the MMG sensors, which detect five muscles' MMG signals. The CNN model is constructed to extract the MMG signals' features. The LSTM model is constructed to estimate acceleration.

2 CNN-LSTM

2.1 CNN

The CNN is specially used to process time series or image data with a similar network structure [14]. CNN includes an automatic feature extractor which consists of a convolutional layer and a subsample layer. Convolution and subsample operations can extract features as much as possible, thereby avoiding the drawbacks of manually extracting features.

During the forward calculation of the convolutional layer, a neuron is only connected to a part of the neighboring neurons. Under the shared weight mechanism, the input features of the layer are convolved, and the result of the convolution is summed with the corresponding offset. Finally, input the sum result into the nonlinear excitation function to get the output features. The calculation process is shown in Eq. 1.

$$F_j^l = f\left(\sum_{i \in M_j} F_i^{l-1} * K_{ij}^l + b_j^l\right) \tag{1}$$

where F_j^1 is the jth feature figure in the lth layer, M_j is the feature map of all inputs in the lth layer; K_{ij}^1 is the jth convolution kernel in the lth layer, $*$ is the convolution operation, b_j^1 is the bias parameter, $f(x)$ is the nonlinear excitation function.

The subsample layer scales the input features of the convolutional layer, extracts main features, and achieves data compression. The calculation process is shown in Eq. 2.

$$F_j^l = f\left(\beta_j^l \mathrm{down}\left(F_j^{l-1}\right) + b_j^l\right) \tag{2}$$

where $\mathrm{down}(x)$ is the subsample operation of the eigenvalues in the x region, β_j^l and b_j^l are the bias parameters; $f(x)$ is the nonlinear excitation function.

2.2 LSTM

Usually, the LSTM is used for processing sequence data [15, 16]. The LSTM can reduce or add the information of unit memory through gate structure, and better control the gradient flow. The unit of the LSTM is composed of the gates, as shown in Fig. 1. The calculation process [17] is shown in (3)–(9).

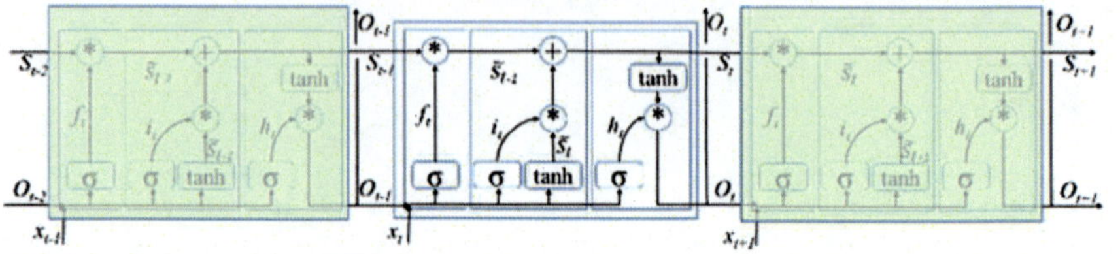

Fig. 1 The component of the LSTM unit

$$f_t = \sigma\left(W_f \cdot [o_{t-1}, x_t] + b_f\right) \tag{3}$$

$$\widetilde{S}_{t-1} = f_t * S_{t-1} \tag{4}$$

$$\mathrm{i}_t = \sigma(W_i \cdot [o_{t-1}, x_t] + b_i) \tag{5}$$

$$\widetilde{S}_t = \tanh(W_c \cdot [o_{t-1}, x_t] + b_c) \tag{6}$$

$$S_t = i_t * \widetilde{S}_t + \widetilde{S}_{t-1} \tag{7}$$

$$h_t = \sigma(W_h \cdot [o_{t-1}, x_t] + b_h) \tag{8}$$

$$O_t = \tanh(S_t) * h_t \tag{9}$$

where σ is the sigmod function, O_{t-1} is the output of the previous layer, x_t is the input at the present moment, W_f, W_i, Wc, and W_h are the weight parameters, b_f, b_i, b_c, and b_h are the bias terms, and S_{t-1} is the unit state output by the previous layer, S_t is the unit state output at the current moment.

2.3 *CNN-LSTM*

The CNN takes the original data directly as input through operations such as multi-layer convolution and subsample, which can adaptively extract more robust features from the training data. The LSTM integrates short-term and long-term time-series-related information to learn time-series information. Figure 2 shows the general CNN-LSTM's component. The original data is inputted into the constructed CNN. According to time series, the fully connected layer's data of the CNN is inputted into the constructed LSTM. Finally, the output is obtained by output layer. The CNN-LSTM can fully cover the features information in the temporal and spatial dimensions. The CNN-LSTM makes up for the insufficiency of CNN in processing time-series data, and makes up for the insufficiency of the LSTM in the feature extraction.

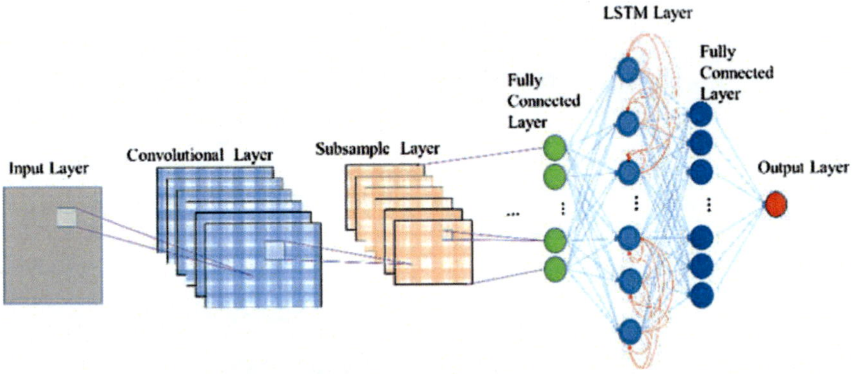

Fig. 2 The component of the CNN-LSTM

3 Process of Experiment

Figure 3 shows the process of knee joint movement acceleration estimation. First, the collected acceleration and MMG signals are processed. The MMG data features are extracted by using the trained CNN network. Finally, the estimated value of acceleration is obtained by using the LSTM model.

3.1 Data Acquisition and Processing

The MMG signals are detected by selecting lap muscles that can control knee joint flexion and extension and play a major role. Figure 4 shows selected muscles.

For estimating the knee joint movement acceleration during the flexion and extension, the thigh body segment is in a static state, which eliminates the acceleration caused by the movement of the lower leg body segment driven by the thigh body segment. The experimental participants are sitting in a chair. The laps are in horizontal state. The participants are approximately $\pi/2$ rad/s (FV) and $\pi/4$ rad/s (SV) speed flexion and extension of the knee movement for 30 s, as shown in Fig. 5.

The ADXL335 acceleration sensor developed by ADI company is used as the MMG sensor. The five MMG sensors are placed horizontally on the outside of the clothes of the five muscles' central part. The MMG sensors are wrapped with elastic straps. The MMG signals were obtained by collecting the output from the Z-axis of the ADXL335 acceleration sensor. In addition, the angular acceleration can be further calculated by measuring the linear acceleration of knee joint motion. The ADXL203EB acceleration sensor developed by ADI company is laid vertically on the outside of the medial ankle point. The ADXL203EB acceleration sensors are wrapped with elastic straps. The X-axis of the ADXL203EB acceleration sensor is

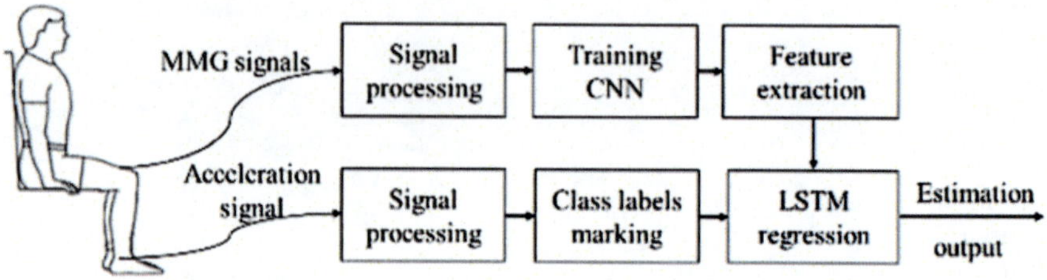

Fig. 3 Process of the model

Fig. 4 Muscle map

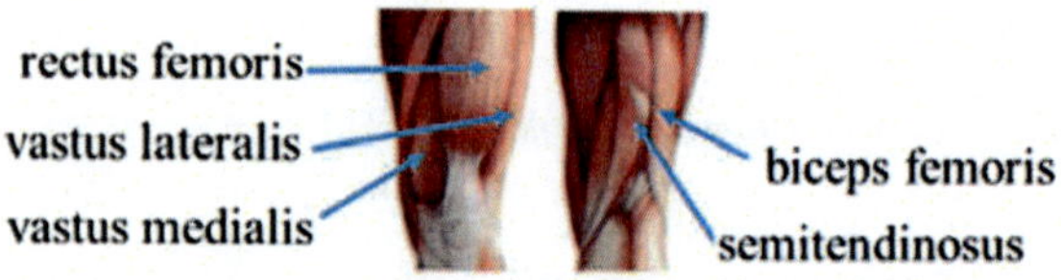

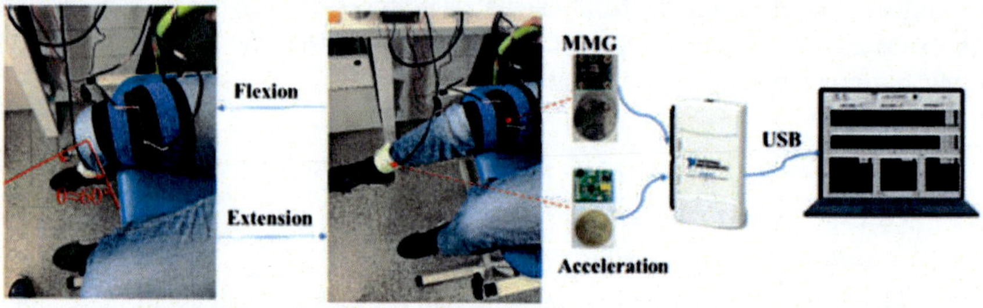

Fig. 5 Movement and collection system of data

perpendicular to the leg, and the X-axis is positive toward the front of the body. The X-axis output from the ADXL203EB acceleration sensor is collected to obtain the acceleration signal. Data from the MMG sensor and acceleration sensor are uploaded to a computer for storage via a portable USB collector (NI USB-6215), as shown in Fig. 5. The sampling frequency is 2000 Hz.

The original MMG signals are filtered by thrird-order Butterworth bandpass filter of 5–100 Hz for filtering human motion artifacts. The original acceleration signal is filtered by fourth-order Butterworth low pass filter of 1.5 Hz for filtering the noise.

3.2 *Construction of the CNN-LSTM Model*

The collected data are divided by sliding window and stepping methods. The combination of 250 ms (500 data points) window and 62.5 ms (150 data points)

Table 1 The component and parameters of the CNN-LSTM

Layer	Operation	Input	Filter	Stride	Padding	Output
1	Conv2D + BN + Leaky Relu	5 * 500	32@2 * 101	[1 1]	[0 1 50 50]	5 * 500
2	Average_Pooling2D	5 * 500	1 * 2	[1 2]	[0 0 0 0]	5 * 250
3	Conv2D + BN + Leaky Relu	5 * 250	16@2 * 101	[1 1]	[0 1 50 50]	5 * 250
4	Conv2D + BN + Leaky Relu	5 * 250	16@2 * 101	[1 1]	[0 1 50 50]	5 * 250
5	Average_Pooling2D	5 * 250	1 * 2	[1 2]	[0 0 0 0]	5 * 125
6	Conv2D + BN + Leaky Relu	5 * 125	8@2 * 31	[1 1]	[0 1 15 15]	5 * 125
7	Average_Pooling2D	5 * 125	1 * 2	[1 2]	[0 0 0 0]	5 * 62
8	Conv2D + BN + Leaky Relu	5 * 62	8@1 * 6	[1 1]	[0 0 2 3]	5 * 62
9	Fully connected	5 * 62	–	–	–	1 * 50
10	LSTM + Leaky Relu	1 * 50	–	–	–	1 * 500
11	Fully connected + Leaky Relu	1 * 500	–	–	–	1 * 100
12	Fully connected	1 * 100	–	–	–	1 * 50
13	Regression	1 * 50	–	–	–	1

step quantity is adopted [9]. At the same time, five groups of fixed length MMG data in the sliding window are used to train the CNN model. The MMG data features are obtained by the trained CNN model. The features as the input of the LSTM model. Finally, the estimated value of acceleration is obtained by the LSTM model. In order to realize the CNN-LSTM, we call Deep Learning Toolbox in MATLAB. Table 1 shows the structure and parameters of CNN-LSTM. For improving the training speed of the model and generalization ability, each convolutional layer processes the feature map generated by the previous layer, including convolution operation (Conv2D), batch normalized operation (BN), and rectified linear units (Leaky ReLU). Finally, the full connection layer in the 9th layer was 50-dimensional MMG features data.

The 50-dimensional MMG feature data is extracted by the CNN network. The feature data as the input of the LSTM model. The LSTM model is constructed as shown the 9th to 13th layers in Table 1.

4 Experiment and Result Analysis

According to the above experimental method, 5 healthy young people are selected as the participants in this experiment. The experiment is conducted in accordance with the Medical Ethics Committee of Hefei Institutes of Physical Sciences. Each

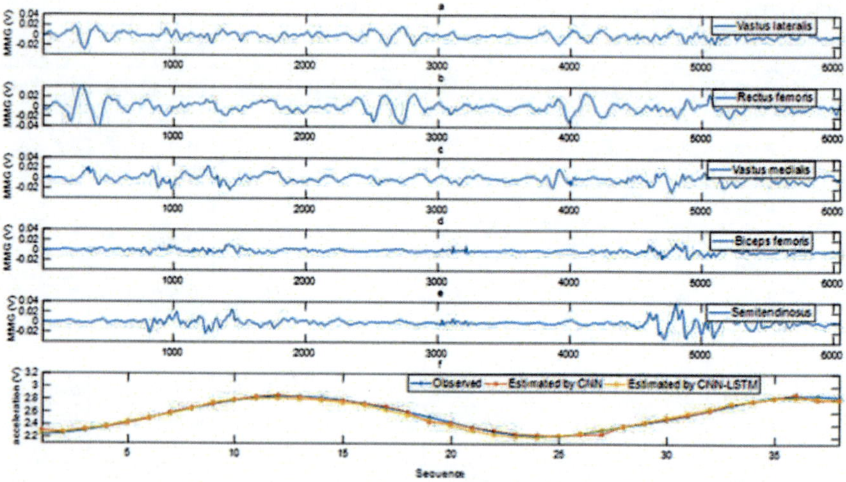

Fig. 6 The MMG signals and the acceleration values

participant can get close to 390 effective data pairs (MMG features-acceleration) in each experiment. When training the CNN model, 90% of the data pairs in each group are used as the training set. The remaining 10% of the data pairs are used as the test set. The performance of the CNN model is tested through tenfold cross-validation. We select the best performance in the validation set as the final CNN model used for features extraction.

Figure 6 shows the filtered MMG signals of the participant's test set at SV, and the estimated value of acceleration during knee flexion and extension. We observe that the relativity between the observed value and the estimated value is higher, when the CNN-LSTM model is used to estimate the acceleration. It proves that the model has better performance and can accurately understand the human motion intention to a certain extent. It provides a reliable signal source for validly controlling the knee joint flexion and extension.

Meanwhile, we use correlation coefficient (R) and root mean square error (RMSE) as evaluation indicators of the built model [17].

Figure 7 show the performance comparison of the first 9 floors CNN model on the test set of 5 participants. The CNN model has a good performance in estimating the knee joint motion acceleration at FV. The R is at the level of $95.69 \pm 2.44\%$. The RMSE is at the level of 0.086 ± 0.043.

New data pairs are formed by the 50-dimensional MMG feature data and acceleration data based on the CNN model. The new data pairs as the input of the LSTM model. When training the LSTM model, the first 90% of the data pairs sorted by time in each group of data pairs are used as the training set. The remaining 10% of the data pairs are used as the test set. The performance comparison of the CNN-LSTM model on the test set of 5 participants are shown in Fig. 8.

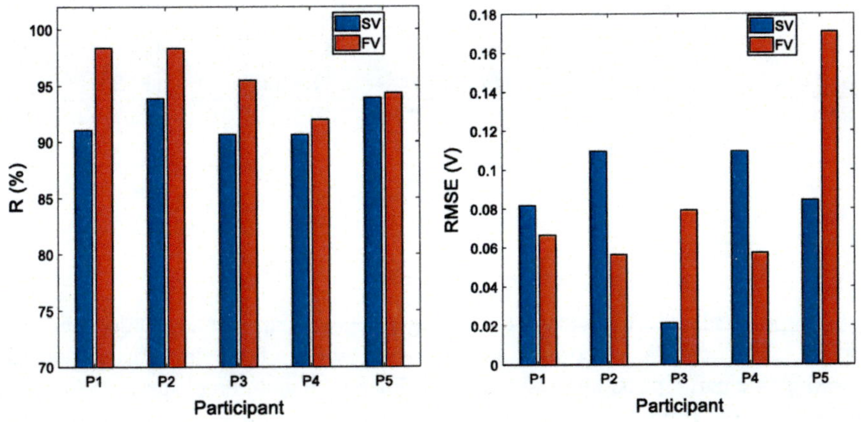

(a) R (b) RMSE

Fig. 7 The R and RMSE of the CNN model

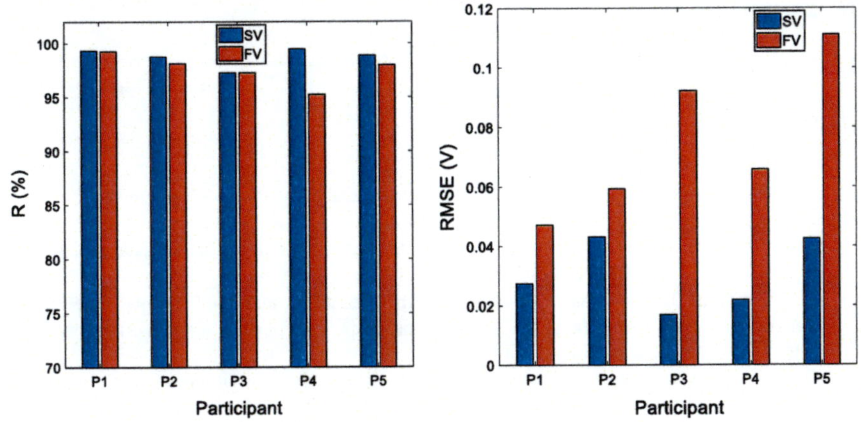

(a) R (b) RMSE

Fig. 8 The R and RMSE of the CNN-LSTM model

Table 2 shows the performance between the CNN model and the CNN-LSTM model. The mean and standard deviation of R and RMSE of the CNN-LSTM model is significantly better than CNN model. The R of the CNN-LSTM model are more stable at FV or SV.

Table 2 The R and RMSE on the test set

	SV R (%)	FV R (%)	SV RMSE	FV RMSE
CNN	92.00 ± 1.53	95.69 ± 2.44	0.08 ± 0.03	0.09 ± 0.04
CNN-LSTM	98.74 ± 0.79	97.56 ± 1.33	0.03 ± 0.01	0.07 ± 0.03

5 Conclusion

This paper explores a method for estimating the movement acceleration of the knee using MMG signals. According to the swing angle of the leg during normal walking, the participants flex and extend the knee joint at different speeds. We build the CNN-LSTM model using the MMG signal, and use the model to estimate the knee joint movement acceleration. The results show that the model can better estimate the knee joint movement acceleration, which will promote the development of human-computer interaction for the WPAR.

Acknowledgements This work was supported by Anhui Jianzhu University's Research Project (No. JZ192012).

References

1. Ding H, Yan J (2016) Research on the measurement and trend of disability rate of the elderly in China. Chin J Popul Sci 3:97–108 + 128
2. Zhao YC (2012) China disabled persons' federation released the latest data of disabled population in China. Disabil Res 1:11–11
3. Yu YP (2016) The research of motion pattern recognition and joint moment analysis of human lower limb based on sEMG. M.S. thesis, Dept. SOMAEE., SOOCHOW Univ., SZ, CHN
4. Wu HF (2018) Study on motion intention recognition of human knee based on mechanomyography and CNN-SVM model. M.S. thesis, Dept. SI., USTC Univ., HF, CHN
5. Orizio C, Liberati D, Locatelli C et al (1996) Surface mechanomyogram reflects muscle fibres twitches summation. J Biomech 29(4):475–481
6. Oster G, Jaffe JS (1980) Low frequency sounds from sustained contraction of human skeletal muscle. Biophys J 30(1):119–127
7. Park J, Kim SJ, Na Y et al (2016) Custom optoelectronic force sensor based ground reaction force (GRF) measurement system for providing absolute force. In: 13th International conference on ubiquitous robots and ambient intelligence (URAI), pp 75–77. IEEE Press, Washington, DC USA
8. Alves N, Chau T (2010) Recognition of forearm muscle activity by continuous classification of multi-site mechanomyogram signals. In: 32nd annual international conference of the IEEE Engineering in Medicine and Biology Society, pp 3531–3534. IEEE Press, Washington, DC USA
9. Wu HF, Wang DQ, Huang Q et al (2018) Real-time continuous recognition of knee motion using multi-channel mechanomyography signals detected on clothes. J Electromyogr Kinesiol 38:94–102

10. Wang DQ, Guo WB, Wu HF et al (2018) An estimation method of quadriceps femoris contraction strength using mechanomyography signal. Chin J Sens Actuators 31(11):84–90
11. Lei KF, Cheng SC, Lee MY et al (2013) Measurement and estimation of muscle contraction strength using mechanomyography based on artificial neural network algorithm. Biomed Eng Appl Basis Commun 25(2):1–8
12. Wilson S, Vaidyanathan R (2017) Upper-limb prosthetic control using wearable multichannel mechanomyography. In: International conference on rehabilitation robotics (ICORR), pp 1293–1298. IEEE Press, Washington, DC USA
13. Huston RL (2007) Fundamentals of biomechanics, 1st edn. Springer, NY, USA
14. Lecun Y, Bengio Y (1995) Convolutional networks for images, speech, and time-series. Handb Brain Theory Neural Netw 2:1–14
15. Hochreiter S, Schmidhuber J (1997) Long short-term memory. Neural Comput 9(8):1735–1780
16. Sutskever I, Vinyals O, Le QV (2014) Sequence to sequence learning with neural networks. In: Advances in neural information processing systems (NIPS), pp. 3104–3112. MIT Press, MA, USA
17. Xie CL, Wu HF, Wang DQ et al (2020) A long short-term memory neural network model for knee joint acceleration estimation using mechanomyography signals. Int J Adv Rob Syst 17 (6):1–10

The Uniformity Analysis of Dual-Sample Relative Movement During Microwave Heating Process

Biao Yang, Hao Gao, and Hongtao Ma

Abstract Microwave heating is a complex process of multiphysics coupling. This research aims to study the influence of relative movement of dual sample on heating uniformity during microwave heating process. First of all, we set up a series of simulation experiments on COMSOL Multiphysics which use FEM method for calculating. Secondly, the comparison and analysis of the heating effect and electric field distribution are carried out on simulation experiments, and the influence of relative movement and different position of the two samples is discussed. Third, the experimental data of the simulation experiments are extracted and used COV and ΔQ to quantitatively analyze the uniformity and heating efficiency, respectively. Finally, according to the above series of discussions and analyses, it is shown that relative movement can improve the uniformity and efficiency of heating two samples simultaneously in a cavity.

Keywords Microwave heating · Dual sample · Relative movement · Uniformity

1 Introduction

Microwave heating has become one of the most common heating methods at present, whether in household or industrial fields, because of its high efficiency, cleanliness and no hysteresis [1, 2]. However, microwave heating also has a problem that cannot be ignored which is the uneven heating [1]. In order to improve the uniformity of heating, researchers have done a lot of research over the years.

The commonly used method to improve the uniformity is to add a mechanical structure in the reaction chamber and use a moving object to change the microwave

B. Yang (✉) · H. Gao · H. Ma
Kunming University of Science and Technology, Kunming Yunnan 650500, China
e-mail: ybiaocn@163.com

B. Yang
Key Laboratory of Unconventional Metallurgy, Ministry of Education,
Kunming Yunnan 650093, China

Z. Yu et al. (eds.), *Advancements in Mechatronics and Intelligent Robotics*,
Advances in Intelligent Systems and Computing 1220,
https://doi.org/10.1007/978-981-16-1843-7_65

absorption or change the distribution of the electromagnetic field in the chamber [2]. The former is like adding a turntable [2, 3]; the latter is like adding a mode stirrer [3]. At the same time, on this basis, the researchers also carried out some related expansions. For example, Ye et al. simulated and analyzed the temperature distribution when a turntable and a mode stirrer with different rotation speeds exist in the same cavity [2]. They also used a turntable made of different materials as a mode stirrer [3]. Meng et al. studied the effect of a rotating metal block in the cavity or a metal spot on the turntable on the heating uniformity [4]. From another point of view, Zhu et al. used a rotating waveguide to improve heating uniformity [5], and He et al. further used two rotating waveguides as research targets [6]. Yvan et al. studied the power absorption when there are multiple materials (water and soy oil) in the same cavity with different proportions [7].

Many of the studies mentioned above have not yet involved the analysis of heating uniformity when two heated objects in the same cavity are in relative motion. In this research, we use COMSOL Multiphysics simulation software to simulate and analyze the heating uniformity of the two materials in static and dynamic heating process.

2 Methodology

2.1 3D Model Description

The simulation model uses a rectangular microwave reaction cavity filled with air. The right side of the cavity is a standard WR340 rectangular waveguide. The material of the cavity and the waveguide is copper. The heated sample is a rectangular potato block (including single and double sample). The detailed geometric of the simulation model is shown in Fig. 1. The meshing of the simulation model is shown in Fig. 2.

Fig. 1 Geometric of simulation model

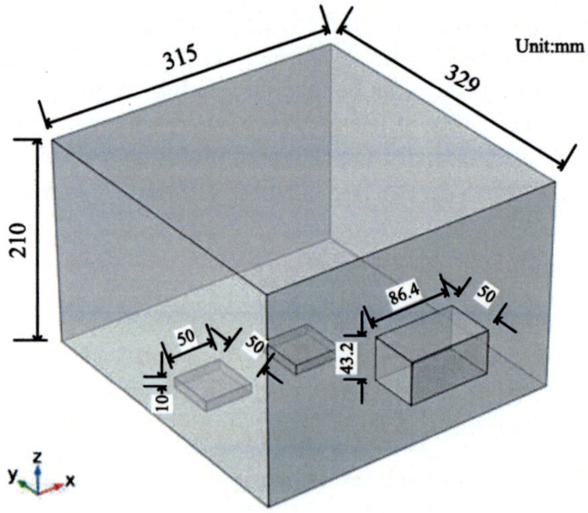

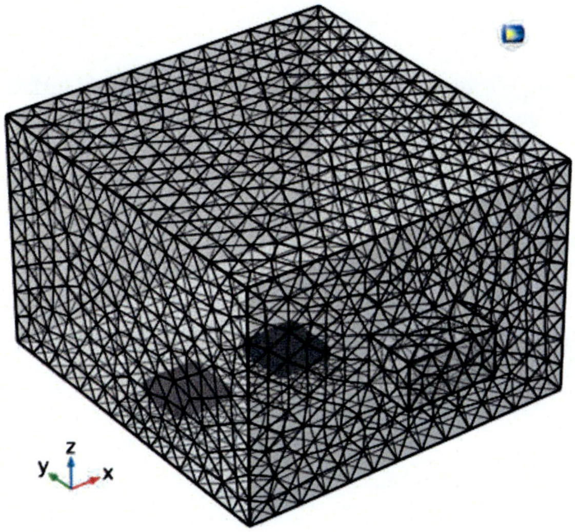

Fig. 2 Meshing of simulation model

2.2 *Governing Equations*

The electric field distribution in the cavity can be solved by Maxwell waveform equations [5] (Formula 1).

$$\nabla \times \mu_r^{-1}(\nabla \times E) - k_0^2\left(\varepsilon - j^{\sigma}/_{\omega\varepsilon_0}\right)E = 0 \quad (1)$$

where E is the electric field intensity; μ_r is the magnetic permeability; k_0 is the wavenumber of free space; ε_0 is the vacuum permittivity; ε is the relative permittivity.

The temperature of the heated sample can be solved by the following heat conduction equation (Formula 2).

$$\rho C_p \frac{\partial T}{\partial t} - k\nabla^2 T = \frac{1}{2}\omega\varepsilon_0\varepsilon''|E|^2 = Q_e \quad (2)$$

where Q_e is the dielectric loss; ω is the frequency of electromagnetic waves; ε'' is the imaginary part of the complex relative permittivity; T is the temperature; ρ is the density of the heated sample; C_p is the heat capacity; k is the thermal conductivity.

2.3 *Boundary Conditions and Initial Conditions*

As mentioned earlier, the material of the entire cavity and waveguide is copper, which is considered a perfect electrical conductor during the entire simulation process. So, the boundary conditions can be expressed as follows (Formula 3):

$$n \times E = 0 \tag{3}$$

where n represents the normal vector of each surface.

In the simulation settings, the initial temperature of each element is 20 °C, the fed microwave mode is TE10, the frequency is 2.45 GHz, and the power is 700 W. The heating time for all simulation experiments is 10 s. In particular, for the accuracy of the simulation results, the relative permittivity of potatoes varies with temperature, and it can be expressed as [2] (Formula 4):

$$\varepsilon = -6.4 \times 10^{3} T^{2} + 2 \times 10^{-1} T + 56.8 + j\left(-10^{-4} T^{2} - 1.08 \times 10^{-1} T + 16.1\right) \tag{4}$$

3 Results and Analysis

3.1 Simulation Results

Static Simulation. In the static simulation, we studied the heating effect of two samples at different relative positions and used the heating effect of a single sample as a comparison experiment. The result is shown in Figs. 3 and 4. It can be seen from Fig. 3 that when a single sample is statically heated, the temperature at the center of the heated sample is lower, while four hot spots appear near the four corners.

As shown in Fig. 4, the two samples in the cavity are A and B, respectively. The sample fixed at the center of the bottom surface of the cavity is B, and the other is A. We simulated the heating effect of sample B, when sample A is on the left (position 1), lower (position 2), right (position 3) and upper (position 4) of sample B. It can be clearly seen from the results that the different placement of sample A will have a very significant impact on the heating effect of sample B. It is worth noting that when sample A is in position 4, the heating rate of sample B is significantly lower than the other three positions.

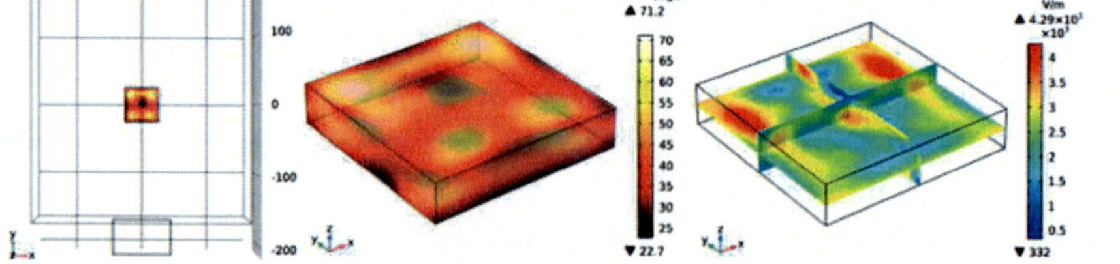

Fig. 3 Result of single-sample static heating

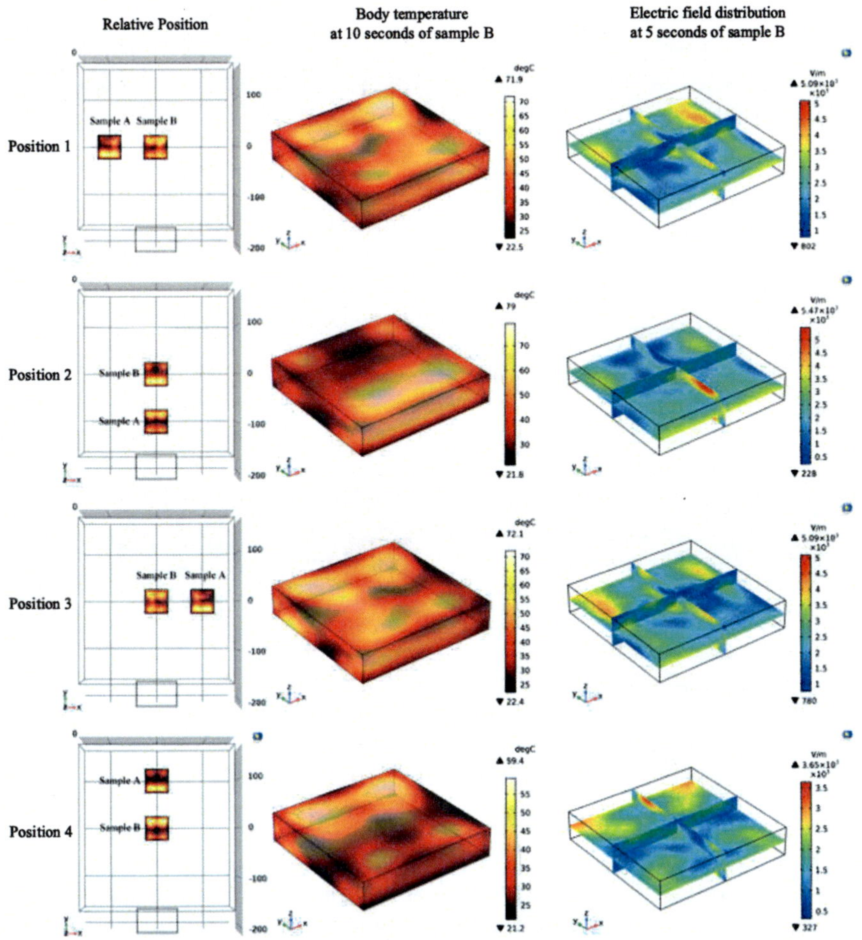

Fig. 4 Result of dual-sample static heating

Dynamic simulation. In the dynamic simulation, sample B is still fixed at the center of the bottom of the cavity, and sample A rotates around sample B at a speed of 6r/min. Heating process is also 10 s (sample A rotates exactly once) as shown in Fig. 5.

Figure 5 also shows the final heating result. It can be seen that the uniformity of heating is significantly improved compared to the static state. Figure 6 shows the electric field distributions at 2.5 s, 5 s, 7.5 s and 10 s, respectively. At these time points, the position of pattern A is exactly the same as the four positions of A during static heating.

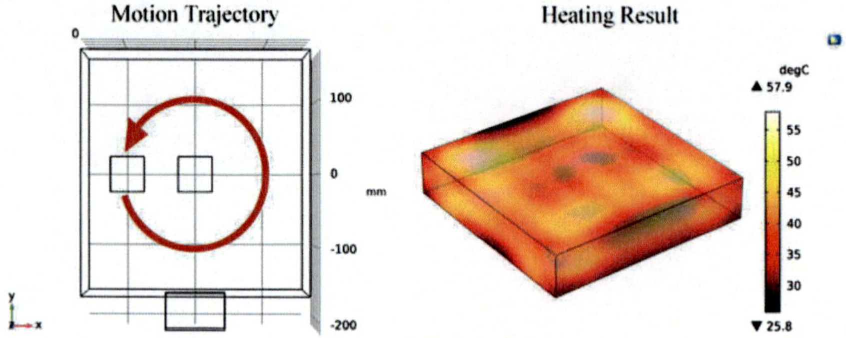

Fig. 5 Motion track and the result of dual-sample dynamic heating

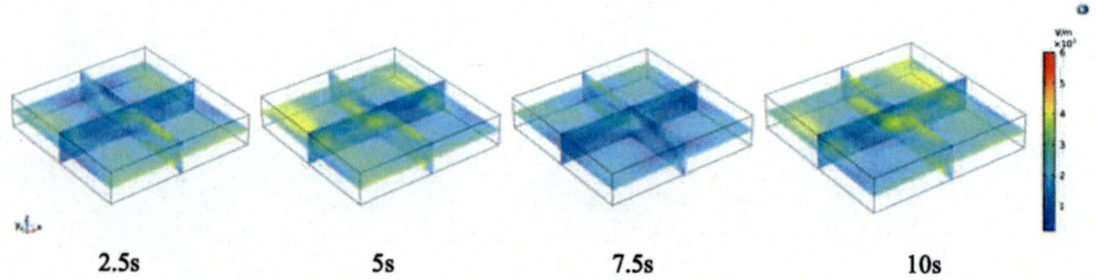

Fig. 6 Electric field distributions of dual-sample dynamic heating

3.2 *Results Analysis*

In order to quantify the heating effect, we use the coefficient of variation (COV) to measure the uniformity, and COV can be expressed by the Formula 5 [6]; we use the integral increase in body temperature of the sample ΔQ to measure the heating efficiency, and ΔQ can be expressed by the Formula 6.

$$\mathrm{COV} = (T_a - T_0)^{-1}\sqrt{n^{-1}\sum(T_i - T_a)^2} \tag{5}$$

$$\Delta Q = Q_t - Q_0 = \iiint\limits_V (t - t_0)\mathrm{d}V \tag{6}$$

where T_a is the average temperature of the sample after heating, T_0 is the initial average temperature, and T_i is the temperature of each sampling point.

When calculating the COV, we sampled nine points on the sample. The positions of these nine points are shown in Fig. 7. Figure 8 shows the COV change of sample B from 0 to 10 s in each model. As can be seen from Fig. 8, the COV of the dual-sample dynamic heating fluctuates greatly during the whole process, but the final moment (10 s) has the best uniformity.

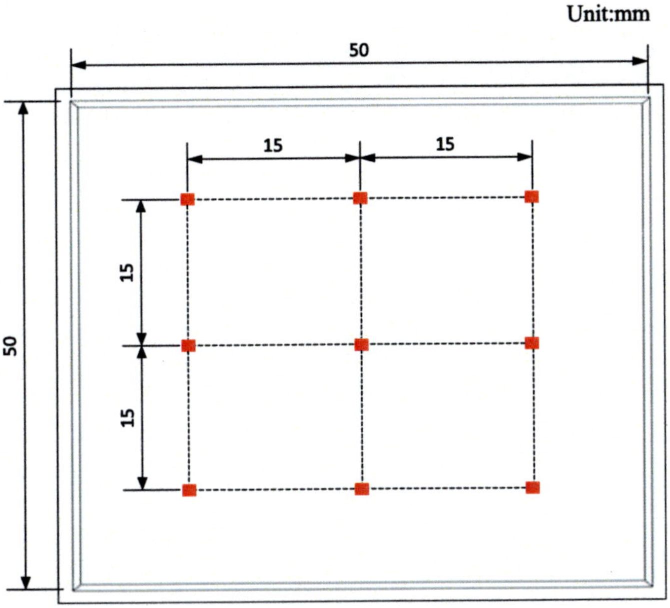

Fig. 7 Position of sample points

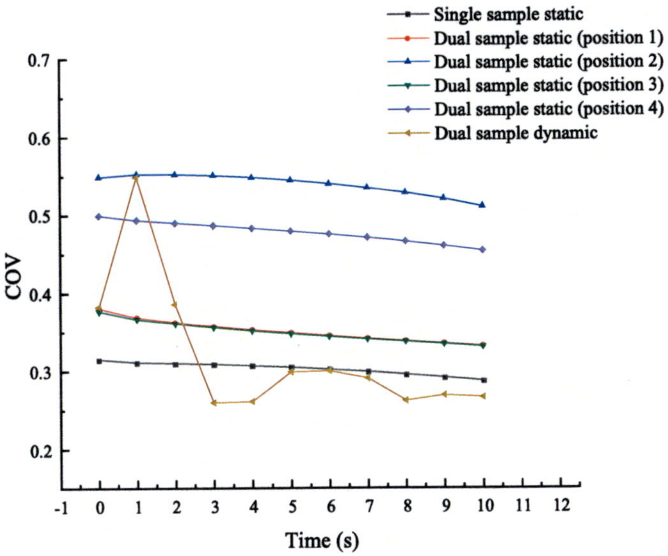

Fig. 8 COV of sample B

Table 1 Heating efficiency of sample B

	Single-sample static	Dual-sample static (position 1)	Dual-sample static (position 2)	Dual-sample static (position 3)	Dual-sample static (position 4)	Dual-sample dynamic
Q_t	0.007802	0.007832	0.007761	0.007831	0.007613	0.007838
Q_0	0.007329	0.007329	0.007329	0.007329	0.007329	0.007329
ΔQ	0.000473	0.000503	0.000432	0.000502	0.000284	0.000509

Unit: $m^3 \cdot K$

It can be seen from Table 1 that the dual-sample dynamic heating has the highest heating efficiency.

4 Conclusion

In order to study the influence of the relative movement of the two samples on the heating uniformity during the microwave heating process, a series of simulation experiments are designed. According to the results and data analysis, the following conclusions can be drawn: When there are two samples need to be heated simultaneously in the cavity and one sample is located at the center of the bottom surface of the cavity, the heating effect (uniformity and heating efficiency) of one sample rotating around the center sample is better than that of both samples in a static state.

Acknowledgements This work is supported by the National Natural Science Foundation of China (No.61863020).

References

1. Yang B, Sun J, Li W et al (2016) Numerical modeling dynamic process of multi-feed microwave heating of industrial solution media. J Cent South Univ 23:3192–3203
2. Ye J, Lan J, Xia Y et al (2019) An approach for simulating the microwave heating process with a slow-rotating sample and a fast-rotating mode stirrer. Int J Heat Mass Transf 140:440–452
3. Ye J, Hong T, Wu Y et al (2017) Model stirrer based on a multi-material turntable for microwave processing materials. Materials 10(2):95–108
4. Meng Q, Lan J, Hong T et al (2018) Effect of the rotating metal patch on microwave heating uniformity. J Microw Power Electromagn Energy 52(2):94–108
5. Zhu H, He J, Hong T et al (2018) A rotary radiation structure for microwave heating uniformity improvement. Appl Thermal Eng 141:648–658
6. He J, Yang Y, Zhu H et al (2020) Microwave heating based on two rotary waveguides to improve efficiency and uniformity by gradient descent method. Appl Thermal Eng 178:115594
7. Yvan L, Daichi K, Mika F et al (2020) Power absorption analysis of two-component materials during microwave thawing and heating: experimental and computer simulation. Innov Food Sci Emerging Technol 66:102479

Temperature Optimization of Thermal Runaway in Microwave Heating Process Based on Sliding Mode Control

Biao Yang, Zhuo Deng, Zhibang Liu, Qihai Mu, and Na Zhu

Abstract Microwave heating is a clean and efficient heating modes, and it heats the materials from inside to outside, so that the temperature of materials increases rapidly, greatly improves heating efficiency and shortens heating time. Although microwave heating has more advantages in heating materials, there are still some problems in the process. And the major challenges are thermal runaway and the non-uniformity of heating. To control the temperature of the heated materials, a reaching law sliding mode control method is proposed in this paper. First, based on the analysis of energy change in heating process and physical properties of materials, the thermodynamic model is established, and the state space expression is derived from the above model. Then, according to the principle of sliding mode control, the appropriate sliding mode and control law are selected to establish the power reaching law to control the material temperature change. Finally, the effectiveness of the proposed sliding mode control strategy is verified by simulations.

Keywords Microwave heating · Thermodynamic model · Sliding mode control · Power reaching law

1 Introduction

Microwave is a green and efficient energy source, and it has the advantage of fast heating, shorter heating time, higher-energy utilization efficiency and ease of operation [1]. Although microwave heating has been increasingly applied in industrial production and daily life, there are still some challenges limiting its

B. Yang (✉) · Z. Deng · Z. Liu · Q. Mu · N. Zhu
Faculty of Information Engineering and Automation, Kunming University of Science and Technology, Kunming 650500, China
e-mail: ybiaocn@163.com

B. Yang
The Key Laboratory of Unconventional Metallurgy, Ministry of Education, Kunming University of Science and Technology, Kunming 650500, China

Z. Yu et al. (eds.), *Advancements in Mechatronics and Intelligent Robotics*,
Advances in Intelligent Systems and Computing 1220,
https://doi.org/10.1007/978-981-16-1843-7_66

development. The most common problem is non-uniform heating in the microwave heating process. In order to control the temperature in the microwave heating process, the state space expression is established in conjunction with the principle of microwave heating, and then the appropriate sliding mode and control law are selected to design the sliding mode controller. The simulation result shows that the reaching law sliding mode controller designed in this study can improve the reaction speed of the system, make the material temperature change along the expected temperature curve and ultimately achieve temperature control.

2 Thermodynamic Model

2.1 Analysis of Energy

Microwave heating process involves the coupling of electromagnetic field and temperature field. In this process, the electromagnetic energy is absorbed by the materials and converted to thermal energy, leading to fast heating materials. According to the energy conservation law and characteristics of material temperature rise, the heat conduction equation can be expressed as follows

$$\rho_m C_p(\partial T/\partial t) = \nabla \cdot (k_t \nabla T) + Q \tag{1}$$

where ρ_m, C_p and k_t are the density, specific heat capacity and thermal conductivity factor of the material, respectively, T is the temperature of heated medium, and Q is the power dissipated during microwave heating and can be expressed as follows:

$$Q = 2\pi f \varepsilon'' |E_{\text{erm}}|^2 + \sigma |E_{\text{erm}}|^2 \tag{2}$$

where f is the frequency, ε'' and σ are the dielectric loss and conductivity of the heated material, respectively, and $|E_{\text{erm}}|$ is denoted as E for simplifying the calculation [2].

Considering that the loss of microwave power is mostly absorbed by the material and converted into heat and assuming that the volume of the material is V, Eq. (1) can be written as follows:

$$\rho_m V C_p(\partial T/\partial t) = V\nabla \cdot (k_t \nabla T) + P \tag{3}$$

where P is the power absorbed by the materials in the heating process.

Considering the heat transfer between the material and the ambient environment, the boundary condition can be written as follows::

$$-k_t \nabla T = h(T - T_a) \tag{4}$$

where h is the heat transfer coefficient; T_a = 22.8 °C is the ambient temperature.

2.2 The Model of Microwave Heating Process

According to the existing literatures, thermal runaway mainly occurs on the surface of the heated material. Thus, the temperature of material surface is selected to study the temperature distribution, and the material is placed on a platform that does not react with microwave. The geometric model of system is shown in Fig. 1.

To simplify the analysis, the following assumptions were made for the study [3, 4]:

Assumption 1: The heated medium is homogeneous and isotropic.
Assumption 2: Only the temperature distribution in the z-axial is considered.
Assumption 3: The volume change and phase change of the material during the heating process are not considered.
Assumption 4: The lower surfaces of the medium are perfectly insulated.

Thus, the initial and boundary conditions are expressed as follows:

$$\begin{cases} T(z,0) = T_{\text{ini}} & t = 0 \\ -k_t(T(z,t))\frac{\partial T}{\partial z} = 0 & z = R_z \\ h(T_a - T(z,t)) = -k_t(T(z,t))\frac{\partial T}{\partial z} & z = 0 \end{cases} \tag{5}$$

where R_z is the height of the material on z-axial, namely the thickness of medium.

The one-dimensional temperature distribution of the material is taken into account in this study, and Eq. (3) can be written as follows:

$$\rho_m(T)VC_p(T)\frac{\partial T}{\partial t} = V\frac{\partial k_t(T)}{\partial z}\frac{\partial T}{\partial z} + Vk_t(T)\frac{\partial^2 T}{\partial z^2} + P \tag{6}$$

From the above analysis, the finite volume method is adopted to analyze the temperature changes of the upper surface, middle part and lower surface of the heated medium. And the material (cube with side length l) is divided into n $(0 \leq n \leq l)$ parts with the same volume, in which the upper surface is denoted as the first layer, the middle part is marked as the ith layer (i = 2, 3, …, $L - 1$), and the

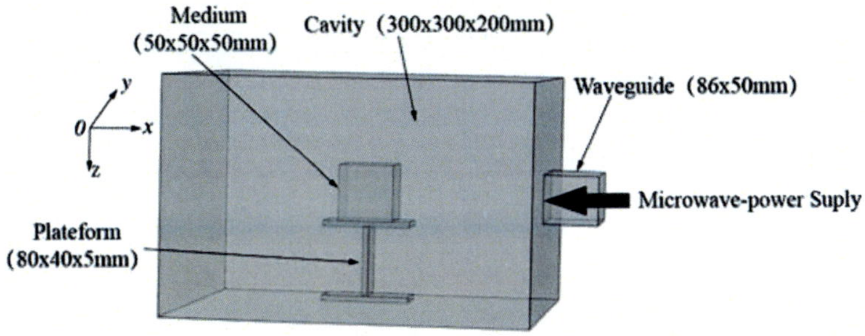

Fig. 1 Geometric model of microwave heating system

lower surface is represented as the l-layer. According to Eq. (6), the heat conduction equations of the upper surface, the middle part and the lower surface can be expressed as follows:

$$\begin{cases} \rho_m(T(1,t))VC_p(T(1,t))\dfrac{\partial T(1,t)}{\partial t} = h(T_e - T(1,t))S + k_t(T(1,t))\left(\dfrac{T(2,t)-T(1,t)}{\Delta z}\right)S + P(1,t) \\ \rho_m(T(i,t))VC_p(T(i,t))\dfrac{\partial T(i,t)}{\partial t} = k_t(T(i,t))\left(\dfrac{T(i-1,t)-T(i,t)}{\Delta z}\right)S + k_t(T(i,t)) \\ \left(\dfrac{T(i+1,t)-T(i,t)}{\Delta z}\right)S + P(i,t) \\ \rho_m(T(n,t))VC_p(T(n,t))\dfrac{\partial T(n,t)}{\partial t} = k_t(T(n,t))\left(\dfrac{T(n-1,t)-T(n,t)}{\Delta z}\right)S + P(n,t) \end{cases} \tag{7}$$

where $T(i,t)$ and $P(i,t)$ represent the temperature and microwave power absorbed by each layer of the material at time t, respectively. S and Δz denote the cross-sectional area and height of each layer, respectively, and the volume of the cube can be expressed as follows $V = n \cdot S\Delta z$.

3 Sliding Mode Control of Microwave Heating Process

3.1 The Design of Sliding Mode Controller

According to the thermodynamic model of microwave heating, with the temperature $T(i,t)$ of each layer as the state variable, the heating process can be described by the following state space expressions:

$$\dot{T}(t) = A + Bu(t), y(t) = cx(t) \tag{8}$$

where $T(t) = [T(1,t), T(2,t), \ldots, T(n,t)]$ is the state vector. $u(t)$ and $y(t)$ are the input vector and output, respectively. A, B and C represent the connection of internal state.

Since thermal runaway is manifested as local hot spots and often occurs on the surface of medium, the temperature change on the upper surface is selected for analysis in this paper. Combined with Eq. (8), A and B can be expressed as follows [3]:

$$A = \frac{k_t(T(1,t))}{\rho_m(T(1,t))C_p(T(1,t))(\Delta z)^2} \cdot \left(-1 - \frac{h\Delta z}{k_t(T(1,t))}\right) \tag{9}$$

$$B = d(1,t)/\left[\rho_m(T(1,t)) \cdot C_p(T(1,t)) \cdot V\right] \tag{10}$$

where $d(1,t)$ is the microwave power coefficient.

And the actual surface temperature and the expected value of upper surface are denoted as Tu and Tm, respectively. Thus, the expected and actual temperature rise curves can be obtained, as shown in Fig. 4. Moreover, the tracking error of temperature is denoted as $e = T_m - T_u$. Thus, the sliding mode function can be designed as follows:

$$s = ce + \dot{e} = c(T_m - T_u) + (\dot{T}_m - \dot{T}_u) \quad c > 0 \tag{11}$$

From the above analysis, an improved exponential reaching law for sliding mode control is proposed in this paper [5]. And the control law is obtained as follows:

$$\dot{s} = -k|s|^{\alpha}\mathrm{sgn}(s) - \varepsilon|T(t)|^{\beta}s \tag{12}$$

where k, ε, α, β are the parameters, and $k > 0$, $0<\alpha<1$, $\varepsilon > 0$, $\beta \geq 1$.

Thus, the control law is obtained, in conjunction with Eqs. (11) and (12), as follows:

$$u(t) = \frac{1}{CB}\left[k|s|^{\alpha}\mathrm{sgn}(s) + \varepsilon|T(t)|^{\beta}s + C\dot{T}_m - CAx(t) + \ddot{T}_m - \ddot{T}_u\right] \tag{13}$$

3.2 *Simulation and Analysis*

Combining Fig. 1 with Eq. (7), SiC ceramic is heated in the microwave cavity, and the constant microwave power is 200 kW. The simulation result and overall temperature rise characteristic curve at 340 s can be obtained as shown in Figs. 2 and 3.

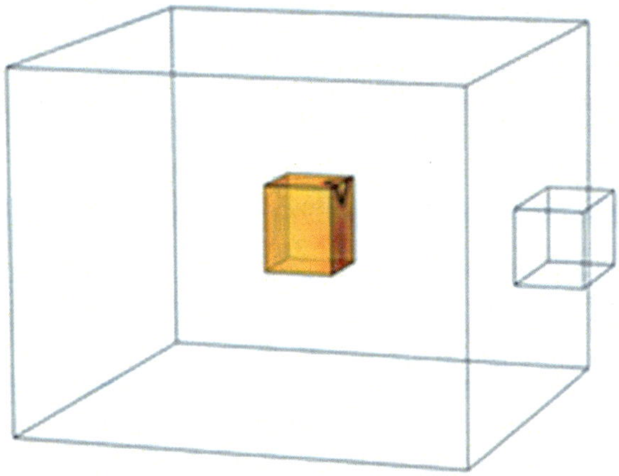

Fig. 2 Temperature distribution of ceramic surface in microwave field

Fig. 3 Temperature rise curve of simulation experiment

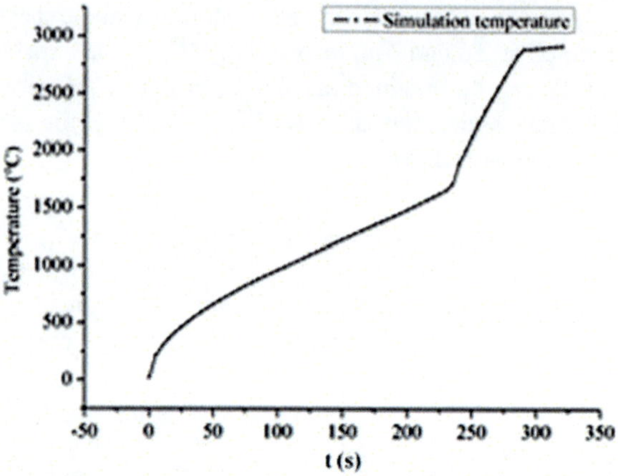

It can be seen that the temperature distribution of the material is not uniform, the temperature rises rapidly when heated to 235 s in some positions and heating to 295 s, and there is a maximum temperature, that is, $T_{\max} = 2900$ °C.

From the analysis of heating process, a piecewise function is developed to represent the expected temperature of the SiC ceramic as Eq. (14), and the expected temperature rise characteristic curve is shown in Fig. 4.

$$T_m = \begin{cases} ve^{-(t/l)} + w & 0 \le t \le 265 \\ T_{\max} & t \ge 265 \end{cases} \tag{14}$$

where $w = 3162$, $v = -3089$ and $l = 108$ in this paper.

The SMC parameters are selected as $A = -7 \times 10^{-3}$, $B = 2 \times 10^{-10}$, $C = 150$, the simulation 1 and simulation 2 are operated with the parameters $k_1 = 100$,

Fig. 4 Expected and simulated temperature rise curves

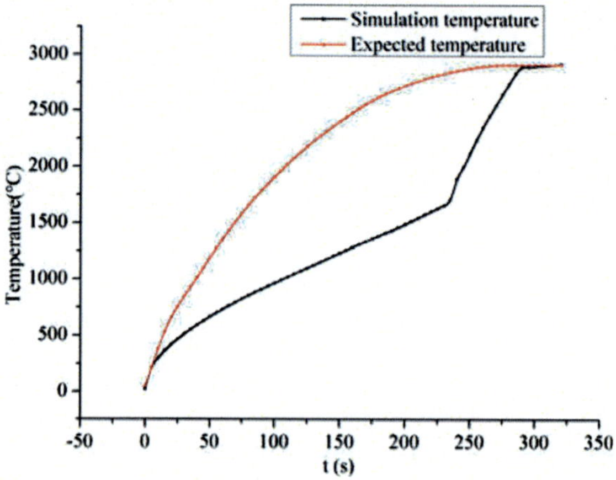

$\alpha_1 = 0.8$, $\beta_1 = 25$ and $k_2 = 10$, $\alpha_2 = 0.5$, $\beta_2 = 2$, respectively. The results are shown in Figs. 5 and 6.

From Figs. 5 and 6, it can be seen that under the effect of the reaching law, the material temperature can increase rapidly, and there is no sudden change occuring in the process. In addition, when relatively large ε and relatively small k are selected, the temperature can converge to the expected temperature rise curve faster, which conforms to the design requirements of reaching law and can realize rapid heating.

Fig. 5 Temperature of simulation 1

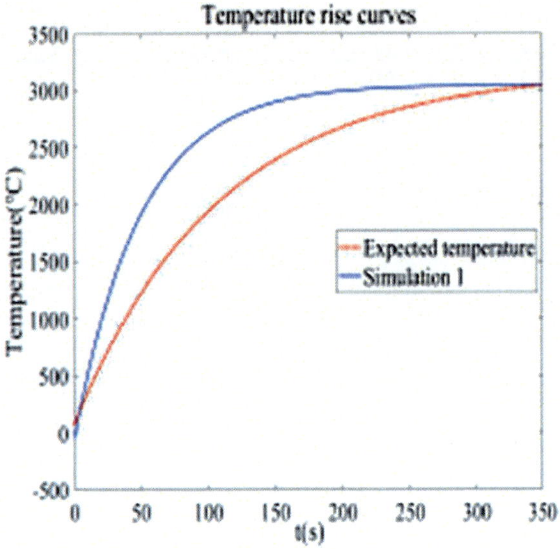

Fig. 6 Temperature of simulation 2

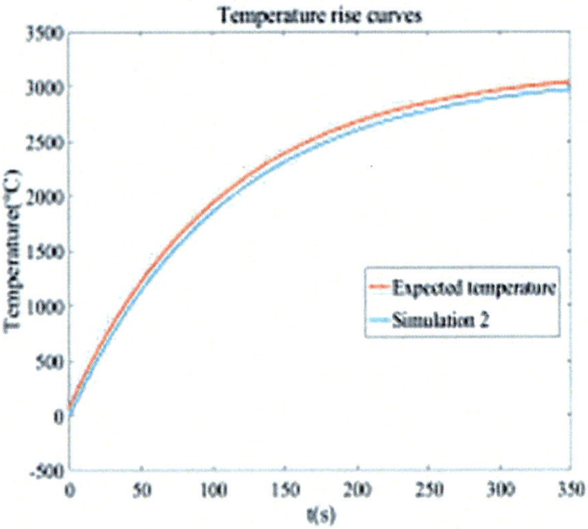

4 Conclusion

In this paper, a modified reaching law for sliding mode control is proposed to improve the temperature rise in the microwave heating process. Through the energy analysis of the process, a thermodynamic model and state space expressions are constructed, and the accuracy of the model is verified by experiment. And then a sliding mode function and reaching law are designed to control the rising trend of temperature in accordance with the temperature rise curve. The results show that the control strategy can change the trend of temperature and accelerate the reaction speed.

Acknowledgements This work is supported by the National Natural Science Foundation of China (No.61863020).

References

1. Guo Q, Sun D, Cheng J (2017) Microwave processing techniques and their recent applications in the food industry. Trends Food Sci Technol 67:236–247
2. Saltiel C, Datta A (1999) Heat and mass transfer in microwave processing. Chem Eng Prog 33(6):1–94
3. Akkari E, Chevallier S, Boillereaux L (2006) Observer-based monitoring of thermal runaway in microwaves food defrosting. J Process Control 16(9):993–1001
4. Yuan Y, Zhong J, Wang Z (2016) Thermodynamics model based temperature tracking control in microwave heating. J Therm Sci Technol 11(1):15
5. Devika K, Thomas S (2018) Sliding mode controller design for MIMO nonlinear systems: a novel power rate reaching law approach for improved performance. J. Frankl Inst. 355(12):17

Visual Aesthetic Arrangement of Chinese Characters

Jiayu Wang and Lin Shi

Abstract The neat arrangement of Chinese characters is an important factor which affected the visual aesthetics of Chinese characters. Chinese characters have many kinds of glyph centers: the center of convex hull (CCH), the center of boundary rectangle (CBR), the center of circumscribed circle (CCC), the physical gravity center (PGC), and the visual center of calligraphy (VCC). Which center is the best reference point of aesthetic arrangement of Chinese characters? Here, we used the font of stone inscription sets handwritten by Zhenqing YAN in the Tang Dynasty (the Yan font) as a dataset in a visual psychophysics experiment to study how different centers could affect visual aesthetic arrangement of Chinese characters. Results showed that (1) various centers showed different statistical distribution in the arrangement of Chinese characters; (2) the variation of centers in horizontal arrangement was smaller than that of centers in vertical arrangement; (3) the center of the boundary rectangle had the smallest variation range; (4) there was a significantly different arrangement between naive subjects in calligraphy and non-naive subjects. These results suggested that the center of the boundary rectangle of the Chinese characters was the best reference point for the visual aesthetic arrangement of Chinese characters.

Keywords Center of boundary rectangle · Center of physical gravity · Center of circumscribed circle · Center of convex hull · Visual center of calligraphy · Visual aesthetics · Chinese characters · Horizontal arrangement · Vertical arrangement

---

J. Wang · L. Shi
School of Information Engineering and Automation, Kunming University of Science and Technology, Kunming 650500, China

L. Shi (✉)
Computer Application Key Laboratory of Yunnan Province, Kunming University of Science and Technology, Kunming 650500, China
e-mail: lin.shi@kust.edu.cn

Z. Yu et al. (eds.), *Advancements in Mechatronics and Intelligent Robotics*,
Advances in Intelligent Systems and Computing 1220,
https://doi.org/10.1007/978-981-16-1843-7_67

1 Introduction

The unique visual characteristics and aesthetic value of Chinese characters occupy an increasingly important position in practice. The commonly used arrangement of Chinese characters has changed from vertical to horizontal. Horizontal and vertical are the issues of great concern in the study of reading psychology in modern times.

Santayana [1] believed that the formal beauty of calligraphy is expressed in the lines of Chinese characters, and Chinese characters are injected with collective subjective will and aesthetic emotions. Cao and Qiu [2] pointed out that the center of gravity was the supporting center point of a word, and the position of the center of gravity was uniform, giving people a neat, stable and aesthetic feeling. Although Chinese characters are square characters, their glyphs are different and there are certain rules [3]. Sun et al. [4] proposed to use the convex hull and rectangle of Chinese characters and the center of gravity of Chinese characters to describe the characteristics of Chinese character alignment and stability. Kai et al. [5] expressed the center of gravity of Chinese characters by calculating the average of the coordinates of black pixels in the stroke area of Chinese characters. Deng et al. [6] proposed a method to calculate the visual center of gravity of Chinese characters by marking the visual center of gravity of sample Chinese character images. Wu [7] explained the application of the center of gravity in the design of printed fonts and the determination of the geometric center of gravity and the visual center of gravity, and the center of glyph. Jia et al. [8] introduced the concepts of centroid and moment of inertia in the field of mechanics and used two parameters, the centroid of font images and the moment of inertia of font structure, to quantitatively describe the concepts of centroid and middle palace in font structure.

Chu [9] pointed out that Chinese characters pay much attention to the visual space formed by their arrangement and combination structure in the specific application. Ai [10] pointed out that the efficiency of horizontal reading was greater than that of vertical reading, which provided a scientific basis for writing Chinese characters mainly horizontally. Wang [11] elaborated on the alignment of the center of gravity in the vertical era and the important role of the center of gravity in the transformation of Chinese character typesetting from vertical to horizontal. Shi and Zhang [12] pointed out that the vertical and horizontal arrangements of English, Chinese and other languages showed different reading characteristics in many aspects.

The alignment of the center of gravity of Chinese characters is particularly important for the arrangement of reading and printing. This paper studied the influence of glyph centers of Chinese characters of the Yan font, the Song and the Kai on the visual aesthetic of Chinese character arrangement.

2 Method

2.1 Chinese Font and Glyph Centers

Image moment is a very useful algorithm in image processing. It can be used to calculate the geometric characteristics such as the centroid, radius, area and direction of the regional image, so it is also called geometric moment [13].

Two common computer fonts were selected: the Song and the Kai, and the Yan font. Number of glyphs in each font: 1635. The selection of candidate Chinese characters was based on the Yan font, and the corresponding Chinese characters in the Song and the Kai are selected—first, generated and collected the glyph images corresponding to the Chinese characters as the experimental objects and secondly, performed image-processing operations such as denoising, smoothing, graying, and edge detection on the image of Chinese characters. Treated Chinese characters as plane figures of different shapes, extract the entire outline of Chinese characters, then collected and calculated the positions of different types of the glyph centers of Chinese characters. The center of the boundary rectangle, the center of the convex hull, and the center of circumscribed circle, respectively, refer to the center point of the boundary rectangle, the geometric center point of the convex hull polygon, and the center point of the circumscribed circle that constitute the boundary points of all strokes of the Chinese character glyph; the physical gravity center refers to the center of gravity calculated by taking the strokes of the Chinese character as a uniform physical rigid body.

Here only the visual center of calligraphy of the Yan font of Chinese characters was measured. The visual center of calligraphy refers to the center obtained through visual psychophysics experiments and subjectively marked by non-naive subjects in calligraphy. Initially, Chinese characters were displayed on the gray background of the computer screen. When the subjects pressed the left mouse button and dragged, they would draw the Mi-zige (a square with double vertical and diagonal lines) with the center of the cursor as the center point of the Mi-zige, as a reference when marking the visual center of calligraphy of Chinese characters. The strokes and structure of the Mi-zige are very consistent with the writing style of Chinese characters. It can be used to check whether each character is symmetrical and the center of gravity is stable. Participants moved the cursor to the position of the visual center of calligraphy of the Chinese character. At this time, the center point of the Mi-zige coincides with the calligraphy visual center. Record the position of the calligraphy visual center.

2.2 The Process and Setting of Chinese Character Arrangement Experiment

Participants: the Yan font (naive subjects), the Yan font (non-naive subjects), the Song and the Kai, 10 people each, male-to-female ratio 1:1, age 20–26 years old.

1. Arrangement, {horizontal, vertical}.
2. Glyph center type, {CCH, CBR, CCC, PGC, VCC}.
3. Font type {the Yan font, the Song, the Kai}.
4. Screen settings: width: 1080 px, height: 1080 px, background: gray square.

Write experimental procedures and invite participants to participate in the experiment. The experiment required ten naive subjects in calligraphy and ten non-naive subjects to use the mouse to drag and drop the given Chinese characters on the display screen to make the visual effect the most aesthetic. In addition to the Yan font, we also conducted experiments on common computer fonts: the Song and Kai.

The experiment included two types of arrangement: horizontal and vertical. Every five Chinese characters are a group. Chinese characters were initially arranged at random positions near the center vertical or horizontal line on a gray square background. Participants were required to use the mouse to aesthetically arrange Chinese characters in the horizontal or vertical direction according to their personal subjective experience, to achieve the most aesthetic effect, and to record the position, serial number, arrangement, and operating time of Chinese characters.

After the experiment, firstly, according to the positions of the Chinese characters recorded in the experiment, combined with the positions of the glyph centers of different types of Chinese characters in the Yan font, the Song, and the Kai to calculate the center position, the mean of center position, and the variation of center position of the Chinese characters in the horizontal and vertical arrangement experiment. The same applies when the Chinese characters are arranged vertically. Secondly, when calculating the vertical and horizontal arrangements, the distance between the glyph centers of Chinese characters and the corresponding centers is divided by the height or width of the corresponding Chinese character boundary rectangle to standardize. Finally, analyze the variation in the center of the Yan font, the Song, and the Kai and the operation time of the horizontal and vertical arrangements of Chinese characters. The t-Location-Scale Distribution was used to fit the distance distribution between the center of the Chinese character glyph and the corresponding mean. And returned the maximum likelihood estimated value of the shape parameter k, the scale parameter and the position parameter mu, the confidence interval with the confidence level of 95%.

3 Results

3.1 *The Distance Between the Center and the Corresponding Mean Value of the Chinese Characters Arranged Horizontally and Vertically*

Comparing the horizontal and vertical arrangements of the same glyph center of Chinese characters, found, whether it was for all subjects, naive subjects or

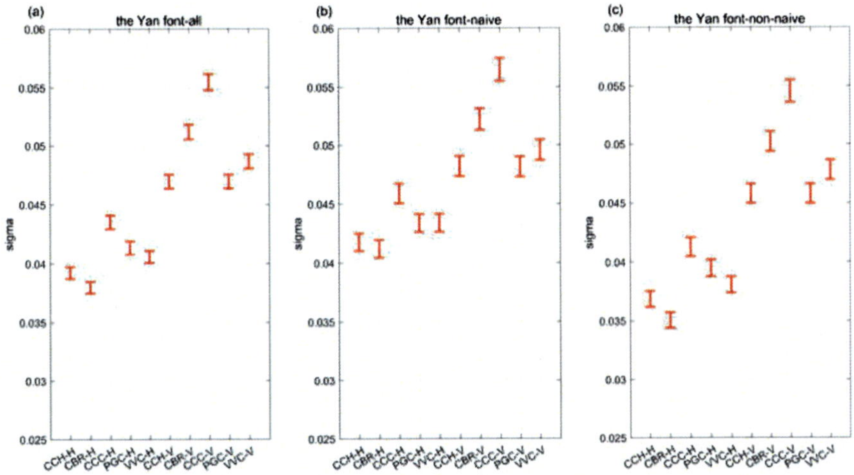

Fig. 1 Confidence intervals of the Yan font (all subjects, naive subjects, non-naive subjects). The confidence interval of the t-Location-Scale Distribution fitting parameter sigma of the distance between the center of the Chinese character glyph and the corresponding mean. The abscissa is different glyph centers in of horizontal and vertical arrangement

non-naive subjects, the confidence intervals of the distance distribution parameter sigma of the various centers of the Chinese characters of the Yan font did not cross each other. Comparing the Yan font (naive subjects) and font (non-naive subjects) found that for the same glyph center, When Chinese characters were arranged horizontally and vertically, the confidence intervals of the distance distribution parameter sigma between various centers and the corresponding mean did not cross each other (Fig. 1) while cases CCC_H v.s. CCC_V and PGC_H v.s. PGC_V were exception (Fig. 2).

It could be shown that the variation of centers of Chinese characters in horizontal arrangement was significantly smaller than that of centers in vertical arrangement, the center of the boundary rectangle (including the center of the convex hull in some cases) had the smallest variation range, which was consistent with the structural characteristics of square Chinese characters. Moreover, there was a significant difference in the horizontal and vertical arrangements of Chinese characters between non-naive subjects and naive subjects, and non-naive subjects had a smaller variation range in various centers when Chinese characters were arranged horizontally and vertically.

When Chinese characters were arranged horizontally and vertically, for the same glyph center of the same font, the confidence interval of the distance distribution parameter sigma of the boundary rectangle center, the convex hull center, the physical gravity center and the corresponding mean value of the Chinese characters of the Song do not cross each other. The confidence intervals of the distance distribution parameter sigma of the boundary rectangle center, the convex hull center, the circumscribed circle center and the corresponding mean value of the

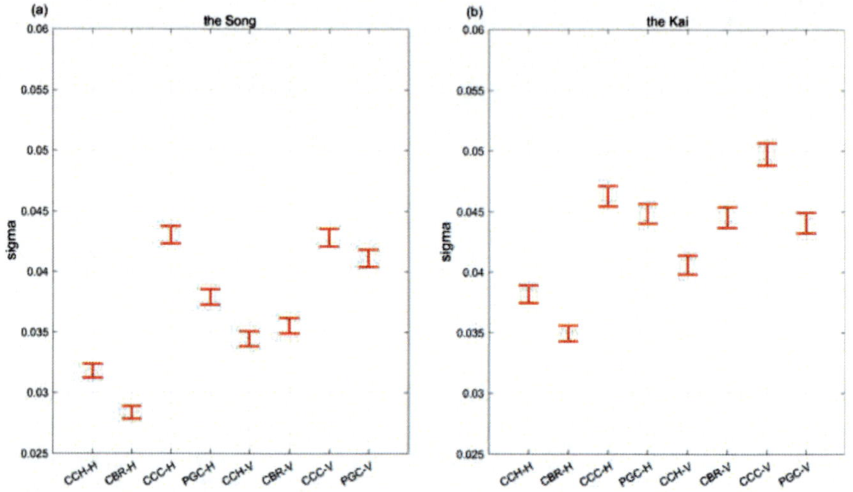

Fig. 2 Confidence intervals of the Song and Kai fonts. The confidence interval of the t-Location-Scale Distribution fitting parameter sigma of the distance between the center of the Chinese character glyph and the corresponding mean. The abscissa is different glyph centers in horizontal and vertical arrangement

Chinese characters of the Kai did not cross each other. The distribution laws of the various types of glyph centers of the Song and the Kai in the arrangement of Chinese characters were consistent with the above-mentioned the Yan font.

In addition, the Generalized Extreme Value Distribution (GEVD) was used to fit the operation time distribution of the horizontal and vertical arrangements of Chinese characters. And returned maximum likelihood estimated of the shape parameter k, the scale parameter sigma and the position parameter mu, the confidence interval with a confidence level of 95%.

3.2 Operation Time of Chinese Characters Arranged Horizontally and Vertically

The operation time of the horizontal and vertical arrangements of Chinese characters was based on the entire experimental process. In order to further explore the difference between the operation time of the horizontal and vertical arrangements of Chinese characters, a rank sum test was carried out on the operation time of the same group of Chinese characters in each font. The results found that for the Yan font (naive subjects), the Song and Kai, the p-values were all less than 0.05. This showed that there was a significant difference in the operation time distribution of the horizontal and vertical arrangements of Chinese characters in the Yan font (naive subjects), the Song and Kai (Fig. 3).

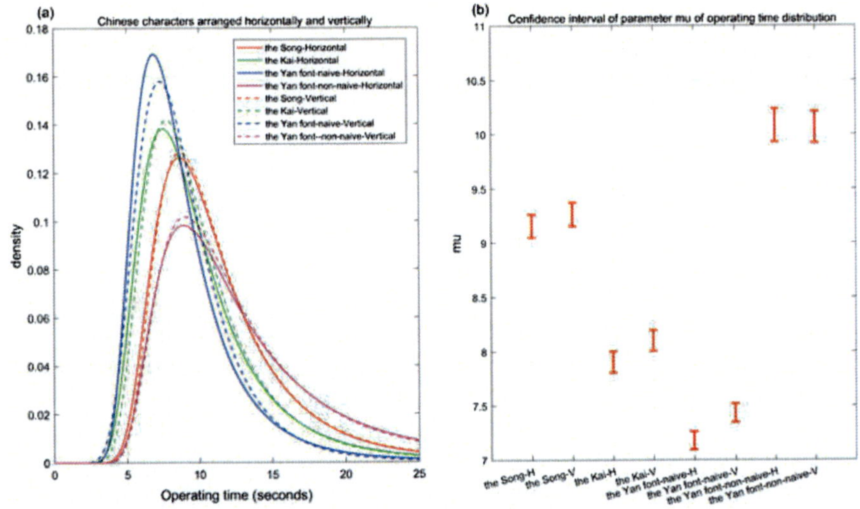

Fig. 3 Distribution functions and confidence intervals. The generalized extreme value distribution fitting of the operation time of the horizontal and vertical arrangement of Chinese characters and the confidence interval of the parameter mu. The abscissa of sub-figure (**b**) is the operation time of Chinese characters in both horizontal and vertical arrangements

4 Discussions

In other words, the center of the boundary rectangle of the Chinese characters was the best reference point for the visual aesthetic arrangement of Chinese characters. And compared to the vertical arrangement, the horizontal arrangement is more conducive to the neat arrangement of Chinese characters. This may be related to the fact that Chinese characters themselves are square characters, and the boundary rectangle and convex hull of Chinese character shapes are consistent with the square, oblate, rectangular, and polygonal structural features of the basic form of the font, and the circumscribed circle is not one of the basic forms of the font. Moreover, the physical gravity center is based on the global operation of the pixel coordinates of the Chinese character image, without in-depth consideration of the influence of the Chinese character structure.

The variation of centers and the operation time of Chinese characters was significantly smaller in horizontal arrangement than that of vertical arrangement. This may be related to people's reading and writing habits or the physiological structure of the left and right horizontal distribution of human eyes, when reading horizontal text, the eyes have a mutual compensation effect, and the reading movement burden of the eyeball is smaller than that of vertical text. In addition, for non-naive subjects, the variation range in various centers of Chinese characters in horizontal or vertical arrangement was smaller, and there was no significant difference in the operation time of Chinese characters in horizontal and vertical arrangements, and this may be because they were more exposed to the vertical arrangement of Chinese characters and were more sensitive to the visual aesthetic of Chinese characters than the naive subjects.

5 Conclusion

We have studied the influence of different types of glyph centers of the Yan font, the Song, and the Kai on the visual aesthetics of Chinese characters neatly arranged horizontally and vertically. Results showed that (1) various centers showed different statistical distribution in the arrangement of Chinese characters; (2) the variation of centers in horizontal arrangement was smaller than that of centers in vertical arrangement; (3) the center of the boundary rectangle had the smallest variation range; (4) there was a significantly different arrangement between naive subjects in calligraphy and non-naive subjects. These results suggested that the center of the boundary rectangle of the Chinese characters was the best reference point for the visual aesthetic arrangement of Chinese characters.

References

1. Santayana G (1955) The sense of beauty. Dover Publications (1955)
2. Cao Z (1994) Handbook of practical typefaces. Printing Industry Press, Beijing
3. Xu X (2013) Exploring the legibility factors of printed fonts. Print Mag 07:53–55
4. Sun R, Lian Z, Tang Y, et al (2015) Aesthetic visual quality evaluation of Chinese handwritings. IJCAI
5. Kai W, Yang YY, Suen CY (1988) Multi-layer projections for the classification of similar Chinese characters. In: International conference on pattern recognition, IEEE
6. Deng X, Li B, Zhang J et al (2015) Visual gravity calculation of Chinese characters based on statistics. J Chinese Inf Process 29(04):159–165
7. Wu Z (2018) The application of gravity in the design of printed fonts. Printing Mag 08:49–54
8. Jia R, Zhao L, Zhang H, et al (2019) A font structure evaluation method based on mechanics. Mech Eng 41(03):337–340 + 319
9. Chu X (2008) On the linear features of Chinese character formation art. J Liaocheng Univ (Social Sciences Edition) 02:363–364
10. Ai W (2013) The problem of Chinese characters. The Commercial Press
11. Wang J (2015) Alignment of the center of gravity-the modernization of Chinese character design. Decoration 12:80–81
12. Shi L, Zhang Y (2016) Vertical and horizontal arrangements of Chinese characters. J Vis Assoc Res Vis Ophthalmol 16(12):1423
13. Hu MK (1962) Visual pattern recognition by moment invariants. Inf Theory IRE Trans 8(2): 179–187

Visual Aesthetics of Chinese Kai-Style Characters Tend to Be "Thin" and "Plump"

Lin Shi, Qianqian Qian, and Xi Xia

Abstract Visual aesthetics of Chinese characters is influenced by many factors, e.g., the rate of height and width of boundary rectangle (RHW), the rate of pixel counts of glyphs and pixel counts of the convex hull (RGC), the rate of pixel count of the convex hull and the boundary rectangle (RCB), the center of gravity position, the vertical symmetry axis position, the component gap, the thickness of strokes, the gap of strokes, and the background. Here, we studied the effect of RHW, RGC, and RCB on visual aesthetics of Chinese Kai-style characters. We selected 2100 characters from *Chinese Calligraphy Dictionary Kai Volume*. 10 naïve subjects in calligraphy and 8 non-naive subjects were asked to score and rank the Chinese characters in visual aesthetics. Results showed that (1) two different presentation, white characters on a black background and black characters on a white background, had significantly different effect on visual aesthetics of Chinese characters; (2) the aesthetic score tended to increase with the increase of RHW and RGC and to decrease with the increase of RCB. Results suggested that the visual aesthetics of Chinese Kai-style characters tended to be "thin" and "plump".

Keywords Visual aesthetics · Chinese Kai-style characters · Thin · Plump

1 Introduction

Different colors of Chinese characters would affect people's subjective visual perception; the rate of height and width of boundary rectangle (RHW) could measure the "thin" degree of the glyph; the rate of pixel counts of glyphs and pixel

L. Shi · Q. Qian · X. Xia
School of Information Engineering and Automation, Kunming University of Science and Technology, Kunming 650500, China

L. Shi (✉)
Computer Application Key Laboratory of Yunnan Province, Kunming University of Science and Technology, Kunming 650500, China
e-mail: lin.shi@kust.edu.cn

Z. Yu et al. (eds.), *Advancements in Mechatronics and Intelligent Robotics*,
Advances in Intelligent Systems and Computing 1220,
https://doi.org/10.1007/978-981-16-1843-7_68

counts of the convex hull (RGC) represented the “plump” degree of the glyph; the rate of pixel count of the convex hull and the boundary rectangle (RCB) could measure the degree of “blank” near the strokes in a glyph. The above metrics played an important role in the process of studying the visual aesthetics of Chinese characters.

The frame structure and stroke structure of Chinese characters would affect the visual aesthetics of Chinese characters. Wei proposed a method to study the aesthetics of the three-dimensional design of Chinese characters in connection with the structure of the Chinese characters [1]. In the structure of Chinese characters, a stable center of gravity was the most basic requirement. The adjustment of the center of gravity of Chinese characters could make the Chinese characters stable [2]. The stable center of gravity of Chinese characters ensured the fluency and comfort of reading [3]. Wang et al. [4] proposed in 2018 that the center of gravity was the weight balance point formed by combining the strokes of Chinese characters according to a certain rule, which could measure the stability of handwritten Chinese characters. Sun et al. [5] used the center of gravity calculation formula to obtain the center of gravity of the handwritten Chinese character in the picture based on the given handwritten Chinese character picture; the aesthetic treatment of Chinese characters also involved visual focus, e.g., Deng et al. [6] used artificially annotated samples of the visual center of gravity of Chinese characters and used statistical analysis methods to summarize the general rules of calculation of the visual center of gravity of Chinese characters in line with the public’s perspective, and pointed out that the visual center of Chinese character shape is in the design and optimization of Chinese character structure. The field had important research significance. In addition to the frame structure, the stroke structure was another factor that affected the appearance of Chinese characters. Cao et al. [7] proposed that the stroke structure and distribution of Chinese characters would also affect the appearance of Chinese characters. In the printed Chinese characters, for the Chinese characters with the left, middle and right structure, the number of strokes of the components was quite different. This kind of distribution was in order to achieve human visual balance and visual aesthetics; For Chinese characters with upper, middle and lower structures, the difference in strokes and relative position ratios were also to achieve better visual effects. In addition, the different colors of Chinese characters would also affect the visual aesthetics. Li used the method of visual perception psychology to study the readability of colored Chinese characters and people’s visual attention to colored Chinese characters [8].

Existing research work mainly described the visual aesthetics of Chinese characters from the objective factors of Chinese characters, while ignoring that the visual aesthetics of Chinese characters was also subject to human vision. In this paper, a large amount of relevant experimental data was obtained by convening naïve and non-naïve subjects to conduct visual aesthetics experiments, and linear regression was performed to analyze the scores of subjects when the three metrics of RHW, RGC and RCB took different values. Trends affecting the visual aesthetics of Chinese characters.

2 Method

2.1 Chinese Character Font Data Set

In this paper, we selected the RHW of the regular Chinese characters in the "Chinese Calligraphy Dictionary Kai Volume" to be about 0.7, 0.8, 0.9, 1, 1.1, 1.2, 1.3, and 300 Chinese characters pictures for each ratio. At the same time, the selected Chinese characters must have the RGC which range was (0.2 0.6), the RCB which range was (0.6, 0.9), and finally 2100 Chinese character pictures were obtained. Next, we merged 2100 black characters on a white background and 2100 white characters on a black background to obtain 4200 Chinese character pictures as our data set.

2.2 Experimental Steps

2.2.1 Three Kinds of Chinese Character Glyph Metrics

This article de-manipulated Chinese character pictures. First of all, the Chinese characters were regarded as a whole, which was convenient to calculate the stroke area contained in the Chinese characters and the convex area of the Chinese character shape, and then calculated the convex boundary rectangle of the Chinese characters. Among them, the convex hull area was the area of an irregular polygon formed by the boundary points of the strokes of the Chinese character, and the convex hull boundary rectangle of the Chinese character was a straight rectangle with the boundary of all black pixels included in the Chinese character. And RHW = H/W, RGC = G/C, and RCB = C/B. Where H was the height of the boundary rectangle, W was the width of the boundary rectangle, G was the pixel counts of glyphs, C was the pixel count of the convex hull, and B was the area of the boundary rectangle.

2.2.2 Aesthetic Score for Chinese Character Sets

After preparing the experimental procedures, the subjects were invited to make a rough score on the Chinese character set. Among the 18 subjects, 10 naïve subjects and 8 non-naive subjects. The age distribution range of the subjects was (20 28). The male-to-female ratio is: 7:11. The experiment required the subjects to press the 1–9 keys on the keyboard to score the Chinese characters according to their preferences on the pictures of Chinese characters that appear randomly. 9 was the most beautiful Chinese character and 1 was the ugly Chinese character. We recorded the number of a single Chinese character, score and operation time, etc.

2.2.3 Sort the Aesthetics of Chinese Character Sets

We took the average value of the subjects' scores on the same Chinese character in the first experiment, arrange the average values from high to low, and rearrange the Chinese character pictures according to the average value, e.g., the subjects gave different scores to "seven" character. Then found the mean. In this paper, the italics with black characters on a white background were transparently processed and after preparing the experimental procedures, the subjects were invited to do visual aesthetic experiments. According to their perception of 10 Chinese characters randomly appearing on the computer screen, the participants dragged the Chinese characters to the scoring position corresponding to their aesthetics according to the different scoring standards of 1–10. Among them, the participants should start with 10 put Chinese characters, 10 means higher score and recorded the Chinese character number, page number, placement position and page operation time, etc.

3 Results

This article first performed a paired T-test on the scores and operating time of the italics fonts of white characters on a black background and black characters on a white background, and then performed linear regression analysis on the scores of Chinese characters. This chapter would analyze the two aspects of the scores of white characters on black background and white characters and black characters, and the linear regression of aesthetic scores and aesthetic rankings.

3.1 White Characters on a Black Background and Black Characters on a White Background

This article performed a paired T-test on the scores of all the subjects under the white characters on a black background and black characters on a white background in *Chinese Calligraphy Dictionary Kai Volume*. The *p*-value of the test was 1.7293e−34. And the *p*-value was less than 0.05, indicating that for all subjects, there was a significant difference between the aesthetic scores of white characters on a black background and black characters on a white background. In addition, we subtracted the difference between the subjects' aesthetic scores in the case of black text on a black background and the subjects' scores in the case of black text on a white background. The difference was less than 0, indicating that the subjects' aesthetic scores in the case of black text on a white background were higher than those with white text on a black background. The results showed that there were significant differences between the two different presentation modes of white characters on a black background and black characters on a white background.

3.2 Linear Regression Analysis of Aesthetic Score

1. As the RHW in sub-figure (a) increased from 0.6 to 1.4, the aesthetic score showed an increasing trend between 5 and 6; As the RGC of strokes in sub-figure (b) increased from 0.1 to 0.7, the aesthetic score showed an increasing trend between 5 and 6.
2. In the process of increasing the RCB in sub-figure (c) from 0.5 to 1, the aesthetic score showed a decreasing trend between 5 and 6.

Table 1: (1) The *p*-values in the fourth column of Table 1 were all less than 0.05, indicating that the aesthetic score tended to increase with the increase of RHW and RGC. At the same time, the *k* values of RHW and RGC are all greater than 0, which fully showed that the aesthetic score tended to increase with the increase of RHW and RGC; In the case of the RCB, the value of k was less than 0, indicating that the aesthetic score tended to decrease as the RCB increases.

(2) This article also performed a linear regression on the aesthetic sense of the naive subjects. Under the condition of RHW, *p*-value was 8.63e−27, *k* value was 0.5254; RGC *p*-value was 7.28e−11, *k* value was 0.6099; under the RCB, the *p*-value was 1.24e−12 and the *k* value was −0.8922, which showed that the aesthetic score tended to increase with the increase of RHW and RGC, and with the decrease as the RCB increases.

(3) This article also performed a linear regression on the aesthetic perception of the non-naive subjects. Under the condition of RHW, *p*-value was 6.62e−43, *k* value was 0.7942; RGC *p*-value was 0.32, *k* value was 0.1098; Under the RCB, *p*-value was 5.91e−21, and the *k* value was −1.3926, which showed that the aesthetic score tended to increase with the increase of RHW and decreases as the RCB increases.

3.3 Aesthetic Ranking Linear Regression Analysis

Since the aesthetic score trend observed from the linear regression graph of the subjects' aesthetic score in 3.2 is not obvious, at the same time, in order to further distinguish the aesthetic differences of Chinese characters with the same score. Next, this article also ranked the Chinese characters by their aesthetic perception, the aesthetic ranking score = the average score of the first experiment * 10 + the ranking of Chinese characters.

Table 1 Linear regression parameter table of all subjects' aesthetic scores

Metrics	R-squared	Adjusted R-squared	*p*-value	*k*	*b*
RHW	0.0151	0.0151	3.92e−127	6.0394	55.9963
RGC	0.00459	0.00456	1.1e−39	6.3518	59.5761
RCB	0.00939	0.00936	1.64e−79	−12.1960	71.0611

Table 2 Linear regression parameter table of all subjects' aesthetic ranking

Metrics	R-squared	Adjusted R-Squared	*p*-value	*k*	*b*
RHW	0.00391	0.00389	2.44e−66	0.6448	5.0723
RGC	0.000388	0.000374	6.15e−08	0.3876	5.5668
RCB	0.00178	0.00177	4.01e−31	−1.1146	6.5419

1. As the RHW in sub-figure (a) gradually increased from 0.6 to 1.4, the aesthetic score showed an increasing trend from 60 to 70. In sub-figure (b), the RGC increased from 0.1 to 1.7. The score showed an increasing trend from 60 to 70.
2. As the RCB in sub-picture (c) increased from 0.5 to 1, the aesthetic score showed an increasing trend from 60 to 70.

Table 2: (1) The *p*-values in the fourth column of Table 2 were all less than 0.05, indicating that the RHW, the RGC, and the RCB had a certain relationship with the aesthetic score. At the same time, the *k* values of the RHW and the RGC were both Greater than 0, which fully indicates that the score of Chinese characters increases with the increase of RHW and RGC; The value of k under the condition of RCB was less than 0, indicating that the score of Chinese characters decreases as the RCB increases.

(2) This article also performed a linear regression on the aesthetic perception of the naïve subjects. Under the condition of RHW, the *p*-value was 1.2e−61, and the *k* value was 5.3368; the RGC was 1.62e−48 and the *k* value was 9.0051; Under the RCB, the *p*-value was 7.72e−25 and the *k* value was −8.5092, which showed that the aesthetic score tended to increase with the increase of RHW and RGC, and decreases as the RCB increases.

(3) This article also performed a linear regression on the aesthetic perception of the non-naive subjects. Under the condition of RHW, *p*-value was 2.58e−69, *k* value was 6.9176; RGC *p*-value was 5.63e−05, *k* value was 3.0351; Under the RCB, the *p*-value was 1.85e−62 and the *k* value was −16.8045, which showed that the aesthetic score tended to increase with the increase of RHW and RGC, and decreases as the RCB increases.

4 Discussions

We selected 2100 Chinese character glyph pictures from the "Chinese Calligraphy Dictionary Kai Volume" and required 10 naïve subjects and 8 non-naive subjects to score and rank the Chinese characters in visual aesthetic. The experimental results showed that there were significant differences in aesthetic scores under the two different presentation modes of white characters on a black background and black characters on a white background, and there are different trends between the RHW, the RGC and the RCB and the aesthetic score. Next, this article discusses the reasons for the above-mentioned significant differences based on calligraphy knowledge and psychological theories.

4.1 The Visual Aesthetics Influence of White Characters on a Black Background and Black Characters on a White Background

In the regular script glyphs, the p-value of white characters on a black background and black characters on a white background is less than 0.05, when the paired T-test is performed, indicating that the two have a statistically significant difference. The black characters on white background score higher, which is consistent with calligraphy knowledge match, primary and middle school students generally choose black characters on a transparent background when practicing copybooks [8]. At the same time, many people prefer black characters on a white background when they appreciate calligraphy.

4.2 The Influence of Metrics on the Visual Aesthetics of Chinese Characters

The aesthetic score tended to increase with the increase of RHW and RGC of all subjects and the naïve subjects, and decreased as the RCB increases, although the p The value is greater than 0.05, its k value is greater than 0, the p-value of the RGC of all subjects is less than 0.05, and the k value is greater than 0, which shows that non-naïve subject experimental data also conforms to the trend of aesthetic score; The aesthetic ranking score also increases as the RHW and RGC of all subjects, naive subjects and non-naive subjects increase, and decreases as the RCB increases; The results of the two experiments consistently show that the aesthetic score tended to increase with the increase of RHW and RGC, and decreases as the RCB increases. The experimental results of subjects with naive and non-naive subjects consistently show that the visual aesthetics of Chinese Kai-style characters tended to be "thin" and "plump".

4.2.1 The Impact of RHW and RGC on the Visual Aesthetics of Chinese Characters

From the sub-graphs (a) of Figs. 1 and 2, we can observe that as the RHW increases from 0.6 to 1.4, although the subjects' aesthetic scores change slowly (the values are all between 5 and 6), However, it still showed an obvious slow increasing trend; Similarly, from the sub-graphs (b) of Figs. 1 and 2, we can also observe that as the

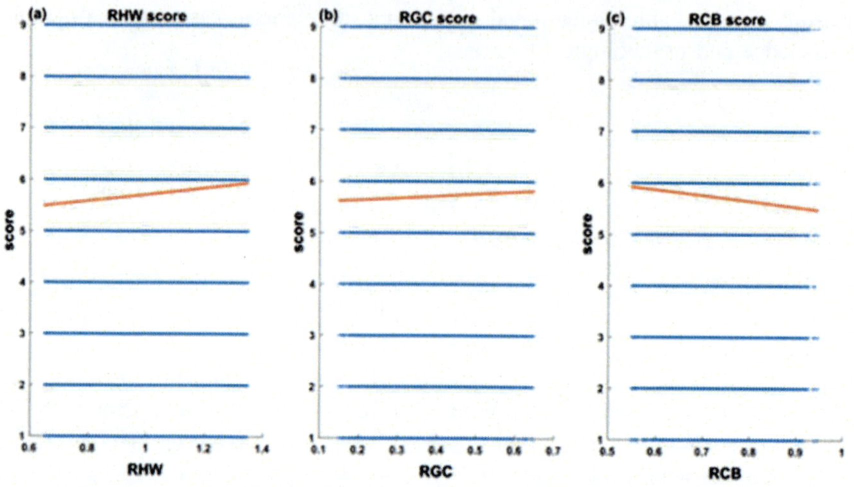

Fig. 1 Linear regression of aesthetic scores. The linear regression graph of aesthetic scores of all subjects. The abscissa is the measurement value of different indicators, and the ordinate is the aesthetic score of the subjects under the corresponding measurement value. The blue dots in the figure represent the number of Chinese character scores, and the red line is the linear regression line

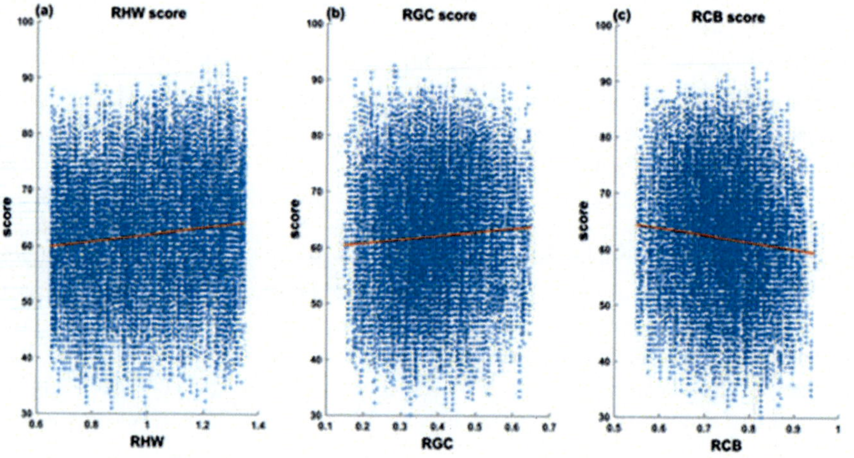

Fig. 2 Linear regression of aesthetic rank scores. The linear regression graph of all subjects' aesthetic ranking. The abscissa is the measurement value of different indicators, and the ordinate is the subject's aesthetic score under the corresponding measurement value. The blue dots in the figure represent the number of Chinese character scores and the red line is the linear regression line

RGC increases from 0.1 to 0.7, the subjects' aesthetic scores also show a slow increase between 5 and 6 trend. The k value under the RHW in Table 1 is 0.6448, and the k value of the RGC is 0.3876; the RHW in Tables 1 and 2 The k values of

RHW and RGC are not the same, but their values are all greater than 0, which shows that the aesthetic score tended to increase with the increase of RHW and RGC.

In summary, we can find that the aesthetic score tended to increase with the increase of RHW and RGC, which are the visual aesthetics of Chinese Kai-style characters tended to be "thin" and "plump". Zhuang [9] mentioned a good-looking Chinese character, which often had the characteristics of "correct", "proportioned" and "plump", which is consistent with our conclusion.

4.2.2 About "Thin" and "Plump"

Generally, when the RHW is greater than 1, the Chinese characters appear "long and thin", and the larger the RGC, the "plumper" the Chinese characters appear. The regular scripts of the Tang Dynasty developed to the extreme, with a sense of horizontal lightness and vertical weight. The ancient characters were mostly vertical and rectangular. The RHW was a golden rectangle with an aspect ratio of 1:0.618, which was considered to be the most beautiful shape in line with the laws of nature [10]. At the same time, the width and height ratio of the entire Chinese character affects the aesthetic evaluation of the Chinese characters [9]; Chinese characters should achieve a symmetrical effect. For characters with few strokes, the lines should be thicker. For printed Chinese characters with more vertical strokes, it was easy to cause horizontal widening. The thickness of printed font strokes was a problem that cannot be ignored in the design of printed fonts. The thickness ratio of horizontal and vertical strokes was 1:2. This kind of Chinese characters more beautiful and beautiful [7]; Sun et al. [5] introduced a new rectangular coordinate system with the center of the convex hull as the origin, and divided the pixel area of handwritten Chinese character strokes into four quadrants, and obtained the pixel distribution by using the ratio of the pixel area of each quadrant to the convex hull area to explore the visual aesthetic quality evaluation of handwritten Chinese characters.

"Thin" mainly refers to the shape of the entire glyph, while "plump" mainly refers to the strokes, e.g., Imitation-Song, the RHW value of this glyph is larger, from an overall point of view, the Imitation-Song characters are relatively slender. The strokes of the Chinese characters are relatively thick. For the "three" character with few strokes, the bottom one is long and the italics are very thick and long. Recognition used for children's books; Song is characterized by horizontal and vertical, horizontal and vertical, thin and vertical, full structure to make the font beautiful. The thickness of the horizontal painting is generally determined to be one-third to one-fifth of the vertical thickness. The horizontal and vertical strokes are of the same thickness in Imitation-Song style, and the horizontal strokes are slightly inclined to the upper right.

4.2.3 The Influence of RCB on the Visual Aesthetics of Chinese Characters

In the sub-graphs (c) of Figs. 1 and 2, we can observe that as the RCB increases from 0.5 to 1, the aesthetic score of the subjects is slowly decreasing between 5 and 6. The k value of the RCB in Table 1 is −1.1146, and the k value in the case of the RCB in Table 2 is −12.1960. Although the k value of the RCB in Tables 1 and 2 is different. And the value both are greater than 0, which indicates that the aesthetic score tended to decrease as the RCB increases.

According to the definition, the RCB and the RGC should have the opposite trend. The result also confirms the aesthetic score tended to decrease as the RCB increases.

5 Conclusion

We studied the effect of the RHW, the RGC, and the RCB on the visual aesthetics of Chinese characters. The results showed that the aesthetic score tended to increase with the increase of RHW and RGC and to decrease with the increase of RCB. Results suggested that the visual aesthetics of Chinese Kai-style characters tended to be "thin" and "plump".

References

1. Wei J (2020) Practical exploration of three-dimensional design of Chinese characters. Packag Eng (06):1–7 (Peking University core)
2. Wang J (2015) Focus on it-the modernization of Chinese character design. 80–81
3. Fan L (2013) Research on visual gravity point extraction algorithm of chinese character image. Comput Appl Softw 09
4. Wang J, Zhu S (2018) Discrimination of the concept of gravity in Chinese font design. Packag Eng 39–42
5. Sun R, Lian Z, Tang Y et al (2015) Aesthetic visual quality evaluation of Chinese handwriting. IJCAI[C]
6. Deng X, Li B, Zhang J (2015) Visual gravity calculation of Chinese characters based on statistics. J Chinese Inf Process 29(4):159–165
7. Cao Z, Qiu C (1994) Handbook of practical printed fonts. Printing Industry Press, Beijing
8. Li X (2018) Research on the readability of colored Chinese characters on digital interface and design application. Southwest University of Science and Technology
9. Zhuang Z (2019) Handwritten Chinese character recognition and aesthetic scoring based on deep learning. Beijing University of Posts and Telecommunications
10. Liang Y (2010) On the architectural visual aesthetics of Chinese character design. Grand View Fine Arts 216–217

Reference Axes of Chinese Characters in Visual Aesthetics

Lin Shi and Wei Hong

Abstract Reference axes are important tools in writing Chinese characters and designing Chinese character fonts. It provides functions such as referencing to the center position and stroke position of the Chinese character shape, referencing to the alignment of Chinese characters, and constraining the character size. In a visual psychophysics experiment, 10 naïve subjects in calligraphy and 10 non-naive subjects adjusted the position and the size of Chinese characters which came from the stone inscription sets handwritten by Zhenqing YAN in the Tang Dynasty (the Yan font) to achieve the best effect in visual aesthetics under various constraints. Which included a single horizontal line (SHL), a single vertical line (SVL), a double horizontal line (DHL), a double vertical line (DVL), a small center circle (SCC), a crossline (CL), a big circle (BC), a square, a square with crossline (Tian), a square with cross and diagonal line (Mi), and a square with # (Jiu). Results showed that (1) the constraints could be classified into four categories: the single horizontal line, the single vertical line, the dot and double line (DADL), and the full encirclement (FE); (2) various categories of constraint had different effects on the diagonal length of the glyph boundary rectangle, the vertical-hortical aspect ratio, and the center position; (3) for non-naive subjects, the diagonal length of the bounding rectangle was the largest case under the constraint of Jiu. The results indicated that reference axes affected the visual aesthetics of Chinese characters and the constraint of Jiu could enlarge the visual size of characters for those who lack calligraphy skills.

Keywords Reference axes · Visual aesthetics · Font size · Fat or thin · Central location

L. Shi · W. Hong
School of Information Engineering and Automation, Kunming University of Science and Technology, Kunming 650500, China

L. Shi (✉)
Computer Application Key Laboratory of Yunnan Province, Kunming University of Science and Technology, Kunming 650500, China
e-mail: lin.shi@kust.edu.cn

Z. Yu et al. (eds.), *Advancements in Mechatronics and Intelligent Robotics*, Advances in Intelligent Systems and Computing 1220,
https://doi.org/10.1007/978-981-16-1843-7_69

1 Introduction

The Chinese calligraphy culture has a long history, and reference axes that as auxiliary tools play very important role in improving the visual aesthetics of Chinese characters. The development of Chinese calligraphy culture and the evolution of reference axes is a complementary process in the long history of culture, with the change of time, Chinese characters significant changes have taken place in the way of restraining the shape of the characters from no restriction to the emergence of reference axes such as a square with crossline (Tian), a square with cross and diagonal line (Mi) and a square with # (Jiu), and it has important application value to the spread of Chinese character culture and the visual aesthetics.

In order to facilitate the initial teaching of Chinese character writing and improve people's ability to achieve the aesthetics of Chinese characters, scholars have made great progress in the visual aesthetics of Chinese characters after years of hard research. It is said that the original Jiu was created by Ouyang [1], a master of calligraphy in the Tang Dynasty, combining the characteristics of Chinese characters. Its purpose was to learn and copy Chinese characters. Yin [2] judged the creation time of the Jiu based on the characteristics of the times. He proposed that it was created no later than the Tang Dynasty, and combined its cultural connotation and significance to discuss its advantages and disadvantages: it promoted the development of calligraphy and restrained the creativity and affection of calligraphers. Zhang et al. [3] described in detail that the eight-direction line segments of the Mi better assisted the movement and teaching of Chinese character strokes. It guided people to achieve the best effect in visual aesthetics. From the background and significance of the Mi, he comprehensively expounded that the Mi was of great significance to the development of calligraphy culture. Guo et al. [4] proposed a new type of reference axes based on the center of gravity of traditional reference axes, which absorbed the advantages of traditional reference axes and combined with the latest research results. The structure of calligraphy showed the charm of Chinese characters. In recent years, new reference axes have appeared constantly, such as the "a square with # according to the proportion" proposed by Qi [5]. Based on the visual aesthetics of Chinese characters in the Jiu, Yang [6] invented the law of writing Chinese characters by means of internal and external conjugation. Based on the recognition that the italics are "it is both square and round, and seen as a circle from the square", Wei [7] proposed a circle with cross and diagonal and so on. History has proved that the function of reference axes is to assist people in writing Chinese characters neatly. The character pattern has an important influence and significance on the visual aesthetics of Chinese character writing.

In order to explore more deeply the influence and effect of reference axes on the visual aesthetics of Chinese characters, this article used the Yan font as an example to conduct experiments on the reference axes and aesthetics of Chinese characters. Since the visual aesthetics of Chinese characters depends on people's subjective judgments, this article invited participants to join in it to obtain experimental data. From the three aspects of the size, fat or thin and center position of the Chinese characters, the influence of different reference axes on the Chinese characters was analyzed. The experimental results showed that for the Yan font, different types of glyph restraints had an effect on the size, the vertical-hortical aspect ratio, and the center position. And there was a significant influence on the center position. Under the condition of the single horizontal line, the subjects adjusted the Chinese characters to smaller and fatter. And under the condition of the single vertical line, the Chinese characters were adjusted to thinner.

2 Method

In order to explore the influence of different reference axes on the aesthetics of Chinese character, this paper selected 1635 Chinese characters of the Yan font as the experimental Chinese character set, combining with people's subjective judgments on the visual aesthetics of Chinese characters to explore the effect of different reference axes on the size, fat or thin, and position.

2.1 Reference Axes

The 11 kinds of reference axes were used as the independent variables of the experiment during the experiment. The subjects used the mouse to adjust the position and size of the single Chinese Character picture (Arbitrarily changed the length and width of it) to achieve the best effect in visual aesthetics.

2.2 Experimental Requirements and Procedures

Experimental requirements: (1) Subjects were required to use the red line segments, borders, circles and other Chinese character shape constraints on the screen as references and constraints that the border range, line segment length, and line position provided the position, size, and fatness of the Chinese characters shape constraint and reference, the subjects adjusted Chinese characters according to this. (2) For the SCC in Fig. 1, the subjects were required to place the visual center of the Chinese character at the small circle. For the CL, the subjects were required to use the length of the line segment as the size constraint and reference and placed the visual center of the Chinese character at the intersection point (Table 1).

Fig. 1 Reference axes. Reference axes included: a single horizontal line (SHL), a single vertical line (SVL), a double horizontal line (DHL), a double vertical line (DVL), a small center circle (SCC), a crossline (CL), a big circle (BC), a square, a square with crossline (Tian), a square with cross and diagonal line (Mi) and a square with # (Jiu). Among them, the side length of the single horizontal line, the single vertical line, the double horizontal line, the double vertical line, the crossline, the square, the Tian, the Mi, the Jiu, and the diameter of the big circle remain equal, and the center of all glyph restraints be consistent

Table 1 The experiment flow chart, the specific flow of the experiment is shown in the following table

Display: Chinese characters placed in the glyph constraint mode (nearby) Output: Data included the number, coordinates, width, height, and constraint method of glyph pictures
1. A picture of Chinese characters and a way of reference axes randomly appeared on the computer screen
2. Using the mouse to adjust the position and size of the Chinese character glyphs of a single of the Yan font (Arbitrarily changed the length and width), making the glyphs the most aesthetic
3. Click "Next"
4. Repeating steps 2 and 3 until the end of the experiment
5. Recording the subject's name, the number, coordinates, width and height, and reference axes
6. Data analysis of the diagonal length of the glyph boundary rectangle, the vertical-hortical aspect ratio, and the center position through the recorded data

2.3 *Experimental Data Processing*

Firstly, we obtained the original Chinese character glyph information of the Yan font. They were the width and height of the Chinese character glyphs, the width and height of the Chinese character glyph pictures, and the five center coordinates of the Chinese character glyphs. The five centers included the visual center of calligraphy (VCC), the center of circumscribed circle of glyph (CCC), the center of convex hull of glyphs (CCH), the physical gravity center (PGC) and the center of boundary rectangle of glyphs (CBR). The visual center of calligraphy was the visual center of Chinese characters marked according to calligraphy knowledge; the center of the circumscribed circle was calculated by drawing the circumscribed circle of the Chinese character shape; the center of the convex hull was calculated by drawing the convex polygon of the Chinese character shape center; the physical gravity was the center calculated by considering the Chinese character shape as the object; the center of the boundary rectangle was the center calculated by drawing the boundary rectangle according to the Chinese character shape. Then, combining the data recorded in the experiment to calculate the diagonal length of the adjusted Chinese character shape. Taking the logarithm of the ratio of the adjusted Chinese character vertical-hortical aspect ratio to the original Chinese character vertical-hortical aspect ratio. Calculating five kinds of center eccentricity of the adjusted Chinese character shape in the Chinese character coordinate.

Creating Chinese character coordinate. Converting the experimentally recorded data into the center eccentricity, the diagonal length of the glyph boundary rectangle, and the vertical-hortical aspect ratio in the Chinese character coordinate. The center eccentricity referred to the distance between the center and the origin of the Chinese character coordinate, and the vertical-hortical aspect ratio referred to the ratio of the height to the width of the glyph boundary rectangle. Taking the midpoint of the reference axes as the origin, and the relevant data was calculated to obtain.

3 Results

A total of 20 subjects were invited to participate in the experiment, including 10 general subjects and 10 calligraphic subjects. Both types of subjects are composed of 5 male subjects and 5 female subjects.

Analyzing the size, fatness and position of Chinese characters based on the data obtained. It mainly focused on the diagonal length of the glyph boundary rectangle, the ratio of the logarithm of the subject to the original vertical-hortical aspect ratio, and the five center eccentricities of the Chinese character shape.

3.1 Diagonal Length of the Glyph Boundary Rectangle

The results showed: (1) sub-pictures (a), (c) showed: under the 95% confidence level, the confidence interval value of the position parameter mu was the smallest in the single horizontal line. (2) Sub-figure (a) showed: under the 95% confidence level, the confidence interval value of the position parameter mu was the largest in the Jiu. (3) According to the difference between the parameters, 11 reference axes could be classified (Fig. 2).

The 11 reference axes are divided into three categories based on their differences. Sub-figures (a), (e), and (i) showed that the three types of test data had the same trend. (1) Under the 95% confidence level, the confidence intervals of the position parameter mu did not overlap in the three cases of the single horizontal line, the single vertical line, and the other cases. It indicated that the differences in these three cases were statistically significant. (2) Under the condition of a single horizontal line, the confidence interval value of the position parameter mu was the smallest. (3) Under the other conditions, the confidence interval value of the position parameter mu was the largest (Fig. 3).

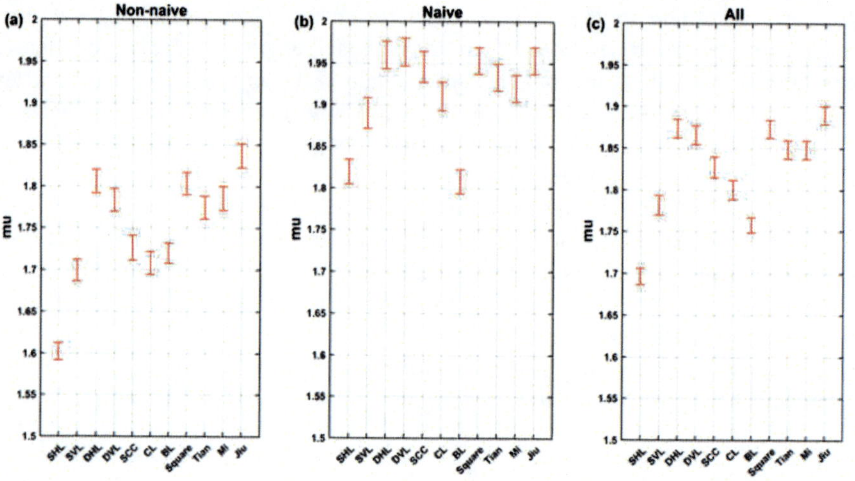

Fig. 2 Confidence intervals of diagonal length. The confidence interval distribution map of the position parameter mu of the diagonal line of the glyph boundary rectangle. The Generalized Extreme Value distribution was fitted to the diagonal length data of the non-naive subjects in calligraphy, the naive subjects, and all subjects. This figure was obtained from the confidence interval distribution of the position parameter mu. And the abscissas corresponded to 11 glyph constraints

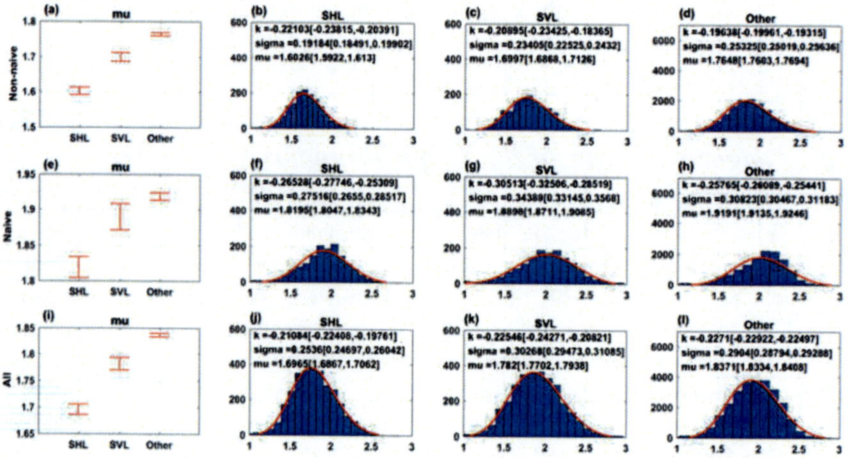

Fig. 3 Confidence intervals and distributions of diagonal length. The Generalized Extreme Value distribution fitting diagram of the diagonal length of the glyph boundary rectangle. The diagonal length data was fitted to the Generalized Extreme Value distribution. According to the difference, reference axes were divided into three categories: the single horizontal line, the single vertical line, the other situations classified into one category. Three rows correspond to three types of data: the non-naive subjects in calligraphy, the naive subjects, and all subjects. The first column is the confidence interval distribution graph of the position parameter mu. The three sub-graphs behind each row are the data graphs for the distribution fitting of the data in the three types of situations

3.2 Vertical-Horizontal Aspect Ratio

After dividing the 11 reference axes into three categories according to their differences, the sub-pictures of (a), (e), and (i) showed: (1) the data of general subjects and all subjects showed that under the 95% confidence level, the confidence intervals of the position parameter mu did not overlap in the single horizontal line, the single vertical line, and the other cases. It indicated that the differences between these three were statistically significant. (2) The data of the subjects with calligraphy literacy showed: at the 95% confidence level, the single horizontal line and the single vertical line, and the confidence intervals of other position parameters mu did not overlap each other. The confidence intervals of the position parameter mu overlapped in the single horizontal line and the other cases, but the confidence intervals of their shape parameter k and scale parameter sigma did not overlap each other. It indicated that the differences in these three situations were statistically significant. (3) All three types of data showed: in the case of a single horizontal line, the confidence interval value of the position parameter mu was the smallest (Fig. 4).

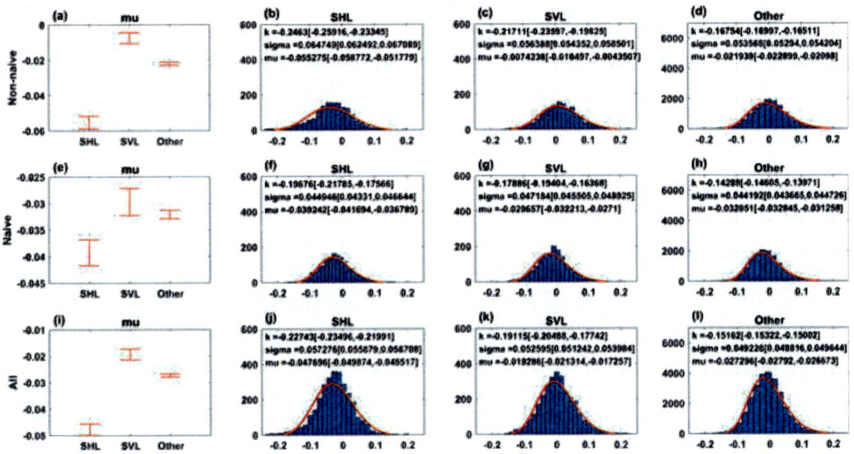

Fig. 4 Confidence intervals and distributions of vertical-horizontal aspect ratio. Taking the logarithm of the ratio between the subjects' adjusted Chinese character vertical-horizontal aspect ratio and the original Chinese character vertical-horizontal aspect ratio, and taking the Generalized Extreme Value distribution fitting graph of the logarithmic vertical-horizontal aspect ratio. According to the difference, reference axes were divided into three categories: the single horizontal line, the single vertical line, the other situations classified into one category. Each row corresponds to three types of data: the non-naive subjects in calligraphy, the naive subjects, and all subjects. The first column is the confidence interval distribution graph of the position parameter mu. The three sub-graphs behind each row are the data graphs for the distribution fitting of the data in the three types of situations

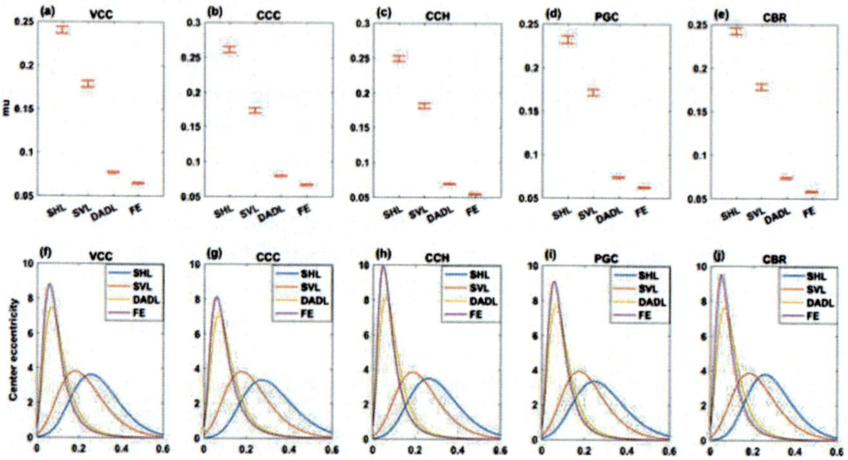

Fig. 5 Confidence intervals and distributions of various centers. The distribution of five kinds of center eccentricity and the confidence interval of position parameter mu. Under the constraints of 11 reference axes, the center eccentricity data of all subjects were fitted to the Generalized Extreme Value distribution. According to the differences, they were divided into four categories: the single horizontal line; the single vertical line; a double horizontal line, a double vertical line, a small center circles and a cross were grouped together, called the dot and double line (DADL); other situations were grouped together, called the full encirclement (FE). The first row is the confidence interval distribution diagram of the position parameter mu, **a–e** correspond to the five types of centers, respectively. The second row is the distribution fitting diagram of the four types of data, **f–j** correspond to the five types of centers

3.3 Center Eccentricity

After dividing the 11 reference axes into four categories based on their differences, the sub-figures (a)–(e) showed that the five types of center data had the same trend. (1) At the 95% confidence level, the confidence intervals of the position parameter mu did not overlap in the four cases of the single horizontal line, the single vertical line, the dot and double line (DADL), and the full encirclement (FE). It indicated that the differences in these four cases were statistically significant. (2) Under the condition of the single horizontal line, the confidence interval value of the position parameter mu was the largest. (3) Under the condition of the full enclosure, the confidence interval value of the position parameter mu was the smallest (Fig. 5).

4 Discussions

As an auxiliary tool for people to write Chinese characters aesthetically, reference axes may have different effects on the shape of Chinese characters. Reference axes can help writers determine the size of the font of the Chinese character, the position of the Chinese character on the paper, and the overall stability of the font [8]. The visual aesthetics of Chinese characters is reflected in people's preferences. Therefore, people's preference is used as the basis for this article's aesthetic evaluation of Chinese characters.

According to the difference of the data, the 11 restriction methods could be divided into 4 categories: the single horizontal line, the single vertical line, the dot and double line, and the full encirclement. Different types of reference axes had significant differences in the diagonal length of the glyph boundary rectangle, the vertical-hortical aspect ratio, and the center position. Among them, there was no significant difference between the diagonal length of the glyph boundary rectangle and the vertical-hortical aspect ratio of the dot and double line and the full encirclement.

4.1 Diagonal Length of the Glyph Boundary Rectangle

This paper used the diagonal length of the glyph boundary rectangle to measure the size of the Chinese character. The longer the diagonal, the larger the Chinese character. For the Yan font: in the three cases of the single horizontal line, the single vertical line and the other situations, the diagonal data of the three types of subjects were significantly different. It indicated that these three types of reference axes had a significant influence on the font size of Chinese characters. The data of non-naive subjects in calligraphy showed that the diagonal length of the glyph bounding rectangle was the largest under the Jiu. This indicated that under the restraint of the Jiu, the non-naive subjects in calligraphy adjusted the font of Chinese characters to a larger size. In the case of the single horizontal line, the confidence interval value of the position parameter mu was the smallest. This indicated: in the case of the single horizontal lines, people liked to adjust the font of Chinese characters to a smaller size. The possible reason is: the single horizontal line only has the horizontal guidance to the Chinese character shape, which makes it easy for people to ignore the development of Chinese characters in the vertical direction. In the other cases, the position parameter mu had the largest confidence interval value. It indicated that people thought that larger Chinese characters were more aesthetically pleasing in the other cases.

4.2 Vertical-Horizontal Aspect Ratio

This paper compared the adjusted vertical-horizontal aspect ratio of Chinese characters, which showed that people liked the fat or thin Chinese characters and reflected people's assessment of the visual aesthetics of Chinese characters. For the Yan font, the vertical-horizontal aspect ratio data of the three types of subjects were significantly different in the single horizontal line, the single vertical line and the other situations. It indicated that these three types of glyph restraint methods had a significant influence on the fatness and thinness of Chinese characters. In the case of the single horizontal line, the confidence interval value of the position parameter mu was the smallest. This indicated: in the case of the single horizontal lines, people liked to adjust the font of Chinese characters to be fatter. The reason for this phenomenon may be: the single horizontal line has a horizontal guide to the shape of Chinese characters, which makes people adjust the Chinese characters to be fatter.

4.3 Center Eccentricity

This article used five kinds of center eccentricity to reflect the position of Chinese characters. Regarding the Chinese character shape of the Yan font, in the four cases of the single horizontal line, the single vertical line, the dot and double line and the full envelopment, the center eccentricity data of the three types of subjects were significantly different. This indicated that these four types of glyph constraint methods had a significant influence on the glyph position of Chinese characters. In the case of the single horizontal line, the confidence interval value of the position parameter mu was the largest, and the opposite was true in the case of full encirclement. This indicated that: in the case of the full encirclement, the center distance of the Chinese character was the smallest. The reason for this phenomenon may be: full envelopment has greater constraints on the shape of Chinese characters, which can provide people with a better reference for writing and be consistent with calligraphy knowledge.

5 Conclusion

Based on the subjective judgments of the subjects, the data showed that different types of reference axes have significant differences in the diagonal length of the glyph boundary rectangle, the vertical-horizontal aspect ratio, and the center position, which showed that different types of reference axes had significant effects on the font size, fatness or thinness, and relative positions of Chinese characters. It also meant that different types of reference axes had an impact on the aesthetics of the font. The non-naive subjects in calligraphy adjusted the Chinese character shape to a larger size under the Jiu, but the naive subjects did not have the same phenomenon. It indicated that the reference axes of the Jiu had the effect of enlarging the Chinese character shape for the non-naïve subjects. Therefore, the conclusion indicated that reference axes had an impact on the aesthetics of the glyphs and the reference axes of the Jiu had the effect of enlarging the size of the glyphs for those lacking in calligraphy.

References

1. Ouyang Z (2008) New calligraphy course. Higher Education
2. Yin Q (2016) On the cultural connotation of the "Jiugongge" calligraphy boundary form. Calligraphy 09:39–42
3. Zhang X, Ma Z, Wu D (2014) A brief discussion on the role and significance of "Mizige" in calligraphy teaching. Neijiang Sci 35(02):120–122
4. Guo N], Wang W (2000) Double palace gravity center—a new type of character pattern that accurately reflects the structure of calligraphy. J Xiaogan Univ 02:79–82
5. Qi G (2005) Qi Gong tells you about calligraphy. Zhonghua Book Company, pp 85–89
6. Yang W (2000) The square embedded with a small square Hard Pen Calligraphy Course. China Academy of Art Press
7. Wei B (2001) Round rice-character copybook. Xiling Yinshe Publishing House
8. Liu C (2008) The effect of the pattern of learning characters on the practice of Chinese characters. Chin Pen Calligr 05:38–39

Implementation and Optimization of FPGA-Based Edge Detection Algorithm

Jinmei Zhang, Zhangyao Zi, Tao Jiang, Chao Zhang, and Yonghang Tai

Abstract Under the premise of the rapid development of FPGA, field-programmable gate arrays are widely used in large-scale data processing fields such as digital image processing because of the characteristics of parallel pipeline structure, their fast data processing speed, and large data processing volume. Based on this, this paper uses the Verilog language to design and optimize the FPGA-based image processing system for the Sobel algorithm to complete the acquisition, storage, and image display of image data, to realize the Sobel edge detection algorithm, and to implement the Sobel edge detection algorithm in Altera's Cyclone IV series. Perform simulation and hardware debugging on the FPGA chip. The simulation results show that the FPGA-based Sobel algorithm is combined with VGA display technology. Compared with the serial processing using software, the speed is much improved and the accuracy of the detection results is guaranteed. This research has a lot of useful value in the field of digital image processing.

Keywords Sobel edge detection algorithm · Field-programmable gate array · Digital image processing

1 Introduction

The most important purpose of edge detection is to find the most obvious point of brightness change in the image. The attributes of the image have undergone obvious changes, which usually reflect important events and changes in the attributes, mainly including depth discontinuities, surface direction discontinuities, changes in material properties, and changes in scene lighting [1]. A large part of the image information usually exists on the edge of the image, and the edge is one of the most essential features of an image. Image edge detection has many uses in life. Its

J. Zhang · Z. Zi · T. Jiang · C. Zhang · Y. Tai (✉)
School of Physics and Electronic Information, Yunnan Normal University, Kunming, China
e-mail: taiyonghang@126.com

Z. Yu et al. (eds.), *Advancements in Mechatronics and Intelligent Robotics*,
Advances in Intelligent Systems and Computing 1220,
https://doi.org/10.1007/978-981-16-1843-7_70

effective solution has a very important impact on our target detection and intelligent recognition. Its successful solution is the basis of many image processing.

When the image data to be processed is relatively large, its processing speed is relatively slow, not a real-time processing method. Therefore, it is necessary to use hardware to implement image processing algorithms. This is because the hardware system is parallel, but the flexibility of the hardware system is very poor. After some functions are successfully configured, it can achieve this function well, but if the functional requirements are one Change, it is very difficult to change this system. The introduction of FPGA technology opens up new possibilities for digital logic. FPGA not only has the parallelism of hardware but also has the flexibility of software. The function of FPGA can be reprogrammed or reconfigured [2], and it is programmable. The logic capacity is large, which can meet many design requirements. It can also solve the problems of computers that cannot run in parallel, and greatly shorten the time required for image processing [3]. As the function of the FPGA chip is getting stronger, the manufacturing process is getting better and the cost is getting lower and lower, so FPGA is bound to become a banner in the field of image processing. In the future, the students will integrate various edge detection image processing algorithms, further integrate and study the advantages of various operators, and combine the advantages of each operator. If the points are combined, then the edge detection technology will be more widely used.

2 Overview

FPGA, its Chinese name is field-programmable gate array. Compared with complex programmable logic devices, it has higher integration, faster calculation speed, and lower power consumption. It can be said that FPGA is an upgraded version of CPLD. At present, the mainstream FPGA has still based on the look-up table (LUT) technology, and the internal structure diagram is shown in Fig. 1. It is mainly composed of five parts: programmable input/output modules, configurable logic units, digital clock management modules, embedded blocks, and wiring resources [4].

(1) Programmable input/output module (IOB): The IOB module is distributed around the chip. This module is used when the FPGA chip is connected to the external circuit. The module can support many interface standards. Due to the voltage and frequency of each interface standard Different, so the standard of the interface can be determined according to the size and frequency of the interface voltage [4]; (2) Configurable logic unit (CLB): It is the basic logic unit of FPGA. Its structure is relative.

The configuration of the LUT can be changed according to the user's design requirements. Configured into combinational logic, shift register, etc. to complete different logic functions; (3) Digital clock management module (DCM): It is a very important device that provides users with a stable clock signal required for system design, mainly composed of a phase-locked loop; 4) Embedded block RAM

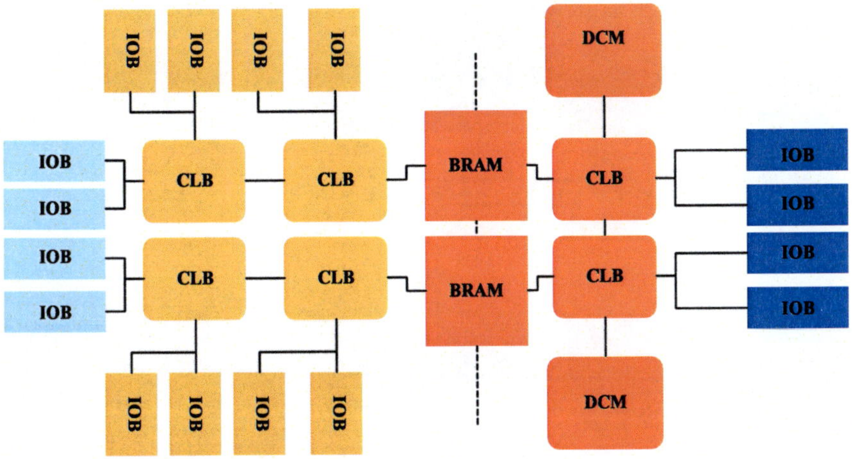

Fig. 1 Internal structure of FPGA

(BRAM): It is the embedded memory embedded in FPGA, which can be configured as single-port, dual-port RAM and FIFO buffer or ROM when in use to facilitate data storage [4]; (5) Programmable Interconnection Resources (ICR): Just like wires in a circuit, its function is to connect resources such as CLB and IOB to form a circuit with functions required by customers. It is thanks to this particular architecture in the diagram that it can perform the tasks required by the project. The operation of these lines is mainly It is realized by programming with hardware description languages Verilog HDL and VHDL like high-level languages, but these two languages have their syntax.

3 Hardware System Framework

This system has five parts in total, namely the clock management module, acquisition module, buffer module, processing module, and display module. The clock management module can output 25 M clock for VGA display, output 24 M clock for image acquisition by the camera, output 100 M clock for SDRAM storage and output −90°100 M clock for Sobel computing module, etc. [5]. First, initialize the camera according to the camera's device manual, and set the camera accordingly to make the camera work in line with the requirements of the system. After the camera configuration is completed, the image can be collected and the collected data can be stored outside the FPGA chip. In the SDRAM chip, the data buffered in the SDRAM is read out to the Sobel operation module for mathematical operation, and finally, the data processed by the operation module is displayed on the computer screen through the VGA interface. The overall framework of the system is shown in Fig. 2.

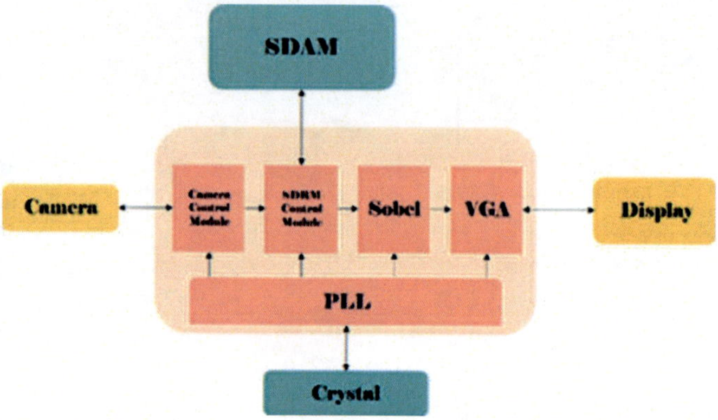

Fig. 2 Overall system diagram

To avoid the respective shortcomings of synchronous preset and asynchronous reset, and to balance each other, the design adopts an asynchronous reset to set the position synchronously [6]. Its basic structure is shown. Load the reset signal into register b when the clock edge arrives, and output the reset signal when the next clock arrives. This method allows the reset signal to be controlled by the clock signal, which not only solves the resource consumption problem caused by synchronous reset but also gets rid of the metastability problem caused by asynchronous reset [6].

4 Experiment

The Sobel operator is mainly used for edge detection, which is used to calculate the approximate value of the degree of the image brightness function. No matter where on the image, using this operator will generate the corresponding gray vector or its normal vector [7]. The Sobel convolution factor is shown in Fig. 3.

This operator consists of two sets of 3 × 3 matrices, which are in the horizontal and vertical directions. By convolving them with the image, respectively, the approximate values of the horizontal and vertical brightness differences can be obtained, respectively. Let A represent the original initial image, Put the values of x

Fig. 3 Sobel convolution factor

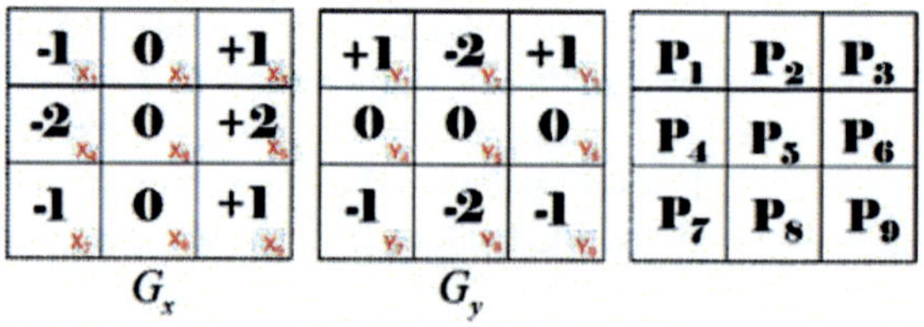

and y into the following formula to calculate the grayscale of this point, In most cases, to improve efficiency, the square root calculation is generally not done, but its approximate value and its formula is shown

$$G_x = \begin{bmatrix} -1 & 0 & +1 \\ -2 & 0 & +2 \\ -1 & 0 & +1 \end{bmatrix} * A \tag{1}$$

$$G_y = \begin{bmatrix} +1 & 2 & +1 \\ 0 & 0 & 0 \\ -1 & -2 & +1 \end{bmatrix} * A \tag{2}$$

$$G = \sqrt{G_x^2 + G_y^2} \tag{3}$$

$$|G| = |G_x| + |G_y| \tag{4}$$

Compare the calculated G value with the threshold. If it is greater than the threshold, set it to 0, and if it is less than the threshold, set the gray value of this point to 1. For the Sobel operator to successfully implement edge detection in FPGA, the following parts are required. They are the image line buffer module, the convolution calculation module, and the threshold processing module.

5 Result

Simply set the threshold size, first input from the outside, temporarily set it to 60, if it is greater than the threshold, it is considered valid, assign 1, otherwise assign 0, and the stimulation waveform is shown in Fig. 4.

After the hardware debugging is completed, as shown in Fig. 5, the original photo is shown. When the threshold is 60, the detection effect is shown in Fig. 6.

Comparing Fig. 5 with Fig. 6, we can see that we have got more edge details. We can see that we have got more delicate edge detection results.

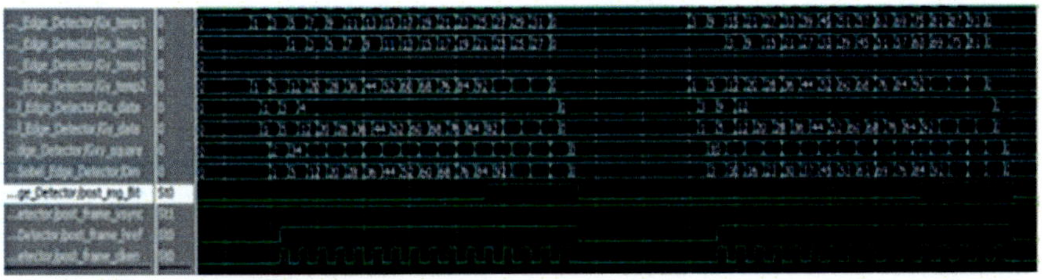

Fig. 4 Simulation waveform at threshold 60

Fig. 5 Original figure

Fig. 6 Detection effect of threshold 30;60;90

6 Conclusion

This paper mainly studies the FPGA implementation of the Sobel edge detection algorithm and proposes a parallel input scheme of image data. The hardware description language is used to transplant the Sobel edge detection algorithm. Compared with the serial input of software, the speed is improved a lot. While completing the edge detection, you can change the Sobel edge detection threshold through the button control method, you can easily find your satisfactory detection results and greatly optimize the design. This design is implemented using Verilog, and simulation results based on Modalism are given, which verifies the feasibility of this design.

References

1. Kirar BS, Agrawal DK (2018) Empirical wavelet transform based pre-processing and entropy feature extraction from glaucomatous digital fundus images. In: 2017 international conference on recent innovations in signal processing and embedded systems (RISE). IEEE
2. Guangyu Z, Ping C (2017) University N D. Design of CMOS industrial camera based on FPGA. Mod Electron Tech
3. Yang L, Li G (2018) Design of a large area array CMOS of camera system that can automatically zoom. In: 2018 IEEE 4th international conference on computer and communications (ICCC). IEEE
4. Hui-Zong F, Ye C, Yang XU (2011) Real-time image acquisition and Sobel edge detection based on FPGA. Transducer Microsyst Technol 30(6):116–118
5. Xiao WH, Wan X, Zhang ZM et al (2011) Design and realization real-time edge detection system based on DM642. Adv Mater Res 216:233–237
6. Yao GX (2016) Design of edge detection algorithm for image sobel based on FPGA. In: 2015 4th international conference on computer science and network technology (ICCSNT). IEEE
7. Guan W, Zheng-Long LI (2018) Driver distraction detection based on reverse binocular recognition. Sci Technol Eng

Study on the Sensing Mechanism of Novel Tactile Sensor

Yuyun Xu, Wei Zhang, Shanhong Li, and Hongqing Pan

Abstract With the deepening integration of information technology and artificial intelligence technology, the application scope of robot is expanding from traditional industrial automation to service and special application scenarios. In different application fields, tactile perception plays an indispensable role when dexterous manipulator completes intelligent and fine tasks. Tactile sensor, as the medium of direct interaction between robot and external environment and target object, is a key component in the development of intelligent robot. In this paper, the sensitive unit of a new type of gasbag tactile sensor is taken as the object. The mathematical model of the sensitive unit of the flexible sensor under the static normal load is constructed by using the theory of material mechanics. The relationship between the external load of the gasbag sensor and the internal pressure of the gasbag is revealed. The influence of the size parameters on the stress and deformation of the new type of gasbag tactile sensor is discussed It provides an important theoretical basis for the structure design and performance optimization of tactile sensor.

Keywords Tactile sensor · Analytic solution · Mechanics of materials

Y. Xu · W. Zhang · H. Pan (✉)
Institute of Intelligent Machines, Hefei Institutes of Physical Science,
Chinese Academy of Sciences, Hefei 230031, China
e-mail: hqpan@iim.ac.cn

Y. Xu · W. Zhang
Department of Science Island, University of Science and Technology of China,
Hefei 230026, China

S. Li
Department of Automation, Hefei University, Hefei 230601, China

Z. Yu et al. (eds.), *Advancements in Mechatronics and Intelligent Robotics*,
Advances in Intelligent Systems and Computing 1220,
https://doi.org/10.1007/978-981-16-1843-7_71

1 Introduction

With the rapid development of automation level in the today's world, intelligent robot technology, as a typical representative of automation level, has been widely concerned in various fields [1]. As robots are more and more widely used in medical, aerospace, military, marine and service fields, unstructured, complex and diverse environment has higher requirements for robots. In the process of precision operation, the tactile perception of intelligent robot's end effector or dexterous hand is very important [2]. Dexterous manipulator can recognize and explore objects and estimate handling stability and control force by uses tactile information [3]. Tactile sensors which conduct important medium between intelligent robot and external objects simulate the perception function of organisms and provide the robot with the information of the surrounding environment and the targets [4].

Because of its softness and flexibility, flexible array sensor can be integrated into any type of robot hand. It can significantly enhance the contact area with the target object, better sense the information of external environment, and complete the complex and fine work [5]. For the static performance of flexible tactile sensor, most scholars qualitatively analyze the deformation of flexible tactile sensor which under external load [6, 7]. The deformation of tactile sensor under static force is calculated and analyzed by finite element analysis software [8]. The relationship between the external load and the sensing information of the sensor is directly obtained by the experimental method [9]. However, the numerical method is only suitable for the determination of sensor structure and size. The experimental method is suitable for the verification stage of the study and cannot provide data support for the study in the early stage.

In this paper, the analytical solution of a new type of gasbag tactile sensor under static normal load is discussed in detail. The sensing mechanism of the gasbag sensitive unit sensing external load is analyzed. The influence with the size parameters of a single gasbag sensitive unit on the internal pressure of the gasbag sensor sensitive unit is discussed in depth. It is helpful to optimize the sensor structure and improve the sensor performance. It also has important theoretical significance and engineering application value.

2 Structure of Sensor Sensitive Unit

In this paper, the lps22hb chip of Italian semiconductor company is selected as the sensitive material. The chip is fully molded package, and the package size is $2 \times 2 \times 0.76$ mm, as shown on the left beside of Fig. 1 [10]. Through the micro holes of the upper silicon wafer, the information of pressure and temperature can be transmitted to the sensitive components. The sensitive unit in the fully molded package is used to sense the external information, and the I2C and SPI communication protocols are used to transmit the sensing data information. The chip can

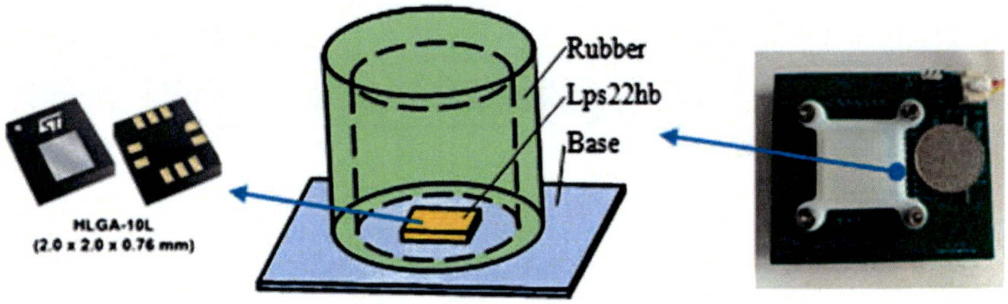

Fig. 1 lps22hb chip, sensing unit and new type flexible tactile sensor

detect the change of pressure in the range of 260–1260 hpa, with the advantages of high precision, ultra-small size, excellent sensing performance, good stability, dust-free and waterproof.

The sensing unit of gasbag sensor is shown in the middle of Fig. 1. The lps22hb chip is placed on the base and inside the sealed rubber gasbag. To ensure the tightness of the inner cavity of the gasbag, the adhesive is used to connect the lower part of the rubber gasbag and the base. Rubber, as a very unique elastic material, has good ductility and elasticity. When the external load acts on the outside of the rubber gasbag, the rubber gasbag will compress, bend or twist. Assuming that the mass of the gas in the sealed gasbag is constant, the change of the volume of the gas in the gasbag will lead to the change of the pressure in the gasbag. The pressure sensitive unit is used to sense the pressure in the rubber gasbag, and then the size and direction of the load are deduced.

The size of the sensitive unit in the micro-air bag array sensor is small, and it can integrate any number of sensitive unit arrays to any shape. As shown on the right side of Fig. 1, the physical figure of the micro-air bag array sensor with 6×6 array is shown. Micro-gasbag array sensor has good flexibility, can be expanded to any shape, and can be integrated on any surface.

3 Mathematical Model of Sensing Unit

In order to quantitatively analyze the deformation of the rubber gasbag, the rubber gasbag is divided into two parts. The upper part is cylindrical, and the lower part is tubular. When the external load F acts on the top of the rubber gasbag, the cylindrical rubber and tubular rubber are elastically deformed. According to Newton's law,

$$F = F_1 = F_2 + F_3 \tag{1}$$

where F_1 is the elastic force of the cylindrical rubber. F_2 is the elastic force of the cylindrical rubber, and F_3 is the pressure force in the rubber gasbag (Fig. 2).

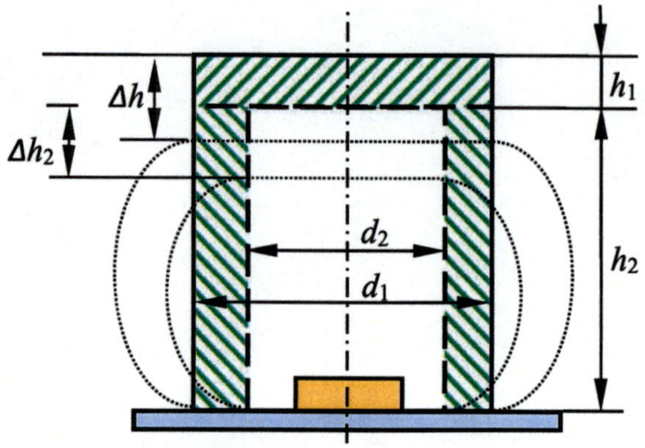

Fig. 2 Dimension diagram of air pressure sensitive unit

When the load F acts on the top of the rubber airbag, the cylindrical rubber and the tubular rubber deform elastically at the same time. In the mechanical relationship, the cylindrical rubber and tubular rubber are regarded as series connection. When the normal load acts on the top of the rubber airbag, the deformation of the rubber airbag is

$$\Delta h = \Delta h_1 + \Delta h_2 \tag{2}$$

where Δh_1 is the compression of the cylindrical, Δh_2 is the compression of the tubular rubber. According to the calculation formula of rubber spring in Table 12–16-3 of Ref. [11], Δh_1 and Δh_2 is

$$\Delta h_1 = \frac{4F_1 h_1}{3.6\left(1 + 1.65\left(\frac{d_1}{4h_1}\right)^2\right) G\pi d_1^2} \tag{3}$$

$$\Delta h_2 = \frac{4F_2 h_2}{3.6\left(1 + 1.65\left(\frac{d_1 - d_2}{4h_2}\right)^2\right) G\pi (d_1^2 - d_2^2)} \tag{4}$$

where d_1 is the diameter of the cylindrical rubber and the external diameter of the tubular rubber, d_2 is the internal diameter of the tubular rubber, h_1 is the height of the whole rubber, and h_2 is the height of the tubular rubber. G is shear modulus of elasticity. On the top surface of the gasbag, the effective acting area of the gas pressure is $\pi d_2^2/4$, so the pressure F_3 acting on the top surface of the gasbag is

$$F_3 = (P - P_a)\pi \frac{d_2^2}{4} = P\prime\pi \frac{d_2^2}{4} \tag{5}$$

where P is the pressure inside the rubber gasbag, P_a is the standard atmospheric pressure outside the rubber gasbag, and P' is the pressure difference between the inside and outside of the rubber airbag. Combining formula (1), (4) and (5), we can get

$$\Delta h_2 = \frac{4h_2}{3.6\left(1+1.659\left(\frac{d_1-d_2}{4h_2}\right)^2\right)G\pi(d_1^2-d_2^2)}\left(F - P\prime\pi\frac{d_2^2}{4}\right) \quad (6)$$

The inner the rubber gasbag is sealed without gas flow, and the total mass of gas in rubber gasbag is certain. According to Boyle's law, the relationship between pressure and volume is as follows.

$$PV = nRT = P_0V_0 = P_0\frac{\pi d_2^2}{4}h_2 \quad (7)$$

where P_0 is the initial air pressure in the rubber gasbag, V_0 is the initial volume of the rubber gasbag, and V is the volume of the deformed rubber gasbag.

According to the circumferential symmetry of the gasbag and the symmetry of the pressure, the cross section of the deformed airbag is circular, which radius is R, and height is $h_2 - \Delta h_2$. The volume V of the deformed rubber airbag can be obtained.

$$V = \int_0^{h_2-\Delta h_2} \pi R^2 dh \quad (8)$$

As shown in Fig. 3, the side of the expanded airbag is rectangular, and the upper and lower sides of the rectangular plate are fixed constraints end. Atmospheric pressure is a uniformly distributed on one side of the rectangular plate, resulting in the deformation of the rectangular plate.

According to the mechanics of materials, the disturbance ω is a function of height h. The diameter R of gasbag can be deduced from the disturbance ω.

$$R(h) = d_2 + 2\omega(h) = d_2 + 2\frac{P\prime\pi d_2 h}{24EI}\left[(h_2-\Delta h_2)^3 - 2(h_2-\Delta h_2)h^2 + h^3\right] \quad (9)$$

Combining formula (7)–(9), we can get

$$\frac{31\pi^2(h_2-\Delta h_2)^9}{90720E^2I^2}P\prime^2P + \frac{\pi(h_2-\Delta h_2)^5}{30EI}P\prime P + (h_2-\Delta h_2)P - \frac{P_0h_2}{4} = 0 \quad (10)$$

Substituting Δh_2 which in formula (6) into formula (10), the mapping relationship $P = f(F)$ between normal load F and the pressure P inside the gasbag can be deduced. Take $E = 100\ \text{N/m}^2$, $h_2 = 2$ cm, $P_0 = 101.325$ kPa, $d_1 = 5$ cm,

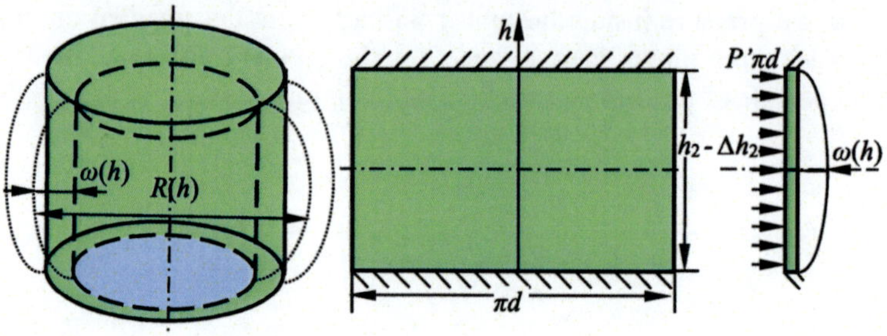

Fig. 3 Stress and deformation diagram of single rubber airbag under normal load

$d_2 = 1$ cm, and draw the relationship curve between normal load F and the pressure P, as shown in Fig. 4.

When the normal load on the upper surface of the rubber airbag increased, the air pressure inside the gasbag increases. However, there are several solutions of pressure P for an external load F. According to the realistic physical meaning, the external load F and the air pressure P are constrained. $F \geq 0$ N and $P \geq P_0 = 101.325$ kPa. The constraint conditions of normal load and pressure are added in the relationship curve, that is, dotted line in Fig. 4. The effective area of the relationship curve between F and P is in the dashed box which in the upper right corner of Fig. 4. From the effective relationship curves of F and P, it can be seen that when the external normal load F is initially loaded, and the pressure P does not change. As the load F increases, the pressure P increases linearly, the linearity between the pressure P and the normal load F is good. Thus, it provides a theoretical basis for the design of the gasbag tactile sensor, and the normal load F can be derived indirectly from the pressure value P.

Fig. 4 Relationship between normal load F and air pressure P in airbag

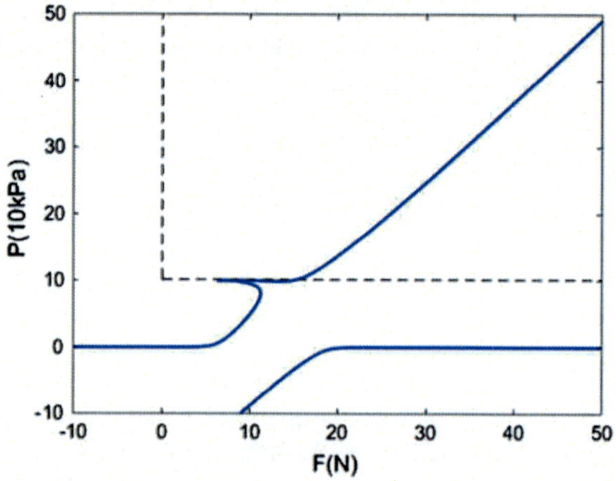

4 Structural Optimization Analysis of Sensitive Element

In order to improve the sensitivity and measurement limit of the flexible tactile sensor, it is necessary to further optimize the size and structure of the rubber gasbag. The dimensions of a single rubber gasbag include whole height h, gasbag height h_2, rubber gasbag outer diameter d_1, and internal diameter d_2. The influence of dimension parameters d_1, d_2 and h_2 on the relationship between load F and pressure P is discussed, respectively.

Firstly, the influence of the external diameter of rubber d_1 on the pressure curve is discussed. The values of the rubber internal diameter d_2 and height h_2 is fixed (d_2 = 1 cm, h_2 = 2 cm). The value of the outer diameter d1 is changed, and the relationship curve between the external normal load F and the pressure P is drawn, as shown in Fig. 5a. When other parameters remain unchanged, the rubber outer diameter d_1 is increased. The slope of the curve between F and P remains unchanged, which means that the sensitivity of flexible tactile sensor unit remains unchanged. The increase of $F|P = P_0$ means that the lower limit of load F is decreased.

Secondly, the values of rubber gasbag internal diameter d1 and height h2 is fixed (d_1 = 5 cm, h_2 = 2 cm), the value of outer diameter d_2 is changed, and the relationship curve between normal load F and pressure P is drawn, as shown in Fig. 5b.

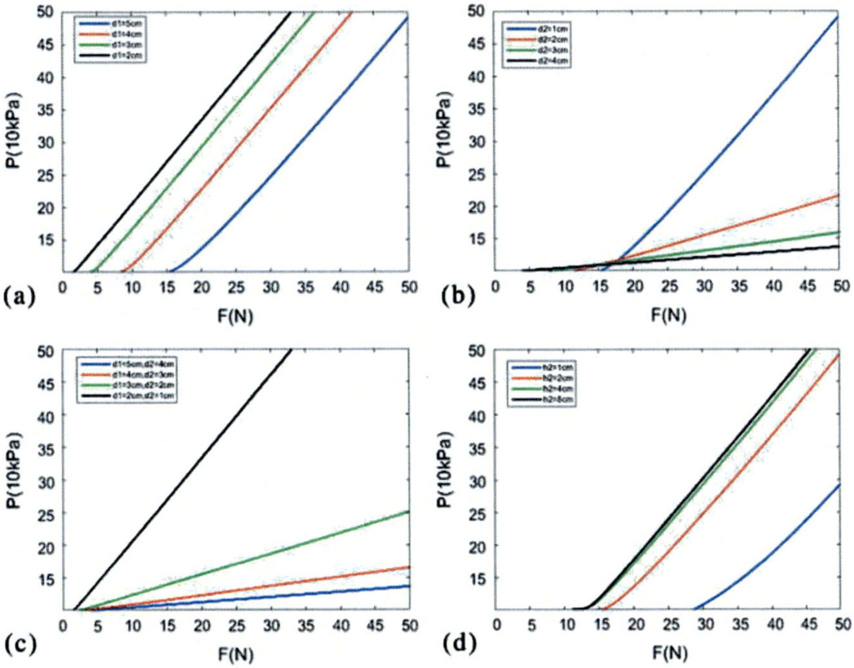

Fig. 5 Relationship between normal load F and pressure P

With the increase of internal diameter d_2, $F|P = P_0$ and the slope of the curve between load F and pressure P becomes smaller. It means that the sensitivity of sensor unit and the lower limit of load F is reduced.

Then, the value of the gasbag height h_2 is fixed ($h_2 = 2$ cm), the values of the tubular rubber outer diameter d_1 and internal diameter d_2 is changed at the same time. Keep the rubber thickness $(d_1 - d_2)/2$ unchanged, and the relationship curve between the normal load F and the pressure P is drawn, as shown in Fig. 5c. As d_1 and d_2 increased at the same time, the slope of the curve between load F and pressure P is abate and $F|P = P_0$ keep invariant. In other words, the sensitivity of sensor unit is descend and the lower limit of load F remains the same.

Finally, the values of tubular rubber internal diameter d_1 and outer diameter d_2 is fixed ($d_1 = 5$ cm, $d_2 = 1$ cm), the value of height h_2 is changed, and the relationship curve between normal load F and pressure P is drawn, as shown in Fig. 5d. When other parameters remain unchanged, the curve slope between load F and pressure P remains unchanged and $F|P = P_0$ is reduced with the increase of h_2. It means that the sensitivity of sensor unit keep hold, and the lower limit of load F is broaden.

To sum up, the influence of the parameter change, gasbag height h_2, the tubular rubber external diameter d_1, the tubular rubber internal diameter d_2 and the tubular rubber thickness $(d_1 - d_2)/2$, on the relationship curve between load F and pressure P is summarized in the following table. Symbol representing ↑ ↓ /: increase, reduce, and unchanged.

It can be seen intuitively from Table 1 that the change trend of curve slope is opposite to the tubular rubber internal diameter d_2. When the internal diameter d_2 of tubular rubber is increased, the curve slope between normal load F and pressure P is decreased. The change trend of $F|_{P=P0}$ was consistent with that of rubber thickness d_1-d_2. $F|_{P=P0}$ of the relationship curve between F and P is rose of with the increase of rubber thickness $d_1 - d_2$. In addition, there is an inverse trend relationship between $F|_{P=P0}$ and tubular rubber height h_2. When the height h_2 is increased, $F|_{P=P0}$ is increases. In summary, it is necessary to reduce the value of internal diameter d_2 to improve the sensitivity of gasbag tactile sensor. In order to extend the lower measuring limit of the external normal load F, it is necessary to reduce the thickness of tubular rubber $d_1 - d_2$ and increase the height of gasbag h_2.

Table 1 Influence of different airbag size parameters on the relationship curve between load and air pressure

d_1	d_2	$d_1 - d_2$	h_2	Slope	$F|_{P=P0}$
↑	/	↑	/	/	↑
/	↑	↓	/	↓	↓
↑	↑	/	/	↓	/
/	/	/	↑	/	↓

5 Conclusion

When the normal external load acts alone, the force state and bending deformation of the gasbag sensor sensitive unit are analyzed in this paper, based on the theory of material mechanics and elastic mechanics. And the relationship between the external load and the pressure is revealed. The conclusion has important theoretical significance and practical application value for the structure optimization and performance improvement of tactile sensor. The study lays a solid foundation for the tactile perception of the intelligent robot. And it realizes the stable control goal of the intelligent robot.

Acknowledgements This work was supported by National Science Foundation of China [No. 91648206, 61673369], Anhui Provincial Key R&D Programmes [No. 202004a07020051, 201904d07020007], Presidential Foundation of Hefei Institutes of Physical Science, Chinese Academy of Sciences [No. YZJJ2020QN17] and Nature Science Research Project of Anhui province [No. 1808085QF195].

References

1. Kappassov Z, Corrales JA, Perdereau V (2015) Tactile sensing in dexterous robot hands—review. Robot Auton Syst 74:195–220
2. Yang T, Xie D, Li Z (2017) Recent advances in wearable tactile sensors: materials, sensing mechanisms, and device performance. Mater Sci Eng R 115:1–37
3. Zou L, Ge C, Wang ZJ et al (2017) Novel tactile sensor technology and smart tactile sensing systems: a review. Sensors 17(11):2653
4. Gu Y, Zhang T, Chen H et al (2019) Mini review on flexible and wearable electronics for monitoring human health information. Nanoscale Res Lett 14(1):1–15
5. Rogers JA, Someya T, Huang Y (2010) Materials and mechanics for stretchable electronics. Science 327(5973):1603–1607
6. Lee HK, Chung J, Chang SI et al (2011) Real-time measurement of the three-axis contact force distribution using a flexible capacitive polymer tactile sensor. J Micromech Microeng 21(3): 035010
7. Pang C, Bao Z et al (2015) Highly skin conformal microhairy sensor for pulse signal amplification. Adv Mater 27(4):634–640
8. Zhu Y, Jiang S, Xiao Y (2018) A flexible three-dimensional force sensor based on PI piezoresistive film. J Mater Sci: Mater Electron 29(23):19830–19839
9. Cheng MY, Lin CL, Lai YT et al (2010) A polymer-based capacitive sensing array for normal and shear force measurement. Sensors 10(11):10211–10225
10. ST. Pressure Sensors. https://www.st.com/zh/mems-and-sensors.html
11. Chen D (2017) Mechanical design handbooks, 6th edn. Chemical Industry Press

Author Index

Z. Yu et al. (eds.), *Advancements in Mechatronics and Intelligent Robotics*, Advances in Intelligent Systems and Computing 1220,
https://doi.org/10.1007/978-981-16-1843-7